OBJECT-ORIENTED DATABASE MANAGEMENT
Applications in Engineering
and Computer Science

ALFONS KEMPER

GUIDO MOERKOTTE

Lehrstuhl für Informatik
Universität Passau
Germany

Fakultät für Informatik
Universität Karlsruhe
Germany

PRENTICE HALL, Englewood Cliffs, New Jersey 07632

Library of Congress Cataloging-in-Publication Data

Kemper, Alfons Heinrich.
    Object-oriented database management : applications in engineering
and computer science / Alfons Kemper, Guido Moerkotte.
        p.   cm.
    Includes bibliographical references and index.
    ISBN 0-13-629239-9
    1. Object-oriented data bases. I. Moerkotte, Guido. II. Title.
QA76.9.D3K455 1994
620'.00285'575--dc20
                                                    93-35995
                                                        CIP

Acquisitions Editor: Bill Zobrist
Production Editor: Bayani Mendoza de Leon
Copy Editor: Brenda Melissaratos
Cover Designer: Design Source
Production Coordinator: Linda Behrens
Editorial Assistant: Phyllis Morgan

 © 1994 by Prentice-Hall, Inc.
A Paramount Communications Company
Englewood Cliffs, New Jersey 07632

The author and publisher of this book have used their best efforts in preparing
this book. These efforts include the development, research, and testing of the
theories and programs to determine their effectiveness. The author and
publisher make no warranty of any kind, expressed or implied, with regard to
these programs or the documentation contained in this book. The author and
publisher shall not be liable in any event for incidental or consequential
damages in connection with, or arising out of, the furnishing, performance, or
use of these programs.

Printed in the United States of America
10 9 8 7 6 5 4 3 2 1

ISBN 0-13-629239-9

Prentice-Hall International (UK) Limited, *London*
Prentice-Hall of Australia Pty. Limited, *Sydney*
Prentice-Hall Canada Inc., *Toronto*
Prentice-Hall Hispanoamericana, S. A., *Mexico*
Prentice-Hall of India Private Limited, *New Delhi*
Prentice-Hall of Japan, Inc., *Tokyo*
Simon & Schuster Asia Pte. Ltd., *Singapore*
Editora Prentice-Hall do Brasil, Ltda., *Rio de Janeiro*

Contents

III Object-Oriented Modeling and Languages   157

7 Objects and Types   159

Contents

Preface

During the Eighties relational database systems have gained a predominant role in the commercial, administrative information management sector. At the same time database researchers began to analyze the information management requirements of other, hitherto neglected application domains, such as mechanical engineering, software engineering, VLSI design, architecture, and science—to name just a few. It turned out that relational database systems do not adequately support these so-called *non-standard* applications. This result led the database researchers to develop more suitable functionality for these "advanced" applications.

The research directions can be distinguished in two categories:

- the *evolutionary* approach: In this research and development work the conventional relational model is taken as a platform for extensions and adaptations.

- the *revolutionary* approach: Here, database research is based on a new data model, the *object-oriented model*.

While this book contains some material on the evolutionary approach—i.e., extensions of the relational model—its main thrust is on describing the object-oriented database technology, i.e., the revolutionary database research direction.

The revolutionary database research direction borrowed ideas from programming language developments—namely the object-oriented paradigm—to develop better database support for such advanced application areas. In the recent past, object-oriented database systems emerged as the "next-generation" database technology—in particular for advanced applications, such as engineering.

As mentioned before, database systems were originally developed for the commercial, administrative sector. Therefore, it is not astonishing that many good textbooks exist that cover this aspect of database technology, i.e., the utilization of conventional DBMSs in business and administration. However, there are no books providing an encompassing discussion of concepts, tools, design methodologies and systems for advanced applications. With this book we attempt to provide a *self-contained* text about object-oriented information management. However, it also includes material on conventional (established) database technology—and extensions thereof—for assessing its strengths and limitations. The development of object-oriented databases was mainly triggered by the needs of the engineering application domains. Consequently, this book covers database concepts from the viewpoint of the engineering community. Therefore, the book can be viewed as an interdisciplinary guide with the goal to bridge the gap between engineering disciplines and database technology.

The book is aimed at two groups of readers:

- For *professionals*—e.g., engineers, software developers—dealing with databases it is a self-contained reference addressing their data modeling and database management needs. The book covers all stages of information management—starting with the conceptual design over implementation design to the physical design. The book provides a thorough description of the state of the art in database development in order to allow engineering users to assess this technology in their area of expertise.

- For *computer scientists* the book serves as an advanced guide to object-oriented database concepts and their use in advanced application domains.

The book is suitable as a text book for an advanced senior or graduate level course on advanced database concepts as well as a self-contained reference guide for professionals. The book is also useful for so-called "technical decision makers" who need to assess forthcoming database technology for their particular application scenarios. Furthermore, it is useful as additional reading in courses involving information processing in technical applications—such as geometric modeling, Computer Aided Design (CAD), Computer Integrated Manufacturing (CIM), VLSI design, software engineering, etc.

The book is self-contained inasmuch as it assumes only basic computer science and engineering prerequisites. The engineering professionals, who want to use this book as a reference in developing database support for technical applications, should have the knowledge of fundamental computer science concepts and a programming language. The computer science reader is familiarized with the database requirements imposed by engineering applications in a dedicated chapter at the beginning.

The book consists of six parts—covering the following topics:

1. Introduction of Basics

2. The Relational Approach

3. Object-Oriented Modeling and Languages

4. Control Concepts

  5. Physical Object Base Design

  6. Sample Systems

In Part I we first motivate the use of database systems in advanced applications, concentrating on mechanical engineering. Then, the basics of conceptual data modeling and engineering database applications are introduced. The material of the remainder of the book is (largely) illustrated by examples introduced in this part.

In Part II the relational database model is surveyed with respect to engineering application support. A critical assessment of relational concepts is followed by a description of extensions of the relational model. This part overviews the "evolutionary" database research and developments.

Part III is dedicated to introduce the modeling concepts of the object-oriented data model. The discussion of this part is based on the generic object model (GOM) which was developed by the authors. Contrary to relational databases, there is no standard object model—as of the time of writing. Therefore, we chose to base our text on a model that is *generic* inasmuch as it unites the most salient features of object-oriented data modeling in one syntactical framework.

Towards the end of Part III we address interface issues to object-oriented databases. It describes three languages for associative object access; that is, retrieval of objects from the database based on selection predicates.

In Part IV control concepts that are integrated in the object-oriented database systems are introduced. Among these we describe the schema management that controls the database consistency, the transaction control concepts that govern the sharing of the database by concurrent users, the version support concepts that control the evolution of objects over time, and the authorization mechanisms that control the access to the stored information.

In Part V the physical object base design is treated. This part is devoted to describing index and object placement methods to enhance the performance of object-oriented database applications.

The last part (Part VI) surveys the existing commercial object-oriented database products. Some systems, which were the forerunners of object-oriented database technology are treated in detail, while others can only be sketched. This part is concluded by a list of influential research prototypes.

In Figure 1 the dependencies among the six parts of the book are depicted. The thick arrows represent strong dependencies whereas the thin arrows indicate a weak dependency in the sense that the material can be read without a detailed knowledge of the parts from which the arrow emanates.

Figure 2 shows a more detailed dependency graph at the chapter level for Parts III – VI. In particular, it indicates that Chapters 11 – 13 can be skipped on first reading without compromising the understanding of the subsequent material.

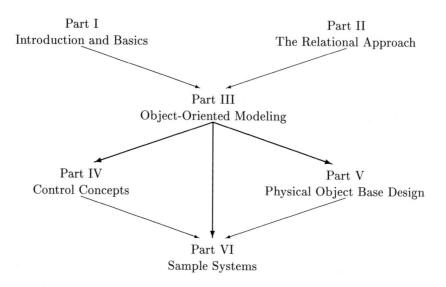

Figure 1: Dependency Graph of the Parts of the Book

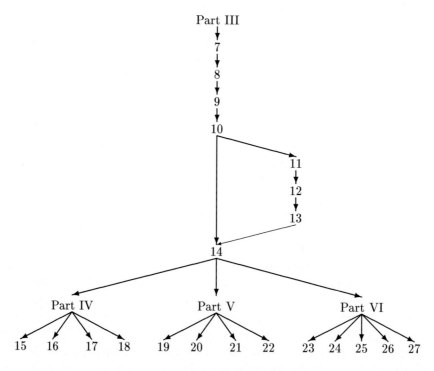

Figure 2: Dependency Graph of Parts III – VI

Acknowledgments

The authors wish to express their gratitude to the many helpful people at the University of Karlsruhe, the Technical University (RWTH) Aachen, and the University of Passau—the three locations where the book was written.

We thank the members of the GOM team (Carsten Gerlhof, Christoph Kilger, Donald Kossmann, Klaus Peithner, Michael Steinbrunn, Hans-Dirk Walter, and Andreas Zachmann) who contributed to the design and, in particular, to the prototype implementation of the generic object model (GOM). Furthermore, we acknowledge the numerous students who participated in the implementation effort.

We are grateful to the many students who attended our courses on object-oriented database systems at the three above mentioned universities. Many improvements evolved from their feedback.

Our students André Eickler and Axel Kotulla deserve our thanks for helping in the testing of programs contained in the text.

Bärbel Kronewetter prepared parts of the manuscript. Thank you for enduring the LATEX "traps" and providing such a well organized secretariat at the RWTH Aachen.

In the final phase of the book two people helped immensely with proof-reading, compiling the index, and final (production-quality) text-formatting: Jens Claußen and Harald Schreiber. Thank you!

Simone Kolb and Alexandra Schmidt provided the (much needed) secretarial and organizational help at the University of Passau during the final production phase.

Peter C. Lockemann, our academic "mentor" at the University of Karlsruhe, deserves our thanks for his continuous support of our research work.

Dennis McLeod continuously encouraged us during the (rather long) writing phase of this book.

Last, not least, the people at Prentice Hall (especially Tom McElwee, Bayani Mendoza DeLeon, Phyllis Morgan, Danielle Robinson, and Bill Zobrist) are thanked for their support and for being patient enough to wait for this book to happen.

Alfons Kemper
Guido Moerkotte

Part I

Introduction

1

New Frontiers for Database Application

1.1 General Remarks about Computerized Information Systems

Our world, that is, our conception of the world, is continually becoming more complex. Every day new information is added to the overall body of existing knowledge. It is estimated that the totally available information about all aspects—technical, scientific, commercial, etc.—of our world is doubling every five years. This huge body of knowledge makes it impossible for an individual to acquire even an encompassing understanding of only small areas of our complex world (model). Even the search for information about a relevant topic is becoming more and more intractable: The Library of Congress contains, for example, nearly 100 million books. Therefore, computerized information systems are becoming an essential part of our professional and, also, our private lives.

Database management systems (DBMSs) are, by now, commonplace in administrative application areas. However, they are still rare in engineering and science applications. This, however, should not be viewed as an indication that DBMSs would play a less important role in engineering or in science. For example, engineering—the entire process from conceiving a product idea, to manufacturing the product, to delivering it to its final destination, and, finally, maintaining it there—consists of a myriad of individual, often complex steps. These steps take place in a purely informational world, such as in design and construction, or in a physical world, such as in fabrication or assembly. But even the physical processes are accompanied by informational processes. Thus, all the engineering steps depend on large volumes of data, and they yield volumes of new data. All these activities are typically scattered across a wide spectrum of physical sites; the information is subjected to and generated by a host of highly specialized software tools on diverse computing hardware. The data are kept on different incompatible storage media in divergent structures and formats that account more for the peculiarities of the tool that generated them than the needs of the tools that will later have to use them.

On the other hand, it is widely agreed that, an integration, as proposed for, e.g., *true* computer-integrated manufacturing (CIM), will only be achieved if all these differences can be overcome and each of the engineering steps can be furnished with a view of the informational world that obscures much of the locational and structural peculiarities of the data. In other words, each of the steps should be provided with a homogeneous, integrated view of the data on a logical level commensurate with the phenomena of the engineering world, all this under severe conditions of technical performance and economy.

1.2   The Integration Problem in CAD/CAM

Today's engineering applications suffer from a severe integration problem. As pointed out before, huge volumes of data are generated by various tools in the course of the life cycle of a technical product. The life cycle of a typical engineering artifact starts with its first conception and proceeds to construction, to manufacturing, and, finally, to marketing and distribution. A very simplistic view of this highly complex and—contrary to the drastically simplified graphic—intertwined life cycle is shown in Figure 1.1.

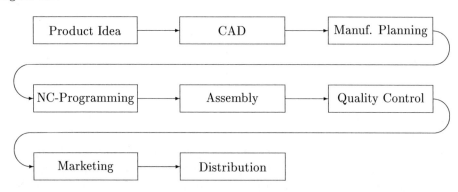

Figure 1.1: Life Cycle of an Engineering Product

All these different phases in the chronological activity chain underlying the product's life cycle can be supported by various computerized tools. However, the current status is that almost no two software modules that support different stages in the life cycle are compatible in such a way that they can exchange data directly—without cumbersome (manual or semi-automatic) transformation. Figure 1.2 depicts graph-ically the current state of the art in CIM: Different computerized tools exist, such as computer-aided design (CAD), support tools for NC-machine[1] programming, tools for computer-aided quality control (CAQ), etc. Unfortunately, the tools interface with their own, customized file structures, which are not accessible by other modules. Even tools that support later stages in the life cycle and are, therefore, dependent on the data generated in earlier stages cannot deal directly with the data structures gen-erated in earlier life-cycle phases of an engineering product. The current state of CIM is characterized by the fact that there are more than 300 commercially available com-puterized tools to support the life cycle of an engineering artifact. But—probably—no two of them are fully compatible such that they can directly work on the same data structures without tedious conversions in between.

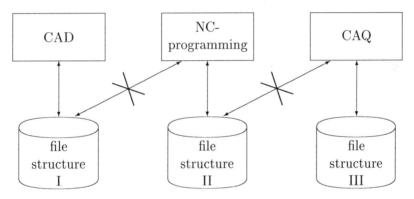

Figure 1.2: Today's Computer-"Integrated" Manufacturing: No Integration

1.3 Databases: An Integration Tool for Engineering Applications

One of the principal challenges in information management in engineering applica-tions is the *integration* of diverse, special-purpose application modules. According to Wedekind [Wed88] we distinguish different integration levels.

As outlined before, the current state of the art in this respect is—unfortunately—characterized by a complete *isolation*, i.e., no integration whatsoever. This state, called integration level 0, is depicted in Figure 1.3. Each of the diverse application modules manages its information in highly customized database (DB) or file system

---

[1]numerically controlled machine

(FS). The customized information structures cannot be accessed by any other application. The best one can do is a manual or semi-automatic transformation from one structure to the other in order to exchange information between different modules.

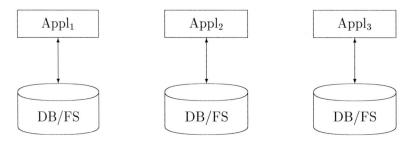

Figure 1.3: Integration Level 0: Isolation

The next step in trying to integrate different modules is depicted in Figure 1.4. In this system architecture the interface of different modules is achieved via special, customized converters that transform the information structure required by one application module to the needs of the other application module. In the diagram the converters are denoted as $Conv_{i,j}$—meaning that the converter transforms the information structure of application $Appl_i$ according to the structural requirements of application $Appl_j$, and vice versa. Unfortunately a great number of converters are needed in order to interface N applications. The number of converters needed is

$$N * (N - 1)$$

For example, for three applications—as exemplified in the figure—six converters are needed. Note that, for example, the converter $Conv_{2,3}$ shown in Figure 1.4 actually constitutes two converters: one from application $Appl_2$ to $Appl_3$ and another one that transforms data structures of $Appl_3$ into those of $Appl_2$.

The next system architecture is called integration level 2 and is graphically shown in Figure 1.5. This overall system architecture reduces the number of converters required. It assumes a *neutral* database or file system that is shared by all applications. This neutral data structure can be viewed as a data exchange standard—such as the ANSI standard IGES (Initial Graphics Exchange Specification) [ANSI81] for mechanical engineering or EDIF (Electronic Design Interchange Format) [EDIF84] for VLSI (Very Large Scale Integration) design data. All the application modules still possess their own customized data structures for local processing and data storage. However, the exchange of the information is achieved via standardized interfaces, which makes it easier to implement the converters—the target of the conversion process is no longer system specific. Also this architecture has a starlike structure and, thus, requires only $2 * N$ converters for N application modules.

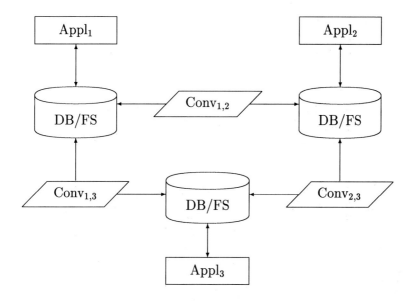

Figure 1.4: Integration Level 1: Converters

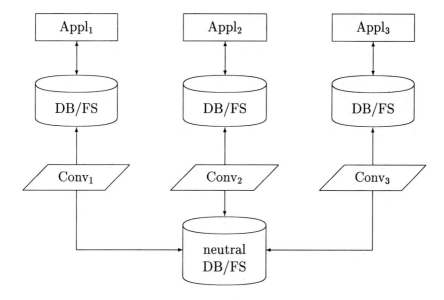

Figure 1.5: Integration Level 2: Neutral File Format

Database management systems have traditionally been considered as one of the most important means for achieving integration by controlling the generation, usage, and exchange of information in such a complex (distributed) environment.

In integration level 3 (Figure 1.6) the database system forms the central "agency" of integration. All application modules interface with a common database.

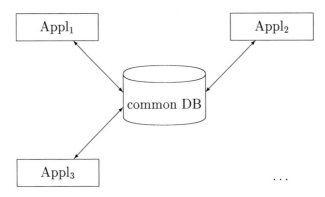

Figure 1.6: Integration Level 3: A Centralized Database

The diagram indicates a global, centralized database that serves as an encompassing model of the production process. Of course, the centralization may be only at the logical level; that is, the information may physically be distributed over several machines. Nevertheless, in current distributed database systems the distribution is transparent to the user; that is, one logically global view of the data is presented to the user of the database system.

Using a database management system in this way—coupled with suitable representation standards of the generated data—promises a drastic increase in efficiency:

- Most of the tedious data conversions become obsolete.

- The data underlying a production process become accessible by all authorized users.

- It becomes easier to retrieve particular data objects and, thus, reuse them in similar (design) work.

- It becomes easier to ensure database consistency because much of the currently encountered redundancy, which is necessitated by the incompatible data structures, can be avoided.

From a database point of view the integration level 3 is the best we can achieve. However, the reality is that many customized CAX^2 modules exist and possess highly

---

[2]CAX: CAD, CAM, CAQ, ...

customized data formats and, therefore, cannot easily be rewritten to interface with a database system. Therefore, we will have to provide means to integrate such *turn-key* systems into an integrated database environment. This can be achieved—analogously to integration level 2—by providing customized converters. Now the database system plays the role of the neutral data format. This is illustrated in Figure 1.7, where a fourth application module, *Appl₄*, which is assumed to be such a *turn-key* system that was developed without knowledge of the integration architecture centered on the database, is integrated via a converter.

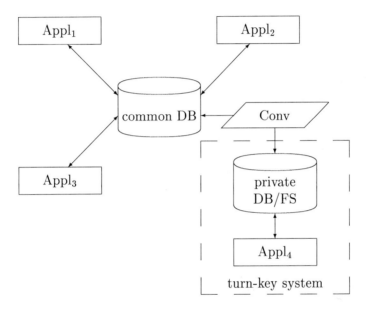

Figure 1.7: Integration Level 4: Integration of Stand-Alone Components

2

Data Modeling Basics

The objective of this chapter is to introduce a semantic data model that is used as a representation tool to model the real (mini) world. The process of mapping the real world, as perceived by the database user, into a schema of the semantic model is called *conceptual design*. The conceptual schema, i.e., the outcome of the conceptual design, is supposed to capture all relevant information units, their interrelationships, and their integrity constraints in one global and consistent representation. It models the database application at the *meta-information* level; that is, it characterizes possible and legal states of the real world but does not define an actual state of the database. The data model used in this process is called a *semantic model* because it typically provides more expressive language constructs than an *implementation model*, such as the relational data model. These expressive language constructs are beneficial to capture the semantics of the information units of the real (mini) world. Since the conceptual database design process involves different groups of users with varying experience in database modeling and usage, it is desirable to have a graphical visualization of the conceptual schema as a basis for communication.

2.1 Database Design Steps

Similarly to "good" software engineering, the task of designing a database application proceeds in a stepwise manner. We distinguish four phases in the life cycle of a database design:

1. Requirement analysis

2. Conceptual design

3. Implementation design

4. Physical design

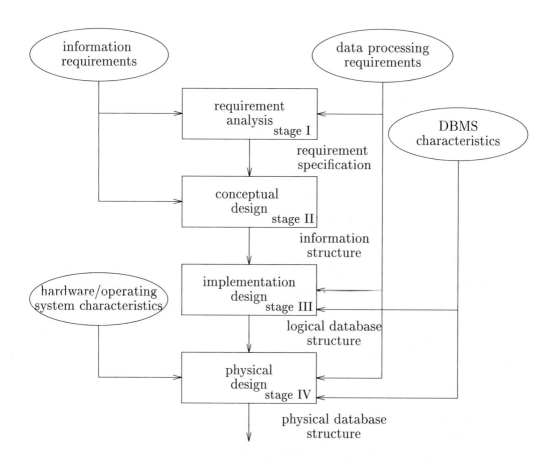

Figure 2.1: Database Design Stages

Graphically, these four stages are depicted in Figure 2.1,[1] where the inputs to the particular life-cycle stages are indicated, as well. It is essential that the output of these four stages are maintained in mutually consistent states. If, for example, a modification of the conceptual design becomes necessary, this modification has to be reflected in the requirement specification, and, of course, it has to be propagated to the implementation and physical design.

2.1.1 The Requirement Analysis

During the requirement analysis the intended functionality and performance of the database application are derived. Thus, the requirement analysis has to take into account the dual aspects of the prospective application:

- The information requirements of the organization, for which the database support is being developed

- The data processing requirements, i.e., the requirements of the programs that interface to the database

This task is typically performed by interviewing the prospective end users and systems programmer who will be using the database under development. It is important to compose a document in a (semi)formal language as a basis for further stages in the database design. This document can later be used to verify that the initially thought-of behavior and performance are actually achieved in the final two stages, the implementation and physical design.

2.1.2 The Conceptual and Implementation Design

The conceptual design is system independent; that is, it makes no assumptions about the database system on which the database application will be installed and its underlying data model. Once the conceptual schema has been completed the *implementation design* can proceed. In this design step the conceptual schema is transformed into the implementation model of the database system being used, e.g., the relational model, the CODASYL network model, or the object-oriented model GOM (generic object model) that is introduced later in this book. The two design steps in generating a database application are summarized in Figure 2.2.

By far the most widely used semantic model for conceptual database design is the entity-relationship (ER) model, which is the central issue of this chapter.

---

[1]This graphic is adapted from [TF82].

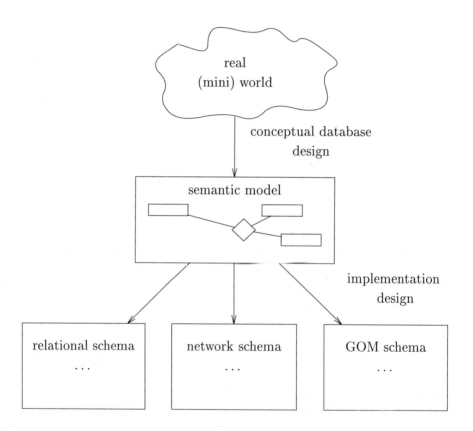

Figure 2.2: Outline of Conceptual and Implementation Database Design Steps

2.2 The Basic Concepts of the ER Model

As the name indicates, the basic structuring features of the entity-relationship model
are the following:

- *Entities*, representing abstract or physical objects in the real world

- *Relationships*, representing interrelations between entities in the real world

- *Attributes*, which describe the properties of entities or relationships

In an ER schema the entities and relationships that exist in the real world—that
is, the part of the real world that is being modeled—are classified into *entity sets* and
relationship types, the latter being defined over entity sets.

2.2.1 Entities and Entity Sets

It is not possible to formally define the notion of entity. Typically, an entity is
characterized as a distinguishable object in the real world and may either be physically
existent or abstract. Prototypical examples of entities are persons and automobiles,
i.e., individual objects like, for example, the person named "Mickey Mouse" or the
"Porsche Speedster" owned by "Jimmy Dean" as an individual automobile.

Because of the large number of entities that exist in even small miniworlds it is not
convenient to deal with individual entities in the conceptual database design. This
would also imply that one needs to know in advance all possible entities that might be
included in the database. Note that this would preclude the insertion of new entities
into and deletion of existing entities from the database. Therefore, similar entities
are grouped into an *entity set*. "Similar," in this context, means that all the entities
share a common collection of attributes describing their characteristics. Furthermore,
entities grouped into a common entity set undergo the same types of relationships
with other entities of (possibly) different entity sets.

Examples of possible entity sets in the CAD/CAM environment are

- All robots

- All engineers

- All products, etc.

The notion of entity set is at the type (or schema) level. It specifies the entities that
could possibly inhabit the respective entity set. In an actual database state, sometimes called a *database instance* or *instantiation*, an entity set contains a particular
subset of all valid entities that could legally be included in the set.

2.2.2   Relationships and Relationship Types

Entities are related in various ways in the real world. For example, the two entities named "Mickey Mouse" and "Mini Mouse," both belonging to the entity set *Persons*, are associated by marriage. Analogous to the classification of entities into entity sets we also classify relationships into *relationship types*. A relationship type describes the valid associations of entities in the real world; e.g., it constrains the sets to which associated entities belong.

Formally, a relationship type R is defined as an ordered list of entity sets E_1, E_2, $\ldots$, E_n, where the E_i are the participating entity sets. In this case R is an n-ary relationship type. It is quite possible that the same entity set participates more than once (at different positions) in a relationship type. An example is the relationship type *Marriage*, which is defined over the participating entity sets *Persons* and *Persons*.

An instantiation of a relationship type is a set of ordered n-tuples. For example,

$$\{[e_{11}, \ldots, e_{n1}], \ldots, [e_{1k}, \ldots, e_{nk}]\}$$

is a valid instantiation of R where $e_{ij} \in E_i$ for $1 \leq i \leq n, 1 \leq j \leq k$. For example, the tuple $[e_{\mathrm{Mickey}}, e_{\mathrm{Mini}}]$ where e_{Mickey} denotes the *Person* entity named "Mickey Mouse" and e_{Mini} the entity named "Mini Mouse" is an element in a possible instantiation of the relationship type *Marriage*.

Very often we will, for simplifying the language, blur the distinction of entity set and entity, and of relationship type and relationship. From the context it should always be obvious whether the instance or the type level is referred to.

2.2.3   Attributes

Entity sets have an associated collection of attributes that describe the properties of individual entities, i.e., members of the entity set. An attribute consists of a name that has to be unique within the entity set and, optionally, an associated *domain*. A domain restricts the possible values that can be assumed by the attribute; therefore, domains are also called *value sets* or *types*. Thus, the entity set with the attribute specifications can be viewed as a template from which individual entities can be molded. An individual entity draws for each attribute a value from the *domain* to which the attribute is constrained. For example, the entity set *Persons* has an attribute *EyeColor*, which, for an individual *Persons* entity, assumes a value of the domain {*blue, green, gray, brown*}.

Attributes may also be associated with relationship types. For example, the relationship type *Marriage* could have an attribute *DateOfMarriage* whose domain is *Date*, i.e., all valid dates.

2.2.4 Key Attributes

The key of an entity set is defined as a subset of the attributes of the entity set such that the values of the attributes in this subset uniquely identify the individual entities within the entity set. Very often, the key is supplied as a single, distinguished numeric attribute whose sole purpose is identification. This is analogous to the real (bureaucratic) world, where we encounter "artificial" key attributes like Social Security number to identify people, license plates to identify automobiles, etc. But, in general, combinations of attributes also may be used when there is no one distinguished attribute that can serve as a key. For example, the two attributes *ManufacturerName* and *ProcessorID* may uniquely identify a member of the entity set *Computers*.

2.3 Graphical Representation of ER Schemas

For the ER model a graphical representation has been introduced using the following icons:

- A *retangle* denotes an entity set.

- A *diamond* represents a relationship type. The diamond is connected by undirected arcs with the participating entity sets.

- A *circle* or *ellipsis* represents attributes that are either associated with an entity set or a relationship type. The association is denoted by an arrow directed from the entity set or the relationship type to the ellipsis.

- Attributes that are part of the key are denoted by *double circles* or *double ellipses*.

An example of an entity-relationship schema is depicted in Figure 2.3. In this conceptual schema we find five entity types: *Divisions*, *Engineers*, *Robots*, *Tools*, and *Products*. There is one ternary relationship type, i.e., *Assembles*, which exists among the types *Robots*, *Products*, and *Tools*. All other relationship types, i.e., *Controls*, *Designs*, *WorksFor*, *Manufactures*, and *Composition*, are *binary*. Except for the relationship type *WorksFor*, which is $1 : N$, all other binary relationships are $N : M$ (which is described in the next section). In our example schema all entity sets have a key consisting of just one attribute, each being marked by a double ellipsis.

Composition is a recursive relationship type. The labels *sub* and *super* of the relationship type *Composition* identify the *role* of an individual entity in the relationship. Roles are attached to relationship types whenever it cannot be deduced from the context which semantics a participating entity has in such a relationship. For example,

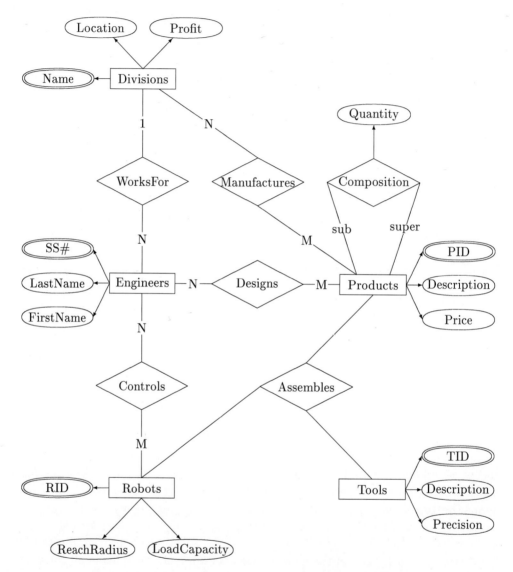

Figure 2.3: A Sample Entity-Relationship Schema

the tuple $[super : e_1, sub : e_2]$ denotes that entity e_1 is the *super* (part) and e_2 is the *sub* (part). An entity of *Products* may participate in the relationship *Composition* in both roles; i.e., it may be a subpart to some entity and also a superpart for another entity.

2.4   Functionality of Relationships

The most common (but not the only possible) kind of relationship is the binary relationship that relates two entities. An abstract example of a binary relationship type is pictured in Figure 2.4.

Figure 2.4: The General Case of a Binary Relationship Type

For binary relationship types it is often useful to characterize the functionality of the association, i.e., in what cardinalities the entities of set E_1 may be associated with entities of set E_2. We distinguish four cases:

1 : 1 relationship Every entity $e_1 \in E_1$ is related with at most one entity $e_2 \in E_2$, and any entity $e_2' \in E_2$ is related with at most one entity $e_1' \in E_1$. In this case R can be viewed as a partial function[2] from E_1 to E_2 or, alternatively, the inverse function from E_2 to E_1.

1 : N relationship If for all entities $e_2 \in E_2$ there exists at most one entity $e_1 \in E_1$ with which e_2 is related via R, then the relationship is called $1 : N$. In this case R can be viewed as a partial function from E_2 to E_1.

N : 1 relationship Analogously defined as above. Now the relationship R corresponds to a partial function from E_1 to E_2.

N : M relationship If there is no limitation imposed on the number of entities with which an entity of either set is associated, then the relationship type is called $N : M$. In this case the relationship R is no longer functional.

Note that this characterization of relationships is at the schema level; that is, it constrains all possible database instantiations to obey the functionalities specified in the ER schema.

The functionality of a binary relationship is typically indicated in the ER schema by labeling the arcs leading into the diamond that represents the corresponding relationship type (see Figure 2.3).

---

[2] A partial function is a function that may not be defined for all elements of the domain.

2.5   Fundamental Work on Abstraction

Abstraction in general denotes a means to reduce the complexity of an application. The concept of abstraction allows the database user or developer to be concerned with one aspect of the system only, while everything else can be treated as a "black box" whose details (e.g., implementation) is hidden in this particular context.

Among the first authors who addressed the concept of abstraction within the context of databases are Smith and Smith [SS77a, SS77b]. They developed the two abstraction concepts of *aggregation* and *generalization* for the relational data model. In order to decrease the complexity of a database application, the model of the real world that is described in the database is decomposed into a hierarchy of abstractions, that is, a combination of aggregations and generalizations. Such an *abstraction hierarchy* allows (possibly different) database users to access the model at different levels of detail, that is, at different abstraction levels. While some user just wants a global overview of the stored data, another user might require very detailed information about parts of the stored real-world model.

2.5.1   Aggregation

Aggregation is a form of abstraction where a relationship between entities is considered and modeled as a (higher-level) entity in its own rights. In this sense, aggregation corresponds to the mathematical notion of Cartesian product. Entities of different types are combined to form another, higher-level aggregate entity. An example in the context of engineering applications is the following: the relationship *assembles* between *Engineers* supervising *Robots*, which assemble certain *Products* using particular *Tools*. Schematically this relationship is depicted in Figure 2.5.[3]

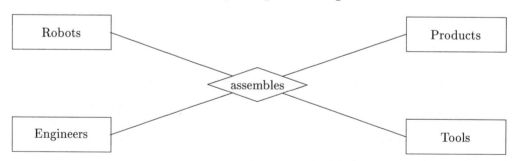

Figure 2.5: Relationship *assembles*

This relationship *assembles* between the entity types *Engineers*, *Products*, *Robots*,

---

[3]Note that this is a conceptual redesign of the relationship type *Assembles* introduced before in our *Company* schema.

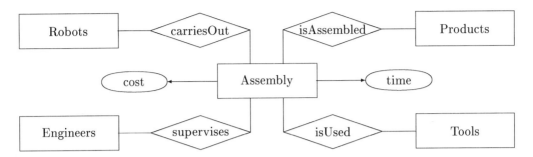

Figure 2.6: Aggregation of the Relationship *Assembly*

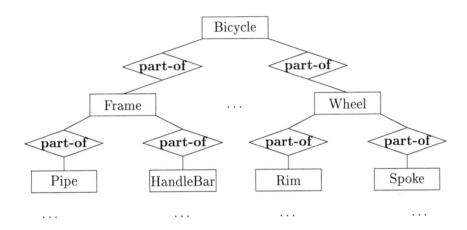

Figure 2.7: Aggregation of a *Bicycle*

and *Tools* could also be modeled as an aggregated entity called *Assembly*, as depicted in Figure 2.6. In this schema *Assembly* is an entity type in its own right. It may have attributes associated to it—such as assembly *cost* and assembly *time*—and it takes part in relationships with other entities, such as *Schedules*, *Engineers*, *Products*, etc.

One of the most important aggregation abstractions is the **part-of** relationship. This relationship exists among a *superpart* and several *subparts*—possibly belonging to different entity types. The subparts are constituent entities of the composite superpart. The **part-of** aggregation often forms *aggregation hierarchies* such as shown in Figure 2.7, in which a *Bicycle* is modeled as an aggregation of a *Frame* and two *Wheels*. The *Wheels* are, in turn, modeled as aggregates of a *Rim*, *Spokes*, etc. Likewise, the *Frames* are themselves aggregate entities that are composed of subparts. Such a **part-of** aggregation may span over many levels—as indicated by the dots in our *Bicycle* example.

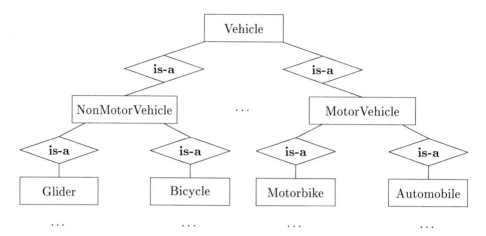

Figure 2.8: Generalization Hierarchy *Vehicle*

2.5.2   Generalization and Specialization

In [SS77b] we find the following definition for *generalization*:

> A generalization is an abstraction which enables a class of individual entities to be thought of generically as a single named entity.

Thus, generalization is used to group together similar entities that can be described by a generic entity type. Generalization is a form of abstraction such that nonrelevant details of the individual entities are omitted in the generic entity type. Or, from the other point of view, the relevant details that all the individual entities share are emphasized by factoring them out as the common attributes in the generic entity type.

In a generalization a collection of individual entities (or entity types) E_{ind}^1, ..., E_{ind}^n is generically described by one entity schema $E_{generic}$. We require that the E_{ind}^i are mutually disjoint entity types, i.e.:

$$E_{ind}^i \cap E_{ind}^j = \emptyset \quad \text{for} \quad 1 \le i, j \le n, i \ne j$$

Each entity of type E_{ind}^i can also be viewed as being an entity of type $E_{generic}$. In this sense, generalization models the relationship **is-a**—meaning an entity e_{ind} of type E_{ind}^i **is-a**n entity of type $E_{generic}$.

Again, the generalization abstraction may span over several levels—thus forming *generalization* or **is-a** *hierarchies*. An example of such a *generalization hierarchy* is shown in Figure 2.8, in which different types of vehicles—such as *Gliders*, *Bicycles*, *Motorbikes*, and *Automobiles*—are generalized in *NonMotorVehicle* and *MotorVehicle*, which themselves are generalized to the entity type *Vehicle*.

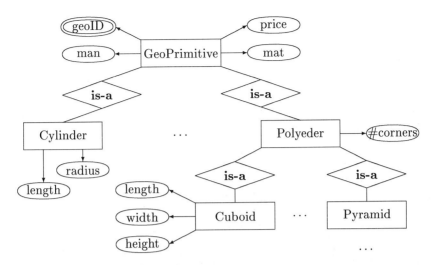

Figure 2.9: Generalization of Geometric Primitives

Inheritance is an integral part of generalization in the way that the specialized entity types inherit the properties of their generalized types. Viewing this in the opposite direction, the common properties of the more specialized entity types are factored out and associated with the more general entity type. Thus, the entity schema $E_{generic}$ comprises all generic properties of the more specialized types E_{ind}^i for $1 \leq i \leq n$. Generic properties are those properties that all individual entities have in common, or more formally stated:

$$Properties(E_{generic}) := \bigcap_{E_{ind} \in \{E_{ind}^1, \dots, E_{ind}^n\}} Properties(E_{ind})$$

This formula states that $E_{generic}$ models all properties that are in the intersection of the properties of the individual entity E_{ind}^i for $1 \leq i \leq n$. Only those properties of the individual entities are modeled within the generic entity that are common to all individuals and that are relevant for the abstraction level at which the generic description is used.

We call the types E_{ind}^i for $1 \leq i \leq n$ *specializations* of the generic entity type $E_{generic}$. The entity schemas of the E_{ind}^i describe those properties that are specific to the individual entities and, therefore, are not covered by the generic entity description.

As an example consider the generalization hierarchy shown in Figure 2.9. In this conceptual schema the geometric primitives—such as *Cylinders*, *Cuboids*, and *Pyramids*—are structured in an **is-a** hierarchy. The top level of the generalization hierarchy is the entity type *GeoPrimitive*. This type has the properties *geoID*, *mat* (material), *man* (manufacturer), and *price*. These four attributes are, by inheritance, common to all geometric primitives—no matter what specialized type they belong to.

The entity type *Cylinder* is a specialization of *GeoPrimitive* with the additional two attributes *radius* and *length*, which account for the specific structural representation of *Cylinders*. A further specialization of *GeoPrimitive* is given by the entity type *Polyeder*. This type has one additional attribute, *#corners*, which determines the number of bounding corners of the particular *Polyeder* entity. The type *Polyeder* is further specialized to *Cuboid* and *Pyramid*. Both types have their own set of specialized attributes. In the case of *Cuboid* these are *length*, *width*, and *height*. Applying the principle of inheritance a *Cuboid* is thus modeled by the following eight attributes:

- *geoID*, *man*, *mat*, and *price*, as introduced in the generic type *GeoPrimitive*

- *#corners*, which is associated with the type *Polyeder*

- *length*, *width*, and *height*, which are associated with the most specialized type *Cuboid*

Analogously, the entity type *Pyramid* has its own specialized attributes.

It should be emphasized that each entity of a specialized type can also be viewed as an entity of a more generic type (along the **is-a** hierarchy)—thus providing different levels of abstraction. For example, a particular entity of type *Cuboid*, say, the entity whose *geoID* attribute equals "cubo#5", is also a *Polyeder*. Likewise, this *Cuboid* entity is also a *GeoPrimitive* entity with just the four relevant attributes *geoID*, *man*, *mat*, and *price*. The level of detail that is required determines which abstraction level has to be accessed.

2.5.3 Combining Aggregation and Generalization

In order to structure the entity types of a complex application, domain, generalization and aggregation abstractions are often applied in conjunction. In this way, an entity type being part of a generalization hierarchy can—at the same time—participate in an aggregation hierarchy. In this, it does not matter whether the entity type forms the root or an intermediate node within the aggregation hierarchy. The use of aggregation and generalization as dual, each other complementing abstraction concepts is demonstrated on our *Vehicle* example in Figure 2.10.

The generalization hierarchy *Vehicle* contains the specialized entity type *Bicycle*. A *Bicycle* is—like all other *Vehicle* types—an aggregation of several constituent entity types. Consequently, the composition of a *Bicycle* is modeled as an aggregation hierarchy in the conceptual schema.

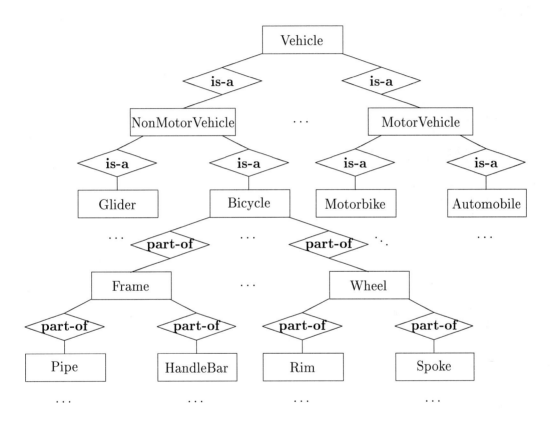

Figure 2.10: Combined Aggregation and Generalization Hierarchies

2.6   Exercises

2.1   Model a *Car* as a conceptual Entity-Relationship schema. Incorporate into the conceptual schema the most essential aspects of the owner, the drivers, and the manufacturer of the car.

In the conceptual design you should apply the abstraction concepts aggregation and generalization—whenever possible. In particular, consider different specialized types of car manufacturers, e.g., American, German, Japanese manufacturers, etc. Aggregation is, naturally, used to model **part-of** relationships— such as an *Engine* being part of a *Car*.

2.2   Model a more complete generalization hierarchy of *Motor Vehicles* which encompasses *Trucks*, *Cars*, *Vans*, etc.

2.3   Model a camera. Pay particular attention to aggregation hierarchies.

2.4   Model a country with cities, rivers, towns, highways, etc.

2.5   Model the train system of your country. Include information about the timetable of the trains.

2.6   Model countries, multinational organizations, political and commercial contracts.

2.7   Design the entity-relationship schema for the robot gripper whose part explosion is depicted in Figure 2.11. Besides modeling the structure of the robot gripper, assume that there are different manufacturers for each component and that each component is supplied by a manufacturer in a certain quantity within a supply. For a later exercise the individuals within each supply must be identifiable and the clients to which a certain robot gripper was delivered must be retrievable.

2.8   So far we have restricted the characterization of relationship types to functionality specifications, i.e., $1:1$, $1:N$, and $N:M$. It is possible to characterize the binary relationships more precisely by the so-called $(\min, \max)$ notation. The specification is graphically denoted as follows:

This schema specifies that each entity e_1 of entity type E_1 participates in the relationship R at least $\min_1$ times and at most $\max_1$ times. Analogously, an

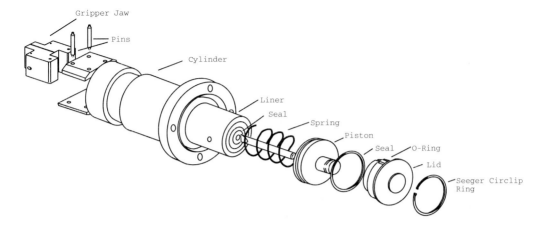

Figure 2.11: Explosion Drawing of a Robot Gripper

entity e_2 of entity type E_2 undergoes a relationship of type R with at least $\min_2$ entities of type E_1 and at most $\max_2$ such relationships. Both, $\min_i$ as well as $\max_i$ can be set to $*$ denoting that the respective entities can participate in an arbitrary number of relationships.

1. Characterize $1 : 1$, $1 : N$, and $N : M$ relationship types in terms of the (min, max) notation.

2. Specify the cardinalities of the relationship types of our conceptual schema in Figure 2.3 using the (min, max) notation.

2.9 Devise an entity-relationship schema modeling entity-relationship schemas. This can be used as a meta schema of the ER model.

2.10 Give a generalization and aggregation hierarchy for a Swiss knife consisting of different blades, scissor, screw driver, corkscrew, etc. A scissor itself should consist of two blades with handle.

2.11 Model the relevant part of a university administration. In particular, apply the abstraction concepts of aggregation—e.g., students enrolling in courses being taught by particular professors—and generalization. For generalization consider the following specialization hierarchy: Students and university staff are a specialization of persons and professors and secretaries are a specialization of university staff.

2.12 Consider computer-aided architectural design (CAAD) as an example application area. Model a house with different rooms—including walls, windows, doors, etc.—for different purposes (like kitchen, living room, etc.) These rooms contain furniture and infrastructure for electricity, water, gas, etc. Give the generalization hierarchy for the different rooms. Model relationships like *adjacency* between rooms.

2.13 Office automation is an advanced application area. There, documents play the predominant role. Documents are also included in every engineering application—the least for documentation. Model documents consisting of headers, sections, title pages, figures, etc. The predominant relationship is *is-included*.

2.14 Software engineering also imposes new challenges upon databases. As with all other advanced application areas, the entities are highly interrelated. Model software programmed in some programming language. Consider entities like modules, procedures, blocks, procedure calls, statements, etc. Important relationships are *uses* and *calls*.

2.7   Annotated Bibliography

The entity-relationship model was proposed in a seminal paper by P. Chen [Che76]. A thorough review of other semantic data models, including the IFO model [AH87a] and the SDM model [HM81], which might, alternatively to the ER model, be used for conceptual design, is presented in [HK87b]. A collection of research papers that describe the state of the art in conceptual modeling languages can be found in [BMS84].

The concepts of aggregation and generalization were first introduced by Smith and Smith [SS77b, SS77a]. These concepts were incorporated into the entity-relationship model in [TYF86]—consequently their model is called the extended entity-relationship (EER) model. A more detailed description of all database design phases can be found in [TF82]. Brachman [Bra83] critically discusses the meaning of **is-a** relationship types.

Recently, some authors have tried to incorporate more object-oriented features—in particular, the behavioral (operational) aspects—into the conceptual design methodology. Most of this work is still based on the entity-relationship model. Booch [Boo91], Rumbaugh et al. [RBP+91], Wirfs-Brock, Wilkerson, and Wiener [WBWW90], and Coad and Yourdan [CY91a, CY91b] propose such methodologies.

Galileo [ACO85] is a strongly typed conceptual data modeling language that incorporates many object-oriented concepts—even though it is more closely related to the semantic data models than to object-oriented models.

3

Application Areas

There are quite a few advanced application areas—within, for example, science and engineering disciplines—for which database support would be very beneficial, as motivated in Chapter 1. In this chapter we will survey just three engineering application areas:

1. Geometric modeling

2. Computer-integrated manufacturing (CIM)

3. Very large scale integration (VLSI) design

These disciplines are investigated with respect to their data modeling requirements. For each of them some (but by far not all) representative entities (objects) are analyzed concerning their structural representation. Furthermore, we investigate the application-specific behavior of these objects in order to analyze the data (object) manipulation requirements an effective database system would have to meet.

3.1 Geometric Modeling

3.1.1 Representation Schemes for Rigid Solids

The basis for (almost) any mechanical engineering application—in particular mechanical computer-aided design (CAD)—is a way to store information about geometric objects in a computer. There are several—quite different—representation methods for solid objects. Some of them are investigated in this section. We do not attempt to give a complete and formal definition of all existing representation schemes for three-dimensional solid objects. Rather, we restrict ourselves to outlining the most important schemes. Only those aspects important to the design of database support for the particular representation are described in detail.

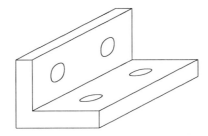

Figure 3.1: A Simple Solid Geometric Object: "Bracket with Four Holes"

There are three representation schemes for which database support is feasible:

1. Primitive instancing

2. Constructive solid geometry (CSG)

3. Boundary representation (BR)

Our presentation is based primarily on the example geometric object of Figure 3.1: a bracket with four holes. Even though this example object is fairly simple, it should suffice to demonstrate the main characteristics of two of the three representation schemes.

Primitive Instancing

In this approach every geometric object is defined as a special instance of a generic primitive object. In relational database terminology this means that one would create a relation for every generic object type. The attributes of the relation would correspond to the parameters that describe the geometric object. Then, each geometric object is stored as a tuple of the relation corresponding to the generic entity type. In this chapter, we use object type as a synonym for generic entity type.

An example of a generic entity type would be brackets with holes. An instance of this object type is the bracket with four holes shown in Figure 3.1.

Now let us consider a generic object type that would describe the object type bracket with a variable number of holes:

>   **generic type** BRACKET (#holes: **integer**)
>   length: real;
>   width: real;
>   height: real;
>   material:{iron, copper, ... };
>   ...

```
    holes: array[1..#holes] of
        begin record
        diameter: real;
        location: array[1..3] of real
        end record
  end generic type BRACKET.
```

A particular object of type *BRACKET* with four holes is instantiated as follows:

create BRACKET(4)

The reader will notice that this is a very simple representation for well-known and highly structured assembly objects, such as brackets, nuts, cog wheels, shafts, etc. Unfortunately, there exist applications where the majority of mechanical objects is produced only in relatively small quantities, i.e., in the order of 5–500. This means that the number of instances of a particular object type is fairly small. On the other hand, there is usually a large number of different generic object types. The primitive instancing approach is not useful in such applications, since it requires the specification of a generic object type for each object type. In database terms this means that we would have to create an abundance of different relations, each consisting of only a small number of tuples. This is the reason why this approach is not always usable in a general purpose CAD system.

Constructive Solid Geometry

The CSG scheme is a volumetric representation of geometric objects. In this approach an object is described as a composition of a few primitive objects. The composition is achieved via motional or combinational operators. Example operators are the (regularized) union, intersection, and difference of two solid objects. Motional operators are, for example, rotate and scale. The description of a geometric object in CSG format is a tree that is defined by the following context-free grammar:

```
⟨mechanical part⟩ ::= ⟨object⟩
⟨object⟩           ::= ⟨primitive⟩|
                      ⟨object⟩ ⟨motion op⟩ ⟨motion arguments⟩|
                      ⟨object⟩ ⟨set operator⟩ ⟨object⟩
⟨primitive⟩        ::= cube | cylinder | cone | ...
⟨motion op⟩        ::= rotate | scale | translate | ...
⟨set operator⟩     ::= union | intersection | difference | ...
```

The description of a geometric object in CSG format forms a tree where the root of the tree represents the solid object being modeled. In Figure 3.2 we show the

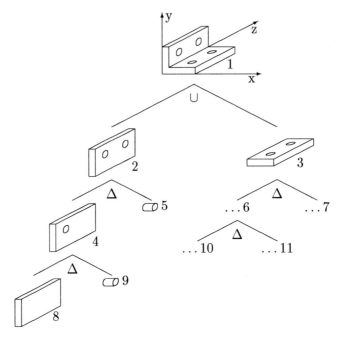

Figure 3.2: CSG Tree of the Bracket

CSG tree for an example object "bracket with four holes". In the CSG tree each
nonterminal node represents an operation, either a rigid motion or a combinational
(set) operator. Terminal nodes either represent a motion parameter or a primitive
object. Each primitive object is described by its parameters, such as length, width,
and height, as well as its relative position. The geometric primitives—the terminals of
the CSG grammar—could be modeled using the primitive instancing representation
scheme introduced in the preceding subsection.

Since a geometric object has to be decomposed into primitive objects (corresponding to the leaves of the tree) the CSG tree of a complex mechanical part can become
very high. This might lead to inefficient data retrieval if there is no suitable data
access support.

Boundary Representation

In the boundary representation scheme a solid object is segmented into its nonoverlapping faces. Each face, in turn, is modeled by its bounding edges and vertices. We
present the representation of a cuboid in Figure 3.3.

From a database point of view, we note that this representation scheme consists
of different abstraction levels, i.e.:

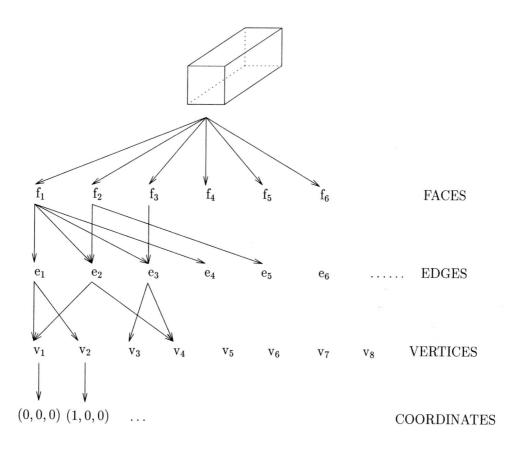

Figure 3.3: Boundary Representation of the Cuboid

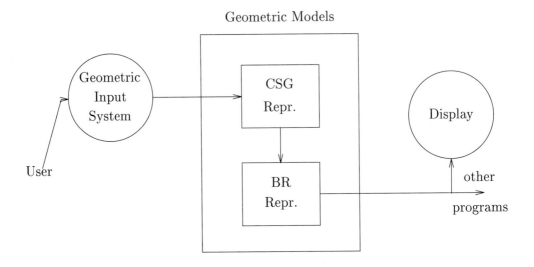

Figure 3.4: Typical Architecture of a Geometric Modeling System

- The *topological* representation, which consists of faces, edges, and vertices

- The *metric* dimension, which is modeled by the coordinates of the bounding vertices

In contrast to the CSG scheme, the height of the tree is constant, that is, 3. A more complex solid object just leads to more nodes in the tree without increasing the height.

The lowest level of the tree stores the metric information, i.e., three-tuples (x_i, y_i, z_i) for vertex v_i. The second level of the tree stores the edges as combinations of vertices. Edge e_i is represented by the tuple (v_{i1}, v_{i2}). In the topmost level of the tree each node describes a *variable* number of edges that represent the boundaries of one face of the rigid object.

Architecture of Typical Geometric Modeling Systems

The CSG scheme is, together with the boundary representation, the most widely used in existing systems. Whereas the boundary representation supports in particular the automatic manipulation of objects, the CSG representation is mostly used to support the input of geometric objects by the user of the geometric modeling system. Figure 3.4 illustrates that the input to the geometric modeling system is usually via the CSG representation. The reason is that this representation scheme is much easier to handle for the user, i.e., the engineer, than the boundary representation. Then, internally, the CSG representation is automatically transformed into the boundary representation. The automatic manipulation of objects is mostly performed on the boundary representation.

3.1.2   Geometrical Transformations

In order to be able to move an object to different locations and to manipulate a particular object in size, one can apply three geometric transformations to the object stored in the geometrical database. The three possible transformations are *translate, scale*, and *rotate*. We will briefly explain how these operations are executed on an example object, say, a cuboid, which is represented by its eight vertices $v_1, v_2, \ldots, v_8$.

Translation

Translation corresponds to moving the geometrical object within the three-dimensional (3-D) coordinate system relative to the origin. The orientation of the object in the 3-D space is not altered by a translation. Manipulating the orientation is achieved by rotations, which will be explained later. A translation is defined by its translation vector $T = (D_x, D_y, D_z)$. A single vertex is translated by adding the translation vector to the vector representing the vertex in the 3-D coordinate system, i.e.:

$$v_i = (x_i, y_i, z_i)$$

$$T = (D_x, D_y, D_z)$$

$$T(v_i) := v_i + T = (x_i + D_x, y_i + D_y, z_i + D_z)$$

To translate a geometrical object represented in boundary representation requires translating all vertices of the object that are stored in the BR schema.

Homogeneous Coordinates

The other two transformation operations, i.e., scaling and rotation, can be naturally defined as multiplications of the vertex (vector) with a corresponding transformation matrix, as we will show below. In order to be able to combine different transformations of the same object, e.g., rotation and translation, one would like to also represent translation as a matrix multiplication. Then we would be able to combine different transformation matrices by multiplying them.

In order to represent the translation also as a matrix multiplication, the concept of homogeneous coordinates has to be employed as is done in many computer graphics systems. This concept requires a vertex to be represented as a four-element vector instead of three elements. Then vertex v_i is represented as

$$v_i = (x_i, y_i, z_i, 1)$$

Now the translation matrix is formed as a 4×4 matrix. The translation of the vertex v_i is then defined as follows:

$$T(v_i) = (x_i, y_i, z_i, 1) * \begin{pmatrix} 1 & 0 & 0 & 0 \\ 0 & 1 & 0 & 0 \\ 0 & 0 & 1 & 0 \\ D_x & D_y & D_z & 1 \end{pmatrix} = (x_i + D_x, y_i + D_y, z_i + D_z, 1)$$

Scaling

An important concept in viewing geometrical objects on a computer display is varying the size in form of scaling (or stretching). Vertices (as end-points of vectors) can be scaled by S_x along the x-axis, S_y along the y-axis, and S_z along the z-axis, according to the scaling matrix S as follows:[1]

$$S(v_i) := v_i * S = (x_i, y_i, z_i, 1) * \begin{pmatrix} S_x & 0 & 0 & 0 \\ 0 & S_y & 0 & 0 \\ 0 & 0 & S_z & 0 \\ 0 & 0 & 0 & 1 \end{pmatrix} = (S_x * x_i, S_y * y_i, S_z * z_i, 1)$$

Rotation

Rotations are used to change the orientation of geometric objects in the three-dimensional space.

In three dimensions we have to distinguish rotations about the three different axes. Here we will show only briefly the definition of rotation about the z-axis. For more details, the interested reader is referred to literature on computer geometry.

A rotation about the z-axis is defined by the rotation angle Φ. Corresponding to this angle the rotation matrix $R_z(\Phi)$ is constructed as:

$$R_z(\Phi) = \begin{pmatrix} cos\Phi & sin\Phi & 0 & 0 \\ -sin\Phi & cos\Phi & 0 & 0 \\ 0 & 0 & 1 & 0 \\ 0 & 0 & 0 & 1 \end{pmatrix}$$

---

[1]From now on we assume homogeneous coordinates.

The rotation of vertex v_i is then given as $v_i * R_z(\Phi)$. Rotation of a geometric object that is stored in boundary representation is carried out analogously to scaling and translation; i.e., each bounding vertex has to be rotated, in turn.

3.2   Computer-Integrated Manufacturing (CIM)

From the vast amount of information that has to be processed in mechanical engineering applications let us just pick one representative example: the *manufacturing cell*. Among other objects a manufacturing cell consists of robots and assembly items, i.e., rigid solid objects. Clearly, a so-called world model database supporting CIM applications has to store computer models of all objects present in a manufacturing cell. Therefore, in this section we will first analyze the data modeling requirements and then investigate the standard operations to manipulate these objects.

We examine the parameters needed to characterize a robot. We shall try to give a classification of these parameters. Since a robot is a highly complex device, we restrict the discussion of details to the robot axes.

3.2.1   Structure and Characteristic Parameters of a Robot

An industrial robot (often also called a manipulator) can, from a mechanical point of view, be divided into three components: base, arm(s), and gripper.

Whereas base and arm are permanent components of a manipulator, a gripper may be exchanged for another one. Each of these components may again be decomposed into smaller components. This is particularly true for the gripper, which may contain sensors that are of further importance to the modeling task. These few remarks already indicate that robot models will turn out to be extremely complex and certainly go beyond what is needed for this presentation. Consequently, we restrict ourselves to the arm and gripper (without sensors). We discuss which information to keep in the database with a model for these components.

In modeling a robot, and hence its components, four aspects must be covered:

- Mechanical structure

- Geometry

- Kinematics

- Dynamics

The different representational dimensions of an industrial robot are depicted in Figure 3.5.

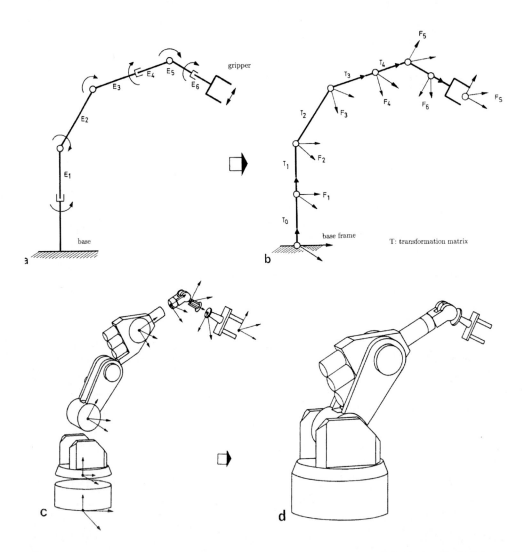

Figure 3.5: Models to Describe a Robot Configuration: (a) Joint Definition, (b) Kinematics Model, (c) Parts Model, and (d) Geometric Representation

Without going into the details, let us sketch the most important representation aspects. The mechanical structure has been outlined above but needs some more refinement. An arm is composed of a number of arm segments (also called axes) that are interconnected via joints. Likewise, a gripper may consist of several axes and, additionally, of one or more tools. Examples of tools are facilities for drilling, welding, or screwing. Gripper fingers can also be considered as tools.

The geometric model of a manipulator or its components is primarily intended for graphic representations. Both constructive solid geometry and boundary representations are suitable. Geometric data, e.g., the axis's dimensions, are also needed in order to plan the robot motions.

The kinematics of a robot are defined by number, types, and mutual position of the axes. The description is usually in the form of kinematic chains, where each axis has its own coordinate system whose position is determined relative to the coordinate system of the preceding axis.

The dynamic model describes the gravitational forces and torques influencing the robot and caused by its intrinsic dynamics, as well as the forces and torques caused by the robot motions. More formally, the dynamic model is a mathematical model consisting of higher-order differential equations. Its parameters can partly be obtained from the kinematic and geometric models.

Figure 3.6 summarizes the discussion in the form of an entity-relationship diagram.

3.2.2   Transformation between Different Coordinate Systems

In computer-integrated manufacturing (CIM) one typically works with a large number of different coordinate systems in order to avoid costly transformations each time an object changes its location within the world model. An example may illustrate this: A manufacturing cell is described within its own, dedicated coordinate system, one that is typically called the world coordinate system. A mobile workbench within the manufacturing cell normally has its own reference coordinate system according to which all relevant points of the workbench are described. Otherwise, the computer modeling system would have to transform all points of the workbench each time it is moved within the manufacturing cell because all points are changed relative to the world coordinate system. This transformation is a very costly process when considering that a complex object may be described by hundreds of boundary vertices.

Rather, the concept of coordinate system transformation is employed. In this case, the coordinate system of the objects within a manufacturing cell, such as the workbench, is assigned its own coordinate system within which the object description remains invariant against change of position within the encompassing object, in this case the manufacturing cell. The object's coordinate system itself is described in

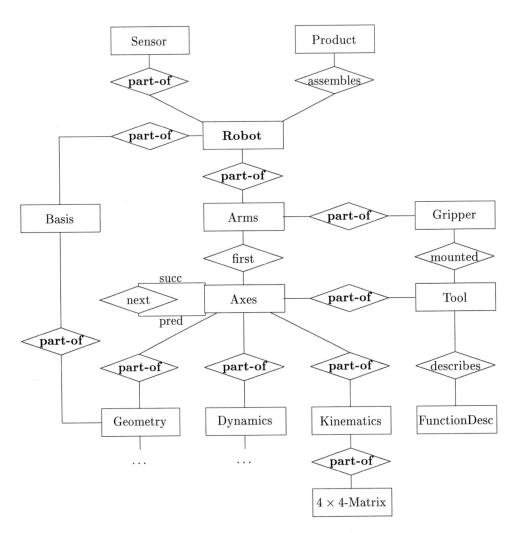

Figure 3.6: Information Structure of a Robot

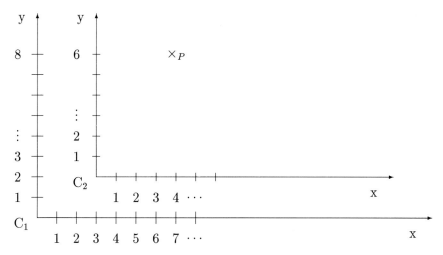

Figure 3.7: Transformation of Coordinate Systems

terms of the world coordinate system. Then we can compute the coordinates of a particular point that is given in one coordinate system within the other coordinate system as follows:

$$C_1(P) = C_2(P) * T_{2,1}$$

where $C_1(P)$ denotes the coordinates of the point P in coordinate system C_1, $C_2(P)$ denotes the same point P in coordinate system C_2, and $T_{2,1}$ is the transformation matrix that would transform coordinate system C_2 into coordinate system C_1.

Let us illustrate this on an example for the two-dimensional case: In this simple example of Figure 3.7, points of coordinate system C_2 are transformed into points of the coordinate system C_1 by translation by the vector $(3, 2)$. Thus, in homogeneous coordinates we have

$$C_2(P) = (4, 6, 1)$$

and

$$T_{2,1} = \begin{pmatrix} 1 & 0 & 0 \\ 0 & 1 & 0 \\ 3 & 2 & 1 \end{pmatrix}$$

Then we compute

$$C_1(P) = C_2(P) * T_{2,1} = (7, 8, 1)$$

3.2.3   Modeling the Kinematics of a Robot Axis

For the remainder we restrict ourselves even further. We examine in more detail how to describe one robot component, an axis, taking the aforementioned four aspects—mechanical structure, geometry, dynamics, and kinematics—into account.

Each axis has its own geometry, i.e., its dedicated coordinate system. In classical robot programming systems the corresponding information is maintained in separate geometry files. Usually, they keep the information redundantly both as a boundary representation for the purpose of internal computations, and in constructive solid representation for communicating with the external world. For efficiency reasons one will often associate with an axis additional attributes such as length, e.g., length of an axis, and volume—although these may be computed from the other parameters.

The kinematics is modeled according to the Denavit-Hartenberg (DH) convention: Place within each arm segment or with each joint a coordinate system (called a motion coordinate system) such that (see Figure 3.8)

1. The z_i-axis of joint i points in direction of the motion axis of joint $i+1$

2. The x_{i+1}-axis is perpendicular to the z_i-axis, pointing away from it

3. The right-hand rule has to be observed which implicitly determines the y-axis

where i and $i+1$ are the sequence numbers of two successive joints.

When the entire kinematics is being modeled, neighboring coordinate systems must be related to one another. Take an object described in coordinate system i. In order to express it in coordinate system $i+1$ the following operations must be performed:

1. Rotation about the z_i-axis with angle θ_i such that axis x_i parallels axis x_{i+1}.

2. Translation along axis z_i by an amount s_i.

3. Translation along axis x_{i+1} by the amount a_i such that coordinate system i shares the origin with coordinate system $i+1$.

4. Rotation about axis x_{i+1} by the angle α_i, such that axes z_i and z_{i+1} agree in position and direction.

In other words, the position of any coordinate system relative to a neighboring one can be described by rotational angles and translation lengths.

Characteristic for the description technique used here is that the kinematic parameters are expressed in so-called robot coordinates, i.e., in values that can directly be interpreted by the robot control system and translated into motions. On the other

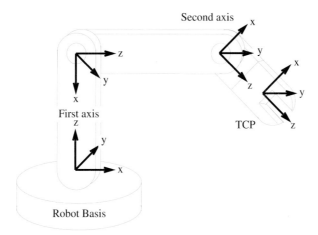

Figure 3.8: Motion Coordinate Systems in Robot Axes

hand, for the programming of robots the objects to be manipulated (picked up) by the robot are expressed in so-called world coordinates. Hence, transformations between robot and world coordinates are necessary.

All transformations may be described in the form of Denavit-Hartenberg matrices. These are 4×4 matrices, where the first three rows and columns contain the information on the orientation of the coordinate system, and the last column describes the position of the origin of the coordinate system (in other words, the last column represents a translation vector with respect to the origin of some other coordinate system). The matrix elements are given by the rotational angles and translation lengths mentioned above.

The matrix $D_{i,i+1}$ denotes the transformation matrix from coordinate system of axis i to the coordinate system of axis $i+1$. $D_{i,i+1}$ can be viewed as being composed of two matrices Z_i and X_{i+1} as $D_{i,i+1} := Z_i * X_{i+1}$. Here Z_i constitutes the transformation matrix of the rotation about and the translation along the z-axis, X_{i+1} constitutes the transformations about/along the x-axis. As homogeneous matrices they have the form

$$Z_i = \begin{pmatrix} cos(\theta_i) & -sin(\theta_i) & 0 & 0 \\ sin(\theta_i) & cos(\theta_i) & 0 & 0 \\ 0 & 0 & 1 & s_i \\ D_x & D_y & D_z & 1 \end{pmatrix}$$

$$X_{i+1} = \begin{pmatrix} 1 & 0 & 0 & a_i \\ 0 & cos(\alpha_i) & -sin(\alpha_i) & 0 \\ 0 & sin(\alpha_i) & cos(\alpha_i) & 0 \\ 0 & 0 & 0 & 1 \end{pmatrix}$$

DH matrices are a particularly elegant way to solve the coordinate transformations in robot programming. Since they are based on the concept of homogeneous coordinate systems, all transformations within the kinematic modeling of an axis may simply be expressed in the form of matrix multiplications.

Thus, to compute the location of some axis n of the robot the coordinate transformations of all axes from the base up to axis n can be multiplied:

$$\prod_{i=0}^{n-1} (Z_i * X_{i+1})$$

Parameters needed to compute the dynamic behavior of an axis are, among others, its mass, its maximum acceleration, and the maximum velocity of the joint.

3.2.4   Simulation of Assembly Operations

As described in [DDKL87] the so-called world model database forms the central part of a robot programming system. One major task in such a system is to simulate off-line robot operations, e.g., assembly operations of the form:

> *mount cog wheel c on shaft s*

The standard geometrical transformations described above are the operations used to model such an operation. We assume that object c (the cog wheel) exists at some location in the world of this robot application; that is, it exists in the world model database. The same holds for object s, the shaft onto which c has to be mounted. Simulating this assembly operation means, in terms of the world model database, changing the location of object c. In our particular case this is achieved by the following (standard) geometric operations:

1. Translate c by T_1 *(pick up object c)*.

2. Rotate c about the z-axis by $R_z(\Phi)$.

3. Rotate c about the x-axis by $R_x(\Theta)$ *(rotate the cog wheel)*.

4. Rotate c about the y-axis by $R_y(\Gamma)$.

5. Translate c again by T_2 *(mount object c on the shaft s)*.

A program fragment that would achieve this transformation is given below:

for all v_i **in** $\{v_1, v_2, \ldots\}$ **do**
    begin
        $v_i := v_i * T_1;$
        $v_i := v_i * R_z(\Phi);$
        $v_i := v_i * R_x(\Theta);$
        $v_i := v_i * R_y(\Gamma);$
        $v_i := v_i * T_2;$
    end

In this program fragment we assumed that the object c is represented by the bounding vertices v_1, v_2, ... This results in altogether $5 * j$ multiplications of a vector with a matrix, where j is the number of vertices used to describe object c. For example, if eight vertices are used to model the object c the above method would result in 40 multiplications of a vector with a matrix. A much more efficient way is to combine the transformation matrices $T_1, R_z(\Phi), R_x(\Theta), R_y(\Gamma)$ into one matrix M.

$$M := T_1 * (R_z(\Phi) * (R_x(\Theta) * (R_y(\Gamma) * T_2)))$$
for all v_i **in** $\{v_1, v_2, \ldots\}$ **do**
    $v_i := v_i * M;$

Besides computing M, this method results in only eight multiplications of a vector with a matrix.

3.3   Computer-Aided VLSI Design

VLSI design is another application area that would benefit substantially from adequate database support. This is due to a number of reasons:

- In VLSI design vast amounts of data are encountered, e.g., geometric layout data.

- A large number of iterations is required until the design is complete and valid. This requires to control the state of design versions.

- There exists a large number of primitive parts (cells) that can be reused in the design of more complex chips. A database of primitive cells would substantially support the reuse of predeveloped library parts.

- The model of a VLSI cell has to span many different representational aspects, each leading to a separate description.

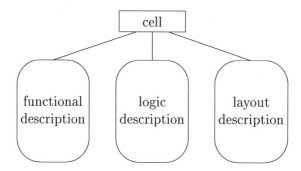

Figure 3.9: Hierarchical Design Approach

A VLSI object can be described by a number of equivalent, but not necessarily isomorphic representations. Each of these representations is complete in the sense that it provides a self-contained description of the VLSI cell. The different representation forms emerge from the distinct viewpoints from which a cell can be modeled:

- *Register transfer description:*
  This viewpoint captures the functional behavior of a chip at an implementation independent level.

- *Logic design:*
  In this representation form the chip's logic is described. For the most fundamental circuits this can be done in the form of logic gate schematics.

- *Layout masks:*
  This representation emphasizes the topological design of the chip, i.e., the placement of its subcomponent, the topological arrangement of wires and pins, etc. From the layout masks the fabrication procedure of the chip is deduced.

Along the natural borderlines of the three different viewpoints of a chip VLSI design is often approached in a hierarchical way. This is outlined in Figure 3.9. The root of the hierarchy represents the VLSI cell under development. A cell is an autonomous unit of the design, e.g., a gate or an aggregate of gates, an adder, an ALU, etc. The root of the subtrees describe the complete cell from the respective viewpoint, e.g., the root of the *functional* hierarchy models the behavior of the cell. The children within a subtree represent major subcomponents in the design of the complete cell. Each subsystem itself has to be fully specified—unless it is a primitive cell that can be extracted from some existing library. The three different subtrees of the design task can often be developed in parallel, once their common interface has been agreed upon.

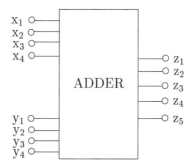

Figure 3.10: A Block Diagram Representation of an Adder

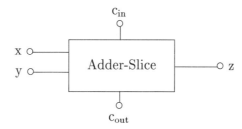

Figure 3.11: Interface of an Adder Slice

3.3.1 The Different Design Dimensions

In accordance with the major representation forms for VLSI chips, three design dimensions are typically distinguished in VLSI CAD. Independent of the three dimensions a chip is—at the very highest level—modeled as a block diagram. For example, the block diagram of a 4-bit adder is pictured in Figure 3.10.

The block diagram representation highlights the chip's interface, which is essential in order to use it as a subcomponent of some other design. The block diagram serves primarily as a documentation and organizational aid whose key information is how subsystems are to be wired together. The wiring specifies how outputs of one component are used as inputs for some other component.

Let us demonstrate the specification of interconnections between subcomponents on our example chip. Figure 3.11 shows the block diagram of an adder slice, which will serve as a basic building block in the realization of a 4-bit adder. This block diagram determines the connection points of an adder slice, but it leaves the realization still unspecified.

Based on the adder slice interface description, Figure 3.12 shows how the wiring of a 4-bit adder has to be done when using adder slices as subcomponents. The adder

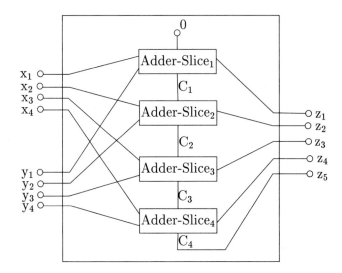

Figure 3.12: Implementation of a 4-Bit Adder

is realized as a "ripple carry through" circuit of four adder slices.

To complete the design process, we show the interface description of a half adder in Figure 3.14 and its implementation as a gate schematics in Figure 3.15.

In this example we developed the integrated circuit in a top-down manner, starting from the highest-level cell and working our way down to the most basic subcomponents. Alternatively, a bottom-up—or even a mixed—approach would be feasible. In large projects, the overall design process is subdivided into smaller tasks that are handled individually by different engineers. This requires that each design part observes the specified interface of the respective subpart.

The Functional Dimension

The functional dimension covers a fairly high-level description of the chip. It emphasizes the behavioral component of the cell. In the case of pure combinational logic the functional description of a chip can be captured in a truth table. For example, the adder can functionally be modeled with the subsequent truth table:

x_1	x_2	x_3	x_4	y_1	y_2	y_3	y_4	z_1	z_2	z_3	z_4	z_5
0	0	0	0	0	0	0	0	0	0	0	0	0
1	0	0	0	0	0	0	0	1	0	0	0	0
	...				...				...			
1	0	0	0	1	0	0	0	0	1	0	0	0
	...				...				...			

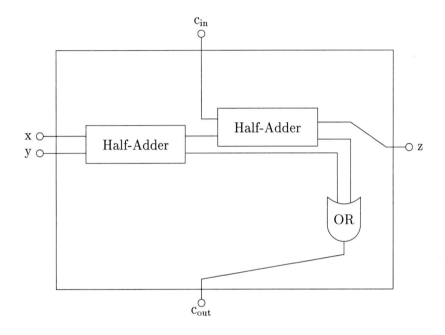

Figure 3.13: Implementation of an Adder Slice

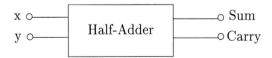

Figure 3.14: Interface of a Half-Adder

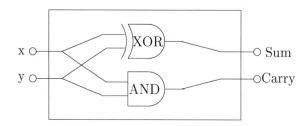

Figure 3.15: Implementation of a Half-Adder

If the circuitry contains storage cells the functionality depends on the internal state of the cell. In this case the behavior may be specified in the form of *state transition tables*. For more complex chips whose behavior cannot be captured easily in the form of transition or truth tables, special hardware description languages have been developed. They allow to model the functionality of a circuit as a program. Such programs are particularly helpful in simulations of a complex chip that is composed of more basic components each of which is described with its own hardware description program.

The Logical Representation

The logic of a simple chip, such as an adder slice, can still be represented as a gate schematic. Figure 3.15 shows the logic of a half adder, a subcomponent in the realization of an adder slice. Based on the half adder an adder slice can be logically realized as shown in Figure 3.13.

Geometrical/Physical Dimension

This dimension of the design process is concerned with the physical layout of a chip. The geometrical design of a chip is viewed as a two-dimensional grid of tiles. The placement of the subcomponents of the complete chip with their interconnection points is called the *floor plan*. Figure 3.16 represents the (simplified) *floor plan* of an adder slice. The floor plan is very similar to the block diagram, except that it contains some more "wiring" information that is usually left out in the block diagram, e.g., the connection points for ground (GND) and power supply (VDD). The floor plan constitutes a topologically correct placing of the input/output points of a chip and its subcomponents. It does not, however, lead to a fabrication method for the chip.

The fabrication procedure of a chip is determined from the *layout* or *mask geometry*. The mask layout distinguishes between the different layers of a chip, i.e.,

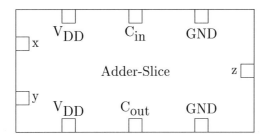

Figure 3.16: Layout Description of an Adder Slice

- Diffusion

- Polysilicon

- Metal

For each of the layers a set of paths is specified. A path is modeled as a rectangle that is described by its width, length, center location, and orientation. Two crossing paths, one in the polysilicon layer and the other in the metal layer, form a transistor. Crossing paths in the diffusion and the polysilicon layer have no functional significance.

Similar to the decomposition of the block diagram representation of a circuit into subcomponents, the geometrical layout design can often be simplified by replicating basic (library) cells in the overall chip layout. This is particularly easy when the circuit has a very homogeneous structure with little "random" wiring between nonadjacent subcells.

For example, to define the layout of the adder, one could first design the geometrical layout of one adder slice and then insert this adder slice four times at certain positions in the layout grid. All paths within one subcomponent layout are to be interpreted as relative to the actual position of the component on the chip. The placement of a subcell on the chip corresponds to the geometric operations *rotate* and *translate*. Most layout tools of VLSI chips allow the rotation only by multiples of 90 degrees.

The layout description of a cell has to observe fabrication-dependent consistency constraints. For example, different fabrication procedures require distinct minimum distances between components (paths) on certain layers. This is contrary to the overall objective to reduce the size of a chip to the minimum.

From the path description a mask is generated and is used to fabricate the wafers, i.e., the essential components of a chip, by etching out the paths in the respective layers.

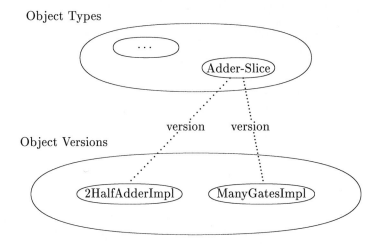

Figure 3.17: Versioning of an Object Type

3.3.2 Versioned Objects

An important aspect in VLSI design is *versioning* of objects. A versioned object is an occurrence of a single object type. The relationship between an object type and its versions is sketched in Figure 3.17.

Versions of the same object type share a common interface and the same functionality. This means in particular that one version of an object can be replaced by another version without compromising the functionality of the chip. For example, the 4-bit adder could be realized with adder slices of either version:

- The adder slice that is built from two half adders and a gate, the one that is called *2HalfAdderImpl* in Figure 3.17

- The adder slice version that is realized as a basic building block with only gates, called *ManyGatesImpl*

The common properties of versioned objects are factored out in the object type. Among these are definitely the functionality and the interface.

3.3.3 Cell Instantiation

The term *cell instantiation* in VLSI design has a different semantic from that in the context of object-oriented models. In the VLSI context instantiation refers to generating a copy of a prespecified cell. This is needed whenever the cell is to be used as a subcomponent of another cell. For example, the adder implementation, shown in Figure 3.12, contains four instantiations of adder slices, each of which is a fixed copy of the adder slice modeled in Figure 3.13. An instantiation in this context cannot be modified; it is a nonmodifiable copy of some existing cell specification.

3.4 Exercises

3.1 For the conceptual schema for modeling robots (cf. Figure 3.6) specify the integrity constraints of the various relationships:

- Using the functionality notion—i.e., $1 : 1$, $1 : N$, and $N : M$

- Using the $(\min, \max)$ notation—as introduced in Exercise 2.8

3.2 A 32-bit adder can be constructed from 32 individual full adders. Each full adder, in turn, is composed of two half adders. A half adder can, for example, be built with one AND gate and two NOR gates.

Convert the preceding (verbal) description of a 32-bit adder into a corresponding entity-relationship schema. Make use of the **is-a** aggregation relationship type. Use the $(\min, \max)$ notation—of Exercise 2.8—to specify the number of constituent subentities that make up the composite object 32-bit adder.

3.3 Outline the data modeling requirements of an engineering application of your own choice. It should be an application domain that you are familiar with— e.g., your area of expertise. Make the requirement specification precise by developing the conceptual schema of (a part of) the application. The conceptual schema should contain the most representative entity types. You should also discuss the data manipulation characteristics of the application domain. In particular, provide a high-level specification of the operations associated with the representative entity types.

3.4 Reconsider Exercises 2.12, 2.13, and 2.14. Revise the initial conceptual schema designs—if necessary.

3.5 Annotated Bibliography

A very thorough review of solid geometric modeling can be found in the survey articles [Req80] and [VR77]. A textbook on this topic was written by M. Mäntylä [Män88]. Another text was written by C. M. Hoffmann [Hof89]. The database aspects of geometric modeling are analyzed in [KW87], [Fis79], and [Mei86]. The design and analysis of very efficient spatial data structures to represent geometric information is described in [Sam90]. The geometric transformations can be studied in more detail in the textbooks [Fv83, PS85]. Lee and Fu [LF83] outline a relational database structure for constructive solid geometry.

There are many good texts on computer-integrated manufacturing. A very comprehensive treatment of computer-integrated manufacturing can be found in [RBD87]. Another comprehensive (though somewhat dated) textbook on CAD/CAM is [GZ84]. The data modeling aspects of CIM are described in various papers:

- [DH86] presents the design of a robot simulation system with emphasis on the data structure definitions.

- [Hei87] reviews the modeling of the robot's kinematics based on the approach proposed originally by Denavit and Hartenberg [DH55].

- Database support for robotics applications—especially simulation of robot assembly operations—is outlined in [KWL86a, KLW87, DKL89, KWL86b].

VLSI design is thoroughly covered in the textbook by Mead and Conway [MC80]. The state of the art in database support for VLSI design applications is illustrated in the monograph by Katz [Kat85]. Another textbook covering information needs of VLSI design is [Rub87]. There are several research papers that cover special aspects of data modeling requirements in VLSI databases:

- [AKMP85] and [BK85] investigate object-oriented modeling of VLSI design data.

- [DL88] treats the versioning problem of data with particular emphasis on VLSI applications.

- [GCG+89] gives an account of using an object-oriented database system, Vbase [Ont87], in design applications.

- Concurrency control requirements of VLSI CAD applications are covered in [KW84, KL84, Kat90].

Part II

The Relational Approach

4

The Relational Database Model

In this chapter the *relational* data model, currently the most popular model, is described. Database management systems (DBMSs) based on the relational data model play a predominant role in today's market. Estimates are that 80% of currently sold DBMS's are based on the relational model. This demonstrates that the relationally organized database systems have almost displaced the other two traditional data models: the *network* and the *hierarchical* model. Today, the hierarchical model is almost propelled into a historical position: Organizations that introduced database technology early on—when no relational systems were available—were forced to settle for a hierarchical system. The enormous cost of restructuring their information system to a relation-based structure might prevent these organization from shifting to relational DBMSs. The network data model, sometimes referred to as the CODASYL model, still occupies a stronger position than the hierarchical systems. This is probably due to the fact that systems based on the network model provide more leverage for optimization than relational systems. Therefore, for a long time performance was the prevailing argument in favor of network-based systems as opposed to relational DBMSs. The enormous efforts invested on optimization in relational systems has—however—almost closed the performance gap to network systems. This suggests that suppliers of relational DBMSs will even increase their share of the market compared to network and hierarchical systems. However, it is expected that the object-oriented DBMS vendors will be able to obtain a growing share of the information technology market.

In this chapter we will first introduce the essential concepts of relational data modeling. Then, the relational data manipulation and query language SQL is discussed—on the basis of the sample relational system ORACLE.

4.1   Definition of the Relational System

The underlying mathematical concept of the relational model is—as the name actually implies—the set-theoretic notion of *relation*. A (mathematical) relation of arity n is defined as a subset of the Cartesian product of n *domains* D_1, ..., D_n. A domain is a set of atomic values, e.g., the set of characters, the set of integers, etc., are valid domains. The Cartesian product, denoted as

$$D_1 \times D_2 \times \cdots \times D_n$$

is the set of n-tuples

$$\{[a_1, \ldots, a_n] \mid a_1 \in D_1, \ldots, a_n \in D_n\}$$

An example of a Cartesian product of two domains might be constructed as follows:

- D_{string} is the set of all strings of length less than 20.

- $D_{natural}$ is the set of natural numbers $\{1, 2, \ldots\}$.

The Cartesian product $D_{string} \times D_{natural}$ includes tuples like

["John Smith", 32], ["Mickey Mouse", 60], and ["X!U#RS?", 923].

A relation is any subset of the Cartesian product of at least one domain. Mathematically, a relation, say, R, is denoted as

$$R \subseteq D_1 \times \cdots \times D_n$$

For example, the set *Persons* consisting of two tuples, i.e.,

Persons = { ["Micky Mouse", 60], ["Mini Mouse", 50] }

constitutes a valid relation over the two domains $D_{string} \times D_{natural}$. The members of a relation are called *tuples*. A relation $R \subseteq D_1 \times \cdots \times D_n$ is said to have *arity* or *degree n*.

It is often easier to view a relation as a table than as a mathematical set. For example, the relation *PhoneBook* based on the Cartesian product

$$PhoneBook \subseteq D_{string} \times D_{string} \times D_{natural}$$

has—in tabular format— three columns and as many rows as elements (tuples) in the set. The columns are named according to the semantics of the respective values. The column names are called *attributes*. Our example relation *PhoneBook* has the three attributes *LastName*, *FirstName*, and *Phone#*, which yield the following table (with three example tuples):

PhoneBook		
LastName	FirstName	Phone#
"Mouse"	"Micky"	213 747 2222
"Duck"	"Donald"	213 749 0007
"Rabbit"	"Roger"	212 123 1130
...	...	...

4.1.1 Relational Schema

A relational database, consisting of a collection of relations, has an underlying schema that can, roughly speaking, be viewed as the type of the relations. For a relation R the schema is denoted as

$$R : \{[A_1 : D_1, \ldots, A_n : D_n]\}$$

This schema defines the name of the relation, R, and the names of the attributes, A_1, ..., A_n, which have to be pairwise distinct. Each attribute A_i is constrained to draw its values from a domain of atomic values, D_i. Note, that the domains D_1, ..., D_n do not have to be distinct, as our example *PhoneBook* indicated. The inner brackets $[\ldots]$ denote the type of a single tuple, the outer braces $\{\ldots\}$ denote the set of the tuples. The set of all relational schemas constitutes the *database schema*. The collection of current values (tuples) stored in a relational database schema is called the relational database.

4.1.2 Relational Key

In the entity-relationship model a key was defined—rather informally—as a set of attributes that uniquely identify an entity. In the relational model a key can be defined more formally. A key K of a relation R is a subset of the attributes of R and satisfies the following two conditions:

1. The relation R will never contain two tuples t_1 and t_2 that agree on all attributes in K.

2. No proper subset of K satisfies property 1.

It is, however, quite possible that a relation R has two (or more) distinct keys K_1 and K_2. Typically, one such key is chosen as the *primary* key.

The *PhoneBook* example has the key $K = \{\text{Phone\#}\}$ under the assumption that no two entries have the same phone number; i.e., a phone number cannot be shared by two people. On the other hand, the set of attributes *NonKey* = {*LastName*, *FirstName*} is not a key because, e.g., some people have more than one telephone number.

4.2   Mapping a Conceptual Schema into a Relational Database Schema

The database schema that is developed in the conceptual design phase of a database application constitutes the basis of the relational schema design. A thorough conceptual design almost automatically leads to a well-defined, consistent relational schema. In the following, transformation patterns are described that transform an entity-relationship schema into a corresponding relational schema. However, it cannot always be guaranteed that the relational schema captures all the semantics that are implicitly contained in the ER schema. This is due to the fact that the relational model—as an implementation model—provides less expressive structuring concepts than the semantic ER model.

The subsequent discussion will be based on the example conceptual schema of Chapter 2 (cf. Figure 2.3), which is redrawn in Figure 4.1. It represents a very simplified model of a company divided into divisions. Each division employs engineers who design products and control robots. The products are assembled by robots that are equipped with appropriate tools. A product may be composed of several subparts.

4.2.1   Representing Entities in Relations

Each entity set that is identified in the conceptual ER schema is modeled in a separate relation. In general, an entity set E with the attributes $A_1, \ldots, A_n$ is mapped to the relational schema

$$E : \{[A_1 : D_{A_1}, \ldots, A_n : D_{A_n}]\}$$

Here the D_{A_i} for $(1 \leq i \leq n)$ denote appropriately chosen domains for the respective attribute A_i.

For each relation representing an entity set a (primary) key is specified according to the key designated in the ER schema. In the relational schema the key is marked by underlining the constituting attribute(s).

Following this procedure for the conceptual schema of Figure 4.1 yields a—still incomplete—relational database schema of five relations:

> Divisions: {[Name: string, Location: string, Profit: money]}
> Products: {[PID: string, Description: string, Price: money]}
> Tools: {[TID: string, Description: string, Precision: real]}
> Robots: {[RID: string, LoadCapacity: real, ReachRadius: real]}
> Engineers: {[SS#: string, LastName: string, FirstName: string]}

A relational schema definition is very similar to a record declaration in programming languages, like Pascal. There is, however, only a very limited collection of legal

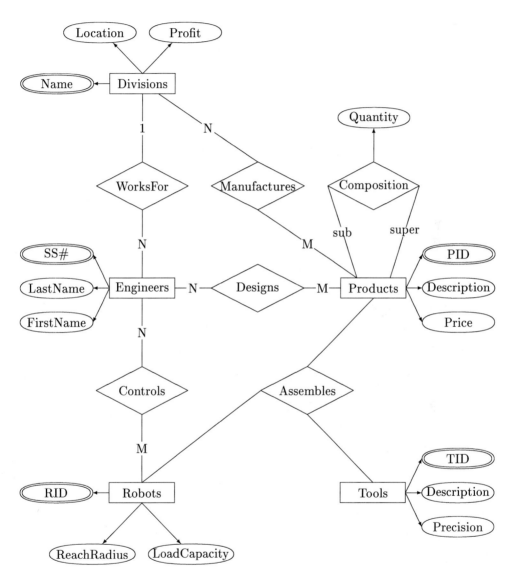

Figure 4.1: Conceptual Model of a Company

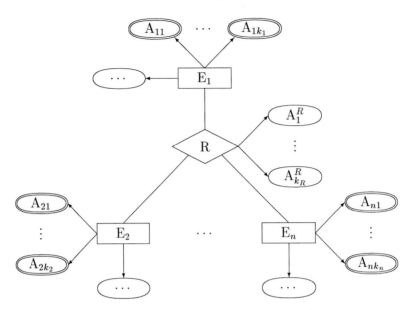

Figure 4.2: General n-ary Relationship

domains (data types) that can be associated with attributes. The relational model requires all attribute domains to be atomic; that is, an attribute value cannot be a composite structure, e.g., a record, itself. This prevented us from defining a *Name* attribute for relation *Engineers* as

Name: [Last: string, First: string]

because it would constitute a composite record-structured attribute.

4.2.2   Representing Relationships in Relations

The entity-relationship model provides two conceptually different structures for modeling data: the *entity set* and the concept of *relationship type* between entity sets. The relational model has only one structuring concept: the relations. This implies that also relationship types that were established in the ER schema have to be mapped to an equivalent relational representation. The most straightforward approach is to model each relationship type as a separate relation. The general transformation rule can be defined on the abstract conceptual schema pictured in Figure 4.2.

The relationship R relates n entity sets $E_1, \ldots, E_n$. The relationship R has itself k_R attributes, denoted $A_1^R, \ldots, A_{k_R}^R$. Let the entity set E_i have a key consisting of the attributes $A_{i1}, \ldots, A_{ik_i}$ for $1 \leq i \leq n$. Without loss of generality, we assume that all the attributes participating in keys of $E_1, \ldots, E_n$ are pairwise distinct—if not, we can always rename any conflicting attributes. Then, the relational schema to model the relationship R is defined as

$$R : \{[\underbrace{A_{11}, \ldots, A_{1k_1}}_{key\ of\ E_1}, \underbrace{A_{21}, \ldots, A_{2k_2}}_{key\ of\ E_2}, \ldots, \underbrace{A_{n1}, \ldots, A_{nk_n}}_{key\ of\ E_n}, A_1^R, \ldots, A_{k_R}^R]\}$$

For the sake of brevity, the domain specification of the attributes is omitted. The relation R can be viewed as a collection of all key attributes of the entity sets related by relationship R. The collection of attributes $A_{i_1}, \ldots, A_{i_{ki}}$ constituting the key of entity set E_i is called a *foreign key*indefKey!foreign of relation R. A foreign key is generally used to refer to a tuple that is described in another relation in which the foreign key is a *primary key*. Later in this chapter we will demonstrate the possible tuples of such a relation that represents a relationship type. Let us first conclude the schema definition for our example application, the model of a company.

In addition to the seven relations for representing entity sets, six relations to pattern the relationship types are defined:

WorksFor: {[DivisionName: string, EngineerSS#: string]}

Strictly obeying our rules the relation would have the following format:

WorksFor: {[Name: string, SS#: string]}

This, however, is less mnemonic than the relation that is obtained by carefully renaming the attributes. The renaming is used to indicate the relation to which a foreign key refers. For example, *DivisionName* refers to a tuple in relation *Divisions* by the *Name* attribute. We have not yet indicated the key of the relation *WorksFor*; let us postpone this discussion to the next section.

Further relational schemas are the following

Designs: {[EngineerSS#: string, ProductID: string]}
Controls: {[EngineerSS#: string, RobotID: string]}
Manufactures: {[DivisionName: string, ProductID: string]}
Assembles: {[RobotID: string, ToolID: string, ProductID: string]}
Composition: {[SubPartID: string, SuperPartID: string, Quantity: integer]}

In relations representing general $N : M$ relationships, such as *Designs, Controls*, and *Manufactures*, all foreign keys in its entirety form the relational key. For example, relation *Designs* has the key {*EngineerSS#, ProductID*}, where *EngineerSS#* is the foreign key referring to relation *Engineers* and *ProductID* refers to relation *Products*. The analogous holds for general ternary $(N : M : P)$ relationship such as *Assembles*.

The last relation, *Composition*, is of particular interest because it models a recursive relationship; that is, it relates entities of the same entity set with each other.

In order to distinguish the role of a participating entity in the relationship, the ER schema contains labels of the two arcs connecting the diamond—representing the relationship—with the rectangle (standing for the entity set). These labels, also called *roles*, are utilized as prefixes to the attribute names in order to distinguish the two entities' role; i.e., *SuperPartID* designates the *Products* entity of which the entity referred to by *SubPartID* is a constituent subpart.

4.2.3   Schema Refinement

The transformation rule that every relationship type is represented by a separate relation is often too restrictive. A particular category of relationships, the binary $1 : N$ relationship type, is typically better represented by including the information within one of the relations that models the participating entity sets. There is one example of a binary $1 : N$ relationship in our company schema: the relationship type *WorksFor* that relates *Engineers* with *Divisions*. Such a $1 : N$ relationship corresponds to a partial function from one entity set to the associated set, in our case:

> WorksFor: Engineers → Divisions

This functional characteristic of the relationship *WorksFor* is also reflected in the relational schema. Since each *Engineer* is associated with at most one *Division*, the *EngineerSS#* is the key of relation *WorksFor*, i.e.:

> WorksFor: {[DivisionName: string, <u>EngineerSS#: string</u>]}

Such a functional relationship can be represented within the relation modeling the entity set corresponding to the domain of the function. In our example, the relation *WorksFor* could be dropped and the relation *Engineers* augmented to:

> Engineers: {[<u>SS#: string</u>, LastName: string,
> FirstName: string, WorksForDivision: string]}

Here, *WorksForDivision* is intended to contain the key attribute value, i.e., the *Name* of the *Division*, in which the respective *Engineer* is employed.

In general, the two relations can be merged because they have the same key—not considering any renaming. By merging them the key attributes of the one relation are ignored while all other attributes are added to the target relation.

We note, that $1 : 1$ relationships between two entity sets E_1 and E_2 can be viewed as a partial function with domain E_1 and range E_2, i.e., $E_1 \rightarrow E_2$, as well as a partial function of the form $E_2 \rightarrow E_1$. Therefore, the relationship may be represented either in the relation modeling E_1 or, analogously, in the relation designed for E_2.

4.3   Sample Relational Database Extension

In Figure 4.3 the extension of the relations introduced above is depicted for a small example database.

4.4   SQL: The Standard Relational Language

In this section the relational data definition, query and manipulation language SQL[1] (structured query language) is discussed. SQL was originally designed in the system R project, a database research project carried out at the IBM Research Center at San Jose, California. Subsequently, SQL was incorporated in the relational products SQL/DS and DB2. In a standardization effort by the American National Standards Institute (ANSI) organization a modified version of the originally proposed SQL, called Standard SQL evolved.

So far, no commercially available DBMS obeys exactly the SQL standard. In order to have a running system as a test vehicle for working out the examples, our following discussion will be based on the "dialect" of SQL that was implemented in the commercial database management system ORACLE. Even though ORACLE's version of SQL differs in some minor details from the proposed standard, we will, nevertheless, often refer to it as just SQL.

4.4.1   Data Definition

In SQL a relation is declared in the following syntax (here, the brackets [...] denote optional parts and the braces {...} denote an arbitrary number of occurrences of the construct inside the braces):

> **create table** ⟨relation name⟩
>> "(" ⟨attribute name⟩ ⟨type⟩ [**not null**]
>> {, ⟨attribute name⟩ ⟨type⟩ [**not null**]} ")"

The ⟨relation name⟩ has to be unique for an individual user only; i.e., different users may have identically named relations. As pointed out before, ⟨attribute name⟩ has to be unique within the relation. The ⟨type⟩ of an attribute is one of the following list of built-in atomic data types[2]:

number   a numeric field that can hold any kind of number; the maximum representation is 40 digits.

---

[1] Often pronounced as SEQUEL.

[2] These data types are ORACLE specific; they may differ substantially in other systems.

Divisions		
Name	Location	Profit
'Big Money'	'Utopia'	50000000
'Lemon'	'Mars'	-10000000
'Wheeler'	'Wheel City'	10000000
'Top Secret'	– –	5000000

Robots		
RID	LoadCapacity	ReachRadius
'R2D2'	500	50
'X4D5'	200	150
'Robi'	150	30

Products		
PID	Description	Price
'Bike'	'mountain bike'	1000
'Wheel'	'bicycle wheel, 26 inch'	50
'Tube'	'tube for bicycle'	5
'Frame'	'aluminum frame'	100
'HAL'	'computer'	10000

Tools		
TID	Description	Precision
'S10'	'screw driver'	0.01
'P01'	'painting'	0.5
'W02'	'welding'	0.001

Engineers			
SS#	LastName	FirstName	WorksForDivision
'007'	'Bond'	'James'	'Top Secret'
'74689'	'Rockefeller'	'Nelson'	'Big Money'
'76591'	'Drais'	'Karl'	'Wheeler'
'78951'	'Spock'	– –	'Big Money'

Controls	
EngineerSS#	RobotID
'007'	'R2D2'
'78951'	'X4D5'

Assembles		
RobotID	ToolID	ProductID
'R2D2'	'S10'	'Bike'
'X4D5'	'W02'	'Frame'

Manufactures	
DivisionName	ProductID
'Wheeler'	'Bike'
'Big Money'	'HAL'

Composition		
SuperPartID	SubPartID	Quantity
'Bike'	'Frame'	1
'Bike'	' Wheel'	2
'Wheel'	'Tube'	1
'Wheel'	'Spoke'	24
'Spoke'	'Screw'	2

Designs	
EngineerSS#	ProductID
'76591'	'Bike'
'78950'	'HAL'

Figure 4.3: Sample Extension of the Relational Schema

number(m,n) decimal number of m digits, n (of the m digits) to the right of the decimal point. The size m cannot exceed 105 digits.

number(m) a numeric field whose maximum number of digits is m; in particular used to constrain numeric attributes whose values are always small in order to preserve space; m may not exceed 105.

float floating point number representation is used; this has an impact on the precision of the stored numbers.

char(n) a character string of fixed length n. Storage space is statically allocated for n characters.

varchar(n) character string of variable length, maximum length is n. Storage is dynamically allocated for attributes of this type.

date valid dates in the range of January 1, 4712 BC to December 31, 4712 AD.

The optional keyword **not null** indicates that the respective attribute has to hold a value of the specified domain (attribute type) at all times. If **not null** is omitted in an attribute specification, a tuple may have an unspecified value for the particular attribute, denoted as **null**. In Figure 4.3 such a **null** value is denoted as $--$.

We can now define a few of the relations of our *Company* database in SQL:

```
create table Divisions
    (Name char(20) not null,
     Location char(20),
     Profit number(14,2))

create table Robots
    (RID char(10) not null,
     LoadCapacity float,
     ReachRadius float)

create table Engineers
    (SS# char(9) not null,
     LastName char(20) not null,
     FirstName char(20),
     WorksForDivision char(20))

create table Controls
    (EngineerSS# char(9) not null,
     RobotID char(10) not null)
```

create table Composition
    (SuperPartID char(10) **not null**,
    SubPartID char(10) **not null**,
    Quantity number(5) **not null**)

Key attributes should always be defined as **not null** because, by the definition of key, it does not make sense to leave a key attribute of a tuple unspecified.

4.4.2   Schema Alteration

It is possible to dynamically augment existing relations by an additional attribute. For example,

    alter table Engineers
      add Salary number(10,2)

incorporates an attribute *Salary* into the relation *Engineers*. The additional attribute may not be specified as **not null**, because existing tuples of the respective relation would, of course, violate this condition.

4.4.3   Indexing

An index in SQL serves two purposes:

- It expedites the search for tuples that have certain attribute values.

- A **unique** index also supports the concept of a key that is not otherwise integrated in the SQL language.

An index is specified in the following syntax:

    create [**unique**] **index** ⟨indexname⟩
      on ⟨relation name⟩
        "(" ⟨attribute name⟩ [**asc** | **desc**]
        {, ⟨attribute name⟩ [**asc** | **desc**]} ")"

The syntax of the index creation indicates that an index may be defined on a concatenation of attributes of a relation. Furthermore, it can be specified whether the index is to be maintained in **asc**ending or **desc**ending order of the attribute values. This is important for sequential processing of the tuples if a particular order of access is required.

The specification of an index triggers internally the creation of an access path structure, typically in the form of a B-tree or a hash table. The particular access path data structure depends on the system implementation; in some systems, e.g., INGRES, the database user can explicitly choose the most appropriate structure according to the projected access profile of the relation.

Modeling a Relational Key by an Index

The characteristic of a key is that the relation cannot contain two (or more) tuples that agree in their values for all attributes constituting the key. This consistency constraint can be enforced in ORACLE SQL[3] by defining a unique index over the attributes of which the key is composed. For example, the key of *Engineers* consists of one attribute, *SS#*.

> **create unique index** KeyOfEngineers
> **on** Engineers (SS#)

On the basis of this index the system would reject the insertion of a new tuple or the modification of an existing tuple, which results in a database state where two tuples with matching *SS#* attributes would be present in the relation *Engineers*.

In general, a key may be composed of more than one attribute. For example, the relation *Controls* has the key consisting of all two attributes, *EngineerSS#* and *RobotID*. The uniqueness of these two attributes conforms to the exclusion of multiple occurrences of tuples with all identical attribute values from the relation:

> **create unique index** NoMultiSet
> **on** Controls (EngineerSS#, RobotID)

We should mention that in Standard SQL the relation declaration language was augmented by a **unique** clause, which allows to specify the relational key directly. In this case the data definition statement for *Engineers* would have the form

> **create table** Engineers
> (SS# char(9) **not null**,
>   . . .
> **unique**(SS#))

Additional (Secondary) Indexes

In addition to the unique indexes that serve as an access path as well as a consistency constraint for preserving the key specification of a relation, one should define indexes on those attributes or attribute combinations over which a search is frequently issued. For example, the field *LastName* of *Engineers* is likely to be a frequent search argument in queries like

> "Find the *Engineer(s)* whose *LastName* is 'Smith'"

The execution of such a query can be expedited by a so-called *secondary index*, which is created by the statement

---

[3]This also holds for the relational systems SQL/DS, DB2, and INGRES.

```
create index FastNameAccess
    on Engineers (LastName)
```

In this case the keyword **unique** is omitted because the relation *Engineers* may very well contain two or more tuples with identical *LastName* attributes.

A combination of index attributes can also be supported, e.g.,

```
create index FastFullNameAccess
    on Engineers (LastName, FirstName)
```

Now a search for, e.g., *Engineers* whose *LastName* is 'Smith' and whose *FirstName* is 'Henry' is optimally supported. The index *FastNameAccess*, on the other hand, would support the query execution partially; i.e., it facilitates fast access to tuples whose *LastName* is 'Smith'. All the qualifying tuples would still have to be inspected in order to look up whether their *FirstName* is 'Henry'. This is avoided by the index *FastFullNameAccess*.

This kind of composite index is meaningful only when many tuples in the respective relation agree on the first index attribute, e.g., if many *Engineers* agree on their *LastName* attribute.

Optimal Number of Indexes

The question about the optimal number of indexes per relation should be raised. This question, however, cannot be answered in general. The decision has to be deduced from the access and update profile of the respective relations. In general, an index expedites the access operations (if they are based on indexed fields) and delays update, insert, and delete operations because index structures have to be updated as well. Therefore, for a static relation—one that is mostly accessed in read operations—a higher number of indexes is appropriate than for a highly dynamic relation. A relation is called highly dynamic if a large portion of the database operations on this relation are modifications, i.e., updates, insertions, and deletions. In general, one should bear in mind that the more dynamic a relation is, the less indexes are advisable. On the other hand, a more static relation, i.e., one that is mostly read as opposed to modified, is more likely to warrant a higher number of indexes. The minimum number of indexes for a relation is 1, i.e., the index that preserves the key characteristics. The maximal possible number is, of course, the number of attributes of the relation—in this case additional indexes based on combinations of attributes provide only marginal additional access support. As a rule of thumb one could state that in addition to the one unique index that preserves the key one should create two or three secondary indexes per relation.

An index that turns out to be too costly or obsolete can be dropped, e.g.:

```
drop index FastFullNameAccess
```

4.4.4   The Syntax of the Query Language

Here we will describe the interactive SQL query language, which provides an inter-active interface for formulating adhoc queries. Later, in Section 4.5 we will discuss the embedding of SQL in a host programming language that accommodates compu-tationally more complex data manipulation tasks.

All SQL database queries obey a syntax that is braced by the keywords "**select** ...**from** ...**where** ..." In detail, the syntactical form of SQL queries is as follows:

(1)   **select** ⟨attribute list⟩
(2)   **from** ⟨relation list⟩
(3)   [**where** ⟨search condition⟩]
(4)   [**group by** ⟨attribute list⟩
(5)   [**having** ⟨search condition⟩]]
(6)   [**order by** ⟨attribute list⟩]

The result of such a query is always itself a (temporary) relation consisting of the attributes listed in line (1). If more than one attribute name follows the **select** key-word, they are separated by commas. The **from** clause of line (2) lists the relation(s) over which the query is formulated—again multiple relation names are separated by commas. Line (3) states a predicate that has to be satisfied by the tuples of the argument relations in order to be included in the result. The **group by** and **order** clause are described later in this section.

A Simple Select

In order to illustrate the most basic query facilities of SQL, let us retrieve the (already mentioned) *Engineers*, whose *LastName* is 'Smith':

> **select** SS#, LastName, FirstName
> **from** Engineers
> **where** LastName = 'Smith'

Every tuple of the relation *Engineers* that satisfies the predicate following the **where** clause is included in the result. The first line of the above query can be abbreviated to:

> **select** *
>   ...

The special symbol * denotes that the result relation should contain all attributes of the argument relation(s).

Ordering

The **order by** clause allows us to retrieve tuples of the result relation in a special, user-defined sequence based on some attribute values. The relation *Engineers* could be retrieved in alphabetical order according to the *LastName* field:

> **select** ∗
> **from** Engineers
> **order by** LastName **asc**

The keyword **asc** is the default—standing for **asc**cending. Alternatively, **desc**ending sorts the tuples in opposite sequence.

4.4.5   The Join Operation

As pointed out earlier, the relational model does not provide any explicit reference or relationship concept that relates tuples with each other. Rather, the association is achieved by so-called foreign keys, that is, user-supplied attribute values that refer to another tuple whose key attributes match the foreign key. In a *join* operation tuples associated in such a way can be combined.

Joining two different Relations

As a first example of a join operation consider the following query:

> "Find the *Engineer(s)* that *Control* the *Robot* with *RobotID* matching 'R2D2'"

> **select** SS#, LastName, FirstName
> **from** Engineers, Controls
> **where** SS# = EngineerSS#
>   **and** ID = 'R2D2'

This query is logically evaluated by first forming the Cartesian product of the two relations *Engineers* and *Controls*, i.e.,

Engineers × Controls

Engineers				Controls	
SS#	LastName	FirstName	WorksForDivision	EngineerSS#	RobotID
'007'	'Bond'	'James'	'Top Secret'	'78951'	'X4D5'
...	...	...	...	...	...
'007'	'Bond'	'James'	'Top Secret'	'007'	'R2D2'
...	...	...	...	...	...

Only a few tuples of this Cartesian product are represented in the diagram above. Altogether, the temporary relation contains

$$\text{cardinality(Engineers)} * \text{cardinality(Controls)}$$

tuples. From this temporary relation those tuples are selected that satisfy the condition specified in the **where** clause, i.e.,

$$\text{SS\#} = \text{EngineerSS\#} \textbf{ and } \text{RobotID} = \text{'R2D2'}$$

This—in our example—leaves us with just one qualifying tuple of which the projection on the three attributes *SS#*, *LastName*, *FirstName* yields the following result:

SS#	LastName	FirstName
'007'	'Bond'	'James'

It should be stressed that the strategy for evaluating the query above is to be viewed as a logical algorithm. In real systems, the query may be executed in an efficient way that avoids the costly derivation of the cross product and takes existing indexes into account.

Joining more than two Relations

A join may be performed over more than two relations. Consider, for example, the following query:

> "Retrieve the *Engineers* who *Control* a *Robot* with a *LoadCapacity* that exceeds 200."

This query can be formulated in SQL as follows:

> **select** SS#, LastName, FirstName
> **from** Engineers, Controls, Robots
> **where** SS# = EngineerSS#
>   **and** RID = RobotID
>   **and** LoadCapacity > 200

Again, logically, the Cartesian product of the three argument relations

$$\text{Engineers} \times \text{Controls} \times \text{Robots}$$

is created. From the resulting tuples only those are selected that satisfy the condition of the **where** clause. From those tuples only the attribute values for *SS#*, *LastName* and *FirstName* are displayed.

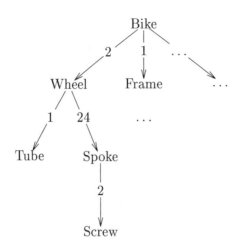

Figure 4.4: Tree Representation of the Relation *Composition*

Join over One Relation

It is sometimes meaningful to join a relation with itself. Consider, for example, the recursive relationship *Composition*, which is represented in the relation of the same name. This relation can be viewed as a collection of trees, one of which is shown in Figure 4.4. This tree corresponds to the sample database given in Figure 4.3.

The arrows in Figure 4.4 indicate the sub-/super-part relationship; i.e., an arrow points from the superpart to the subpart. Each arrow is labeled by the respective value of the *Quantity* attribute in relation *Composition*. Suppose we want to retrieve the nodes at level two of the tree rooting at the *Product* 'Bike'. This conforms to finding all those *Products* identifiers of which the (immediate) subparts of 'Bike' are composed. Let us also output the total quantity of the second-level subparts of the Product 'Bike'. This is computed as the quantity of the respective first-level subpart multiplied by the quantity of the second-level subpart. For example, the 'Wheel' is a first-level subpart of 'Bike' that is needed in a quantity of 2. Then 'Spoke' is a second-level subpart of 'Bike' that is needed in *Quantity* $2 * 24 = 48$ to assemble the bike. The query is stated as follows:

> **select** super.SuperPartID, sub.SubPartID,
>   super.Quantity $*$ sub.Quantity
> **from** Composition super, Composition sub
> **where** super.SuperPartID = 'Bike'
>   **and** sub.SuperPartID = super.SubPartID

The identifiers *super* and *sub* are called *aliases* or tuple identifiers or range variables. The query is logically executed by forming the Cartesian product of relation

Composition with itself. The tuple identifiers *super* and *sub* are needed to refer to otherwise identically named attributes. This situation is represented in the table below:

Composition × Composition

super			sub		
SuperPartID	SubPartID	Quantity	SuperPartID	SubPartID	Quantity
. . .	. . .	. . .	. . .	. . .	. . .
'Bike'	'Wheel'	2	'Wheel'	'Tube'	1
'Bike'	'Wheel'	2	'Wheel'	'Spoke'	24
. . .	. . .	. . .	. . .	. . .	. . .

The pictured table also shows the (two) qualifying tuples for which the search predicate is fulfilled. The dots stand for those tuples that are in the Cartesian product but do not satisfy the **where** clause.

Thus, for our example database extension of Figure 4.3 the result of the query has the form

super.SuperPartID	sub.SubPartID	super.Quantity * sub.Quantity
'Bike'	'Tube'	2
'Bike'	'Spoke'	48

4.4.6 Transitive Closure

The preceding query gave an indication that it is sometimes desirable to compute the transitive closure of a relation representing a tree (or forest), such as *Composition*. For example, the operation to retrieve all subparts of some product like 'Bike' demands the computation of the transitive closure of the relation *Composition*.

Let us denote the transitive closure of a relation R for two attributes A and B as $trans_{A,B}(R)$. The $trans_{A,B}(R)$ is a relation of the form

$$trans_{A,B}(R): \{[A: \ldots, B: \ldots]\}$$

The domains of A and B coincide with the respective domains in R. Formally, for a given extension of R the relation $trans_{A,B}(R)$ contains the following set of tuples:

$$trans_{A,B}(R) := \{t \mid \exists t_1, \ldots, t_k \in R \text{ for } (k \geq 1) \text{ with}$$
$$t.A = t_1.A \text{ and}$$
$$t_i.A = t_{i+1}.B \text{ for } (1 \leq i \leq k-1) \text{ and}$$
$$t_k.B = t.B\}$$

As an example consider the following extension of *Composition*, which is characterized as a tree:

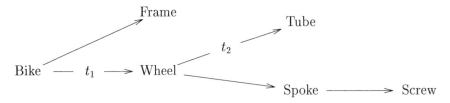

In the above figure two tuples, labeled

$$t_1 = (Bike, Wheel) \text{ and } t_2 = (Wheel, Tube)$$

are highlighted, which match our (mathematical) definition of transitive closure. These two tuples give rise to a third tuple $t_3 = (Bike, Tube)$, which is to be included in the transitive closure of relation *Composition*.

Graphically, the transitive closure $trans_{SubPartID, SuperPartID}(Composition)$ contains—in addition to the base relation *Composition*—the arrows $Bike \rightarrow Tube$, $Bike \rightarrow Spoke$, $Bike \rightarrow Screw$, and $Wheel \rightarrow Screw$. This yields the directed acyclic graph:

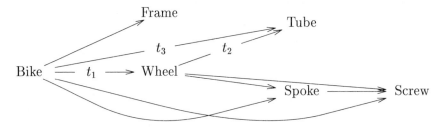

Unfortunately, it can be shown that the SQL language is not powerful enough to express this query. The best one can achieve in SQL is to compute the tree up to a fixed maximum depth. This is done by unioning queries that are analogous to the one that was formulated for finding the second-level subparts of 'Bike'. For fixed n $(2 \leq n)$, the query to find the n-th level subparts is formulated over relation *Composition* as follows:

Q_n: **select** $\cdots$
    from Composition var$_1$, ..., Composition var$_n$
    where var$_1$.SubPartID = var$_2$.SuperPartID
        and var$_2$.SubPartID = var$_3$.SuperPartID
        $\ddots$
            and var$_{n-1}$.SubPartID = var$_n$.SuperPartID

Finding all subparts up to the n-th level involves the union of n queries:

> (Q_1) **union** (Q_2) **union** ... **union** (Q_n)

where Q_1 denotes the query returning the first two attributes of relation *Composition*.

But the formulation of a query that creates the tree up to an arbitrary depth is impossible in SQL. To achieve this one has to retreat to an embedding of SQL in a programming language (cf. Section 4.5).

4.4.7   Aggregate Operators

SQL supplies a collection of *aggregate operators*, operators that are invoked on a (multi) set of values. The aggregate operators on numeric values are:

- **avg** computes the *average* of a set of numbers.

- **sum** computes the *sum* of a set of numbers.

- **min** returns the *minimum* number out of a set of numeric values.

- **max** returns the *maximum* number out of a set of numeric values.

Let us illustrate the use of these operators on just a simple example query: "retrieve the total *Profit* of all *Divisions*."

> **select sum**(Profit)
> **from** Divisions

Another aggregate operator is **count**, which is used to count the number of tuples in a (result) relation. For example, to count the number of *Divisions* one could issue the query

> **select** count(∗)
> **from** Divisions

The aggregate operator **count** may be followed by the keyword **distinct** if one wants to count the number of distinct elements. For example, in order to count the number of different *Locations* in which *Divisions* are located we could formulate the subsequent query

> **select count** (**distinct** Location)
> **from** Divisions

This way duplicates are eliminated before the **count** operator is applied.

4.4.8   Nested Queries: Set Operations in the where Clause

So far we have only used very primitive predicates in the **where** clause, which were based on attribute comparison with a constant or another attribute. SQL also allows sets as operands in the **where** clause. There are four operators on sets:

- **in** is satisfied if a particular value is contained in a set.

- **not in** is satisfied if a particular value is *not* contained in a set.

- **any** is used for existential quantification.

- **all** is used as a universal quantifier.

A set is determined by an SQL query that is to be nested within the where clause. Let us illustrate the nesting of **select** ... **from** ... **where** ... expressions on the following query:

> "Find the *Robots* that are controlled by an *Engineer* who works for a *Division* located in 'Utopia'."

This query can be decomposed into four parts. Part $(i + 1)$ utilizes a set of values retrieved by part (i) for $(1 \leq i \leq 3)$.

1. Retrieve the set S_1 of *Names* of *Divisions* whose *Location* is 'Utopia'.

2. Retrieve the set S_2 of *SS#*'s of *Engineers* whose *WorksForDivision* attribute is contained in S_1.

3. Retrieve the set S_3 of *RobotID*'s from relation *Controls* whose *EngineerSS#* field is contained in S_2.

4. Retrieve all attributes of the tuples in *Robots* whose *RID* attribute value is contained in S_3.

These four independently conceived subqueries can be assembled to one SFW (select from where) expression with multiple nested **where** clauses:

```
select *
from Robots
where RID in
        (select RobotID        ## S₃ ##
          from Controls
          where EngineerSS# in
              (select SS#        ## S₂ ##
                from Engineers
```

>   **where** WorksForDivision **in**
>   (**select** Name   ## S_1 ##
>   **from** Divisions
>   **where** Location = 'Utopia')))

The advantages of this query formulation as opposed to a corresponding join query concern updatability, when embedded in a general-purpose programming language (cf. Section 4.5). Join queries are generally not modifiable because it is impossible to determine how to propagate the updates to the base relations of which the joined result was materialized.

To illustrate universal quantification using the **all** operator, let us formulate the query

>   "retrieve the most profitable *Division(s)*"

in SQL. This query involves the comparison operator ">=" standing for "greater than or equal".

>   **select** *
>   **from** Divisions
>   **where** Profit >= **all**
>   (**select** Profit
>   **from** Divisions)

An alternative (and more readable) formulation would utilize the aggregate **max**:

>   **select** *
>   **from** Divisions
>   **where** Profit =
>   (**select** **max**(Profit)
>   **from** Divisions)

4.4.9   Group Operations

The **group by** operator of SQL facilitates the partition of a relation into subgroups. This is necessary whenever an aggregate operator is to be applied to each individual partition of the argument relation. For this purpose the relation is segmented into the desired groups and then the aggregate operator can be employed on each partition. Again, let us illustrate the use of the **group by** operator by way of an example:

>   "retrieve the total number of *Engineers* that are employed by the individual *Divisions*"

The corresponding SQL query partitions the relation *Engineers* into groups with identical *WorksForDivision* attributes. For each such group the *WorksForDivision* attribute value and the **count** of the tuples in that partition are returned as a result:

> **select** WorksForDivision, **count**(∗)
> **from** Engineers
> **group by** WorksForDivision

The **group by** clause may optionally be augmented by a condition that has to be satisfied by the respective partition in order to be included in the result. Such a predicate is stated in the **having** clause. As an example let us modify the previously expressed query:

> "Retrieve all those *Divisions'* names and the number of their employed engineers for those *Divisions* that employ at least 10 *Engineers*"

> **select** WorksForDivision, **count**(∗)
> **from** Engineers
> **group by** WorksForDivision
>   **having count**(∗) ≥ 10

The **having** clause is analogous to the **where** clause. While the **where** clause specifies a condition that has to be satisfied by particular tuples in order to be included in the further evaluation, the **having** clause evaluates the associated condition on each partition, individually. Either the whole partition satisfies the predicate, or it is excluded from further consideration as a whole.

4.4.10   Views

The SQL *view* construct provides a means for defining local, user-specific views of the relational database. A view is, in many respects, similar to a stored relation. The main difference is that a view is dynamically derived from the base relations over which it is defined.

Let us demonstrate the principal approach of defining and using a view on an example. Suppose we want to give some outside users access to some of the data stored in relation *Divisions*. But we want to limit the access in such a way that the *Profit* attribute is erased and also the information concerning the division named 'Top Secret' is left out. For this purpose we could define the view *DullDivisions* as follows:

> **create view** DullDivisions
>   **as select** Name, Location
>   **from** Divisions
>   **where not**(Name = 'Top Secret')

This view can, subsequently, be accessed like a normal (base) relation. A major difference, however, concerns modification operations. Not all views can be modified; especially those views that were derived using a join operation are generally not updatable.

4.4.11 Insertion, Deletion, and Update of Data

Tuples are inserted into a declared relation using the **insert** statement. For example,

> **insert into** Robots
> **values** ('R2D2', 500, 50)

inserts a new tuple into *Robots*. The attribute values have to obey the same order in which the attributes were declared in the **create table** definition.

Tuples are deleted from a declared relation using the **delete** statement. For example,

> **delete from** Robots
> **where** RID = 'R2D2'

deletes all tuples with *RID = 'R2D2'* from the *Robots* relation. Here, the **where** clause allows the same flexibility as the **where** clause of the select statement when restricted to one relation.

Tuples within a declared relation are updated using the **update** statement. For example:

> **update** Robots
> **set** LoadCapacity = LoadCapacity * 2
> **where** RID = 'X4D5'

doubles the *LoadCapacity* of robot *'X4D5'*.

4.5 A Programming Language Interface of SQL

The interactive SQL language described in the preceding section is suited for formulating adhoc query and data manipulation statements. However, most data manipulation tasks in engineering applications are too complex to be handled by such an interactive tool. Typically, data are generated and manipulated in computationally complex application programs that, therefore, need an interface to the database system. Such an interface is provided in the form of embedded SQL, an embedding of SQL into various programming languages, such as Pascal, C, Cobol, Fortran, PL/I, etc. Since all these interfaces exhibit comparable functionality, we will representatively describe SQL embedded in C here. Since the embedding of SQL in general-purpose programming languages varies for different systems in some details, we will

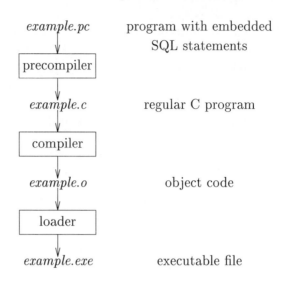

Figure 4.5: Outline of Generating a PRO*C Application

base our discussion here on the ORACLE interface called PRO*C. The emphasis is placed on the description of the database interface—not on introducing the programming language C. Only very few, mostly self-explanatory C language constructs are used in this section.

4.5.1   General Concepts

PRO*C programs are precompiled into regular C programs that can subsequently be compiled and linked like any regular C program. Figure 4.5 summarizes these steps.

Let us now turn to the structure of a PRO*C program. In general, (almost) any SQL statement can be embedded in the C program. In order to make SQL statements recognizable by the precompiler, they are always preceded by the keywords **exec sql**.

4.5.2   Host Language and Communication Variables

Embedded SQL statements may contain host language variables for two purposes:

1. To use the value of the variable as a constant in the **where** clause for comparison predicates

2. To transfer attribute values retrieved from a database relation into the variables

Such a host language variable that is to be used in a subsequent SQL statement has to be declared within a particularly marked declaration block. An example, which will be reused throughout this section, is given as follows:

exec sql begin declare section;   /∗ beginning of declaration ∗/
    varchar SomeSSNo[9], FName[20], LName[20];
    float Sal;
exec sql end declare section   /∗ end of variable declaration ∗/

The variables *SomeSSNo*, *FName*, *LName*, and *Sal* can now be used as regular C variables and also within an SQL statement. Within the SQL statement they have to be prefixed by a colon, i.e., *:SomeSSNo*, *:FName*, *:LName*, and *:Sal* to differentiate them from database variables.

In addition to the host language variables that are used to relay data between the database and the C application program, one needs to provide some status variables, which are used to pass status information concerning the database access to the application program. By including the line

exec sql include SQLCA.H;

in your C program, an appropriate collection of status variables is automatically copied into your application program by the PRO∗C preprocessor.

4.5.3   SQL Statements without Host Variables

These statements constitute the most straightforward way of invoking a database operation from within a C program. For example, one could delete a particular *Engineers* tuple—due to firing of an employee—from the database:

    . . .
exec sql
    delete
    from Engineers
    where SS# = '007';
    . . .

This statement initiates the deletion of the (in this case unique) tuple from the relation. However, each time the program is run, it attempts to delete the same tuple from the relation *Engineers*. A more meaningful "firing" program would read in the Social Security number of the engineer to be fired and then delete the tuple from the relation *Engineers* that has a matching *SS#* attribute. In order to achieve this, one has to use host language variables within the **delete** statement.

4.5.4   Using Host Language Variables

Let us demonstrate the principle on our "firing" program. This rather trivial program performs the following steps:

1. Prompt the user for the Social Security number of the engineer to be fired. If the user inputs "END," the program terminates.

2. Delete the respective tuple from the database relation *Engineers*.

3. Go back to step 1.

This functionality is coded in the following program fragment:

```
exec sql begin declare section;
    varchar SomeSSNo[9];
exec sql end declare section;
exec sql include SQLCA.H;          /* predefined status variables */
main ( )
{
    exec sql connect ...;          /* connects to the database */
    while (1)        /* keep on firing */
    {  printf ("Enter employee's SS# (or END to end): ");
       scanf ("%s", SomeSSNo.arr);        /* read the SS# */
       SomeSSNo.len= strlen(SomeSSNo.arr);
       if (strcmp(SomeSSNo.arr, "END"))
           break;        /* no more firings, leave the while loop */
       exec sql
            delete from Engineers
            where SS# = :SomeSSNo;
    }
    exec sql commit work release;
}
```

For social reasons, we should mention that analogously to deleting tuples one could also insert new tuples into a relation, e.g., to implement a "hiring" program, and also updates could be performed in this way.

4.5.5   Embedded Queries

We distinguish embedded SQL queries in two categories based on the number of tuples the query returns:

- Queries returning just one tuple

• Queries that return a relation, i.e., a set of an unknown number of tuples

This distinction is needed because C is a programming language that may be characterized as processing one record at a time. On the other hand, SQL is a set-oriented language that naturally operates on sets of records (tuples). Some authors have called this problem of coupling a set-oriented query language with a record-at-a-time-oriented programming language the *impedance mismatch* between the two systems.

Queries Returning One Tuple

Let us first deal with the simple case of a query returning just one tuple that may, nevertheless, be composed of attributes from different relations by formulating a join query. As an example let us retrieve some attributes of the tuple from the relation *Engineers* whose *SS#* attribute matches some user-supplied value.

```
exec sql begin declare section;
    varchar SomeDivisionName[20], LName[20], FName[20];
    SomeSSNo[9];
exec sql end declare section;
exec sql include SQLCA.H;        /* predefined status variables */
    ...                 /* read the social security */
                        /* number into SomeSSNo */
exec sql
    select LastName, FirstName, WorksForDivision
    into :LName, :FName, :SomeDivisionName
    from Engineers,
    where SS# = :SomeSSNo;
    ...            /* further processing of the tuple */
```

In this example program the attribute values of the uniquely identified tuple are retrieved into the host language variables *LName, FName,* and *SomeDivisionName.* Following the **select** statement the information could be further processed in the C program.

Queries Returning Multiple Tuples

The more general case of a query embedded in a C program returns a set of tuples, i.e., a relation. Since the programming paradigm of the conventional programming languages permit to work on only one tuple at a time a special mechanism called *cursor* is used. A cursor is basically a temporary work area in which the results of a query are stored for further processing in the application program. A cursor can be perceived as a pointer into the result set returned by the query. This pointer (cursor) can be moved forward one position whenever the processing of the tuple at the current position of the cursor is finished.

Cursor Declaration   Before a cursor can be used it has to be declared in the application program; that is, it has to be associated with a query. An example cursor over a set of *Engineers* tuples is declared as follows:

> **exec sql declare** Engs **cursor for**
>   **select** LastName, FirstName, Salary
>   **from** Engineers
>   **where** WorksForDivision = :SomeDivisionName

The declaration of the cursor does not, however, initiate the execution of the associated query. This is triggered by the **open** statement.

Opening a Cursor   Let us demonstrate the opening of the cursor *Engs* on a small program fragment. Again, one could interactively read in a division name into the host language variable *SomeDivisionName*, e.g.:

> scanf ("%[^\n]", SomeDivisionName.arr)
> SomeDivisionName.len= **strlen**(SomeDivisionName.arr)

The cursor *Engs* could then be opened:

> **exec sql open** Engs;

The **open** statement leads to the evaluation of the query and associates all qualifying tuples of the *Engineers* relation with the cursor *Engs*.

Fetching Tuples from the Cursor   Initially the pointer (cursor) is positioned in front of the first tuple in the result set. One could now transfer the attribute values of the first tuple in the opened cursor *Engs* into appropriately declared host variables using the **fetch** command:

> **exec sql fetch** Engs **into** :LName, :FName, :Sal;

We note that, when using a cursor, the query may not contain an **into** clause; this is associated with the **fetch** command.

All **fetch** statements move the cursor one position ahead in the result set. When the last position of the cursor is reached, a subsequent **fetch** command raises the SQL error **not found**.

It is not possible to move the cursor back to a position that it has passed already. The only way to re-access a tuple over which the cursor passed is by closing the cursor and re-**open**-ing it again.

Closing a Cursor   Finally a cursor is closed by the **close** statement:

> **exec sql close** Engs;

Only after a cursor has been explicitly closed it may be opened again.

Complete Example Program Based on the program fragments supplied above we want to formulate a (more) complete program which is listed in Figure 4.6. In this program the user is prompted to input a division name (lines 11, 12, 13) for which the engineers' salaries are to be adjusted. The cursor *Engs* is opened (line 19) for the query that retrieves all tuples of *Engineers* who work for the user-supplied division. The statement in line 20 specifies that whenever a fetch on the cursor results in a **not found** error, i.e., the end position has been reached, control should go to the statement labeled *"finish."*

In this example application the **fetch** command is nested into an infinite surrounding loop that is left by the **whenever** statement. Within this loop for result tuple the *FName, LName*, and the *Sal* values are displayed and the user is prompted to assign a new salary. In lines 28–30 the *Salary* attribute of the relation *Engineers* is updated. The clause **where current of** *Engs* refers to the tuple currently referenced by the cursor. Finally, in line 34 the cursor is closed and, then, the work is committed. The **release** indicates that the connection to the database should be relinquished.

4.6   Exercises

4.1   Translate the conceptual schema for modeling cars—developed in Exercise 2.1 in Chapter 2—into a relational schema and define the according relations by SQL **create table** statements. In addition, create the indexes that you deem useful for typical applications of this database schema.

4.2   Based on the database schema given in the text formulate the following SQL queries:

   • Retrieve all divisions which generate losses.

   • Retrieve all robots that have ever painted.

   • Retrieve all engineers working for divisions in 'Utopia'.

   • Retrieve all products that were assembled with the screwdriver identified 'S10'.

4.3   Outline a PRO*C program that computes the transitive closure of the relation *Composition*.

4.4   Augment this program developed in Exercise 4.3 such that it also computes the total cost of all subparts of which a particular (super) *Product* is composed.

4.5   Outline a PRO*C program that determines the top-level *Product* in which a given sub-*Product* is used.

```
1      exec sql begin declare section;
2          char LName[20], SomeSSNo[9], FName[20];
3          varchar SomeDivisionName[20];
4          float Sal, NewSal;
5      exec sql end declare section;
6      exec sql include SQLCA.H;          /* predefined status variables */
7      float TotalSal = 0;
8      main( )
9      {
10         exec sql connect ··· ;
11         printf ("Enter Division's Name: ");
12         scanf ("%[^\n]", SomeDivisionName.arr);
13         SomeDivisionName.len= strlen(SomeDivisionName.arr);
14         exec sql declare Engs cursor for
15             select LastName, FirstName, Salary, SS#
16             from Engineers
17             where WorksForDivision = :SomeDivisionName
18             for update of Salary;
19         exec sql open Engs;
20         exec sql whenever not found goto finish;
21         while (1)       /* infinite loop */
22         {
23             exec sql fetch Engs
24                 into :LName, :FName, :Sal, :SomeSSNo;
25             printf ("%-20s\t %-20s\t %6.2f\n", FName, LName, Sal);
26             printf ("Enter new Salary: ");
27             scanf ("%f", &NewSal);
28             exec sql update Engineers
29                 set Salary = :NewSal
30                 where current of Engs; /* the current tuple */
31             TotalSal = TotalSal + NewSal;
32         }
33     finish:
34         exec sql close Engs;
35         exec sql commit work release;
36         printf("\n Total salary of %s is %8.2f\n", SomeDivisionName.arr,TotalSal);
37     }
```

Figure 4.6: Program to Adjust and Total the Salaries of a Division

4.6   Devise and analyze several alternative relational representations of the CSG
geometric modeling. Discuss the problems of the relational model with respect
to the recursive structure of CSG.

4.7   Devise a relational representation of the conceptual robot model presented in
Figure 3.6. Discuss the shortcomings of the relational modeling.

4.8   Transform the conceptual schema for modeling robot grippers that was developed in Exercise 2.7 into a relational schema.

4.9   A client of the robot gripper factory reports a failure of a spring in a robot gripper with identifier 'RoGripp#1'. Give a select statement retrieving all robot
grippers together with the clients they were delivered to which use springs
from the same supply.

4.10   Give the relational schema of the ER diagram in Figure 2.3.

4.11   Give a relational schema modeling the Swiss knife example for which the conceptual schema was developed in Exercise 2.10.

4.12   Reconsider Exercise 2.12. Convert the ER schema into a relational schema.
Additionally, state the following queries in SQL:

1. Which rooms are adjacent to the kitchen of a given house?
2. How many restrooms of a given house are there?
3. Is there any afternoon sun on the balcony of a given house?
4. Retrieve all houses with two kitchens and three restrooms.
5. What is the area of a given house?
6. How many rooms besides kitchens and restrooms does a given house possess?
7. Is there any house designed by Gaudy?
8. What is the average number of floors of houses designed by Wright?
9. What is the maximum number of floors of houses designed by Wright?

If you cannot give an SQL counterpart of a certain query, reconsider your ER
schema and then the relational schema.

4.13   For the ER diagram designed in Exercise 2.13, give the according relational
schema. Additionally, state the following queries in SQL:

1. How many documents do exist?

   2. Retrieve the total number of pages of all documents.

   3. How many figures are included in a given document?

   4. Is there a subsection entitled "ER Modeling" in a given document?

   5. Retrieve all documents concerned with traveling costs.

   6. What are the documents processed by Mister X?

   7. What is the average time needed to manipulate a document of a certain type?

   8. Is there any document showing an explosion drawing of a robot gripper?

If you cannot give an SQL counterpart of a certain query, reconsider your ER schema and then the relational schema.

4.14 For the ER diagram designed in Exercise 2.14, derive the corresponding relational schema. Additionally, state the following queries in SQL:

   1. How many calls are there to a given procedure?

   2. Where do these calls occur?

   3. What are the procedures implemented in a certain module?

   4. Does module X use a procedure of module Y?

   5. How many lines of code does module X possess?

   6. Who is the designer of the interface of module X?

   7. How long did it take to design the interface of module X?

   8. What is the average time needed to implement a procedure consisting of 40–50 lines of code?

   9. Retrieve the programmer who is involved in the highest number of modules.

  10. Retrieve all programmers together with the number of lines of code they produced back in 1966.

If you cannot derive an SQL formulation of a certain query, reconsider your ER schema and then the relational schema.

4.7   Annotated Bibliography

The relational data model was introduced in a landmark paper by E. Codd in 1970
[Cod70]. Two early prototype implementations had a strong influence on the technical
advances that made the efficient implementation of the relational model possible:

- *System R*, a relational database research project at the IBM Research Laboratory San Jose, California [ABC+76]

- INGRES, a relational database prototype that was developed at the University
  of California, Berkeley under the leadership of M. Stonebraker and E. Wong
  [Sto85]

Both research projects eventually lead to commercial products. System R gave rise
to two commercial DBMSs distributed by IBM: SQL/DS and DB2. The INGRES
prototype, which was first distributed as public domain software with the Berkeley
UNIX operating system, was redeveloped by Relational Technology, Inc. (RTI), a
company that was founded by the leading researchers of the UC Berkeley project.
Many other commercial database systems based on the relational model have been
developed in the mean time. Among those are ORACLE [Ora86], developed by a
company of the same name, RDB by Digital Equipment Corp. All of these systems
support a dialect of the SQL language—not always identical to the Standard SQL,
but the differences are typically limited to minor details. Standard SQL was developed by a consortium of database researchers and defined in an American National
Standards Institute (ANSI) document [ANSI86]. Another source for the Standard
SQL description is [Dat87].

   This presentation is based on the ORACLE product, which is described in [Ora86].
The C interface, called Pro*C, is described in [Ora85].

   View updates are discussed by Dayal and Bernstein in [DB82].

   Further textbooks that cover the relational model in more detail than we could
invest on this database model are [EN89, KS91, Ull88, Dat90].

5

Assessment of Conventional Database Technology for Engineering Applications

After having introduced the relational database model and the standard relational language SQL in the preceding chapter, we will now assess the applicability of relational databases to engineering applications. The relational model was conceived as an information structure for administrative, business-oriented data. Originally, there was no intention to also meet the data modeling and data manipulation requirements of the so-called non-standard applications, e.g., engineering and science applications.

5.1 A Representative Engineering Example

In order to back up and to animate our discussion, we will consider the following representative data modeling task: boundary representation (cf. Section 3.1). The conceptual schema of the most simple boundary representation is graphically depicted in Figure 5.1.

The schema consists of four entity sets: *Polyeder* modeling the highest level abstraction of a solid geometric object; *Faces* approximating the outer hull of a *Polyeder* in the form of polygons; *Edges*, which represent the boundaries of the polygons; and finally *Vertices*, which contain the metric information in the form of X, Y, Z coordinates. The four entity sets are associated by three $N : M$ relationship types: *Hull*, *Boundaries*, and *StartEnd*. We assume that distinct *Polyeders* may be connected via a common *Face*, which makes the relationship *Hull* many-to-many ($N : M$).

Actually one could specify the cardinalities of these relationships more precisely (see Exercise 5.5). For example, every edge of a polyeder bounds exactly *two* faces;

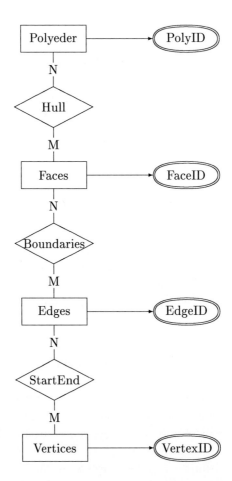

Figure 5.1: Conceptual Model of Boundary Representation

every vertex is associated with at least three edges; and every edge is bounded by two vertices.

Applying the transformation rules for converting a conceptual ER schema into a corresponding relational schema (cf. Section 4.2.2) yields the relational database schema shown in Figure 5.2. The depicted database extension includes some of the tuples representing a geometric object of type cuboid, which is identified within the *Polyeder* relation by "*cubo#5*".

See Exercise 5.6 for an alternative relational modeling that reduces the number of base relations by exploiting the more precise cardinality specification derived in Exercise 5.5. However, this reduced relational representation induces different problems making several commonly formulated queries extremely complex (see Exercise 5.7).

5.2   Weak Points of the Relational Model

On the basis of the relational schema of Figure 5.2 we will analyze the shortcomings of the relational model with respect to representing technical data.

5.2.1   Segmentation

One of the most severe drawbacks of the relational model is the need to decompose of logically coherent application objects over several base relations. This is obviously demonstrated in the boundary representation of Figure 5.2, in which the object "*cubo#5*" is decomposed into many information pieces that are segmented over seven different relations.

The segmentation induced by the relational modeling disassociates *external* objects—as encountered in the application domain—from the *database-internal* objects that model the external objects. Thus, there is no one-to-one correspondence between an internal database structure and the external objects represented in that structure.

Let us animate this problem on our example application, geometric modeling. From the geometric modeling perspective, the object "*cubo#5*" forms *one* composite object that should really be viewed as one coherent unit. This is also reflected in the application-specific operations that apply to such a geometric object, e. g.:

- *Rotate* the cuboid identified by "*cubo#5*"

- *Scale* the cuboid identified by "*cubo#5*"

- *Delete* the object named "*cubo#5*"

All these operations are associated with the geometric object as a unit, not with partitions thereof that are stored in the various relations.

Polyeder

PolyID	Volume	...
...	...	...
"cubo#5"	1000.00	...
...	...	...

Hull

PolyID	FaceID
"cubo#5"	"f1"
"cubo#5"	"f2"
...	...
"cubo#5"	"f6"
...	...

Faces

FaceID	Circumference	...
...	...	...
"f1"	40.00	...
"f2"	40.00	...
...	...	...
"f6"	40.00	...
...	...	...

Boundaries

FaceID	EdgeID
...	...
"f1"	"e1"
"f1"	"e2"
"f1"	"e3"
"f1"	"e4"
...	...
"f6"	...
...	...

Edges

EdgeID	Length	...
...	...	...
"e1"	10.00	...
"e2"	10.00	...
...	...	...
"e12"	10.00	...
...	...	...

StartEnd

EdgeID	VertexID
...	...
"e1"	"v1"
"e1"	"v2"
"e2"	"v2"
"e2"	"v3"
...	...
"e12"	...
...	...

Vertices

VertexID	X	Y	Z
...	...	...	...
"v1"	0.00	0.00	0.00
"v2"	10.00	0.00	0.00
...	...	...	...
"v8"	...	...	...
...	...	...	...

Figure 5.2: Relational Database of Boundary Representation

From the application viewpoint the constituent subparts of which such an object is constructed are often of no interest. For many operations one wants to abstract the subparts to one logical unit. Unfortunately, this is not supported by the relational model. Rather, one has to explicitly reconstruct a complex application object from the segments distributed over the relational schema. This often results in unnaturally complex queries with highly redundant results (see also the exercises). Consider, for example, the query that is formulated to find the *PolyID* and the X, Y, and Z coordinates of all bounding *Vertices* of *Polyeders* whose *Volume* exceeds 10.

```
select distinct p.PolyID, v.VertexID, v.X, v.Y, v.Z
from Polyeder p,
     Hull h,
     Boundaries b,
     StartEnd s,
     Vertices v
where p.Volume > 10 and
      p.PolyID = h.PolyID and
      h.FaceID = b.FaceID and
      b.EdgeID = s.EdgeID and
      s.VertexID = v.VertexID
```

This query consists of a costly join of five relations, each containing some pieces of the information needed to reconstruct the qualifying *Polyeder*. (Note that this query could, alternatively, be formulated using nested **where** clauses—if we omitted the attribute *PolyID* from our result list.[1] However, this would not reduce the complexity of the resulting statement.)

There is yet another problem with a query that reconstructs a complex object, as demonstrated above. In general, a view that was constructed by a join query cannot be updated. For more details on this problem see Section 5.3.

5.2.2   Identifier Attributes

Tuples are uniquely identified by their key attribute values. There are numerous problems with using attributes that model the state of an object (tuple) as a key. For example, consider the tempting choice of using the X, Y, Z attributes of relation *Vertices* as a key. This would imply that a vertex is referred to by its coordinate values, e.g., from relation *Edges*. A geometric transformation resulting in the modification of the coordinates would thus assign a new identity to the respective vertex. Thus, all references to the particular vertex become invalid.

---

[1] SQL does not allow to access attributes from a nested query at an outer level.

To avoid the problems of intermingling object identity with object state, one usually incorporates some artificial attribute in the relation whose sole purpose is identification. The identifying attribute values, e.g., *PolyID*, are user-supplied. These identifying attributes serve as "key" in the relation representing the entities, e. g., *Polyeder*, and as foreign keys in the relations that represent relationship types in which the entities participate.

References to tuples are maintained by foreign keys, as explained in Section 4.2.2. Such a reference is always needed when modeling a relationship in the relational model. For example, the tuple ["*cubo#5*", "*f1*"] of relation *Hull* represents the fact that the *Polyeder* identified by the *PolyID* attribute value "*cubo#5*" has a bounding *Face*, which is identified by the *FaceID* attribute value "*f1*". Thus, relation *Hull* consists of merely two foreign keys relating the tuples of relations *Polyeder* and *Faces* without any additional information.

The following problems occur with these identifier attributes:

Uniqueness   The attribute values have to be chosen unique. It is not sufficient to guarantee uniqueness within one complex application object, because different application objects are mapped onto the same relational schema. For example, the *FaceID* values have to be unique within the entire relation *Faces*; i.e., no other faces may assume the identifier values "*f1*", ..., "*f6*", which are assigned for the *Faces* belonging to the Polyeder "*cubo#5*". The burden of guaranteeing uniqueness of the identifier attributes is placed on the shoulders of the end user. While this may seem reasonable in a single-user environment, it becomes increasingly more difficult to guarantee uniqueness under concurrent use of the same relational database by many users—all being required to assign globally unique key values.

Referential Integrity   The term *referential integrity* denotes that the tuple to which a foreign key refers actually exists. Consider, for example, the insertion of a tuple ["*cubo#5*", "*xyz*"] into relation *Hull*. There is no guarantee that a tuple with *FaceID* "*xyz*" does exist in the current extension of relation *Faces*. If not, the referential integrity is violated because we encounter a (dangling) reference to a nonexistent tuple. Currently, database implementors are working on incorporating concepts to ensure referential integrity in relational database management systems.

5.2.3   Lack of Data Abstraction

The relational model has only one very simple structuring concept, the relation. In application domains of the business and administrative sector, this spartanic data structure seems sufficient because all data objects exhibit a very homogeneous structure. Thus, they can easily be mapped into relational representations.

On the other hand, engineering objects possess a more heterogeneous structure. A complex object may be composed of a variety of differently structured subobjects as, for example, a *Polyeder* in boundary representation. A natural (and user-friendly) representation of such complex objects demands more sophisticated abstraction mechanisms than the relational model offers.

Two such abstraction concepts that were already introduced in this book are the following:

- *Aggregation*, which abstracts composite objects composed of subobjects to aggregate objects.

- *Generalization* and *specialization*, where generalization denotes the classification of similar objects to a generic object type. Specialization is the complementary concept by which an object type can be refined to account for more specialized instances.

5.2.4 Only Built-in Data Types

Today's commercially available relational DBMSs have a fixed collection of built-in data types that are "hard-wired" into the system. It is not possible to augment this data type collection by application-specific types, e. g., a data type *vector* consisting of three components that could represent vertices in 3-D.

5.2.5 Modeling Object Behavior

From an application programmer's point of view, an external object, i.e., one that can be identified in the application domain, consists of two dual dimensions:

1. The structural representation, which models the current state of the external object

2. The behavioral specification

While the first representational dimension of application-specific objects can be mapped to a relational schema (with the aforementioned disadvantages), the behavioral dimension of objects is entirely missing in relational databases.

The behavioral dimension of an object type constitutes the semantically valid operations that are applicable to the individual objects. The set of valid operations is highly application specific. For example, the operations that are meaningful on geometric objects in the context of geometric modeling include the following:

- *Scaling* of the objects by user-supplied scale factors

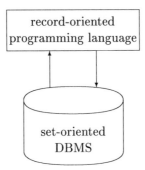

Figure 5.3: Coupling a DBMS with a Programming Language

- *Translation* of the objects by user-supplied X, Y, and Z offsets

- *Rotation* of the objects about a particular axis by a user-supplied angle

The relational DBMSs only provide generic language constructs that can be invoked on any kind of relational schema—disregarding the semantics of the modeled objects. There is no way of associating, for example, the three geometric transformations with the geometric objects modeled in the relational schema of Figure 5.2. This implicates that a great deal of the application-specific semantics of external objects is not recovered in the relational database. The behavioral characteristics of an object type can, at best, be represented in application programs that interface with the relational DBMS. Thus, the behavioral dimension is really divorced from the representational one, which, as a consequence, leads to two independent partial models of the same object.

5.3   Programming Language Interface: The "Impedance Mismatch"

As explained earlier, SQL is an adhoc query language that is useful for interactive queries to retrieve the database state. But the interactive database language is inappropriate for computationally complex data manipulations. For this purpose an interface between the DBMS and general-purpose programming languages has been devised, in a form shown in Figure 5.3.

The programming language interface developed for the standard SQL language suffers from a severe mismatch. It is tried to couple two systems that are based on entirely different execution paradigms:

- General-purpose programming languages, such as C, Pascal, Fortran, etc., execute data in a mode that can be termed *"record-at-a-time."*

- Relational database systems handle data in a *set-oriented* fashion.

Therefore, any embedded database operation that returns a set of tuples, i.e., a relation, as a result causes problems because none of the widely used programming languages possesses set-oriented operators.

To overcome the mismatch between set-oriented and record-oriented execution models, the *cursor* control structure was devised as a "bridge" between the two systems. The cursor is used to temporarily maintain references to the result tuples returned by a database query. Within the application program one can thus iterate over this set of results by fetching the tuples one after the other into some local data structure of the application program.

Even though this approach seems to work well for relatively simple applications as encountered in, e.g., business applications, it has severe limitations in the context of engineering applications. These are caused by the limited functionality of the cursor modifications, which are analyzed below.

5.3.1 One-Way Motion: Sequential Access

Within the cursor one can only move in one direction; i.e., one can only move from the currently accessed result tuple to the next one. One can never reaccess a tuple that was already seen before the current tuple was fetched—unless the cursor is closed and reopened. While this constraint may seem reasonable in batch-mode processing of data, it is highly restrictive for complex applications. In a batch-oriented application one typically accessed each data item just once to perform some modifications. In more complex applications, e.g., an interactive geometric modeling system, objects are more often randomly accessed, for example, on the basis of user selection. In this context the sequential access requirement of cursors is highly restrictive because it results in frequent and costly reopening of the same cursor.

5.3.2 Nonupdatable Joins

Not all declared cursors are updatable. If the cursor is associated with a join-query, the resulting tuple set cannot be updated, because it is generally undecidable how updates of the joined query are to be propagated to the base relations. For example, the following cursor *EdgeVertices* joins the relations *StartEnd* and *Vertices*:

```
exec sql declare EdgeVertices cursor for
      select s.EdgeID, v.VertexID, v.X, v.Y, v.Z
      from StartEnd s, Vertices v
      where s.VertexID = v.VertexID;
```

However, it is impossible to perform updates, like, for example, geometric transformations on this cursor. To achieve this, one has to explicitly reprogram the join in the form of a nested loop in the application program by iterating over cursors that are defined on the base relations. For our example, one would declare a cursor *EdgeToVertex*, which retrieves the *EdgeID* and the associated *VertexID* attributes. Then, in a nested loop one could for each *VertexID* perform the retrieval and update directly on the base relation *Vertices*. The control flow program is outlined below:

```
exec sql declare EdgeToVertex cursor for
    select *
    from StartEnd;

exec sql open EdgeToVertex;
while(1)        /* Iterate over the EdgeToVertex cursor */
{
   exec sql fetch EdgeToVertex into :E#, :V#;
   exec sql select X, Y, Z    /* read the coordinates */
           into :Xvar, :Yvar, :Zvar
           from Vertices
           where VertexID = :V#;
   . . .
   exec sql update Vertices set . . . ;    /* perform modifications */
}
```

5.4   Discussion and Outlook

The previous discussion concerning the *impedance mismatch* between the relational data model and the application programming language is graphically emphasized in Figure 5.4.

The graphic highlights that the database schema is not aware of the application-specific behavior that is associated with the objects; e.g., the operation *rotate* can only be modeled as part of the application program. This has the disadvantage that the database system cannot serve as a repository for the operations. This makes the sharing of application-specific operations among different applications, say, applications *A* and *B*, difficult. In reality one often experiences that the same operations are multiply coded by different application programmers, as exemplified in the graphic.

In order to implement such operations it is typically necessary to reconstruct the application objects in data structures of the programming language. For this purpose tedious transformations have to be coded that retrieve the information concerning one external object from the base relations—where this information is segmented. Again, these transformations are very often replicated by different applications as

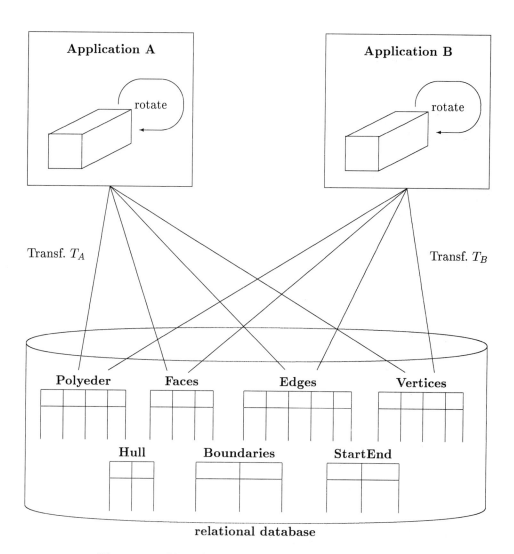

Figure 5.4: Visualization of the Impedance Mismatch

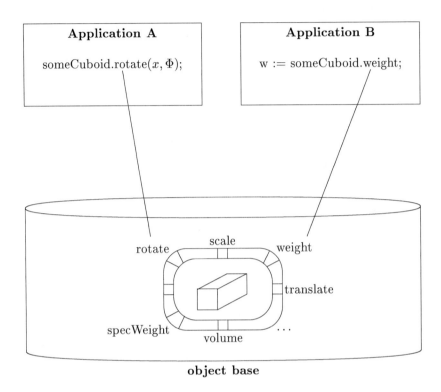

Figure 5.5: Visualization of the Advantages of Object-Oriented Data Modeling

exemplified by the two transformation procedures T_A and T_B in the graphic. Also this transformation process can be very time-consuming, since it typically involves a join over many base relations—as in our boundary representation example.

In the core of this book we propose object-oriented database technology as a promising data model, especially for non-standard applications, such as engineering applications. In fact, the object-oriented modeling of our boundary representation model would overcome the problems discussed so far. This is graphically highlighted in Figure 5.5.

Without going into the details of object-oriented modeling—this is done in Part III of the book—we should briefly explain the graphic. In object-oriented data models the *structural* and the *behavioral* components of objects are represented in a uniform schema. Thus, the operations naturally associated with application objects are an integral part of the database schema. Thereby, the sharing of operations among different applications is directly supported by the data model.

By the principle of *information hiding*, an object, such as our cuboid, is shielded from arbitrary manipulations. Only the predefined operations, such as *rotate*, *scale*, etc., are applicable.

The segmentation of an application object is not visible in the object-oriented representation, because the semantically rich operations are associated with an object and hide the internal structural representation from the user.

Since the programming language, in which the operations are coded, is integrated in the object model, the transformation process—to transform the objects into data structures of the application programming language—that is needed in the relational model for coding computationally complex operations becomes obsolete.

5.5 Exercises

5.1 Insert a cuboid with identifier "cubo#1" including its faces, edges, and vertices together with the corresponding relations into the *Polyeder* database.

5.2 Formulate an SQL query that retrieves all information about "cubo#1".

5.3 Construct the result relation for the above query, count the number of distinct values occurring in each column, and compute the percentage of redundancy.

5.4 Outline PRO*C application programs that perform scale, translate, and rotate operations on a given cuboid.

5.5 Use the (min, max) notation introduced in Exercise 2.8 to specify the cardinalities of the relationships of the boundary representation more precisely.

5.6 Exploit the more precise cardinality specifications of the relationships and transform the conceptual schema into a relational schema that minimizes the number of base relations.

Hint: The relationship *StartEnd*, for example, can be modeled by two attributes *Start* and *End* in the relation *Edges*.

5.7 Analyze the alternative relational boundary representation with respect to query formulation. Formulate, for example, the query: Find all *Polyeders* that have one vertex in the origin of the coordinate system. Why is this query more complex under the new relational schema?

5.8 Demonstrate the very limited support that the relational model provides for representing generalization hierarchies—such as the one developed in Exercise 2.11.

Investigate different alternatives for modeling the generalization hierarchy shown in Figure 5.6 (the upward directed arrows denote a generalization).

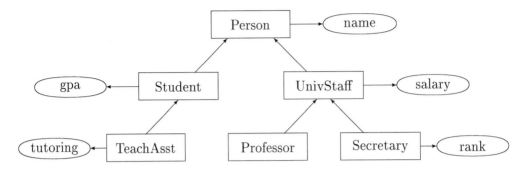

Figure 5.6: Generalization Hierarchy of a University Administration

In particular, the view concept of the relational database model should be investigated with respect to modeling the generalization hierarchy.

5.9   In Exercises 4.12, 4.13, and 4.14, conceptual ER schemas had to be transformed into relational schema. Redo these exercises under consideration of the discussion of this chapter.

5.6   Annotated Bibliography

The problems of conventional database systems relying on the relational model for supporting VLSI design applications are highlighted in [DMB+87]. A general discussion of the weaknesses of the record-oriented data models can be found in [Ken79]. Codd, the designer of the relational model, proposed some enhancements of the pure relational model to augment its expressive power in [Cod79]. A thorough discussion of the requirements of CAD/CAM applications on database systems is supplied in [LAB+85]. An analysis of data modeling requirements for geometric modeling is carried out in [KW87, Kem86]. The performance bottlenecks of conventional DBMSs with respect to engineering applications are investigated in [Mai89]. The shortcomings of the conventional short-duration transaction model for engineering design transaction are investigated in [BKK85, KLMP84, DK88].

The problems induced by the incompatibility of the interface between (relational) database management system and programming language was one of the catalysts for developing so-called database programming language. The basic idea is to extend the programming language by constructs that provide for a more natural (set-oriented) access to persistent data items. An early example is Pascal-R designed by J. Schmidt [Sch77]. Atkinson and Buneman [AB87] extensively survey these approaches.

6

Extensions of Relational DBMSs

In the preceding chapter we concluded that the (classical) relational model does not adequately support advanced database applications. When this was first found out in the early 1980s many projects were initiated to enhance the pure relational model by concepts that allow a more natural modeling of complex applications. Since these projects took the existing relational model as a basis for their developments, they are called *evolutionary* approaches—as opposed to the *revolutionary* approach of designing a new data model, such as the object-oriented model. In this chapter we will discuss several extensions to the relational model that were designed to enrich its semantics.

6.1 Modeling Generalization and Aggregation in the Relational Model

In Chapter 2 the basic abstraction concepts *generalization* and *aggregation* were introduced. While the relational model provides—at least some—support for modeling aggregate objects, the representation of generalization hierarchies is not supported in the pure relational model. Here we will describe some straightforward extension of the relational model to better support generalization and specialization.

In the following discussion we will use a notion of generalization and specialization that is semantically slightly richer than the one introduced in Chapter 2. To explain this semantic extension, we use the abstract generalization/specialization hierarchy—consisting of just two levels—shown in Figure 6.1. The generic entity type R is specialized into the entity types $R_{11}, \ldots, R_{1p_1}, \ldots, R_{mp_m}$. Now we consider so-called specialization dimensions; in our example schema the m specialization dimensions are called $s_{k1}, \ldots, s_{km}$. These s_{ki} form a subset of all the properties $s_1, \ldots, s_n$ describing the entities of type R. Each specialization dimension s_{ki} specifies a particular prop-

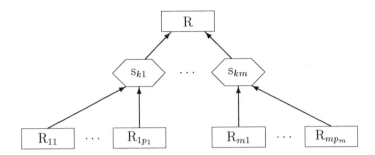

Figure 6.1: Specialization along m Dimensions $s_{k1}, \ldots, s_{km}$

erty of the entities of type R based on which they can be grouped into the disjoint specialization subsets $R_{i1}, \ldots, R_{ip_i}$. That is, the sets $R_{i1}, \ldots, R_{ip_i}$ are mutually disjoint. Furthermore, we assume that each entity of type R belongs to exactly one of the sets $R_{i1}, \ldots, R_{ip_i}$—depending on its s_{ki} property. Thus, each entity of type R is specialized along m different dimensions and, therefore, belongs to exactly m different specialized entity types.

In our graphical notation (cf. Figure 6.1) the specialization dimensions are denoted by an angled box. The specialized, mutually disjoint subsets of the particular specialization dimension are connected by an arrow (or just an upward directed line) leading into the angled box. Furthermore, an arrow (line) from the angled box leads to the generic type that is being specialized.

Let us illustrate this concept of specialization on a more intuitive example. Consider the entity type *Person* with the three specialization dimensions:

- *sex*, which distinguishes the *Persons* into *Females* and *Males*

- *nationality*, which groups all the *Persons* into the numerous national groups—e.g., Canadians, Austrians, etc.

- *maritalStatus*, which groups the *Persons* into the specialized subsets *Singles*, *Married*, *Divorced*, and *Widowed*

This generalization/specialization of *Persons* is shown schematically in Figure 6.2. Note that the three specialization dimensions are completely independent of each other. For example, the *sex* has no influence on the *maritalStatus* or the *nationality* of a *Person*.

6.1.1   Syntactic Extensions of the Relational Model

In their paper Smith and Smith [SS77b] proposed the following general syntax for modeling abstraction hierarchies within the relational model:

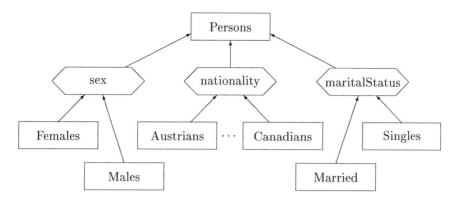

Figure 6.2: Specialization of *Persons*

var R [: **generic**

$\quad\quad s_{k1} = (R_{11}, ..., R_{1p_1});$

$\quad\quad$...

$\quad\quad s_{km} = (R_{m1}, ..., R_{mp_m});$

$\quad$**of**]

$\quad$**aggregate** [keylist]

$\quad\quad s_1: [\textbf{key}] \ R_1;$

$\quad\quad$...

$\quad\quad s_n: [\textbf{key}] \ R_n;$

$\quad$**end**

This abstract schema corresponds to the description of object type R in the conceptual schema of Figure 6.1. Objects, i.e., tuples, of type R constitute aggregates that are represented by the attributes s_1, ..., s_n. Some of these attributes are references to other objects (of the same or another relation). We see that—according to Figure 6.1—the objects of R are specialized along m (specialization) dimensions: $s_{k1}, ..., s_{km}$. For each specialization dimension s_{ki} ($1 \le i \le m$), we have the mutually disjoint object sets $R_{i1}, ..., R_{ip_i}$, to which the elements of R belong.

There are several rules that have to be satisfied for such a schema definition to be a valid generic specification:

1. R_i ($1 \le i \le n$) is either a generic identifier (then **key** must appear in the definition) or a type identifier (then **key** must not appear).

2. **keylist** is a subset of $\{s_1, ..., s_n\}$.

3. Each R_{ij} $(1 \leq i \leq m, 1 \leq j \leq p_i)$ is a generic identifier whose key domains are the same as those of R.

4. Each s_{ki} $(1 \leq i \leq m)$ is the same as some s_j $(1 \leq j \leq n)$,

5. If s_{ki} is the same as s_j, then the type **key** R_j is the range $(R_{i1}, \ldots, R_{ip_i})$.

Let us very briefly discuss these syntactic requirements concerning their semantics for the relational database schema. Requirement (1) states that an attribute s_i is either

- A reference to another object in which case **key** must appear

- An atomic value of type R_i, in which case R_i is a type identifier

Each attribute s_{ki} $(1 \leq i \leq m)$ constitutes a disjoint partitioning of the objects of R into the object classes $R_{i1}, \ldots, R_{ip_i}$, where R_{ij} itself stands for a type definition. Thus, the value of attribute s_{ki} is drawn from the range $R_{i1}, \ldots, R_{ip_i}$.

6.1.2   An Example Application of Generalization and Aggregation

The discussion of abstraction support in the (extended) relational model has been rather abstract so far. Therefore, let us now specify a conceptual model of some specific engineering applications to illustrate these concepts on a more intuitive level. For this purpose we consider the CSG (constructive solid geometry) representation of rigid solid objects as introduced in Section 3.1. This representation model naturally incorporates the concept of aggregation to describe particular components of a rigid solid and generalization in the way the solid objects are constructed, e.g., composed or moved objects. In Figure 6.3 we show a model of this representation method in the form of an extended entity-relationship digram. We have two generalizations: the entity set *mech_obj* is a generalization of *comp_obj* (composed objects), *prim_obj* (primitive objects), and *mot_obj* (moved objects). The partition of *mech_obj* is controlled by different values of the attribute *TYPE*, which belongs to *mech_obj* and is represented using the specialangled box. Another generalization has the entity set *prim_obj* as father object. Primitive objects are partitioned into the classes *cuboid*, *cylinder*, etc. The classification attribute is *PTYPE*.

6.1.3   Schema Definition

We are now ready to convert such an extended entity-relationship schema into the augmented relational schema. The first entity type to be represented is *mech_obj*.

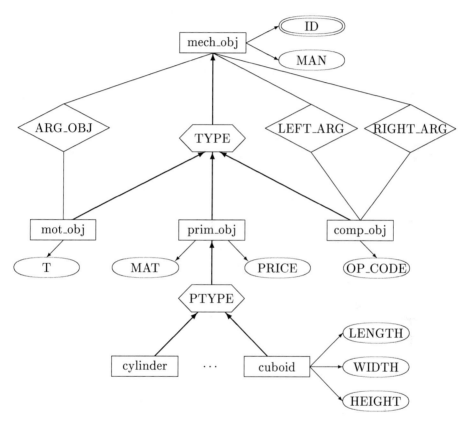

Figure 6.3: Extended Entity-Relationship Model of CSG

```
(1)     var mech_obj: generic
(2)           TYPE = (prim_obj, mot_obj, comp_obj);
(3)        of
(4)        aggregate [ID]
(5)           ID: identifier;
(6)           MAN: manufacturer;
(7)           TYPE: structural_comp
(8)        end
```

Line (1) of the schema definition states that *mech_obj* is a *generic* type, which is partitioned into the disjoint subsets labeled *prim_obj*, *mot_obj*, and *comp_obj* as specified in line (2). The controlling attribute is named *TYPE*. The type *mech_obj* is furthermore an aggregate of the attributes *ID* of type *identifier*, *M* of type *manufacturer* and *TYPE*, the specialization attribute. Line (4) defines *ID* to be the key.

The definition of the generic (relation) type *prim_obj* is analogous:

```
        var prim_obj: generic
              PTYPE = (cylinder, ..., cuboid);
           of
           aggregate [ID]
              ID: identifier;
              MAN: manufacturer;
              MAT: material;
              PRICE: money;
              PTYPE: geom_form;
           end
```

The type *prim_obj* is—as pointed out above—a specialization of the type *mech_obj*. This requires that the attributes of *mech_obj*, except for the attribute *TYPE* that controls the specialization, are replicated in the type definition of *prim_obj* in order to model *inheritance*. This induces redundancy, which will be discussed later.

The type *cuboid* is not generic; it merely constitutes an aggregate of the attributes *ID* (the key attribute), *MAN*, *MAT*, *PRICE*, *LENGTH*, *WIDTH*, and *HEIGHT*:

```
        var cuboid
           aggregate [ID]
              ID: identifier;
              MAN: manufacturer;
              MAT: material;
              PRICE: money;
              LENGTH: real;
              WIDTH: real;
              HEIGHT: real;
           end
```

The type *mot_obj* contains an attribute, *ARG_OBJ*, which is specified as "**key** *mech_obj*". This means that this attribute assumes references to tuples of the relation *mech_obj* as values. These references are maintained in the form of foreign keys—compare with Section 4.2.2—in the relational model; i.e., the attribute contains the key (in our case *ID*) of the associated tuple in *mech_obj*.

```
var mot_obj
    aggregate [ID]
        ID: identifier;
        MAN: manufacturer;
        ARG_OBJ: key mech_obj;
        T: matrix;
    end
```

The relation *comp_obj* contains two such key attributes, called *LEFT_ARG_OBJ* and *RIGHT_ARG_OBJ*, both referring to a tuple in *mech_obj*.

```
var comp_obj
    aggregate [ID]
        ID: identifier;
        MAN: manufacturer;
        LEFT_ARG_OBJ: key mech_obj;
        RIGHT_ARG_OBJ: key mech_obj;
        OP_CODE: (union, difference, intersection, . . . );
    end
```

In Figure 6.4 the extension of the relational schema is shown for a few data items corresponding to the CSG representation of a bracket. From relation *mech_obj* it can be deduced that the object identified by "*bracket#1*" is a *comp_obj*. The description of "*bracket#1*" in *mech_obj* constitutes the highest abstraction level; merely the manufacturer (*MAN*) and the *TYPE* are specified at this abstraction level. A more refined (or specialized) description of "*bracket#1*" can be found in the relation *comp-obj*. Here, the information contained in *mech_obj* is duplicated and, in addition, the *LEFT_ARG_OBJ* and *RIGHT_ARG_OBJ* attributes are specified. As defined in the schema, they refer to objects of relation *mech_obj*, in this case the two objects identified by "*plate#1*" and "*plate#2*". These two objects are described at three abstraction levels:

1. The tuple in *mech_obj*, constituting the highest level abstraction—i.e., the most general level.

2. The tuple in *prim_obj*, where an intermediate abstraction level is maintained.

3. The tuple in *cuboid*, constituting the most detailed description of the particular object. This is the most specialized (and encompassing) description of the object.

mech_obj		
ID	MAN	TYPE
. . .	. . .	. . .
"bracket#1"	"Steal, Inc."	"comp_obj"
"plate#1"	"Steal, Inc."	"prim_obj"
"plate#2"	"Steal, Inc."	"prim_obj"
. . .	. . .	. . .

comp_obj				
ID	MAN	LEFT_ARG_OBJ	RIGHT_ARG_OBJ	OP_CODE
. . .	. . .	. . .	. . .	. . .
"bracket#1"	"Steal, Inc."	"plate#1"	"plate#2"	"union"
. . .	. . .	. . .	. . .	. . .

prim_obj				
ID	MAN	MAT	PRICE	PTYPE
. . .	. . .	. . .	. . .	. . .
"plate#1"	"Steal, Inc."	"iron"	20.00	"cuboid"
"plate#2"	"Steal, Inc."	"iron"	10.00	"cuboid"
. . .	. . .	. . .	. . .	. . .

cuboid						
ID	MAN	MAT	PRICE	LENGTH	WIDTH	HEIGHT
. . .	. . .	. . .	. . .	. . .	. . .	. . .
"plate#1"	"Steal, Inc."	"iron"	20.00	10	5	0.5
"plate#2"	"Steal, Inc."	"iron"	10.00	10	2	0.5
. . .	. . .	. . .	. . .	. . .	. . .	. . .

Figure 6.4: Relational Extension of a "bracket with no holes" in CSG Representation

Each lower (more specialized) level of abstraction *inherits* the attributes of the higher-level abstraction.

6.1.4   Data Manipulation

So far we have only discussed the data modeling issues related to the concepts of generalization and aggregation within the augmented relational model. Of course, particular care has to be taken concerning the data manipulation because the schema—as exemplified in Figure 6.4—contains redundancy. This redundancy is caused by the duplication of inherited attributes in the more specialized object types.

It has to be ensured that the update of an attribute A within some relation R is reflected in all relations that are generalizations of R and contain A and also in all relations that are specializations of R and—obviously—contain attribute A. This so-called *cascaded updating* should, of course, be automatically performed by the database system in order to prevent consistency violations within the database. For example, an update of the attribute *PRICE* of the tuple identified by "plate#1" within relation *cuboid* has to be reflected in relation *prim-obj* in order to guarantee a consistent database state. In general, the update operations may cascade—upward and downward—over many levels of the generalization hierarchy. This depends on the number of levels over which the updated attribute has been inherited (upward cascading) and the number of levels over which the attribute "is handed down" to specialized types (downward cascading).

6.1.5   Discussion

This approach of extending the relational model to support generalization abstraction suffers from several shortcomings:

- Replication of data when mapped into the relational model.

- No behavioral object orientation; that is, the model does not allow to incorporate application-specific operations.

- The semantics of the underlying object structure is not easily seen when one has only a limited view of the application.

- Costly update operations to guarantee the consistency of the replicated information (cascaded updates). This may easily lead to performance degradation if large generalization hierarchies are encountered.

6.2   Molecular Objects

Batory and Buchmann [BB84] extend the work on aggregation by Smith and Smith [SS77a]. The work by Smith and Smith deals only with aggregation as an abstraction of a *single* relationship into a higher-level entity. This notion is extended to the concept of *molecular object*. A molecular object is formed by molecular aggregation, which is an abstraction of a set of entities (of possibly different types) and their relationships into a higher-level entity.

The concept of molecular object is especially useful for engineering database applications, which require to model complex, heterogeneously structured entities that are composed of a variety of subentities. At the highest abstraction level a molecular object is represented by just a single record (or tuple), analogously to the atomic aggregation hierarchies of Smith and Smith. Only at lower abstraction levels are the constituent parts of the molecular object visible: the so-called *atoms*. In this context atoms correspond to records (or tuples) of different types, which are interconnected by relationships to form the encompassing molecular object.

Molecular objects are classified into four categories:

1. Disjoint, nonrecursive

2. Nondisjoint, nonrecursive

3. Disjoint, recursive

4. Nondisjoint, recursive

Disjoint/nondisjoint refers to whether constituent parts (i.e., atoms or submolecules) of a molecular object are shared (= nondisjoint) by other molecules of the same type. *Recursive/nonrecursive* denotes whether the object's structure is recursively defined, that is, whether the structure is defined utilizing the molecular object definition of the same type.

This four-level classification scheme serves two distinct purposes. (1) It supports the database designer in modeling advanced database applications. This is achieved by applying transformation rules to map molecular objects that were appropriately identified in the application area to the database model, e.g., the relational model. (2) The classification scheme allows to evaluate existing and newly developed data models according to their level of support for the particular class of molecular objects.

6.2.1   Disjoint, Nonrecursive Molecules

The easiest form of a molecular object structure is the *disjoint, nonrecursive* structure. In this case no two molecules share components with each other; i.e., all the molecules of the particular type are pairwise disjoint—or *nonoverlapping* concerning their internal structure. Furthermore, the schema of this class of molecular objects is nonrecursively defined. An example of such a disjoint, nonrecursive molecular structure is our *Bicycle* schema developed in Chapter 2—cf. Figure 2.7.

6.2.2   Nondisjoint, Nonrecursive Molecular Objects

This class of molecular objects is distinguished from disjoint molecules in the way that two molecules may share some constituent components; i.e., the extension of two molecules may *overlap* in some dependent subobjects. An example of an overlapping molecular object from the computer geometry domain is the *boundary representation* of geometric objects. Two geometric objects may share the same face—e.g., if the two solid objects are welded together at the particular face.

6.2.3   Recursive, Disjoint Molecular Objects

So far, we have only treated nonrecursive molecular object structures, i.e., structure definitions that do not, as a subcomponent, contain objects of the same type as the molecular object. Recursive molecular objects (recursively) reuse the molecular structure definition in the internal representation.

Let us demonstrate this on a concrete example, again taken from computer geometry applications. The *constructive solid geometry* representation of geometric objects is a very typical example of a recursive molecular object. It actually constitutes a binary tree in which each node represents a solid object that is used to construct the next higher level solid object. The nodes of such a binary tree can, however, be of a different type: They either represent *moved objects*, *primitive objects*, or *composed objects*. *Primitive* objects are again distinguished as *cylinders*, *cuboids*, or other geometric primitives. This molecular object structure is schematically shown in Figure 6.5. The box delimits the molecular object schema.

6.2.4   Recursive, Nondisjoint Molecular Object Types

In the preceding subsection we have assumed that CSG models of solid geometric objects are nonoverlapping, i.e., that an (external or internal) CSG object is not used twice in the representation of a solid object. If we bend the specification of CSG representation slightly in the way that an object, which itself is a CSG object, can be used as an argument for constructing another CSG subcomponent an arbitrary number of times, the CSG representation degenerates to a *directed acyclic graph* (DAG).

The extension of two such overlapping CSG-objects, CSG_1 and CSG_2, is shown in

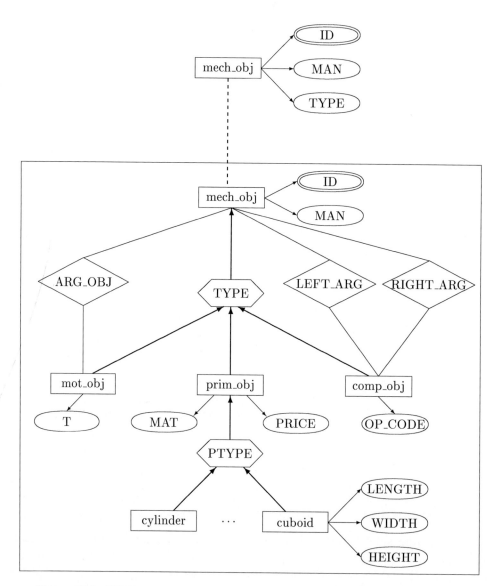

Figure 6.5: ER Representation of Recursive, Disjoint Molecular Objects

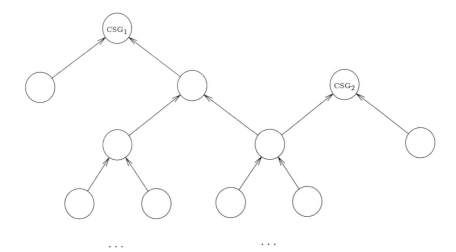

Figure 6.6: CSG Representation Forming a DAG

Figure 6.6. Then the CSG representation constitutes a recursive, nondisjoint molecular structure.

6.2.5   Discussion of Molecular Objects

The notion of molecular objects achieves a classification of object structures that may serve two purposes:

1. It can be utilized as a methodology to classify object types of an application domain, especially within engineering disciplines, in order to aid the conceptual database design. In particular, one can devise general transformation rules that map the various molecule types to flat relational schemas which serve this purpose.

2. It serves as a yardstick for database systems in the way that it allows one to measure the data modeling constructs of the DBMS with respect to the support they achieve in representing the various forms of molecular structures.

As of now, no relational database system extension achieves an adequate support of all four classes of molecular objects.

6.3   A General Reference Mechanism

The pure relational data model associates tuples—that comprise an external object—of possibly different relations by join attributes that are maintained by the system user. In order to access an object in its entirety these tuples have to be explicitly joined in the database query. The extensions of the relational model that we will describe now is designed to reduce the burden of explicitly joining the relations that store constituent tuples of an external object. Rather, the association of tuples is encoded in particular attributes of the relations: attributes that contain a query, expressed in the relational query language QUEL. This principle led to the name "QUEL as a Datatype."

The corresponding join-attribute consists of a QUEL query that retrieves as a result tuples from one or more different relations. The purpose of this extension is to provide a very general referencing mechanism. The database designer could define new objects, such as vectors, cubes, arrays, etc., in separate relations and access them from the parent relation via an attribute of type QUEL. Unlike other referencing mechanisms, no pointers to referenced tuples are visibly[1] maintained; rather, this method represents a dynamic mechanism by executing the query that relates dependent objects to their parent tuple only at the time of access to the particular attribute in the parent relation.

6.3.1   Modeling Constructs

Let us explain this approach on a small example taken from the aggregation hierarchy of the boundary representation. Assume we have a general $N : M$ relationship *boundary*, such as:

This schema relates *mechanical_part* entities to *faces* entities, where a geometrical object is described by its bounding faces. Faces may belong to more than one geometrical object in case the objects touch. This schema could be modeled in normalized relations as shown in Figure 6.7. In order to retrieve a mechanical part, e.g., the one with ID = "cube#1", including all its faces, we would have to join the three (flat) relations *mechanical_part*, *faces* and *boundary* using the following QUEL statement:

> **range of** m **is** mechanical_part
> **range of** f **is** faces
> **range of** b **is** boundary
> **retrieve** m.**all**, f.**all**

---

[1]Actually, the reference to tuples may be precomputed by the system and stored in the form of direct pointers to the corresponding tuples. But this is not visible to the user.

mechanical_part				boundary			faces		
ID	NAME	...		GEO_ID	FACE_ID		ID	SURFACE	...
"cube#1"	...	...		"cube#1"	"f1"		"f1"	...	...
...	...	...		...	...		"f2"	...	...
"cube#2"	...	...		"cube#1"	"f6"		...	...	...
...	...	...		...	...		"f6"	...	...
"cube#3"	...	...		"cube#2"	"f1"		...	...	...
...	...	...		...	...		...	...	...

Figure 6.7: Normalized Relations to Model the Relationship *boundary*

> **where** m.ID = "cube#1"
> **and** m.ID = b.GEO_ID
> **and** f.ID = b.FACE_ID

Using "QUEL as a Datatype," one could encode the join as well as the join relation *boundary* in attributes of the relations *mechanical_part* and *faces*, respectively. The information originally stored in the relation *boundary* is now encoded in the QUEL-attributes *FACE_JOIN* of the relation *mechanical_part* and *GEO_JOIN* of the relation *faces*—as shown in Figure 6.8.

The access to the faces of a *mechanical_part* is now carried out by accessing the attribute *FACE_JOIN*, which triggers the execution of the stored query.

> **range of** m **is** mechanical_part
> **retrieve** m.all, m.FACE_JOIN
> **where** m.ID = "cube#1"

Similarly, we could retrieve the *mechanical_part*(s) that are bounded by a particular face, say "f1":

> **range of** f **is** faces
> **retrieve** f.GEO_JOIN
> **where** f.ID = "f1"

6.3.2   Extended Query Language

Using these attributes of type QUEL in a query facilitates the formation of so-called *reference chains*—by chaining up the accesses to QUEL attributes. For this purpose, the original set oriented query language QUEL has been extended by some more navigational language constructs. A tuple variable t can be appended by a path $A_1.A_2.\cdots.A_n$, where A_1 denotes an attribute of type QUEL and A_2 is an attribute in

mechanical_part			
ID	NAME	...	FACE_JOIN
"cube#1"	...	...	"**range of** f **is** faces **retrieve** f.all **where** f.ID **in** { "f1", ... }"
... "cube#2"	... ...	... ...	... "**range of** ..."

faces			
ID	SURFACE	...	GEO_JOIN
"f1"	10.0	...	"**range of** m **is** mechanical_part **retrieve** m.all **where** m.ID **in** {"cube#1", "cube#2"}"
"f2"	15.0	...	...
... "f6"	... ...	... ...	...

Figure 6.8: "QUEL as a Datatype" Relations to Model the Relationship *boundary*

the result relation returned by executing $t.A_1$. In general, A_i has to be an attribute in $t.A_1. \ldots .A_{i-1}$. Note that only the last attribute, A_n, within such a reference chain could possibly be atomic.

To give the reader an idea of the extended query capabilities of "QUEL as a Datatype", let us consider the following rather simple query.

> Find the *SURFACE* values of all *faces* bounding the *mechanical_part* identified "cube#1".

range of m **is** mechanical_part

retrieve m.FACE_JOIN.SURFACE

where m.ID = "cube#1"

The subclause *m.FACE_JOIN* in the **retrieve** clause references a set of tuples of the relation *faces*. The clause *m.FACE_JOIN.SURFACE* finally retrieves the values of the corresponding SURFACE attributes.

We see that the "." operator is overloaded in this extended query language. The ability to reference tuples of different relations via an attribute of type QUEL results in significantly easier queries. This same query would have involved an explicit join in the traditional relational model—in the case of longer reference chains a correspondingly large number of explicit joins is avoided. For a more detailed discussion of reference chains in object-oriented query languages we refer the reader to Chapter 14.

6.3.3   Lack of Type Safety

"QUEL as a Datatype" provides an extremely general referencing mechanism within the context of the relational model. However, the generality has to be paid for in terms of lack of type safety. For example, in our sample database extension of Figure 6.8 the user could have injected the query

> **range of** f **is** other_faces
>
> **retrieve** f.all
> **where** f.ID **in** {"f20",..., "f25"}

in the tuple identified by "cube#3" of relation *mechanical_part*. Now assume that the relation *other_faces* has the following structure:

other_faces			
ID	SURFACE	...	GEO_JOIN
"f20"	"smooth"	...	"**range of** m **is** mechanical_part **retrieve** m.all **where** m.ID **in** {"cube#3"}"
"f21"	"smooth"	...	...
...	...	...	...
"f25"	"rough"	...	...

In this relational schema of *other_faces* the attribute *SURFACE* has a different type (i.e., *string*) as opposed to the relation *faces* (i.e., *float*).

Then the following query is correct:

> **range of** m **is** mechanical_part
>
> **retrieve sum**(m.FACE_JOIN.SURFACE)
> **where** m.ID = "cube#1"

whereas the next query

> **range of** m **is** mechanical_part
>
> **retrieve sum**(m.FACE_JOIN.SURFACE)
> **where** m.ID = "cube#3"

leads to a type error—because of the attempt to sum attributes of type *string*. This type error cannot be detected at compile time, since it depends on stored attribute values; i.e., it depends on the particular encoding of the QUEL attributes.

6.4   Abstract Data Types in the Relational Model

ADT-INGRES provides a facility that allows the user to define his or her own application-specific data types. The representation of the new data type has to be specified in the programming language C.

Let us now consider an example. We want to define a relation to store cuboids, basically in terms of their eight bounding vertices. Rather than repeating for each of the vertices the fact that a vertex consists of three decimal numbers, i.e., the X, Y, and Z coordinates, we specify once and for all an ADT *vertex_type* and then use it in the definition of the relation *cuboids* as follows:

```
cuboids ( id: integer,
          material: char(10),
          description: char(20),
          V1: ADT: vertex_type,
          V2: ADT: vertex_type,
                 . . .
                 . . .
          V8: ADT: vertex_type )
```

An example query using the ADT attribute *V1* of type *vertex_type* would look as follows:

```
range of c is cuboids
retrieve (c.material, c.description, c.V1)
where c.id = 5
```

The output of this query could then—depending on the implementation of the ADT—look as shown below:

material	description	V1		
		X	Y	Z
copper	massive	1.0	3.5	2.0

And a possible append command could look as follows:

```
append to cuboids
        ( id = 5,
          material = "copper",
          description = "massive",
          V1 = (1.0, 3.5, 2.0),
          V2 = (. . .),
                 . . .
          V8 = (. . .) )
```

The user has to supply the implementation of such an abstract domain. For our example this implementation is made known to the database system as follows:

define ADT
      (typename = "vertex_type",
        bytesin = 9,
        bytesout = 9,
        inputfunc = "to_internal_vertex",
        outputfunc = "to_external_vertex",
        filename = "/usr/ingres/.../vertex")

Inputfunc and *outputfunc* are C subroutines that convert the data type to internal and external representation. These routines are found in the file "/usr/ingres/.../vertex" as specified in the *filename* clause. For example, the *outputfunc to_external_vertex*, would extract the X, Y, and Z coordinates from the internal representation and output them in the format shown above. For the implementation of these routines an intrinsic knowledge of the programming language C is required.

An obvious disadvantage of ADT-INGRES is that each abstract data type has to be mapped onto one attribute—usually an attribute of variable length. In our case this means that the three coordinates are mapped onto an attribute of type string. This is a very unnatural mapping. It would be much more convenient (and natural) to map the coordinates onto three attributes of type float.

Schematically the ADT mapping for our data type *vertex_type* is shown below:

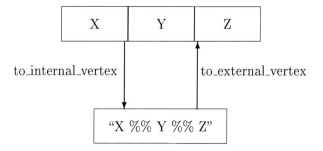

In addition to such a data type, the ADT-INGRES user can define corresponding operations on these domains. As an example let us present the framework of the operation "R_y", which takes as an argument a vertex and an angle. It returns a vertex that is rotated about the y-axis by the given angle. The implementation would look as follows:

define adtop
      (opname = "R_y"
        funcname = "rotate_about_y"
        filename = "/usr/ingres/.../rotate_y",

result = ADT: vertex_type,
arg1 = ADT: vertex_type,
arg2 = ADT: angle_type,
prec like "+")

Once again the file *rotate_y* must contain a C program implementing this operation.

Similarly one can define the other possible geometric transformations scaling, translation, and rotation about the other two axes.

Let us now demonstrate a database operation that rotates a previously inserted copper cuboid:

range of c **is** cuboids
replace c (V1 = R_y(c.V1,PHI),
              V2 = R_y(c.V2,PHI),
                   ...,
              V8 = R_y(c.V8,PHI))
where c.id = 5

ADT-INGRES provides a novel way of specifying new data types and corresponding operations in a database management system. The advantage of this approach lies in the fact that the operations can be arbitrarily complex. For example, we showed the framework for all the geometric transformations on 3-D objects, i.e., scaling, translation, and rotation.

But the additional flexibility of the system also has its penalty. The new data types have to be entirely specified in the programming language C. Thus, the user has to be familiar with two quite different systems: (1) the database language QUEL and (2) the programming language C.

Another shortcoming of this approach is inherent in the underlying data model of the database management system INGRES. ADT-INGRES does not allow to map an ADT onto different tuples (or relations). It requires to map each ADT completely onto one attribute. Thus, the internal representation of engineering objects does not reflect the external structure (as the user perceives it) of the object. This usually results in a fairly tedious transformation process from external to internal representation, and vice versa. For example, the ADT *vertex_type* had to be mapped into a character string rather than onto three attributes of type float, which would have been a much more natural mapping. The reason for this unnatural mapping is that ADT-INGRES does not provide any support to handle hierarchical data structures, but precisely these occur frequently in engineering applications.

ADT-INGRES provides some support for behavioral object orientation by allowing the database user to define application-specific ADT operations. But due to the fact that the underlying data model is pure relational—and, therefore, does not support structurally complex objects—these operations are quite tedious to implement.

6.5   Hierarchical Object Support in XSQL

System R [ABC+76] is a relational database management system developed at the IBM research laboratory in San Jose and is the basis of the commercial products SQL/DS and DB2. In the XSQL [KLMP84, LP83] project efforts were undertaken to enhance the relational data model and data manipulation language to support technical applications. The new type *long field* and the notion of *complex object* were introduced. However, long fields, which are useful to store unstructured data such as text, are only of minor interest for computer-integrated manufacturing (CIM) databases. The data generated in CIM, e.g., CSG and the boundary representation (BR), constitute highly structured data schemes. Of course, one could store this data in long fields and then retrieve it using appropriately defined operations on the long fields. This would resemble the domain ADT approach of ADT-INGRES. This approach of storing structured data in a long field, however, bears the disadvantage that accessing subparts of a long field requires to retrieve and evaluate the whole information before the relevant part can be obtained.

6.5.1   The Modeling Constructs for Complex Objects

In the XSQL terminology a *complex object* constitutes an entity that is composed hierarchically of several subentities that themselves may be composed of subentities via a hierarchical relationship. The notion of *complex object* is suitable to group several tuples of (possibly) different relations together. One of the tuples is the so-called *root* tuple and the others are the dependent tuples (subobjects), which together form the complete description of some complex object. In this respect, a complex object constitutes an aggregate object (cf. Chapter 2).

The basic extensions of XSQL over the pure relational model are described below.

Surrogates

In Chapter 5 we concluded that one of the shortcomings of the original relational model was the lack of system-generated identifiers for the tuples. In XSQL so-called surrogates are introduced to uniquely identify objects (tuples)—for a more elaborate discussion of object identifiers see Section 7.5. These identifiers are called surrogates in the literature, e.g., [Cod79], because they constitute placeholders for the corresponding objects. Surrogates should be independent of contents and physical location of the object. A surrogate remains invariant throughout the lifetime of an object. Thus, surrogates can be read by the database user in order to access the objects being identified by the corresponding surrogate. But surrogates are protected against user update.

One possible method for generating surrogates is as follows: Take the *processor id* and concatenate it with the *date* and *time* the object was created, i.e.:

processor id	date&time

This method guarantees that the generated surrogates are "world-wide" unique because within the same processor they are distinguishable by the creation date and time. And among different computer systems the surrogates can be differentiated by their *processor id* portion. This is particularly important when databases from different sites have to be merged or when engineers work independently on different parts of a large design over a local area network.

Component-Of Attributes

The *component-of* attributes in XSQL are used to model hierarchical relationships between one parent tuple and its direct child tuples. These tuples may belong to the same or to different relations. Thus, this concept is useful for modeling $1 : N$ relationships of the following form—recall the (min, max) notation introduced in Exercise 2.8:

Here, an entity e_2 of the entity type $E2$ is related to exactly one entity e_1 of $E1$ via the relationship $R1$. On the other hand, any e_1 of entity set $E1$ may be associated with several entities of $E2$. Using the *component-of* construct, one can model the relationship $R1$ by the following (extended) relational schema:

    create table E1(E1-ID: identifier,
                E1-DATA: ...)

    create table E2(E2-ID: identifier,
                E2-FATHER: component-of(E1)
                E2-DATA: ...)

The attributes *E1-ID* and *E2-ID* are of type *identifier*; that is, upon insertion of a new tuple these attributes are automatically assigned a surrogate that is generated by the system. The attributes *E1-DATA* and *E2-DATA* stand, representatively, for the remaining properties of entities of type *E1* and *E2*, respectively.

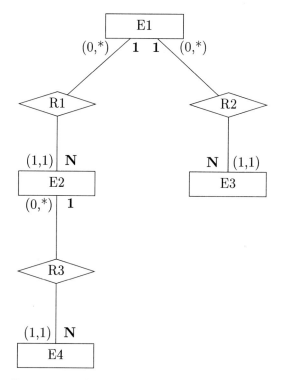

Figure 6.9: A General Hierarchical Schema

The hierarchical relationship *R1* is modeled by the attribute *E2-FATHER*, which is of type *component-of(E1)*. For an entity e_2 of *E2*, this attribute, logically, contains a reference to the entity e_1 of *E1* to which e_2 is related by the relationship *R1*, i.e., to the parent entity. This reference is maintained in the form of the surrogate of e_1, i.e., the system-assigned value of *E1-ID*.

Sor far we have only treated a hierarchical relationship over one level. In general, the *component-of* construct can be employed to handle large hierarchical structures that span over many levels, such as shown in Figure 6.9.

Analogously to the discussion above, this conceptual schema can be mapped into the following XSQL relational schema:

```
create table E1 (E1-ID: identifier,
            E1-DATA: ... )

create table E2 (E2-ID: identifier,
            E2-FATHER: component-of(E1)
            E2-DATA: ... )
```

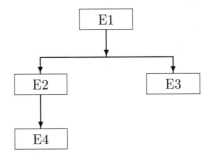

Figure 6.10: Notation of Hierarchical Relationships as a Complex Object

E1	
E1-ID	E1-DATA
$589000100	...
$589000200	...
...	...

E2		
E2-ID	E2-FATHER	E2-DATA
$589000101	$589000100	...
$589000103	$589000100	...
...	...	...
...	...	...

E3		
E3-ID	E3-FATHER	E3-DATA
$589000102	$589000100	...
$589000104	$589000100	...
...	...	...
...	...	...

E4		
E4-ID	E4-FATHER	E4-DATA
$589000105	$589000101	...
...	...	...
...	...	...

Figure 6.11: Instantiation of a Complex Object

create table E3 (E3-ID: identifier,
                     E3-FATHER: component-of(E1)
                     E3-DATA: ...)

create table E4 (E4-ID: identifier,
                     E4-FATHER: component-of(E2)
                     E4-DATA: ...)

In order to convert to the conventions used in XSQL, we will, from now on, denote the direct hierarchical relationships that are modeled by *component-of* attributes by using a thick arrow from the father entity to the dependent entity. For the example used in Figure 6.9 above, this is shown in Figure 6.10.

The example instance of the database in Figure 6.11 contains two complex objects, represented by the root tuples of relation *E1* with the surrogates *$589000100* and

$589000200. Only the dependent tuples of the first complex object are shown in the other relations *E2, E3,* and *E4.* A complex object has a surrogate consisting of nine digits where the first seven identify the root of the complex object and the last two digits are used to identify the dependent subtuples within a complex object. Thus, tuples that belong to the same complex object have the same prefix in their surrogate attribute; only the last two digits differentiate them. It can be seen in Figure 6.11 that the complex object represented by the root tuple *$589000100* has the following sub-objects:

- $589000**101** and $589000**103** of relation *E2*

- $589000**102** and $589000**104** of relation *E3*

- $589000**105** of relation *E4*

$589000**105** of relation E4 is not a direct component of *$589000100*, the root tuple. It is a component of the tuple *$589000101* of relation *E2*. For each complex object in such a schema the system maintains a so-called *map* that contains pointers to the storage location of the tuples in the form of tuple identifiers (TIDs).

6.5.2   Attributes of Type *reference*

The *component-of* attributes are useful to model hierarchical relationships, i.e., $1 : 1$ and $1 : N$ relationships. XSQL also provides limited support for modeling $N : M$ relationship types. For this purpose we can utilize the so-called *reference* attributes, which maintain a reference from one tuple to another tuple—of a different or of the same relation. Let us consider a general $N : M$ relationship type, e.g.:

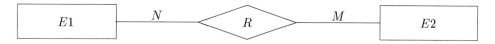

This relationship type can be modeled as a separate relation with references to the participating relations *E1* and *E2*, i.e.:

create table E1 (E1-ID: identifier,
                     E1-DATA: ...)

create table E2 (E2-ID: identifier,
                     E2-DATA: ...)

create table R (R-ID: identifier,
                    RtoE1: reference(E1)
                    RtoE2: reference(E2))

```
create table MP(
        MP_ID: identifier,
        NAME: ...,
        ...)

create table FACES(
        F_ID: identifier,
        F_COMP: component_of(MP),
        ...)

create table EDGES(
        E_ID: identifier,
        E_COMP: component_of(FACES),
        ...)

create table VERTICES(
        V_ID: identifier,
        V_COMP: component_of(EDGES),
        ...)
```

Figure 6.12: BR Schema in XSQL

The attributes *RtoE1* and *RtoE2* contain references to the particular tuples in *E1* and *E2*, respectively.

Unfortunately, XSQL provides less system support for these *reference* attributes than for the *component-of* attributes. While it is possible to retrieve an entire complex object whose structure is given by the *component-of* relationships, it is not possible to implicitly follow a chain of *reference* attributes. An object that is decomposed along *reference* attributes has to be explicitly reconstructed—very much like a join in the pure relational model.

6.5.3   Boundary Representation in XSQL

The problem with the *component-of* concept is that each tuple has exactly one parent tuple in the parent relation. This is a severe shortcoming for modeling the boundary representation of geometric objects. A possible BR schema is shown in Figure 6.12.

We note that the same edge can belong to more than one face. And a vertex usually belongs to many edges. Using the above schema one would have to include a lot of redundant information, since, for example, a vertex has to be stored for each edge that it is an endpoint of. It might even make data manipulation algorithms very complex. Consider, for example, the query:

> Find all faces that have vertex v_i in common.

Because of the data redundancy this query involves an exhaustive search for all vertices $v_j = (x_j, y_j, z_j)$ such that $x_j = x_i$, $y_j = y_i$, and $z_j = z_i$ where (x_i, y_i, z_i) are the coordinates of vertex v_i.

6.5.4   Evaluation of the XSQL Model

The main contribution of the XSQL research project are twofold: First, the incorporation of a robust identification concept in the form of system-generated surrogates was achieved. Second, the relational model was enriched by constructs to model hierarchical relationships among (possibly) different entity types.

In terms of the molecular object notion of Section 6.2 the XSQL model provides the following support for the various kinds of molecular objects:

- Nonrecursive:

    - Disjoint: Supported by the *component-of* attributes
    - Nondisjoint: Very limited support by the *reference* attributes

- Recursive:

    - Disjoint: Also supported by the *component-of* attributes
    - Nondisjoint: Very limited support

A particularly important concept in engineering applications is the *shared (sub)-object*. This notion refers to a (complex) object that is used as a constituent part in the definition of more than one comprising object. Shared objects fall under the category nondisjoint (recursive/nonrecursive) molecule in the classification of Section 6.2. In XSQL shared subobjects can be modeled as autonomous complex objects, which themselves own all the properties of a complex object in its own right. The encompassing objects, of which the shared subobject is a constituent part, contain references to this shared subobject. Graphically, on the instance level, this is depicted in Figure 6.13, in which the complex objects on the top share the object on the bottom. Note that in general the objects referencing a shared subobject are of different type.

The designers of XSQL have in some way solved one of the shortcomings of "QUEL as a Datatype." They have devised a strategy to use system-generated identifiers to reference tuples of different relations. This is a very helpful mechanism to build up abstraction hierarchies without back references. In summary one could say that they

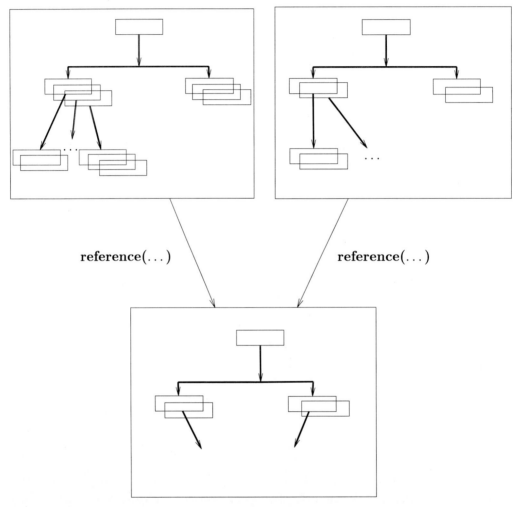

Figure 6.13: Shared Subobjects in XSQL

enriched the relational model to capture the semantics of the hierarchical model as well.

The object structure is maintained via references between the tuples, that is, between children and parent tuples of the hierarchical structure. The main advantage of this approach is that it requires only a very limited extension of a pure relational system. Thus, it can be seen as a relatively pragmatic approach to provide structural object orientation within the framework of the relational model. A severe disadvantage that may result from the XSQL layout of complex objects concerns the physical level of the resulting database management system. The model, per se, does not provide any implicit clustering of complex objects, because the tuples constituting a complex object may belong to a large number of different relations. Thus, it is the database administrator's responsibility to perform a sophisticated physical database design to achieve a clustering of these heterogeneous tuples on the same page. This is required in order to aid object-oriented access to a complex object in its entirety without compromising efficiency by having to access too many pages.

The reason for this potential problem lies in the fact that XSQL does not impose clear-cut object boundaries. As it is nicely said in [HS87] objects are considered a "luxury item" in XSQL, which does not receive the same extensive system support as the built-in structures *relation* and *tuples*. The boundaries of a relation and—even more so—of a tuple are much more predominant than those of a complex object. Even in the schema definition a complex object structure can be identified only by searching all relations for *component-of* attributes that may contain a back reference to the parent relation. Within the extension of such a schema the object boundaries get totally obscured because all subobjects, i.e., dependent tuples, of a complex object can either be accessed via the parent tuple or, independently, via the relation to which they belong. The object's structure is preserved only by the internally stored complex object table, the so-called *map*, which is not visible to the database user. The *map* is only accessible by the system in order to reassemble complex, hierarchically structured objects.

6.6   The Nested Relational Model NF^2

In the evaluation of the XSQL model it was concluded that hierarchically composed object do not possess distinctive object boundaries. One approach to provide more clear-cut object boundaries for hierarchically structured objects constitutes the *nested relational data model*. Object boundaries are maintained in the nested relational model by nesting the structure of the dependent subobject within the parent object. This leads *inherently* to a clustered logical (and, depending on the implementation, also physical) representation of a complex object. This basic idea of nesting relations within each other gave rise to an alternative name of the model: **non first normal** form (NF^2) model.[2]

6.6.1   The Basic Modeling Constructs of the Nested Relational Model

In this section we will contrast the basic modeling constructs of the nested relational model with the pure relational model. Let us first recall the structuring concepts of the relational model: *relations* are sets of equally structured *tuples* consisting of *attributes* that draw their values from *atomic* domains. This is schematically depicted in Figure 6.14.

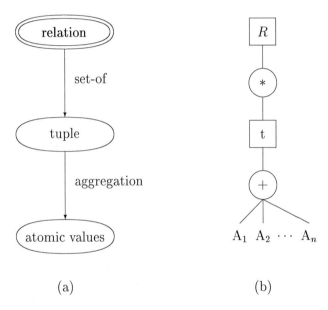

(a)   (b)

Figure 6.14: Modeling Constructs of the Pure Relational Model: (a) Structuring Concepts and (b) Sample Schema

On the left-hand side of the diagram the generic structuring concepts of the relational model are shown. The only valid object structure within the relational model is the relation; therefore, this has been indicated in the diagram by a double ellipsis. The right-hand side of the diagram sketches how a relation R is actually composed by the *set-of* operation of tuples of type t. The *set-of* operation, denoted as "$*$", is sometimes referred to as the *association* operation. A tuple of type t constitutes an aggregation, denoted as $+$, of attributes $A_1, \ldots, A_n$ each of which draws its values

---

[2]The first normal form of relational data modeling demands that all domains of a relation be atomic. This requirement is abandoned in the nested relational model.

from a domain of atomic values. Now the pure relational model—from a structural point of view—is characterized by the fact that the *set-of* and the *aggregation* operation are each applied exactly once in the above-described sequence.

The NF² data model extends the structuring concepts of the pure relational model in the way that the two generic structuring concepts *set-of* and *aggregation* can be applied several times, that is, an arbitrary, though *predefined* number of times. Thus, an attribute A_i of a relation R can be of type relation, i.e., a composite type. This provides for nesting relations R_i that model dependent subobjects within the parent relation R. Schematically, the structuring concepts of the NF² model are shown in Figure 6.15.

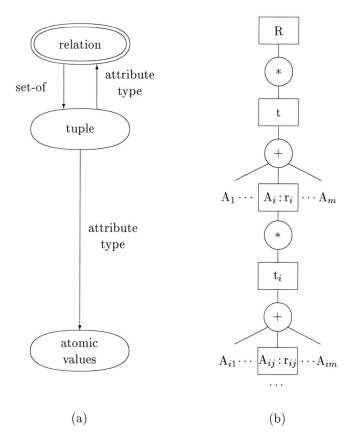

(a)   (b)

Figure 6.15: Modeling Constructs of the Original NF² Model: (a) Structuring Concepts and (b) Sample Schema

Again, the only valid object structure that can exist as an autonomous database construct is the relation. A relation consists of a set of equally structured tuples that are composed of a number of attributes. Now, the attributes of a tuple are either *atomic-valued* (AV) or *relation-valued* (RV). This is depicted on the left-hand side of Figure 6.15.

On the right-hand side of the diagram we show an application of the structural facilities of the model by constructing a sample relation R. This relation consists of tuples composed of n attributes, denoted A_1, ..., A_n. All of these attributes are atomic-valued except for A_i, which is relation-valued. This means that A_i itself is a relation, i.e., a set of tuples with attributes A_{i1}, ..., A_{im} of which A_{ij} is again a relation-valued attribute. We note that the nesting of relations in the NF^2 model can be arbitrarily deep. The only requirement is that the nesting is of *fixed depth* because the schema of a relation, such as R, has to be *finite*. In other words, the tree of Figure 6.15(b) has to have a fixed depth where all attributes A_i ($1 \leq i \leq n$) of R eventually lead to atomic-valued attributes in the tree, which are recognized as leaf nodes.

6.6.2   Formal Definition of NF^2 Relations

In the pure relational model a relation R is defined as a subset of the Cartesian product of domains D_i of atomic values, such as

$$R \subseteq D_1 \times D_2 \times \ldots \times D_n$$

The *first normal form* of the relational theory states that all domains D_i contain just atomic (i.e., noncomposite) values.

In the definition of the NF^2 model we also start with given sets D_1, ..., D_n of atomic values, now called the atomic domains. An NF^2 relation R_{NF^2} is then defined as a subset of the Cartesian product of complex domains C_1, ..., C_k:

$$R_{NF^2} \subseteq C_1 \times \ldots \times C_k$$

where C_i corresponds to some D_j for ($1 \leq i \leq k$, $1 \leq j \leq n$), if C_i is an atomic domain of the relation. Alternatively, C_i could be a relation-valued domain, in which case C_i is constructed as the powerset of a Cartesian product of sets C'_1, ..., C'_r, where the C'_j ($1 \leq j \leq r$) are, again, either atomic domains or themselves complex domains constructed as powersets of Cartesian products of (atomic or complex) domains.

As an example we could construct the two complex domains C_1 and C_2 as

$$C_1 \quad := \quad 2^{(D_1 \times D_2)}$$

$$C_2 \quad := \quad 2^{(D_2 \times D_3) \times 2^{(D_4 \times D_5)}}$$

We denote the powerset of a set S by 2^S. C_1 is constructed as the powerset of the Cartesian products of the atomic domains D_1 and D_2. C_2 is more deeply nested. It is the powerset of the Cartesian product of $D_2 \times D_3 \times C_3$, where C_3 itself is a powerset of the Cartesian product of—now atomic—domains $D_4 \times D_5$.

Then

$$R_{NF^2} \subseteq C_1 \times C_2 \times D_6$$

is a ternary relation consisting of one atomic-valued domain D_6 and two relation-valued domains C_1 and C_2. This relation is represented in a tree format in Figure 6.16 below.

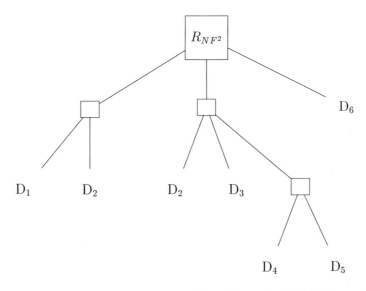

Figure 6.16: Graphical Representation of the Example NF^2 Relation R_{NF^2}

6.6.3   Representation of NF^2 Relations by Attributes

For simplicity, we have so far used unique domain names $C_1, \ldots, C_n$ because we have not made a distinction between attribute names and domains. A formal definition of associating attribute names with complex domains of nested relations is presented in [SS86]. For our purposes it suffices to view each attribute of an NF^2 relation to draw its values either from an atomic domain, in which case the attribute is called an atomic attribute, or from a complex domain, in which case it is a complex (composite) attribute. A complex domain is formed from atomic domains by repeated application of the Cartesian product and the powerset operation.

In-Line Representation of Nested Relations

This leads us to a linear representation of nested relations by nesting the linear representation of a subrelation of its attributes within the superrelation. For example:

$$R(A_1(A_{11}, A_{12}), A_2, A_3(A_{31}, A_{32}(A_{321}, A_{322})))$$

is the relation with the three attributes A_1, A_2, and A_3. A_2 is an atomic-valued attribute, whereas A_1 and A_3 are relation-valued. A_1 corresponds to a subrelation with the attributes A_{11} and A_{12}, which must be atomic-valued attributes because they are not expanded any further in the linear representation.

Tree Representation

This example schema is graphically represented as a tree in Figure 6.17. Here, relation-valued attributes are represented as intermediate nodes in the tree, which we denote by a box. Atomic-valued attributes can exist only as leaf nodes in the tree.

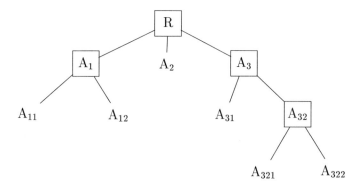

Figure 6.17: Tree Representation of the Example NF^2 Relation R

Tabular Representation

The extension of a nested relation is best represented in *tabular* format. For our example schema this is shown in Figure 6.18.

In this representation relation-valued attributes are represented by curly brackets, such as $\{A_1\}$ and $\{A_3\}$. The (tabular) representation of a nested subrelation is contained within the table of the parent relation. Subrelations are denoted by their attribute names. For example, $\{A_{32}\}$ stands for the subrelation that contains tuples with the attributes A_{321} and A_{322}.

Figure 6.18 also contains some sample tuples; i.e., it contains two tuples of the relation R. The first tuple, say, t_1, contains a set of n tuples $[v_{11}^i, v_{12}^i]$ $(1 \leq i \leq n)$ as

{R}					
{A_1}		A_2	{A_3}		
A_{11}	A_{12}		A_{31}	{A_{32}}	
				A_{321}	A_{322}
v_{11}^1	v_{12}^1			v_{321}^1	v_{322}^1
v_{11}^2	...			v_{321}^2	v_{322}^2
v_{11}^3	...	v_2^1	v_{31}^1	...	...
...	...			v_{321}^m	v_{322}^m
v_{11}^n	v_{12}^n		v_{31}^2	...	...
				...	...
...	...	...	...	...	...
...	...	...	...	...	...
...	...	v_2^2	...	...	...
...	...	...	...	...	...

Figure 6.18: Tabular Representation of the Example NF^2 Relation R

a value for attribute A_1. Attribute A_2 has the value v_2^1 assigned. And the relation-valued attribute A_3 contains two tuples, which we will denote t_{31} and t_{32}. The tuple t_{31} contains as values for attribute A_{31} v_{31}^1 and A_{32} is assigned a set containing the m tuples $[v_{321}^i, v_{322}^i]$ $(1 \leq i \leq m)$, where v_{32j}^i corresponds to the value assigned to attribute A_{32j}. The tuple t_{32} of relation $\{A_3\}$ is analogously constructed.

6.6.4 The Extended NF^2 Model

The *extended* NF^2 model incorporates structuring concepts into the relational model that go beyond nesting of relations. The extended NF^2 model is the basis of the database system prototype AIM-P that was developed at the IBM Scientific Center, Heidelberg. The designers of the extended model [PA86, DKA$^+$86] motivate the extensions to the originally proposed NF^2 model by stating the following goals:

- Provide a model in which all object structures are homogeneously treated; i.e., tuples, relations, atomic values, and lists can autonomously exist within the database.

- Extend the capabilities of the NF^2 model in the way that also an ordering concept is provided. Tuples in a relation are not ordered—that is, the system does not guarantee any order in which tuples will be stored and, subsequently, be accessed during manipulation.

- Allow the combination of any object constructors. The originally proposed NF^2 model allows only relations as sets of tuples, which again may contain set-valued attributes. In the extended NF^2 model one also allows sets of atomic values, sets of sets, sets of lists, and so on. Thus, the extended NF^2 model should be fully orthogonal in its treatment of structuring concepts in the way that it allows to construct new types by combining any of the structuring concepts.

The modeling constructs of the extended NF^2 model are graphically shown in Figure 6.19. As can be seen from the diagram the model provides the following type

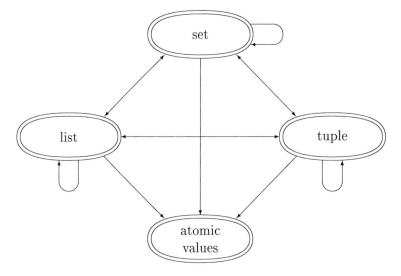

Figure 6.19: Structuring Concepts of the Extended NF^2 Model

constructors:

Atomic Values   *Atomic values* are of numeric, character, or Boolean type. The numeric types include integer and floating point number domains.

Tuples   *Tuples* are aggregations of attributes. Attributes can be atomic-valued, relation-valued, or—which goes beyond the original NF^2 model—*list-valued*. Another distinction to the original NF^2 model is that a tuple can exist as an autonomous entity within the database. A tuple is denoted in the syntax of the AIM-P database system by brackets.

Thus,

$$T : [A_1\text{: } A_1\text{-Type},\dots, A_n\text{: } A_n\text{-Type}]$$

defines a tuple T that aggregates the n attributes $A_1, \dots, A_n$.

Sets *Sets* contain equally structured elements. The elements are either tuples, atomic values, lists, or itself sets. In the original model sets could contain only tuples; such a set is called a relation.

Sets are denoted in the syntax by curly brackets in the form

$$S : \{\text{S-Type}\}$$

where S is declared as a set of elements of type *S-Type*.

It is possible to specify the maximum or the exact cardinality that a set has to obey at any time. For example,

$$S_{max} : \{\text{n S-Type}\}$$

specifies that the set S_{max} contains at most n elements. And

$$S_{exact} : \{\text{n FIX S-Type}\}$$

constrains the set S_{exact} to contain exactly n elements of type *S-Type*.

Lists A *list* provides the ordering concept within the extended NF^2 model. Lists are very similar to sets in the respect that they contain equally structured elements. The distinction is that lists preserve the order which is assigned to inserted elements. Also, lists may contain duplicates of elements.

A list L of elements of type *L-Type* is declared in AIM by using angle brackets as follows:

$$L : <\text{L-Type}>$$

Analogously to sets, one can also constrain the cardinality of a list by either specifying the maximum or the exact number of elements that the particular list contains at any time.

6.6.5 Data Definition Language

We now demonstrate a nested schema definition for the boundary representation of a mechanical part.

```
create object MECHANICAL_PART
    {[ ID: string(10),
       NAME: string(20),
```

```
        FACES:
          {[ ID: string(4),
              EDGES:
                {[ ID: string(4),
                    VERTICES:
                      {[ ID: string(4),
                          LOCATION:
                              [ X: real,
                                Y: real,
                                Z: real ]
                      ]}
                ]}
          ]}
    ]}
```

MECHANICAL_PART is a relation with the atomic-valued attributes *ID* and *NAME* and the relation-valued attribute *FACES*. The relation *FACES* has the atomic-valued attribute *ID* and the relation-valued attribute *EDGES*, which again is split up into *ID* and the relation *VERTICES*. The relation *VERTICES* consists of an attribute *ID* and an attribute of type tuple, *LOCATION*, which consists of the three coordinates *X*, *Y*, and *Z*.

An *extension* of the relation *MECHANICAL_PART* for an example geometric object in the form of a cuboid is shown in Figure 6.20.

The nested relational model supports very well hierarchical relationships, i.e., $1 : N$ and $1 : 1$ relationship types. However, the relation *MECHANICAL_PART* demonstrates the problem of modeling $N : M$ relationships in a nested relation. A straightforward schema design, as carried out in the preceding section, very often induces a severe redundancy problem in the resulting database model of the real world.

Let us demonstrate this on an example object, a simple *cuboid*, that is stored in the relation *MECHANICAL_PART*. In reality, a cuboid consists of 6 faces, 12 edges, and 8 bounding vertices. When a cuboid is inserted in the *MECHANICAL_PART* relation, a lot of information is stored redundantly. The schema leads to the following description of a cuboid:

$$\text{cuboid} \overset{6}{\Longrightarrow} \text{faces} \overset{4}{\Longrightarrow} \text{edges} \overset{2}{\Longrightarrow} \text{vertices}$$

Thus, a cuboid is described by separately modeling its six faces, each of which is modeled by its four bounding edges. Each edge, again, is separately described by its two end-vertices. This leads to $6 * 4 * 2 = 48$ stored *VERTICES* tuples containing the *LOCATION* tuple of *X*, *Y*, and *Z* values instead of the eight actually existing vertices. This means each vertex is redundantly stored six times. The cause of the

{MECHANICAL_PART}							
ID	NAME	{FACES}					
		ID	{EDGES}				
			ID	{VERTICES}			
				ID	LOCATION		
					X	Y	Z
...	...	...	...	...	...	...	...
'cubo#5'	'cuboid'	'f1'	'e1'	'v1'	0	0	0
				'v2'	1	0	0
			'e2'	'v2'	1	0	0
				'v3'	1	1	0
			'e3'	'v3'	1	1	0
				'v4'	0	1	0
			'e4'	'v4'	0	1	0
				'v1'	0	0	0
		'f2'	'e5'	'v5'	...	...	...
				'v6'	...	...	...
			'e6'	...	...	...	...
		...	...	...	...	...	...
...	...	...	...	...	...	...	...

Figure 6.20: Tabular Representation of the Relation MECHANICAL_PART

problem is the lack of modeling support for shared subobjects. This is often poorly remedied by duplicating the shared subobjects.

Unfortunately, this duplication of shared subobjects creates several problems:

increased storage space: The replication of data leads to an increase in the storage space to store an object. This problem is somewhat mitigated by the fact that the nested relational schema typically reduces the number of join attributes that are needed in a flat relational model.

efficiency problem: *Controlled* redundancy of data leads to more efficient data manipulation in some cases. But it may also cause the opposite effect: Consider, for example, the typical geometric transformation of a *cuboid* that is performed by multiplying each bounding vertex by some transformation matrix T. This operation results in eight multiplications of a vector v, representing the vertex, with the matrix T if no redundant information has to be taken care of. In our case, however, the operation results in 48 multiplications—that is, all redundantly stored vertices have to be modified. Thus, we can expect a performance decrease by a factor 6.

consistency problem: *Uncontrolled* redundancy typically leads to consistency problems. When, for example, the vertex v_i is replicated six times in the database, say, $v_i^1, \ldots, v_i^6$, it has to be guaranteed that an update of vertex v_i is carried out on all six replicated copies $v_i^1, \ldots, v_i^6$ of v_i.

6.6.6   Modeling Boundary Representation without Redundant Vertices

Let us now devise a BR schema that mitigates the problems incurred by redundancy somewhat. The following NF^2 schema models geometric objects in boundary representation without redundantly storing metric information within one object description. Only the redundancy within the topological representation remains. Since the topology of a geometric object is not as frequently modified as the other information, this schema is definitely superior to the preceding schema.

```
create object BR_GEOMETRY
        {[GEO_ID : string(10),
          FACES:
                  {[FACE_ID: string(10),
                    EDGES:
                            {[EDGE_ID: string(10),
                              VERTICES: {2 FIX [V_ID: string(10)]}
          METRIC:
```

```
{[VERTEX_ID: string(10),
  X: real,
  Y: real,
  Z: real]}
]}
```

The tabular representation of the nested relation *BR_GEOMETRY* is shown in Figure 6.21. In this schema we have gone "one step back" toward the pure relational model. We have separated the *topological* representation, i.e., subrelation *FACES*, from the *metric* information, i.e., subrelation *METRIC*. Thus, we separated the rather static aspects of a geometrical representation from the more dynamic parts. The coordinates of a geometric object are much more likely to change—due to geometric transformations—than the topological structure. The association between the two components is achieved by "ordinary" join attributes (or foreign keys); i.e., the attribute *V_ID* of subrelation *VERTICES* relates tuples of the subrelation *METRIC* via their *VERTEX_ID* attribute values.

6.6.7 Query Language

For the AIM-P system the extended SQL query language HDBL (*Heidelberg database language*) was proposed. The query language still uses the basic SQL

$$\textbf{select} \ldots \textbf{from} \ldots \textbf{where} \ldots$$

construct. But in order to facilitate queries over nested relations, the **select** ... **from** ... **where** ... construct can be nested in HDBL. Our example queries given below are stated over the relations *MECHANICAL_PART* of Figure 6.20 and *BR_GEOMETRY* of Figure 6.21.

Simple Selection Query Let us first demonstrate some simple query that just selects a complete complex object from the relation *BR_GEOMETRY*:

```
select  geo_obj
from    geo_obj in BR_GEOMETRY
where   geo_obj.GEO_ID = 'cubo#1'
```

The result of this query is just the table shown in Figure 6.21, except that only the information belonging to the complex object identified by 'cubo#1' is selected.

Simple Projection Query If we are only interested in retrieving the metric information of 'cubo#1', e.g., in order to perform a geometric transformation such as a translation of the solid object, we could state the following *projection* query:

{BR_GEOMETRY}							
GEO_ID	{FACES}			{METRIC}			
	FACE_ID	{EDGES}		VERTEX_ID	X	Y	Z
		EDGE_ID	{VERTICES}				
			V_ID				
'cubo#1'	'f1'	'e1'	'v1'	'v1'	0.0	0.0	0.0
			'v2'	'v2'	1.0	0.0	0.0
		'e2'	'v2'	'v3'	1.0	1.0	0.0
			'v3'	'v4'	0.0	1.0	0.0
		'e3'	'v3'	'v5'	0.0	0.0	1.0
			'v4'	'v6'	1.0	0.0	1.0
		'e4'	'v4'	'v7'	1.0	1.0	1.0
			'v1'	'v8'	0.0	1.0	1.0
	'f2'	'e5'	'v2'				
			'v6'				
		'e6'	'v6'				
			'v7'				
		'e7'	'v7'				
			'v3'				
		'e2'	'v2'				
			'v3'				
	'f3'	'e8'	'v5'				
			'v6'				
		'e6'	'v6'				
			'v7'				
		'e9'	'v8'				
			'v7'				
		'e10'	'v8'				
			'v5'				
	'f4'	'e11'	'v1'				
	...	...	...				
'cubo#2'	...	...	...	...	...	...	...

Figure 6.21: Tabular Representation of Relation BR_GEOMETRY

```
    select m
    from   m in b.METRIC,
           b  in BR_GEOMETRY
    where b.GEO_ID = 'cubo#1'
```

The result of this query corresponds to the subrelation *METRIC* of the parent relation *BR_GEOMETRY*, i.e., the right-hand side of the table in Figure 6.21.

Nested Query   In one of the preceding queries we have already shown how a complete tuple corresponding to a complex object is retrieved from a nested relation by merely referencing the uppermost level of the nested relational structure. Implicitly, the subrelations that are contained within the parent tuple are also retrieved. This method is only applicable when one wishes

- To retrieve the nested tuple in its entirety

- To maintain the object structure as defined in the schema of the relation

Alternatively, one can access and restructure lower levels of a nested relational tuple. In the following query, we merely retrieve some parts of the tuple corresponding to *'cubo#5'* of relation *MECHANICAL_PART* without any restructuring. But we will not rely on the implicit retrieval of subrelations. Rather, we explicitly retrieve each subrelation by nesting the **select** ... **from** ... **where** ... clause. The query stated in HDBL below finds the vertices of edge *'e1'* of the face *'f1'* that belongs to the cuboid *'cubo#5'* in the relation *MECHANICAL_PART*.

```
    select [ m.NAME,
       (select [ f.ID,
          (select [e.ID,
             (select [ v.ID,
                       v.LOCATION]
             from v in e.VERTICES)
             ]
          from e in f.EDGES
          where e.ID = 'e1')
          ]
       from f in m.FACES
       where f.ID = 'f1')
       ]
    from m in MECHANICAL_PART
    where m.ID = 'cubo#5'
```

The result tuple of this query would then look as follows:

```
{[ 'bracket',
       {[ 'f1',
              {[ 'e1',
                      {[ 'v1', [1, 0, 2] ] ]
                       [ 'v2', [0, 1, 0] ]}
                      }
              ]}
       ]}
]}
```

This same query in a pure relational model and a relational schema analogous to that described in Chapter 5 would have required about the same number of joins that the NF^2 query has nesting depth. However, we should note that such a deep nesting of NF^2 queries is only necessary when one wants to restructure the nested (hierarchical) objects.

6.6.8 Discussion

The NF^2 data model implicitly incorporates hierarchical relationships between tuples of different relations. Thus, it is really a hybrid of the relational and the hierarchical data model. Again we note that there is a problem with data redundancy; a nested relation (or tuple/list) can be nested within only one parent tuple. In this respect the nested object is *exclusively owned* by the parent object. *Sharing* of objects is not supported.

In the case of a purely hierarchical data structure the NF^2 schema is extremely concise because we can nest relations arbitrarily deep. Even though aggregation hierarchies are an important issue in engineering applications, we note that the NF^2 model does not necessarily lead to a more concise representation in all applications. For example, the CSG representation would still require most of the relations of a pure relational schema. This is due to the fact that CSG incorporates generalization hierarchies for which the NF^2 data model provides no direct support as opposed to aggregation hierarchies.

The few example queries that we presented show that the extended SQL query language of the NF^2 model is nontrivial because of its nested nature. But complex NF^2 queries would be at least as complex in the pure relational model. For some of our example queries this is quite obvious, since they would involve a join over several different flat relations.

In summary, the NF^2 model provides structural support for hierarchically composed objects. Unlike the System R extensions, called XSQL, this is achieved by physically (and logically) clustering such complex objects via nested subrelations.

6.7 Exercises

Aggregation and Generalization Extensions

6.1   Model the boundary representation scheme utilizing the extensions of the relational model to support aggregation and generalization hierarchies.

6.2   Model the *Vehicle* generalization hierarchy that was introduced in Section 2.5 (cf. Figure 2.8).

Augment the conceptual schema by representative attributes and show how the concept of inheritance along the generalization/specialization hierarchy is applied on these attributes.

6.3   Convert the conceptual schema modeling a university administration that was designed in Exercise 2.11 into the relational representation extended with aggregation and generalization support.

6.4   Modeling generalization in the relational model induces redundancy into the schema. This redundancy has to be controlled by the system—as described in the text. In particular, an update of an attribute may have to be propagated upward and downward along the generalization hierarchy.

Outline the system programs that realize the control of the redundant attributes.

6.5   Estimate the cost of redundancy in the extended relational model supporting aggregation and generalization in terms of

1. Additional storage space

2. Performance overhead due to cascaded updates

for some sample database occurrences.

6.6   Devise ways to support generalization hierarchies in the pure relational model utilizing the view concept. Discuss the

- Advantages using view definitions with respect to avoiding redundancy

- Disadvantages concerning the limitations of updating the virtual relations, i.e., the views

6.7   Where can you use the aggregation and generalization concepts of this chapter when considering the domains of Exercises 2.12, 2.13, and 2.14.

Molecular Object Structures

6.8   Find a sample object schema from your area of expertise for each kind of
molecular object structure, i.e.,

1. Disjoint and nonrecursive

2. Disjoint and recursive

3. Nondisjoint and nonrecursive

4. Nondisjoint and recursive

6.9   Devise the transformation rules that transform general molecular objects into
a relational representation. Again, distinguish between the four categories

1. Disjoint and nonrecursive

2. Disjoint and recursive

3. Nondisjoint and nonrecursive

4. Nondisjoint and recursive

Apply these general transformation rules on the sample molecular object structures designed in Exercise 6.8.

6.10   Consider the domains of Exercises 2.12, 2.13, and 2.14. Where can you use
the molecular objects?

General Reference Mechanism

6.11   Develop a relational CSG representation exploiting the general reference mechanism called "QUEL as a Datatype."

On the basis of this example application analyze the virtues (e.g., flexibility,
expressiveness) and the disadvantages (e.g., lack of type safety, performance
problems) of the model.

6.12   The so-called *soft* references—which have to be computed anew each time the
"QUEL as a Datatype" attribute is accessed—impose a severe performance
penalty. One way to mitigate the problem is the materialization of these soft
references; that is, the query is precomputed and stored in the database. Of
course, the materialized queries have to be maintained in a consistent form;
that is, the database system has to automatically invalidate and rematerialize
the query results whenever an update in the database occurs.

Discuss this kind of optimization method. In particular address these questions:

- What rematerialization overhead is to be expected?
- In what kind of database applications is materialization useful?
- Where should the materialized query results be stored?

6.13 Consider the domains of Exercises 2.12, 2.13, and 2.14. Where can you use the "QUEL as a Datatype" reference mechanism?

Abstract Datatypes

6.14 Consider your area of expertise and find some commonly used data types that are candidates for abstract data types in the relational model. Devise their realization in ADT-INGRES. Discuss the problems encountered in the implementation of these data types—if any.

6.15 In the text it was pointed out that the ADT-INGRES data abstraction approach suffers from the requirement that ADTs have to be mapped to a single flat domain, typically a string domain. Very often this leads to rather unnatural internal representations of the abstract datatypes.

One way to overcome this problem is to take the nested relational model as the basis for such an ADT facility. Discuss the nested relational model with respect to incorporating data abstraction. The nested relational model allows the nesting of relations (and lists) and, therefore, should enable a much more structured internal representation of domain ADTs.

6.16 Consider the domains of Exercises 2.12, 2.13, and 2.14. Where can you use abstract datatypes?

XSQL

6.17 Model the aggregation hierarchy representing a *Bicycle*—cf. Figure 2.7 in Section 2.5—in XSQL. Exploit the *component-of* construct as much as possible. Show a sample extension of the database schema (for a small part of a bicycle model).

6.18 XSQL supports disjoint, recursive molecular object structures quite well. Consider our *Company* conceptual database schema (cf. Figure 2.3): The relationship *Composition* between *Products* and *Products* forms such a recursive (one-level) molecular object. Represent this part of the conceptual schema in XSQL. Show a sample extension—maybe modeling a bicycle.

6.19   Why is the *component-of* construct not suitable for supporting shared subobjects?

Nested Relational Model

6.20   Model the conceptual *Robot* schema designed in Section 3.2 (Figure 3.6) as a nested relational schema.

   The **part-of** relationships should be modeled by nesting the corresponding relations.

6.21   Discuss the virtues of the nested relational model with respect to supporting aggregation and generalization abstractions.

6.22   Consider the CSG representation. Is the nested relational model amenable to supporting this data structure?

6.23   Consider the following—somewhat contrived—conceptual schema relating the types *Women*, *Men*, and *Children*:

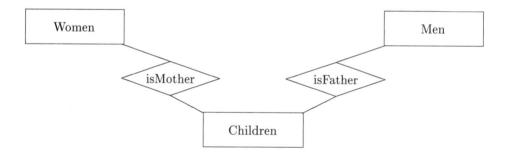

   Model the above conceptual schema as a nested relational schema. Address the problems caused by the fact that a child is a shared subobject of a woman and a man.

6.24   In a foregoing exercise (cf. Exercise 6.18) it was observed that XSQL supports disjoint, recursive molecular structures quite well. How about the nested relational model? Substantiate your analysis on the same *Products/Composition* example.

6.25   Consider the domains of Exercises 2.12, 2.13, and 2.14. Are nested relations useful?

6.8 Annotated Bibliography

The extensions of the relational model to support aggregation and generalization were invented by Smith and Smith [SS77b, SS77a]. Some of the ideas are also incorporated in the augmented relational model RM/T that was proposed by Codd [Cod79]. Recently Codd devoted an entire book [Cod90] on this new version (as he calls it) of the relational model.

The extensions of the relational model to support *complex objects* were proposed by Haskin and Lorie [HL82], Lorie and Plouffe [LP83], and Lorie et al. [LKM$^+$85].

The notion of molecular object was first developed by Batory and Buchmann [BB84]. The molecule structures form the basis of a data model called MAD (molecule atom model) that was realized at the University of Kaiserslautern, Germany. The MAD model is described in [HMWMS87, Mit89].

The general idea behind "QUEL as a Datatype" was first proposed by Zaniolo [Zan83] in the context of the database prototype development called GEM. "QUEL as a Datatype" [SAHR84, SAH87a] was then exploited by Stonebraker et al. as an extension to the database management system INGRES [SWKH76, Sto85]. In a more recent proposal the modeling concepts of this approach have been incorporated in the database system POSTGRES [SR86], the successor of INGRES.

ADT-INGRES was first proposed by Stonebraker in [SAH87a, SRG83]. Most of the data abstraction concepts were subsequently incorporated in the database system POSTGRES [SR86], as well.

Abiteboul and Hull [AH87a] introduced the graphical notation that was used to introduce the NF^2 structuring concepts. One of the first nested relational models proposed in the literature is the NF^2 model, which was originally presented in [SP82]. The underlying theory is described in [SS86]. Several research papers are collected in the book [AFS87]. A reference for the extended NF^2 model can be found in [DKA$^+$86]. The above reference also outlines the architecture of the AIM-P database prototype. Another prototypical database system, DASDBS, based on the (original) NF^2 model was developed by a group headed by H. Schek at the University of Darmstadt [PSS$^+$87, Sch87]. The HDBL language that is based on the extended NF^2 model is contained in [PA86] and, more thoroughly, in [PT86]. Another language for nested relational database along the lines of the standard relational language SQL is described in [RKB87]. A logic-based language for complex objects, called *COL*, was developed by Abiteboul and Grumbach [AG88].

Further related topics are presented in the following original papers:

- [AB84] outlines the database project VERSO, carried out at the research institute INRIA, France.

- A more theoretical overview of operations and properties of nested relations is worked out in [AM83].

- [FT83] is an early account of work on nested relations by Fischer and Thomas.

- [OY87] develops new normal forms to support the hitherto little addressed problem of logical database design for nested relations.

Kemper, Lockemann, and Wallrath [KLW87] report on a nested relational database system—called R^2D^2—that incorporates data abstraction. The realization of these data abstraction concepts is further elaborated by Linnemann et al. [LKD+88].

Part III

Object-Oriented Modeling and Languages

The description of this part is based on an object model called GOM (generic object model) that was developed by the authors. GOM was chosen as the "demonstration vehicle" because of the lack of a widely agreed upon *standard* object model. GOM is generic inasmuch as it provides the most salient features of object-oriented data models in one coherent syntactical framework. We believe that, at this stage of the book, the reader is served better by discussing modeling concepts at a system-independent level rather than dealing with peculiarities of actual systems. Actual systems are discussed at the end of the book—after having thoroughly explained the underlying concepts.

7

Objects and Types

In this Chapter, we will "set the stage" for the discussion of the object-oriented data model GOM (generic object model). This chapter is devoted to the basic data structuring concepts: objects, values, object types, and sorts. The predominant use of object types is to implement the entity types of the entity-relationship model, as described in Chapter 2.

7.1   Distinction between Objects and Values

We distinguish between *objects* and *values*: the former being elements of *object types* and the latter being elements of so-called *sorts*. *Objects* constitute complex entities that are built from other objects and/or from values. The most severe distinction between objects and values concerns the *state*: Objects possess an internal state that is subject to modification. On the other hand, values, such as the integer 10, never change. Values can be seen as "God-given," eternally invariant entities.

Objects, on the other hand, are created by the user. They are derived from previously defined *object types* by instantiation. An object type describes the *structural* and *behavioral characteristics* of a set (class) of "similar" objects. Individual objects are then derived from this object type by an explicit *create* operation.

The subsequent diagram highlights the terminology used throughout the book:

Types		Instances
sorts	$\longrightarrow$	values
object types	$\longrightarrow$	objects

Despite the vast differences between sorts and object types we will often blur the distinction and simply refer to types. For example, we say that a variable (or

159

attribute) is constrained to a particular type, meaning that the variable is constrained to a particular object type or a particular sort.

There are *predefined sorts* in GOM, which include the following:

Sort	Domain
bool	{*true, false*}
int	integer numbers
float	floating point numbers
char	character
string	sequence of characters

These predefined sorts can be used by any object base application without any further sort definition. For the time being, we will only use these built-in sorts; however, in GOM one can also define new, application-specific sorts—as we will describe in Section 7.12.

7.2   Object Types

Similar objects are described by a common object type definition. In this respect, a GOM object type corresponds to an entity type in the entity-relationship model, which was also used to group and describe a set of similar objects. Based on the built-in sorts and the built-in type constructors, the user can define arbitrary complex object types. The syntactical conventions for defining new object types are outlined below.

7.2.1   The Type Definition Frame

For specifying a new object type the so-called *type definition frame* is used. It obeys the syntax shown in Figure 7.1. Parts enclosed by [...] are optional; i.e., they may be omitted in a valid type definition.

Without going into the details of the type definition—this will be done in subsequent sections—let us, at least, outline the meaning of the most essential constructs.

The identifier *TypeName* of the newly defined type has to be unique; that is, there must not be any other type definition with the same name. For the time being, a GOM object type has exactly one supertype, which is either explicitly specified using the **supertype** clause or is implicit. If the **supertype** clause has been omitted, the supertype *ANY* is implicitly assumed. The type named *SuperTypeName* is called the *supertype*; consequently the newly defined type named *TypeName* is called the *subtype*. The subtype *inherits* all properties, i.e., the structural representation and the operations, from its supertype. Thus, GOM supports the concept of *single* inheritance (cf. Chapter 10).

```
[persistent] type TypeName [ supertype SuperTypeName] is
    [public OperationList]
    body TypeStructure
    [operations
        OperationSignature;
        . . .
        OperationSignature;
    implementation
        OperationImplementation;
        . . .
        OperationImplementation;]
end type TypeName;
```

Figure 7.1: The Syntax of the Type Definition Frame

In the **public** clause, the operations are listed that are supplied for the *clients*—that is, users—of this type. These operations provide the only means for accessing and modifying the (internal) state of objects of the respective type. The **public** clause specifies the interface of objects of type *TypeName* to the outside world. This way *encapsulation* of objects is achieved.

Within the **body** clause the internal structure of objects of this type is specified. The structural representation is used to maintain the internal state of objects in between successive operation invocations—or, in the case of persistent objects, even in between different program runs. We distinguish between *tuple structured* object types whose internal representation consists of a collection of named *attributes* and *collection types*. The collection types include *set* and *list structured* types.

The **operations** clause specifies the abstract signatures of the operations that are associated with this type—called *type associated operations*. An *OperationSignature* specifies the name of the operation, which has to be unique within the type[1], the types of the argument objects, and the result type returned upon invocation of the operation. In the optional **code** clause of the *OperationSignature*, we can specify the name of the implementation of the operation, in which case the actual coding of the operation can be found under that particular name within the **implementation** section. If the **code** clause is missing, it is implicitly assumed that the operation name coincides with the implementation name.

In the **implementation** part of the type definition frame, we have to provide the coding of all operations whose signature is specified in the **operations** part. The implementation is coded in the programming language GOMpl (GOM programming

---

[1]This condition will be relaxed when we introduce *overloading* (cf. Section 8.7) later on.

language), whose basic control concepts resemble those of the programming language C. In addition, however, GOM provides many advanced concepts that allow for object-oriented program development (cf. Chapter 8).

From the above (cursory) explanations it becomes evident that an object type consists of two, each other complementing dimensions:

1. *Structural representation*

   The structural representation determines the internal structure of objects of the respective type. The structure is utilized to maintain the internal state of the objects. The structural representation is specified by the **body** clause and, in addition, by the **supertype** clause, since the subtype inherits the structural representation of its supertype.

2. *Behavioral specification*

   The *behavior* of a type is determined by the type-associated operations that are listed in the **public** clause. Again, the behavior is also inherited from the supertype. The operations are further detailed in the **operations** clause, where their *signature*, i.e., their legal invocation pattern, is specified. The **implementation** clause contains the coding of user-defined operations—certain operations are built-in and, therefore, do not require any implementation.

In this chapter, we want to concentrate on the structural representation of object types. Nevertheless, we will also discuss operations that are tightly associated with the structural representation of objects, e.g., operations to read and modify attribute values or operations to insert elements into sets or remove elements from sets.

7.3   Tuple-Structured Object Types

Tuple structured object types contain a **body** clause that specifies an arbitrary, though fixed number of *attributes*—note that now the brackets denote the tuple constructor as opposed to optional parts in the syntax:

> **type** *TypeName* **is**
>   **public** ...
>   **body** [$attr_1$: $Type_1$;
>   ...;
> $attr_n$: $Type_n$;]
>   **operations**
>   ...
>   **implementation**
>   ...
>   **end type** *TypeName*;

In this case, the attribute names $attr_i$ for $(1 \leq i \leq n)$ have to be pairwise distinct, the $Type_i$ have to be—not necessarily distinct—type names. The type $Type_i$ denotes either a sort or an object type. We say that the attribute $attr_i$ is *constrained to* type $Type_i$ or simply $attr_i$ is of type $Type_i$. There are no restrictions concerning recursive object structures; it is even possible that one of the attributes is constrained to the same object type in which it is defined. An illustrative example is the object type *Person* containing itself an attribute *spouse* that is constrained to type *Person*:

```
type Person is
    body [name: string;
          age: int;
          spouse: Person;]
    end type Person;
```

This object type definition states that all *Person* objects are described by three attributes: *name* constrainted to sort *string*, *age* constrainted to sort *int*, and *spouse* constrainted to object type *Person*,

For every attribute $attr_i$, two built-in operations, denoted $attr_i \rightarrow$ and $attr_i \leftarrow$ are provided. These operations have the following implicit signatures (cf. Section 8.8):

```
declare attri: → Typei;
declare attri: ← Typei;
```

The operation $attr_i \rightarrow$ is used to read the current assignment of $attr_i$ and $attr_i \leftarrow$ is invoked to modify the current value (state) of $attr_i$. These two operations are implicitly defined for each attribute of a tuple structured type; i.e., their signature as well as their implementation are built-in. For example, for an object of object type *Person* $age \rightarrow$ reads the *age* attribute and $age \leftarrow$ modifies the attribute. Since the arrows may be tedious to write, they can be skipped as follows. Whenever there is an assignment statement (an assignment is denoted by ":="), the attribute name always refers to the modifying operation on the left-hand side whereas it refers to the read operation on the right-hand side of the assignment. Hence, in $p.age := p.age + 1$, the left occurrence of *age* refers to $age \leftarrow$ whereas the right occurrence of *age* refers to $age \rightarrow$. The statement is equivalent to $p.age \leftarrow (p.age \rightarrow + 1)$ and, hence, increases the age of a *Person* object referred to by p by one.

These attribute access and attribute assignment operations are called VTO (*value returning operation*) and VCO (*value receiving operation*), respectively. More details on these are provided in Section 8.8.

Although the attribute access and attribute assignment operations are implicitly available, it is still the choice of the object type designer whether these operations become *visible* to clients of the type—in this case they are to be included in the **public** clause.

Let us now illustrate the definition of object types on another example: the object type *Vertex*. Objects of type *Vertex* represent vertices in the three-dimensional space; i.e., their state is characterized by the three x-, y-, and z-coordinates. Consequently, the object type *Vertex* is tuple structured with those three attributes x, y, and z, all being constrained to the sort *float*.

> **persistent type** Vertex **is**
>   **public** ...
>   **body** [x, y, z: float;]
>   **operations**
>   . . .
>   **implementation**
>   . . .
> **end type** Vertex;

The type definition of *Vertex* is preceded by the keyword **persistent**, which specifies that this type definition should be included in the object base schema. More detailed information about the persistence of type definitions, variables, and objects can be found in Section 7.10 and in Chapter 18.

In the above type definition, we have used an abbreviated form for specifying the type constraints of the three attributes. By separating the attribute names x, y, and z by commas, they are all constrained to the same type, the sort *float* in this case.

To proceed with our toy example application, we need one more (preliminary) type definition, the object type *Material*. The objects of type *Material* are modeled by just two attributes: *name* of type *string* and *specWeight* of type *float*. The former models the common name of the respective *Material* object—e.g., "iron" or "copper"—the latter its specific weight. The type definition frame for *Material* is as follows:

> **persistent type** Material **is**
>   **public** ...
>   **body** [name: string;
>   specWeight: float;]
> **end type** Material;

In this example, we have consciously omitted the optional **operations** and **implementation** clauses because, for this object type, we only need the predefined attribute access and assignment operations—which are built-in and, therefore, don't require any signature declaration or coding.

We could abbreviate the above type definition of *Material* even further—to a form that resembles tuple definition in the relational model or record definition in classical programming languages. The utmost abbreviated type definition of *Material* looks as follows:

    persistent type Material **is**
     [name: string;
      specWeight: float;]];

Note that no hiding of any of the attributes is possible in this abbreviated type definition; that is, access and assignment to all attributes are implicitly allowed.

We are now ready to define a somewhat more complex object type, called *Cuboid*, that is used to model three-dimensional geometric objects of cuboidal shape. The representation of cuboids in our object base is according to the diagram in Figure 7.2. It is important to obey the shown topological ordering of the bounding vertices of the cuboid representation—this same ordering will be assumed in subsequently introduced operations to compute the width, height, and length of *Cuboid* objects. The type definition frame for *Cuboid* looks as follows:

    persistent type Cuboid **is**
      public ...
      body [v1, v2, v3, v4, v5, v6, v7, v8: Vertex;
            mat: Material;
            value: float;]
      operations
        ...
      implementation
        ...
    end type Cuboid;

According to the type definition, a *Cuboid* object is represented by its eight bounding vertices, which are given as the eight attributes *v1*, *v2*, ..., *v8*—all being constrained to the previously introduced object type *Vertex*. In addition to the eight bounding vertices, the *Cuboid* type definition specifies an attribute *mat* that is constrained to type *Material*. Obviously, this attribute models the material of which the geometric object is composed. The last attribute, called *value*, is of type *float*, and is used to store the monetary value of the geometric object of type *Cuboid*.

7.4   Instantiation

Objects are "born" by instantiating an object type. Therefore, individual objects are often called *instances* of the respective type. In this sense, an object type can be viewed as a "cooky frame" from which an arbitrary number of "cookies"—all having the same shape—can be generated.

For instantiating an object type, the built-in operation *create* is used. Applied on a tuple structured object type this *primitive constructor* operation supplies a new tuple structured object whose attributes are initialized as follows:

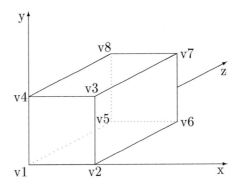

Figure 7.2: The Topological Representation of a *Cuboid*

- *int* attributes are initialized to 0.

- *float* attributes are initially set to 0.0.

- *char* attributes are preset to the character '\0'.

- *bool* attributes are set to *false*.

- Attributes constrained to a complex object type are initialized to *NULL*, a special attribute value denoting that no reference exists.

In the following program fragment, we will generate an instance of type *Cuboid*:

    var myCuboid: Cuboid;

(1)   myCuboid := Cuboid$create;   !! or, equivalently:
(2)   myCuboid.create;

A double exclamation mark ("!!") denotes the beginning of a comment that ends at the end of the corresponding line. Thus, everything between the first "!!" and the end of the line has no influence on the actual program semantics.

In the above program fragment, we used a concept that has not yet been introduced: *variables*. In GOM, variables have to be declared prior to their use in the **var** clause. Analogously to attributes, variables are constrained to a particular type; meaning that the variable can only be associated with an instance of the type to which it is constrained.

The above program fragment shows two statements, (1) and (2), which are equivalent—(2) is an abbreviation of (1). In statement (1), a new instance of *Cuboid* is generated by invoking *Cuboid$create* and, subsequently, this new instance is assigned to the variable *myCuboid*. The assignment is implicit in statement (2). Furthermore,

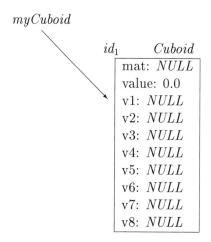

Figure 7.3: The "Skeleton" of a New *Cuboid* Instance

it is implicitly deduced what type of object is to be created from the type constraint of the variable. In this case, since *myCuboid* is constrained to *Cuboid*, a new *Cuboid* instance is generated.

The initial state of the newly created *Cuboid* instance is shown in Figure 7.3. It is obvious from the diagram that the built-in *create* operation merely generates the empty "hull" of an object, which is, of course, not yet meaningful for actual applications.

Furthermore, we note that every newly generated object is associated with a unique *object identifier (OID)*. In this case, the object identifier[2] is logically denoted as id_1. We will return to the very important issue of object identification in Section 7.5.

Let us now complete our example object base. Our goal is to make the *Cuboid* identified by id_1 the unit cube whose *v1*-corner lies in the origin of the coordinate system.

For this purpose we have to create eight new *Vertex* instances—all of which have to be properly initialized.

Furthermore, a new *Material* object for modeling iron is created and associated with the *mat* attribute of the *Cuboid*. Later on we will see that this *Material* object can be referenced by other *Cuboid* instances—leading to a so-called *shared subobject*, i.e., an object that is referenced by more than one object.

---

[2]The object identifiers shown in this presentation denote logical symbols, not the actual system identifiers—which are not accessible by the user.

The following program fragment will do this (tedious) job:

```
myCuboid.v1.create;          !! create a new Vertex
myCuboid.v1.x := 0.0;
myCuboid.v1.y := 0.0;
myCuboid.v1.z := 0.0;
myCuboid.v2.create;
...                          !! creation and initialization of the remaining vertices
myCuboid.v8.create;
myCuboid.v8.x := 0.0;
myCuboid.v8.y := 1.0;
myCuboid.v8.z := 1.0;
myCuboid.value := 39.99;

myCuboid.mat.create;
myCuboid.mat.name := "Iron";
myCuboid.mat.specWeight := 0.89;
```

The same effect induced by the first statement, i.e.,

```
myCuboid.v1.create;
```

on *myCuboid* can be achieved by

```
var aVertex: Vertex;
...
aVertex.create;
myCuboid.v1 := aVertex;
```

The direct application of the *create* operation on the expression *myCuboid.v1* eliminates the need to declare a separate temporary variable *aVertex*—which could possibly also be "abused" to circumvent strict encapsulation—and the subsequent assignment of the newly generated *Vertex* object to the attribute *v1*. This assignment is implicitly contained in *myCuboid.v1.create*.

The object base occurrence resulting from the above program fragment is schematically shown in Figure 7.4. The object base contains 10 individual objects:

- One *Cuboid* instance with object identifier (OID) id_1

- Eight *Vertex* instances with OIDs $id_{11}, \ldots, id_{18}$

- One *Material* instance with OID id_{77}

The states of the *Vertex* instances, i.e., their x, y, and z attribute values, are easily seen in Figure 7.4. The state of the *Cuboid* instance is more interesting because this object contains nine attributes that are constrained to object types—as opposed to

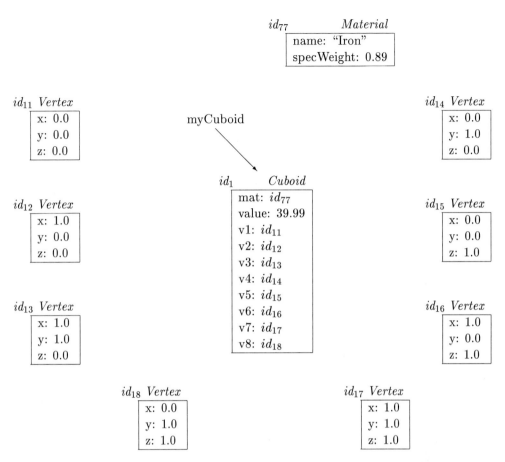

Figure 7.4: Object Base with One *Cuboid* Instance

a sort. The attribute *mat* is associated with the *Material* instance identified by id_{77}. In the case of complex attributes, the association is maintained by storing the respective object identifier—in this case id_{77}—within the attribute. Furthermore, we consider the eight attributes *v1*, ..., *v8*, which are constrained to *Vertex* objects. Their current state is—as shown in the diagram—such that they refer to the *Vertex* instances id_{11}, ..., id_{18}.

7.5   Object Identity

In preceding sections, we already touched the issue of object identification. Since it is one of the key concepts of object-oriented models, we will discuss object identification in more detail in this section.

7.5.1   Motivation for Object Identity

Identity is that property of an object that distinguishes it from all other objects. Unfortunately, this property is often mixed with other properties of an object. Traditional database systems often use the contents (state) of an object to specify its identity. In the pure relational model, for example, identity is achieved by assigning unique values to special attributes, i.e., the key attributes. In the modeled world, the key attributes distinguish entities from each other. This approach is therefore called identity through *contents*. Programming languages, on the other hand, use the *location* within the memory for identification. Both approaches lead to problems:

Identity Through Contents The first problem of this approach is that it does not allow expression of different meanings of equality. $A = B$ always means equality in terms of contents (state). There is no possibility to express equality in terms of identity.

A second problem is the possibility of changing the contents. This is due to the fact that database designers are inclined to use semantically meaningful properties to identify objects within the database. For a relation that stores employees, for example, one might use the attributes *name* and *homeAddress* as identifying attributes. Then the modification of this *homeAddress* gives the person a new identifier and, hence, a new identity. This may lead to severe consistency problems concerning the referential integrity of the database because other entities might refer to this object that now has changed its identity. Therefore, any old references become invalid and have to be changed as well.

Identity Through Location Object identification through the location—typically an address in memory—of the object is mostly employed in programming languages. The memory location uniquely identifies an object from all other objects being generated in an application. These location dependent identifiers are called *physical object identifiers*.

If an object is deleted, the memory formerly occupied by it should be reused. Thus, it may happen, that a new object is created and occupies the same space as the deleted object occupied before. Then all still existing references to the old object, which where invalid after deletion, now become valid again, but to the wrong (new) object. The result of this may be disastrous.

Further, object identification through location has the severe drawback that the moving of an object from one location to another invalidates the identity. This is particularly problematic in distributed systems, in which one may want to migrate objects from one site to another.

Furthermore, in tuning the performance of a database application it is often necessary to restructure the storage area. This, of course, necessitates that objects are moved from one storage location to another—for example, in order to improve the clustering of logically related objects (cf. Chapter 19).

These examples make quite obvious that there is a need for an independent identification mechanism. This requires every object to get a unique identifier at creation time. Moreover, it has to be guaranteed that this identifier remains invariant and unique throughout the lifetime of the object. This requirement leads to so-called *logical object identifiers*—as discussed below.

7.5.2 Logical Object Identity

Every GOM object o can be viewed as a triple of the form:

$$o = (id_\#, \mathit{Type}, \mathit{Rep})$$

The three parts have the following meaning:

- $id_\#$ denotes the system generated unique identity—often called the (logical) object identifier (OID)—of the object o.

- *Type* specifies the object type of which the object was instantiated.

- *Rep* denotes the internal state—i.e., the current value of the internal structure—of the object o.

In the case of a tuple structured object o, *Rep* denotes the current values of the attributes; in the case of set or list structured objects *Rep* denotes the current elements (and the relative order of the elements in the case of a list) of the object o—see Section 7.8 for a detailed discussion of collection types.

As an example let us look at the *Cuboid* object depicted in Figure 7.4: There is one object, identified by id_1, of type *Cuboid*; eight objects, identified by $id_{11}, \ldots, id_{18}$ of type *Vertex*; and one object with identifier id_{77} of type *Material*. The state *Rep* of the object id_1 is given by the current values of the 10 attributes *mat*, *value*, and *v1*, ..., *v8*.

In this presentation we will generally denote object identifiers as id_1, id_2, ... Every object is assigned an OID at the time it is created, i.e., instantiated. The OID remains invariant throughout the existence of the object. Therefore, the object identity remains invariant with respect to the following:

- The storage place, where the object is stored on disk

- The state of the object

An object identifier possessing these two properties is called a *logical object identifier*. This is a basic difference between GOM and other, more value-oriented models, such as the relational model. An OID in GOM will always remain the same no matter what modification the object is subjected to. Further, it is guaranteed, that an OID of a deleted object will never be reused, avoiding the problem indicated above.

7.6 Shared Subobjects

The so-called *shared subobjects* are a "hot" issue in database research. In Chapter 6, we observed that several recently developed data models, like the nested relational model NF^2, do not adequately support this concept. Many of the relational extensions support only strictly hierarchical object composition, where subobjects (or, more precisely, part-objects) are exclusively associated with only one parent object.

It turned out, however, that the data structures in many of the advanced database application areas require general network structures, as opposed to hierarchical structures. In object-oriented models, like GOM, there are no limitations on the number of references pointing to an individual object. Thus, it is easy to model network structures with shared subobjects, i.e., objects that are referenced by more than one other object. We will demonstrate this concept on a very simple object base occurrence that is generated by the following program fragment—assuming that we start with the object base state of Figure 7.4.

```
var anotherCuboid: Cuboid;
   . . .
anotherCuboid.create;
anotherCuboid.mat := myCuboid.mat;
```

The resulting object base is (partially) depicted in Figure 7.5. The *Material* object with the OID id_{77} is now a shared subobject, since it is referenced twice: once by the *mat* attribute of the *Cuboid* instance id_1 and a second time by the *mat* attribute of the *Cuboid* instance id_2. In this way, an object can be referenced an arbitrary number of times. We will later see that the same object can even be an element of several different sets.

Of course, object modifications that are issued via one of the several references become visible via all other references, as well. This is something an inexperienced user of object-oriented systems has to get used to. For example, the updates

```
myCuboid.mat.name := "Copper";
myCuboid.mat.specWeight := 0.90;
```

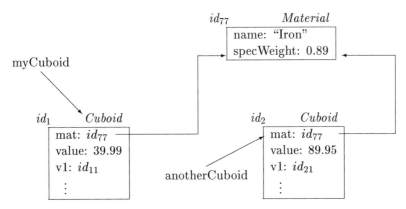

Figure 7.5: Schematical Description of a Shared Subobject

become also visible via the reference *anotherCuboid.mat*, such that the following equalities hold after the foregoing assignment:

anotherCuboid.mat.name = "Copper"
anotherCuboid.mat.specWeight = 0.90;

This indicates that the so-called sideeffects of object updates are rather difficult to foresee in the presence of shared subobjects.

7.7   Referencing and Dereferencing

In the above discussion, we have already seen that the assignment of an object to a variable or an attribute leads to storing a reference to that assigned object. In this respect, object-oriented models employ a so-called *reference semantics* when dealing with objects. On the other hand, values are copied upon assignment—this is called *copy semantics*. Attributes and variables that are constrained to object types are automatically dereferenced, and objects are implicitly referenced upon assignment. We demonstrate the implicit dereferencing and referencing on an example program:

       var someMaterial: Material;
          w: float;
          myCuboid: Cuboid;
          . . .

(1)   someMaterial.create;   !! *someMaterial* holds OID id_{88}, say
(2)   someMaterial.name := "Carbon";
(3)   someMaterial.specWeight := 0.75;
(4)   myCuboid.mat := someMaterial;   !! *myCuboid* holds OID id_1, as before
(5)   w := myCuboid.mat.specWeight;

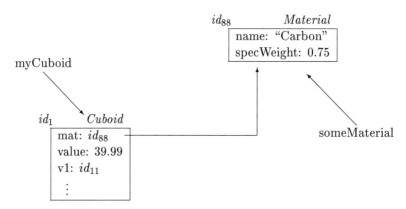

Figure 7.6: Object Base to Demonstrate Referencing and Dereferencing

The statements (1), (2), and (3) generate a new *Material* instance and initialize this object to the desired state; i.e., the *name* attribute is set to "Carbon" and the *specWeight* is set to 0.75. This new *Material* instance obtains a (logical) object identifier—say id_{88}. This identifier is stored in the variable *someMaterial*. The assignment (4) associates this *Material* object id_{88} with the attribute *mat* of the *Cuboid* id_1—which is referenced by the variable *myCuboid*. Implicitly, a reference to the *Material* object is generated and stored within the *Cuboid* object; this is achieved by storing the OID id_{88} within the attribute *mat* of the object id_1. Figure 7.6 shows the resulting state of the object base.

In statement (5), we see an example of automatic dereferencing: Starting with the object identifier id_1 stored in the variable *myCuboid*, the (multiple dot) expression *myCuboid.mat.specWeight* first obtains the *Material* instance with OID id_{88} from which the atomic attribute *specWeight* is then read and assigned to the variable *w*. Note that this assignment is based on copy semantics, since *w* (and the attribute *specWeight*) are constrained to a sort—*float* in this case.

A further implicit (and hidden) referencing occurs during the instantiation of the *Material* instance. In statement (1), *create* was invoked on the variable *someMaterial* whereupon a reference of the newly generated *Material* instance was automatically assigned to the variable *someMaterial*.

7.8   Collection Types

In addition to the tuple structured types, GOM provides two built-in type constructors for collection types:

- The *set constructor*

- The *list constructor*

7.8.1   Set-Structured Object Types

A set structured object type is declared as follows:

> **type** *SetTypeName* **is**
>   **public** ...
>   **body**
>   {*ElementType*}
>   ...
> **end type** *SetTypeName*;

The *ElementType* can be either an object type or a sort—predefined or user-defined. Let us introduce a simple example of a set structured object type:

> **type** TelephoneNumbers **is**
>   {int};

or, equivalently, the "long" version:

> **type** TelephoneNumbers **is**
>   **public** ...
>   **body**
>   {int}
>   **operations**
>   ...
>   **implementation**
>   ...
> **end type** TelephoneNumbers;

We could then declare a variable, called *guidosTelephoneNumbers*, of the set structured type *TelephoneNumbers* and associate it with a set object as follows:

> **var** guidosTelephoneNumbers: TelephoneNumbers;
>   ...
> (1)   guidosTelephoneNumbers.create;   !! creates an empty set
> (2)   guidosTelephoneNumbers.insert(6082080);
> (3)   guidosTelephoneNumbers.insert(28833);

Statement (1) creates an empty set object of type *TelephoneNumbers*. In statements (2) and (3) the two phone numbers 6082080 and 28833 are inserted into the set referenced by *guidosTelephoneNumbers*. Note that the insertion of values into a set obeys the copy semantics—we will see an example of inserting objects into a set according to the reference semantics later on.

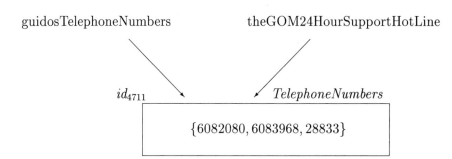

Figure 7.7: Object Base Extension with a Shared Set Object

It is quite possible to have several variables that refer to the same set object—in effect making the particular set structured object a *shared subobject*. This is demonstrated in the following program:

var theGOM24HourSupportHotLine: TelephoneNumbers;
. . .
theGOM24HourSupportHotLine := guidosTelephoneNumbers;

Then, modifications of the set object id_{4711} via one of the two variables, say, *guidosTelephoneNumbers*, become visible through the other variable referring to the same set, as well. An example is shown below:

theGOM24HourSupportHotLine.insert(6083968);

The resulting object base state is depicted in Figure 7.7. Set structured objects are "first-class citizens" in GOM. They are treated analogously to tuple structured objects. In particular, they possess

- An identity, in our case id_{4711}

- A type, in our case *TelephoneNumbers*

Also, they can be used to constrain attribute types.

In the above example, the elements of the set object were atomic values. This, however, is not a requirement, as the following type *CuboidSet* demonstrates:

type CuboidSet **is**
    { Cuboid };

On the basis of this type definition, we can declare two variables, *valuableCuboids* and *workPieceCuboids*, both being constrained to *CuboidSet*. After assigning a newly instantiated *CuboidSet* object to the variables we can then "inhabit" the two sets using the built-in operation *insert*:

> **var** workPieceCuboids: CuboidSet;
>   valuableCuboids: CuboidSet;
>
>   . . .
> workPieceCuboids.insert(myCuboid);
>
>   . . .
> valuableCuboids.insert(. . .);

This program fragment leads to an object base state as shown in Figure 7.8. We notice that the set objects now contain references to the (complex) element objects. This is analogous to the automatic referencing and dereferencing of attributes that are constrained to object types. It thus becomes possible that two set objects, like the sets *workPieceCuboids* with the OID id_{59} and *valuableCuboids* with the OID id_{60}, contain the same object, in our case the *Cuboid* instance id_3. However, a set cannot contain the same object (or the same value) more than once—GOM thus supports the mathematical notion of set as opposed to a so-called *multiset* or *bag*.

7.8.2   Object Type Extensions as Implicit Sets

Aside from user-maintained sets, i.e., instances of set-structured object types, GOM facilitates to maintain implicit sets of all instances of a particular object type. This implicit set is called *type extension* or *extension*, for short. If the type designer desires to maintain the type extension he or she has to indicate this in the type schema using the **with extension** clause, e.g.:

> **type** Cuboid
>   **with extension is**
>
>   . . .
> **end type** Cuboid;

Newly instantiated *Cuboids* are then automatically inserted into the extension.

The extension of the type *Cuboid* can be accessed as *ext(Cuboid)*. The only functionality this distinguished set object allows is to iterate through all the elements of the extension. Explicit modifications of the extension, e.g., removing an element from the extension or inserting an element into the extension, are not allowed—insertion and removal are implicitly carried out when a *Cuboid* is instantiated or deleted, respectively. Extensions often form the basis for queries as will be seen in Chapter 14.

7.8.3   List-Structured Object Types

Lists differ from sets in two respects:

1. The elements of a list are ordered.

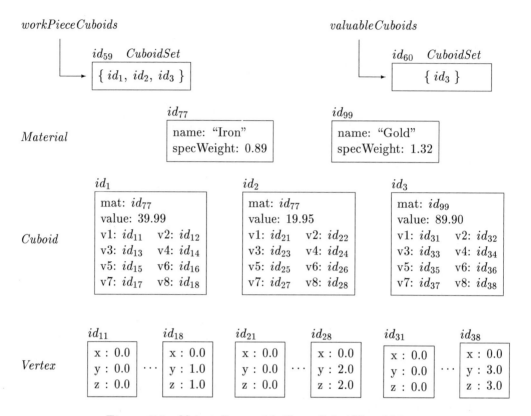

Figure 7.8: Object Base with Two *CuboidSet* Objects

2. A list may contain the same element more than once—at different positions within the list.

List types are defined in GOM with a **body** clause that contains

$$<ElementType>$$

as the structural representation—the angle brackets $<\dots>$ denote the list constructor.

Just like set-structured types list-structured types can be defined for any element type, i.e., for sorts as well as for object types.

An example of a list type whose elements are constrained to be objects of type *Vertex* is the following:

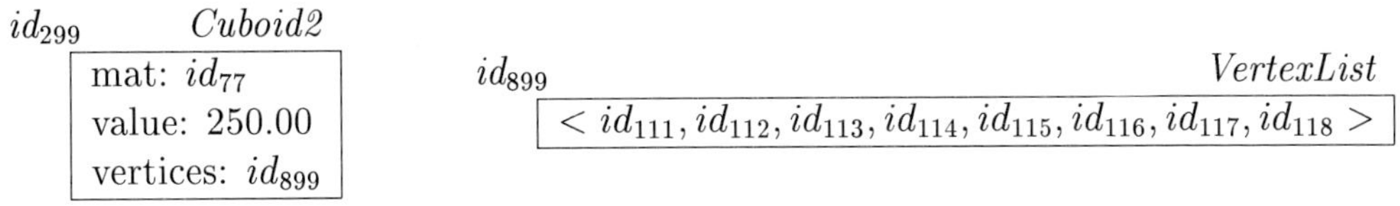

id_{299} *Cuboid2*

| mat: id_{77} |
| value: 250.00 |
| vertices: id_{899} |

id_{899} *VertexList*

$< id_{111}, id_{112}, id_{113}, id_{114}, id_{115}, id_{116}, id_{117}, id_{118} >$

Figure 7.9: Representation of a *Cuboid2* Instance

```
type VertexList is
    public ...
    body
        <Vertex>
    operations
        . . .
    implementation
        . . .
end type VertexList;
```

Based on the type *VertexList*, we could then redefine our example type *Cuboid*—now called *Cuboid2*—with a slightly different internal representation. Instead of explicitly naming the eight attributes *v1*, ..., *v8* we could maintain the bounding vertices with their relative topological order—according to Figure 7.2—as a list containing references to eight *Vertex* instances. We use the previously defined *VertexList* in the specification of *Cuboid2*—thereby making *Cuboid2* a client of *VertexList*:

```
type Cuboid2 is
    public ...
    body [mat: Material;
          value: float;
          vertices: VertexList;]
    operations
        . . .
    implementation
        . . .
end type Cuboid2;
```

A possible instance of the *Cuboid2* type is shown in Figure 7.9. We assume that the attributes *mat*, *value*, and *vertices* have been initialized in a way conforming to the state shown in Figure 7.9. The object identifiers id_{111}, ..., id_{118} are assumed to refer to objects of type *Vertex*, as specified in the type definition of *VertexList*.

7.9   Typing

This section discusses some typing issues relevant for type safe languages. We distinguish between *type safe* data models and *non-type safe* models. The former have the advantage that they guarantee that all type inconsistencies can be detected at compile time. Thus, the compiler rejects any programs that may possibly generate a run-time error due to a type inconsistency. An example of such an error is the access to an attribute that is not present in the respective object, e.g., the attempted access to the attribute *spouse*

    myCuboid.spouse

of an object of type *Cuboid*. In a *type safe* model, this error is already detected at compile time. The question arises why non-type safe models do exist at all. The reason is that these models offer higher expressiveness and more flexibility than the type safe models. In order to guarantee type safety at compile time, the type safe models have to restrict the statements and expressions to those that cannot possibly cause a type error under *any* possible database state. On the other hand, the non-type safe models delegate the responsibility of ensuring type consistency to the programmer. Unfortunately, type inconsistencies are very hard to detect because the occurrence of such an inconsistency depends on the state of the object base. Thus, a program may have run correctly for a long time and may still run into a type error because somebody changed the state of the object base. In non-type safe models, the database components, e.g., attributes, variables, list and set elements, are not constrained to a particular type. This, of course, yields the flexibility that the database user can assign arbitrary objects to these components. For example, in a non-type safe model the following statement

```
var myCuboid;
...
if (...)
    myCuboid := Vertex$create
else
    myCuboid := Cuboid$create;
myCuboid.value := 3.5;    !! potential type error
...
```

could possibly lead to a type error—depending on the branch of the **if** clause that is executed: In case the condition of the **if** clause evaluates to *true*, the *Vertex* instance to which *myCuboid* may refer has no *value* attribute. However, it cannot be determined at compile time which branch of the **if** clause will eventually be executed. This can only be detected at run-time.

Type safe models are also called *statically type safe* in order to indicate that type errors are prevented statically at compile time. Opposedly, *dynamically typed* object models or languages are those in which type errors are not prevented at compile time but have to be discovered dynamically at run-time.

Type safe models constrain data components to particular types. Therefore, these models are called *strongly typed*. GOM is such a *strongly typed* object model. This offers several advantages:

1. *Type safety*

   The GOM compiler verifies the type safety. This ensures that less run-time errors occur; in particular, all run-time errors due to type inconsistencies—one of the main sources of severe errors in non-type safe models—are totally avoided.

2. *Efficiency*

   Languages and object models that do not impose type constraints are inherently less efficient than strongly typed models. This is caused by the lack of type specificity of these languages at compile time. On the other hand, strongly typed models allow to determine the type of all expressions at compile time. This abandons the necessity of dynamic type checks at run-time and provides a much higher potential for optimization—as we will see in subsequent chapters.

3. *Program structuring*

   All data components—i.e., attributes, list and set elements, and variables—are constrained to a particular type. This, in general, leads to a better structuring and readability of the object base schema in comparison to non-typed components. In dynamically typed models the anticipated type of the component is, at best, indicated by the name as, e.g., in *aPerson*, *aDog*, etc.

As already mentioned—and obeyed to in the preceding examples—the database designer has to specify type constraints for all data components. Data components are the following:

- *Attributes*, which constitute the internal representation of tuple structured types

- *Variables*, which may be transient in an application program or persistent in the object base schema

- *Parameters* of operations

- *Elements* of set and list types

Rather than specifying the typing rules formally, we will illustrate the typing rules on an example taken from the administrative application domain. Let us define the schema shown in Figure 7.10.

Utilizing these type definitions, we can now generate a small object base with three *Persons* and one *City*.

```
type Person is                          type City is
    public ...                              public ...
    body [name: string;                     body [name: string;
          age: int;                               mayor: Person;
          spouse: Person;                         inhabitants: PersonSet;]
          livesIn: City;]                   operations
    operations                                  ...
        ...                                 implementation
    implementation                              ...
        ...                             end type City;
end type Person;

type PersonSet is
    public ...
    body {Person}
    operations
        ...
    implementation
        ...
end type PersonSet;
```

Figure 7.10: A *Person* and *City* Type Schema

```
var cityOfLA: City;
    mickey, mini, donald: Person;
    ...    !! initializations
```

Based on these variable declarations, the object base could assume the state shown in Figure 7.11. In this object base, all type constraints are obeyed. For example, the variables *mickey* and *donald*, which are constrained to type *Person*, do actually refer to *Person* objects. Furthermore, the attributes of this database state refer to objects (or are assigned values in the case of sorts) that conform to the type constraints stated in the schema (type definitions). There is one exception, though: The attribute *spouse* of the object referred to by *donald* has the value *NULL*, indicating that it currently has no reference. This is not outruled by the type consistency rules.

Also, the attribute *inhabitants* of the *City* object with OID id_{571} is properly typed: As imposed by the type constraints it refers to an object of the set structured type *PersonSet*. The object id_{115} of type *PersonSet* contains three elements, all of which are *Person* instances—as required by the type constraint.

In a strongly typed language, the compiler is responsible for the verification of the type constraints. Thus, any potential violations are detected statically—at compile time—and are rejected. Furthermore, the compiler verifies the type consistency of all

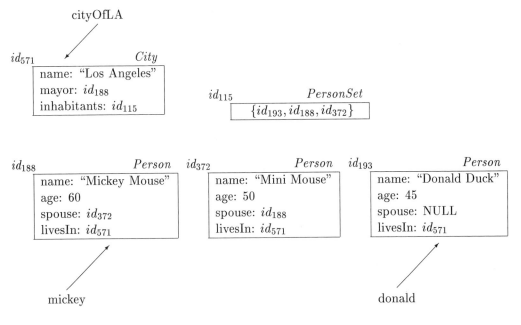

Figure 7.11: Object Base with *Person*s and a *City*

expressions contained in a program.

We will illustrate this on a few example expressions that are based on the object base of Figure 7.11.

```
var totalAge, ageOfSomeBody: int;
    anyBody: Person;
    name: string;
    ...
(1)  ageOfSomeBody := cityOfLA.mayor.spouse.age;
(2)  foreach (anyBody in cityOfLA.inhabitants)
        totalAge := totalAge + anyBody.age;
```

In the first statement, the variable *ageOfSomeBody* is assigned the value determined by the expression *cityOfLA.mayor.spouse.age*. To verify the type consistency of the statement, the compiler has to determine that the types of the expression to the left of the ":=" sign and the expression to the right of the assignment sign match. This match has to be verified based on the type constraints of the data components.

For the two statements of the above example program fragment this can be sketched as follows:

$$\underbrace{\text{ageOfSomeBody}}_{\text{int}} := \underbrace{\underbrace{\underbrace{\underbrace{\text{cityOfLA}}_{\text{City}} .\text{mayor}}_{\text{Person}} .\text{spouse}}_{\text{Person}} .\text{age};}_{\text{int}}$$

The second expression, i.e., the **foreach** loop, is verified type consistent as follows:

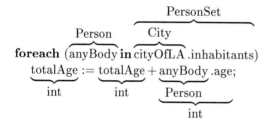

First, the type of the iteration variable *anyBody* is determined as *Person*, since the elements of *cityOfLA.inhabitants* are *Person* objects. Then the assignment statement in the loop body is verified as type safe.

7.10   Persistence

Persistence, in general, means that certain program components "survive" the termination of the program. For this purpose these components have to be stored permanently on secondary storage.

In GOM, we distinguish three different components that can be made persistent:

- Types, i.e., object types as well as sorts

- Objects

- Variables

Subsequently, we will to explain the initiation and implications of persistence of these three components in turn.

7.10.1   Persistence of Types

Object type and sort definitions can be made persistent by preceding the type definition by the key word **persistent**. We already showed some examples of persistent types in the preceding sections, e.g.:

persistent type Vertex **is**
  . . .
end type Vertex;

Thereby, the object type *Vertex* is transferred to the GOM schema manager. From
then on, other application programs can utilize this type definition. But no other
type may be named *Vertex*—neither persistent nor transient. (This will be refined in
Chapter 18.)

There are certain dependencies that have to be taken into account:

- In the case of a tuple structured type, all types to which attributes are constrained have to be persistent as well. Likewise the constrained element type of set or list types have to be persistent.

- Within the type hierarchy (cf. Chapter 10) all supertypes of a persistent type have to be persistent as well.

7.10.2   Persistence of Objects

Every GOM object that was created by instantiating a type can potentially be made
persistent. Objects understand the operation *persistent*, which initiates their permanent storage on secondary memory. As long as this operation has not been invoked
on a potentially persistent object, the object remains transient and disappears after
termination of the program. Examples of making objects persistent are

aVertex.persistent;
myVertices.persistent;

In order to enforce the persistence of all instances of a particular type, the operation
persistent could be invoked in the initialization phase of the objects (cf. Section 8.5).

However, it is not ensured that the object instances being referenced by a persistent object are persistent themselves. For example, one could insert nonpersistent
(i.e., transient) objects into the above-shown persistent *VertexSet* object, which is
referenced by the variable *myVertices*. The referentiation of transient objects from
persistent ones leads to *dangling references* after termination of the program because
the transient objects disappear while the reference remains in the persistent object.
Good programmers will, therefore, avoid referencing transient objects from persistent
ones wherever possible.

7.10.3   Persistence of Variables

Variables can be made persistent by preceding their declaration with the key word
persistent. The prerequisites are:

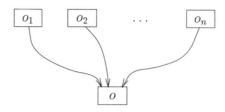

Figure 7.12: Sharing of an Object

- The variable is constrained to a persistent type.

- There is no other persistent variable of the same name. (This will be refined in Chapter 18.)

After termination of a program in which persistent variables are declared, the persistent variables are maintained by the schema manager. Persistent variables can then be reused by the same program in a subsequent program run or may also be used by different programs.

Examples of persistent variable declarations are

    persistent var myVertices: VertexSet;
                aVertex: Vertex;

Again, the declaration of a variable as being persistent does not automatically ensure that the referenced object is persistent. If, at termination time of the program, a persistent variable references a transient object, its value becomes undefined until the next assignment; i.e., again we have a dangling reference problem that should be taken care of by programmers.

7.11   Garbage Collection

7.11.1   Inaccessible Objects

The removal of an object from a collection or the disassociation of an object from an attribute by an assignment statement does not induce the deletion of the respective object from the database. This is not meaningful in an object-oriented system, because in general more than one link to the object may exist—due to object sharing. Consider, for example, the database configuration pictured in Figure 7.12.

The object o is shared by n different objects $o_1, o_2, \ldots, o_n$. The removal of the association between o_1 and o should not affect the remaining $n-1$ references. Therefore, the deletion of an object from the database should be carefully controlled by the system.

Only those objects that are *inaccessible* should be automatically deleted from the database. An object is called inaccessible if no link exists by which a database query could possibly reach the object. In particular, an object o that is neither a member in any set or list nor associated with an attribute or a variable is effectively inaccessible. Thus, all objects on which no references exist are inaccessible and, hence, should be automatically deleted. Nevertheless, as will be shown shortly, also objects that do have an incoming reference may really be inaccessible due to an inaccessible cycle of objects.

7.11.2 Detection of Inaccessible Objects

The automatic deletion of inaccessible objects from the database is called *garbage collection*. The garbage collection could be implemented by associating a reference count with each object in the database. The reference count is increased each time a new reference on the respective object is established by an insertion into a set or list or by an assignment to an attribute or a variable. The removal of a reference on the object automatically decreases the reference count. A reference count of 0 initiates the deletion of the object from the database because it is no longer accessible.

As elegant as this algorithm seems, it has one severe flaw, though. It cannot catch inaccessible cycles, like:

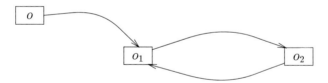

The removal of the reference from o to o_1 decreases the reference count of o_1 to 1 because of the reference from o_2 to o_1. Nevertheless, the two objects o_1 and o_2 are inaccessible as a whole—after removing the link from o to o_1. In order to detect inaccessible cycles, more sophisticated garbage collection strategies have to be utilized. Typically, it suffices to periodically invoke them at low database usage periods in order to reclaim the storage space that is occupied by inaccessible cycles.

7.11.3 Explicit Deletion of Objects from the Database

The automatic garbage collection may—particularly in large databases—be prohibitively time-consuming. Especially the detection of inaccessible cycles requires to scan the whole database, a penalty that can be tolerated only in small databases and/or in large time intervals. The cost of detecting single, inaccessible objects with a reference count of 0 is less expensive: The penalty for this method consists of the extra storage space to maintain a counter for each object and the overhead to increase the counter each time the particular object is associated with some database component.

An alternative to automatic garbage collection is the explicit deletion of objects from the database by the user. For this purpose the *delete* operation can be invoked on any object—*delete* is a built-in operation that is predefined for any object type. Thus,

    yourCuboid.delete;

causes the deletion of the *Cuboid* instance referred to by *yourCuboid*. However, this *delete* statement only destroys the outermost object; it does not delete any other objects, in our case the vertices or the *Material* object referenced by the *Cuboid*. If desired, these objects would have to be explicitly deleted as well, e.g.:

    yourCuboid.v1.delete;
    . . .
    yourCuboid.v8.delete;
    yourCuboid.delete;

In Section 8.6, we will describe a facility that allows to refine the *delete* operations such that the referenced subobjects (e.g., the vertices in our example) are implicitly deleted when *delete* is invoked on the outer superobject (e.g., the *Cuboid*).

The explicit deletion of objects may be dangerous, however, if the user does not take into account, that the deleted object is potentially referenced by other objects as well—i.e., if it is a shared subobject. Let us review the situation sketched in Figure 7.12. In this case, the deletion of the object o leaves the references from o_i to o (for $1 \leq i \leq n$) *dangling* because the object o to which they (should) point is no longer present in the database. A subsequent access attempt to o via such a dangling reference results in a run-time error.

This demonstrates that explicit deletion of objects from the database may be dangerous if objects are inadvertently deleted that are shared by many other objects. The *delete* operation purposely delegates the responsibility of controlling that a deleted object is not referenced by other objects, which still need access to the respective instance, to the user. Nevertheless, it appears that the potential danger of this approach often is accepted for the sake of increased performance by avoiding costly garbage collection.

7.12   Sorts

A sort is similar to an object type with the only exception that the "instances" of sorts are values and not objects; i.e., they do not possess an identity. The syntax for defining a sort is the same as the syntax for defining an object type. The only exception is that the keyword **type** has to be replaced by the keyword **sort** within the type definition frame, which now has to be called *sort definition frame*.

7.12.1   Complex Sorts

Complex sorts are those that are constructed using a tuple-, a set-, or a list-constructor. A typical, simple example sort is *date* which could be defined as

> **sort** date **is**
>   [day: int;
>   month: int;
>   year: int;]];

A variable of type sort can be declared by

> **var** d1: date;

and an actual date can be assigned to *d1* by executing

> d1.day := 1;
> d1.month := 1;
> d1.year := 1980;

Note that we did not *create* (instantiate) a value for *d1*. This is not necessary—and not even possible—since a *date* is a value of a sort and the space occupied by this value is provided at the declaration time of the variable *d1*.

The fact that values do not possess an identity is illustrated by the following small example. Assume that a second variable *d2* of sort *date* has been declared. Then

> d2 := d1;

results in *d2.day* being equal to 1, *d2.month* being equal to 1, and *d2.year* being equal 1980. Increasing the value of *d1.day* by 1, i.e.:

> d1.day := d1.day + 1;

does not change the value of *d2.day*, which still equals 1, whereas *d1.day* now equals 2. This illustrates the intrinsic difference between an object assignment—resulting in a reference—and a value assignment—resulting in a new copy of the value. The former is referred to as reference semantics, the latter is called copy semantics.

As for types, operations can be associated with sorts. For *date*, a useful operation is *nextDay*, which increases the day by one or, if necessary, increases the *month* and starts with the first day of the next month or even increases the *year* as implied by the Gregorian calendar. The declaration and implementation of operations associated with sorts is analogous to object type associated operations—which will be discussed in the next chapter.

It is possible—and quite common—to combine type and sort definitions. An example is

> **type** Car **is**
> [creationDate: date;
> chassis: Chassis;];

where the sort *date* is used within the definition of the object type *Car*. By naming conventions, type names start with an upper-case letter, whereas sort names start with a lower-case letter.

A restriction in the use of sorts as type constraints is that a sort may not appear recursively—directly or indirectly—in the definition of itself without having at least one type definition in the cycle. Thus, the sort definitions

> **sort** s **is** [a: u;];
> **sort** t **is** [b: s;];
> **sort** u **is** [c: t;];

are illegal, whereas

> **sort** s **is** [a: u;];
> **type** T **is** [b: s;];
> **sort** u **is** [c: T;];

are legal.

The above-defined complex sorts are tuple structured. Of course, there exist also set and list structured complex sorts. An example of a set structured sort is

> **sort** childSet **is** {Person};

where the values of *childSet* are defined as sets of *Person*s objects. This sort can now be used within the definition of the type *Person*:

> **type** Person **is** [age: int; spouse: Person; children: childSet;];

A value for a built-in sort can be explicitly stated; e.g., 1 denotes the value one of sort *int*. Though useful, this is not yet possible for complex values. Therefore, a mechanism is introduced and allows one to explicitly denote a certain complex value. As an example, the above initialization of the date value associated with the variable *d1* consisting of three assignments is equivalent to

> d1 := [day: 1; month: 1; year: 1980;];

The brackets "[...]" are used to denote a tuple structured value. Analogously, "{...}" and "< ... >" can be used to denote set structured and list structured values, respectively. Nesting of value denotations is also allowed, e.g.,

$$\{[a:1; b:2; c:3;]; [a:2; b:3; c:4;]\}$$

is a value that on the outer level is set structured and contains two elements that are tuple structured.

7.12.2   Enumeration Sorts

There is another distinguished kind of sorts called *enumeration sorts*. Enumeration sorts model a finite set of elementary values. Their definition obeys the following syntactic frame:

    sort *sortName* **is enum** (*EnumerationLiterals*);

For example, we can model the traffic light colors *TrafColors*—consisting of the colors *red*, *green*, and *yellow*—as follows:

    sort trafColors **is enum** (*red*, *yellow*, *green*);

7.13   Exercises

7.1   Discuss the modeling of aggregation hierarchies in GOM. Illustrate your discussion on the example aggregation hierarchy *bicycle* (cf. Figure 2.7 in Chapter 2).

7.2   Based on our *Person* and *City* object base of Figure 7.11, verify the type safety (or point out potential type errors) of the following statements:

        donald.spouse := cityOfLA.mayor;
        cityOfLA.mayor.spouse := cityOfLA.mayor;
        cityOfLA.name := cityOfLA.mayor.spouse.livesIn;
        cityOfLA.name := cityOfLA.mayor.spouse.name;

Note that you are asked to verify the type safety—and not to analyze whether the statements are semantically meaningful.

7.3   Consider, again, the object base of Figure 7.11. Verify the expression

        donald.spouse.spouse.age

Is this expression type consistent? What happens at run-time of a program containing this expression?

7.4   In the text, we explained the difference between the copy semantics of values and the reference semantics of objects. In our object base of Figure 7.11, the *City* named "Los Angeles" is a shared object that is referenced by all three *Person* objects via their *livesIn* attribute. Discuss the effects of the following program fragment:

```
        . . .
        donald.livesIn.mayor := donald;
        print(donald.livesIn.mayor.name);
        . . .
        mickey.livesIn.mayor := mickey;
        print(donald.livesIn.mayor.name);
```

Why is the effect substantially different from the characteristics of the following program fragment?

```
        donald.age := mickey.age;
        print(donald.age);
        . . .
        mickey.age := 70;
        print(donald.age);
```

7.5   Assume we want to ensure that *mickey* and *donald* always have the same *age*. How can this be achieved? Hint: You could define a new type *AgeType* and share an instance of *AgeType* among *mickey* and *donald*.

7.6   In the text we chose a very simple representation of *Cuboids*. Devise a boundary representation schema in GOM that allows to model arbitrary *Polyeders*. Illustrate your schema design on a sample cuboid object; i.e., show at least the most important parts of the resulting object base.

7.7   Based on the object base developed in Exercise 7.6, discuss the following concepts:

- Shared subobjects
- Implicit referencing
- Implicit dereferencing
- Reference versus copy semantics

7.8   Give an example of a set object being a shared subobject. Hint: Consider the type *Person* with an additional attribute *children* of type *PersonSet*. It may make sense that husband und wife share the same *PersonSet*. Under what condition is this meaningful?

7.9   Discuss the difference between the sort *personSet* and the object type *Person-Set*. Illustrate your analysis on the following two object types:

<table>
<tr><td>

type Person **is**
   [...;
     children: PersonSet;]

type Person′ **is**
   [...;
     children: personSet;]

</td><td>

type PersonSet **is**
   {Person};

sort personSet **is**
   {Person′};

</td></tr>
</table>

Furthermore, consider the six variables:

    var mary, joe, littleJoe: Person;
        betty, jim, jimbo: Person′;

Now describe the difference between the statements (1), (2) and (3), (4):

(0)   joe.children.create;
(1)   mary.children := joe.children;
(2)   mary.children.insert(littleJoe);
(3)   betty.children := jim.children;
(4)   betty.children.insert(jimbo);

Illustrate your discussion by diagrams showing the states of the resulting object bases.

7.10   Illustrate the problems that occur in a cyclic sort definition, as the abstract one introduced in the text:

    sort s **is** [a: u];
    sort t **is** [b: s];
    sort u **is** [c: t];

Focus on storage allocation for the values of the sorts, for example, if a variable of a sort *a* is defined.

7.11   Implement the Swiss knife example of Exercise 2.10 in GOM.

7.12   Model the objects of the GOM object model within GOM. That is, your object types should include tuple structured types, set structured types, list structured types, attributes, sorts, etc.

7.13   Implement the entity types of the conceptual model of Exercise 2.12 in GOM.

7.14   Implement the entity types of the conceptual model of Exercise 2.13 in GOM.

7.15   Implement the entity types of the conceptual model of Exercise 2.14 in GOM.

7.14   Annotated Bibliography

There is still no agreement concerning object-oriented models. Though, recently there have been attempts in identifying the salient features of object-oriented data models—most notably the so-called Manifesto [ABD⁺89]. An overview of object-oriented concepts is presented by Bancilhon [Ban88].

The generic object model (GOM) relies heavily on concepts borrowed from other data models. The strict object identity mechanism that is essential for the GOM reference concept is adopted from the FAD data model [BBKV87]. A detailed discussion of object identity can be found in [KC86]. Other than in FAD the atomic data objects are considered *values* and do not possess an object identity in GOM. Thus, they are not considered autonomous database entities. The constructors for structured types, i.e., tuple, set, and list constructors, are taken from the extended NF² model [DKA⁺86]. They can also be found in the models Extra [CDV88] and Orion [KCB88].

In GOM we distinguish between objects (which may generally be shared) and sorts (which are generally exclusively associated with a data component). Some authors discuss more specialized super-/subobject relationships. For example, the Extra data model [CDV88] facilitates so-called *own* attributes, and Orion has different kinds of composite objects that are described in [KBC⁺87, KBG89].

Furthermore, the influence of the functional data models, most notably the model DAPLEX [Shi81], should be acknowledged. The object-valued attributes in GOM are directly related to single-valued functions from one entity set into another entity set. Set-valued attributes, whose members are (references to) objects correspond to multivalued functions in DAPLEX. Some similarities to POSTGRES' attributes of type POSTQUEL [SR86] should also be stressed. But the references in GOM are strongly typed, whereas the type of POSTGRES references in the form of POSTQUEL queries as attribute values can only be determined at run-time. The reference concept of GOM is, as far as strong typing is concerned, more closely related to the MAD model [HMWMS87]. The *frame* (sometimes called *unit*) constructs of knowledge representation languages such as KRL [BW77, BW79] are also similar to the GOM model referencing other objects. A *slot* in a frame can contain either an atomic value or a reference to another frame. Also the forerunner of most "true" object-oriented programming languages, Simula-67 [DMN70] exhibits many similarities with GOM. Also, Smalltalk-80 [GR83], has a similar reference concept as GOM, except that in Smalltalk even atomic values—except for small integers—are conceptually treated as identifiable objects.

There are some books on object-oriented data base models, most of which are collections of research papers or descriptions of specific systems, though:

- Zdonik and Maier [ZM89] compiled a collection of research papers, many of which are concerned with implementation issues of object-oriented databases.

- Kim and Lochovsky [KL89b] also collected papers and edited them into a reference book.

- Kim, in addition, describes in [Kim90] the model and the implementation aspects of the Orion object database system project, which he headed at the MCC in Austin, Texas.

- Cardenas and McLeod [CM90] edited a set of papers from the areas of semantic and object-oriented data modeling. Their main emphasis is on the data model aspects.

- Cattell [Cat91] gives an introduction to object-oriented databases and emphasizes implementation issues.

- In [BDK92] Bancilhon, Delobel, and Kanellakis describe the implementation aspects of the O_2 object-oriented database system.

- Gray, Kulkarni, and Paton cover object-oriented data modeling from the viewpoint of semantic data models in [GKP92].

- Heuer [Heu92] wrote an introductory book (in German) on object-oriented and extended relational database system concepts.

Object-oriented analysis and design are discussed in detail in several text books, e.g., [Boo91, Bud91, CY91a, CY91b].

Shriver and Wegner [SW87] compiled a book on research directions in object-oriented programming, primarily covering the language aspects. The recent advances in object-oriented software engineering are covered in [MM92].

8

Specifying the Behavior of Objects

We already noted that each object type has an associated collection of operations that constitute the user interface by which an instance (object) of the respective type can be manipulated or by which the state of the instance can be queried. The usage of predefined operations accomplishes *information hiding* to the extent that the object can be viewed as an abstract data type that can be used without detailed knowledge of the internal realization. Once an object type has been specified, the user does not have to be concerned with the internal representation or the implementation of the operations. All one has to know is the interface specification, which consists of the *signatures* of the operations.

8.1 Classification of Operations

In order to be a useful building block for constructing complex applications, an object type has to provide all the operations that are *natural* to create instances, modify the instances' internal state, and access the instances' state. In designing the interface, the object type implementor has to find the right blend between an interface that is too lean, i.e., one that does not provide enough functionality for the object type to be useful, and one that is too abundant. A rather rich interface bears the danger that the usage of the object type becomes too complex.

Depending on their semantics, we distinguish different classes of operations on object type instances. The four classifications for operations are:

1. *Primitive constructors*
   This kind of operation is exclusively used to create a new instance of the respective object type. An example of a primitive constructor is *matrix$create(i)*, which would return the unity matrix of dimension $i \times i$.

197

2. *Constructors*

   The constructor operations are distinguished from the primitive constructors in the way that they create a new instance by performing some calculations. For example, *m.inverse* may create a new *matrix* instance, which is derived from the existing matrix m as the multiplicative inverse.

3. *Observer functions*

   Observers are functions that return information concerning the internal state of the object instances to which they are applied. An example of an observer function is the operation *elem*, which takes the three arguments m (a *matrix*) and i, j (two *integers*) and returns the element at position (i, j) of the *matrix* instance m. The invocation looks as follows: $m.elem(i, j)$.

4. *Mutators*

   Mutators are operations that change the internal state of the object instance on which they are invoked. For example, the operation $m_1.add(m_2)$ adds the matrix m_2 to the matrix m_1 on which *add* is invoked. Of course, it is required that the two matrices m_1 and m_2 are of compatible dimension.

The first three classes of operations are commonly called *functions* in programming languages because they return a result and are free of side effects. By this we mean that their invocation leaves the state of the database object invariant.

The mutators are operations, often called *procedures*, that actually change the internal state of some instance in the database. According to the presence of *mutator* operations, we call an object type either *mutable*, if at least one mutator is defined, or *immutable*, if no mutator operation is defined on the respective object type.

8.2   Type-Associated Operations

The behavior of objects is modeled by operations that are associated with the object type. Thereby, the object type definition encompasses the two dimensions: the structural representation and the behavioral specification of its instances.

So far we have only used the predefined operations, such as *create*, *persistent*, attribute access, etc. Of course, these built-in operations are by no means sufficient to capture the particular behavioral patterns of application-specific object types.

For example, when modeling a *Cuboid* it is obvious that we should provide observers like *volume*, *weight*, *surface*, and mutators that correspond to the standard geometric transformations *translate*, *rotate*, and *scale*. But before we define the object type *Cuboid*, we will first extend the definition of the object type *Vertex*, which will then serve as a major building block for defining more complex geometric types, such as *Cuboid*, *Cylinder*, etc. The type definition—now expanded with application-specific operations—is shown in Figure 8.1.

It can be observed that the definition of an operation is separated into two parts:

1. The *operation declaration*, where the abstract *signature* of the operation is specified

2. The *operation definition*, which supplies the code for implementing the operation

For better modularity these two parts are separated under two distinct headings, called **operations** and **implementation**, within the type definition frame.

We have included three sample mutators, i.e., *translate*, *scale*, and *rotate*, and two sample observers, i.e., *inOrigin* and *distance*. The function *inOrigin* determines whether the *Vertex* instance—on which it is invoked—lies in the origin of the coordinate system or, more precisely, very close to the origin. The observer *distance* determines the distance of the *Vertex* on which it is invoked to the parameter *Vertex*, called *otherVertex* in the implementation.

We trust that the reader has an intuitive understanding of the behavioral specification of *Vertex* objects—even though the details have to be covered in subsequent sections.

8.2.1   Operation Declaration

The operation declaration specifies the abstract *signature* of an operation, which determines the legal invocation pattern of the operation. Except for predefined operations, the signature of all type-associated operations has to be supplied in the **operations** clause of the type definition frame. Its syntactical structure is as follows:

> **declare** ⟨OperationName⟩ : ⟨ParameterTypeList⟩ → ⟨ResultType⟩
>   [**code** ⟨CodeName⟩] ;

The various parts of the abstract operation declaration have the following meaning:

- The ⟨*OperationName*⟩ uniquely identifies the operation within *one* object type definition. Note that different types provide distinct name spaces, e.g., the types *Vertex* and *Cuboid* could provide an identically named operation *translate*. We will later see that even within the same type, we can reuse an operation name by overloading (cf. Section 8.7)—within certain limits, however.

- The ⟨*ParameterTypeList*⟩ specifies the number, the relative ordering, and the types of the parameters with which the operation has to be invoked. The ⟨*ParameterTypeList*⟩ may be empty, in which case no additional parameters—besides the receiver object on which the operation is invoked—are needed in an invocation of the operation.

persistent type Vertex **is**
   public ...
   body [x, y, z: float;]
   operations
      declare translate: Vertex → **void**;
      declare scale: Vertex → **void**;
      declare rotate: float, char → **void**;
      declare distance: Vertex → float;
      declare inOrigin: → bool;
      ...

   implementation
      define translate(t) **is**
         begin
            self.x := **self**.x + t.x;
            self.y := **self**.y + t.y;
            self.z := **self**.z + t.z;
         end define translate;
      define scale(s) **is**
         ...
      define rotate(angle, axis) **is**
         ...
      define inOrigin **is**
         return
            (**self**.x $= 0.0 \pm \epsilon$ **and**   !! abbreviation for
             self.y $= 0.0 \pm \epsilon$ **and**   !! (**self**.y $\leq \epsilon$ **and self**.y $\geq -\epsilon$)
             self.z $= 0.0 \pm \epsilon$);
      define distance(otherVertex) **is**
         var dx, dy, dz: float;
         begin
            dx := **self**.x − otherVertex.x;
            dy := **self**.y − otherVertex.y;
            dz := **self**.z − otherVertex.z;
            return sqrt(dx ∗ dx + dy ∗ dy + dz ∗ dz);
         end define distance;
     ...
end type Vertex;

Figure 8.1: Definition of the Object Type *Vertex* with Operations

- The ⟨*ResultType*⟩ determines the type of the result object that is returned upon invocation of the operation. If none is returned, we have to specify **void** here.

- The (optional) specification of the ⟨*CodeName*⟩ determines the implementation name under which the operation is provided in the **implementation** part of the type definition frame. If it is omitted, the implementation has to be provided under the same name as the operation.

8.2.2   Operation Definition

The **implementation** section of the type definition frame contains the coding of user-defined operations. For each operation declared in the **operations** section, an associated implementation has to be provided. The syntactical structure is as follows:

> **define** ⟨CodeName⟩ [“(”⟨ParameterList⟩“)”] **is**
> ⟨VariableSection⟩
> ⟨OperationBody⟩

The first line represents the *head* of the operation implementation.

Operation Head

The operation head associates by means of the ⟨*CodeName*⟩ the coding with the declaration of the operation. If the optional **code** clause is missing in the operation declaration, the ⟨*CodeName*⟩ has to be identical to the ⟨*OperationName*⟩. Following the code name, the parameters—specified prior by number and type in the declaration—are named in the ⟨*ParameterList*⟩. Of course, the relative ordering and the number of parameter names have to coincide with the declaration of the operations—this is verified by the compiler.

Within the operation definition, parameters are treated like local variables. The type of the parameter is determined by the position of the parameter name in the ⟨*ParameterList*⟩ of the signature of the operation. For operations without parameters the ⟨*ParameterList*⟩ is omitted.

Variable Section

The ⟨*VariableSection*⟩ follows the operation head. Here, all additionally needed local variables are declared; i.e., their name and type have to be defined.

If no local variables are needed, the ⟨*VariableSection*⟩ may be omitted—as in the implementation of the operations *inOrigin* and *translate* of the object type *Vertex* (in Figure 8.1).

Operation Body

The ⟨*OperationBody*⟩ constitutes the actual coding of the operation. In the easiest case, the body contains just one statement. If it contains more than one statement, it has to be bordered by **begin** and **end**. Inside the **begin** ... **end** brackets an arbitrary number of statements, separated by semicolons, can be specified. The **begin** ... **end** can only be omitted if the ⟨*VariableSection*⟩ is omitted as well—otherwise, the **begin** ... **end** bracketing has to be used even if the body contains merely one statement.

8.2.3   Invocation of Type-Associated Operations

The invocation of a type-associated operation has to specify the so-called *receiver object*. The receiver object is separated from the operation name with the "dot" notation. For example, we could initiate the translation of a particular *Vertex* object that is referenced by the variable *myVertex* as follows:

```
var myVertex: Vertex;
    translationVertex: Vertex;
        ...
myVertex.translate(translationVertex);
```

The object referred to by *myVertex* is called the receiver object; sometimes we will also refer to the *receiver type* of an operation that corresponds to the type of the possible receiver objects (in this case the receiver type is *Vertex*). The object referred to by *translationVertex* is called the argument of the operation invocation—incidentally the argument is also an instance of type *Vertex* in this example.

Within the operation definition, the receiver object is referred to by **self**. Thus, when executing the *translation* code for the above call, **self** references the *Vertex* referred to by *myVertex*. Hence, **self** can be seen as an implicitly defined parameter. It is defined for all type-associated operations, and its type is always the receiver type, that is, the type the operation is associated with.

The dot notation may determine the receiver object via an arbitrarily long reference chain by chaining dot expressions. An example of such a reference chain is as follows:

```
myCuboid.v1.translate(translationVertex);
```

Here, the receiver object of the invocation *translate* is determined as the *Vertex* object referenced by the attribute *v1* of the *Cuboid* object referred to by the variable *myCuboid*.

8.3   Bottom-Up Type Definition

Object-oriented data modeling can be characterized as *bottom-up* structuring the application domain. That is, the most basic object types are developed first. Based on the primitive object types, such as *Vertex* in the geometric modeling domain, more advanced and complex object types can be realized.

Let us now demonstrate how the type *Vertex* can be utilized to expand the type *Cuboid* with the behavioral dimension. The augmented type definition of *Cuboid* is shown in Figure 8.2.

We have included the sample mutators *translate*, *rotate*, and *scale* and the observer functions *length*, *width*, *height*, *volume*, and *weight*. The meaning of these operations should be obvious. More interesting is the way the object type *Cuboid* has been implemented. It heavily utilizes the functionality of the prior-defined type *Vertex*. Consider, for example, the observer *length*, which is realized by delegating the computation of the distance between the boundary vertices *v1* and *v2* to the *Vertex* instance referenced by *v1*. The functions *width* and *height* are analogously implemented. Likewise, the mutator *translate* is implemented by delegating the translation to each of the respective *Cuboid*'s bounding vertices.

8.4   Encapsulation

Object-oriented models enforce *encapsulation* and *information hiding*. That is, the state of objects can be manipulated and read only by invoking operations that are specified within the type definition and made visible through the **public** clause.

Let us motivate the concept of encapsulation on our *Cuboid* example. According to the type definition specified thus far, it is still possible to circumvent the user-defined operations *translate*, *rotate*, and *scale*, which should be the only legal mutators for modifying a *Cuboid*'s geometric state. Under the current type definition, it would still be possible to execute the following program fragment:

```
      var myStrangeCuboid: Cuboid;
         ...
      myStrangeCuboid.create;
         ...      !! initialization of the bounding vertices to form a unit cube
(1)   myStrangeCuboid.v1.x := 0.5;
(2)   myStrangeCuboid.v1.y := 0.5;
(3)   myStrangeCuboid.v1.z := 0.5;
```

This program fragment would transform an initially well-formed *Cuboid* referred to by *myStrangeCuboid* into some semantically inconsistent state—as shown in Figure 8.3.

```
persistent type Cuboid is
  public ...
  body [v1, v2, v3, v4, v5, v6, v7, v8: Vertex;
        mat: Material; value: float;]
  operations
    declare length: → float;
    declare width: → float;
    declare height: → float;
    declare volume: → float;
    declare weight: → float;
    declare translate: Vertex → void
      code translateVertexCode;
    declare scale: Vertex → void
      code scaleCube;
    declare rotate: float, char → void   !! rotation angle and axis as parameters
      code rotateCube;
  implementation
    define length is
      return self.v1.distance(self.v2);   !! delegate the computation to Vertex v1
    define width is
      return self.v1.distance(self.v4);
    define height is
      return self.v1.distance(self.v5);
    define volume is
      return self.length * self.width * self.height;
    define weight is
      return self.volume * self.mat.specWeight;
    define translateVertexCode(t) is
      begin
        self.v1.translate(t);   !! delegate translate to the Vertex instance v1
        ...;
        self.v8.translate(t);   !! delegate translate to the Vertex instance v8
      end define translateVertexCode;
    ...
end type Cuboid;
```

Figure 8.2: Incorporating Operations in the Type Definition *Cuboid*

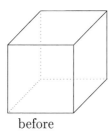 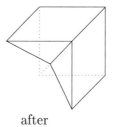

before   after

Figure 8.3: Semantically Inconsistent Transformation of a *Cuboid* Object

On the left-hand side of Figure 8.3, we show the state of the *Cuboid* directly after initialization, and before statements (1), (2), and (3) are performed. The right-hand side illustrates the semantically inconsistent state that was induced by the statements (1), (2), and (3).

The semantic inconsistency is caused by the fact that only one bounding *Vertex* of the *Cuboid* was modified—without observing the rules for correct geometric transformations. In order to guarantee a consistent state, we have to ensure that *all* bounding vertices are simultaneously transformed in a conforming manner. This is guaranteed if we restrict the manipulations to the well-defined three geometric transformations *rotate*, *translate*, and *scale*—which we assume to be correctly defined by an experienced database designer.

For the purpose of encapsulation, our object model GOM provides the **public** clause. The **public** clause determines the collection of operations that can legally be invoked "from the outside" on instances of the respective type. For our *Cuboid* example, the inconsistency problem was caused by allowing access to the bounding *Vertex* referenced by *v1*, which enabled the isolated transformation of a single bounding *Vertex*.

Let us now define the meaning of *inner* and *outer* operation definitions, which is needed for a detailed illustration of encapsulation. A type contains the following classes of operations:

- *Built-in operations*
  For a tuple structured type, these operations consist of the predefined operations to access and write attributes. For set structured types, GOM provides built-in operations for inserting new elements, removing elements, and counting the number of elements, i.e., determining the *cardinality* of the set object. For list structured types, the position of list elements is additionally taken into account.

- *User-defined operations*
  All operations declared in the **operations** section and implemented in the

implementation section of the type definition frame make up the user-defined operations.

The set of *inner operations* of a type consists of all the user-defined and built-in operations of the respective type. The set of *outer operations* is determined by the **public** clause. The **public** clause has to specify a subset of the operations contained in the set of the inner operations of the type. If we want to make all operations public, we can simply omit this clause. Otherwise, the **public** clause contains a list (separated by commas) of operation names.

We call the signatures of all the inner operations of a type the *inner signature* of the type. Likewise, the set of signatures of all those operation made public is called the *outer signature* of the type. Note that the outer signature of a type is always a subset of the inner signature.

The inner signature determines the set of operations—and their invocation patterns—that may be invoked in the implementation of operations within the **implementation** part of the type definition frame. More specifically, the operations included in the inner signature but not in the outer signature may be invoked only on the receiver specified by **self**, i.e., in the **implementation** part of the object type. On the other hand, the outer signature determines the set of operations (and their invocation requirements) that may legally be invoked from the outside of the type implementation.

Let us now define the **public** clause of the *Cuboid* definition that avoids the "nasty" problems of inconsistent geometric modifications:

```
persistent type Cuboid is
    public length, width, height, volume, weight, translate, scale, rotate
    body
        . . .
    operations
        . . .
    implementation
        . . .
end type Cuboid;
```

As clients of the type, we can invoke only the operations *length*, ..., *rotate*. Thus, the program fragment shown before, where the bounding *Vertex v1* was modified separately, becomes impossible because the access to the attribute *v1* has been ruled out—read and write access to *v1* are excluded from the **public** clause. However, reading (and writing) *v1* is still contained in the inner signature of *Cuboid* such that the implementation of *translate* of Figure 8.2 remains valid. The encapsulation of *Cuboid* objects achieved by the redefined **public** clause is depicted in Figure 8.4.

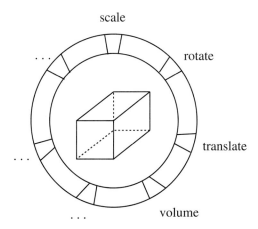

scale

rotate

translate

volume

Figure 8.4: Schematic Characterization of *Cuboid* Encapsulation

In summary we want to tabularize the access right to the inner and outer signature of an object:

access to	from	
	inside	outside
inner signature	okay	illegal
outer signature	okay	okay

8.5 Initialization

In Section 7.4, we showed how objects are instantiated by invoking the predefined operation *create*. The state of the generated instances is thereby initialized to default values—for nonatomic attributes this is generally *Null*. Thereafter, one has to initialize the instantiated object to the desired state by applying corresponding operations. This has two drawbacks:

1. *Violation of encapsulation*
   Again, let us illustrate this problem on our *Cuboid* example. If one has to create the bounding vertices individually, as shown in Section 7.4, then we cannot outrule the possibility that a reference to some of the bounding vertices is maintained outside the instantiated *Cuboid*. This reference could then be (ab)used to individually modify this *Vertex* leading to a semantically inconsistent *Cuboid* object—as outlined in the preceding section.

2. *Tedious procedure*

   If one has to frequently create objects of a particular type then the initialization may be a very tedious procedure—as demonstrated for the *Cuboid* example in Section 7.4. There we had to create the bounding vertices one by one, initialize them to the desired coordinates and assign them to the corresponding attributes $v1, \ldots, v8$.

A better way to initialize newly generated objects is to combine instantiation and initialization. For this purpose, the object model GOM facilitates the definition of so-called *initializers* within the type definition. The initializer has to be defined the same way as any other type-associated operation. The only difference is that its name is fixed by a convention. The name of an *initializer* is the same as the type whose instances it is supposed to initialize. Hence, for the object type *Cuboid*, the initializer is called *Cuboid*.

Let us demonstrate the definition of this initializer. Assume that we want to always initialize *Cuboid* objects to the unit cube at a particular position in the three-dimensional coordinate system. In subsequent geometric transformations, i.e., invocations of *scale*, *translate*, and *rotate*, we can generate any desired size, location, and orientation of the *Cuboid* instance. Furthermore, we want to initialize the *mat* attribute of the newly generated *Cuboid* to a *Material* instance that is passed as an argument to the initializer. Then the initializer *Cuboid* is defined as follows:

```
persistent type Cuboid is
   public ...
   body
      ...
   operations
      declare Cuboid: Material → void;
      ...
   implementation
   define Cuboid(m) is
      begin    !! generate the unit cube made of m
         self.mat := m;
         self.v1.create;
         self.v1.x := 0.0;
         self.v1.y := 0.0;    !! v1 = (0, 0, 0)
         self.v1.z := 0.0;
         self.v2.create;      !! v2 = (1, 0, 0)
         ...
         self.v8.create;
         self.v8.x := 0.0;
         self.v8.y := 1.0;    !! v8 = (0, 1, 1)
```

> **self**.v8.z := 1.0;
> **end define** Cuboid;
> . . .
> **end type** Cuboid;

If the compiler encounters an initializer, it invokes it immediately after each *create* invocation on the newly created object. Thus, two steps are performed upon invoking *create*:

1. Instantiation of the new object

2. Invocation of the initializer on the newly created object immediately afterwards

Of course, if the initializer requires arguments, these have to be passed with the invocation of the *create* operation, as illustrated below for our *Cuboid* example:

> **var** myCuboid: Cuboid;
> gold: Material;
> . . .
> myCuboid.create(gold);
> . . .

This program fragment leads to the creation of a new unit cube—an object of type *Cuboid*, though—whose *mat* attribute references the *Material* instance to which *gold* refers (presumably, a *Material* instance modeling gold).

8.6   Refining the *delete* Operation by Destructors

In Section 7.11, we introduced the *delete* operation, which is applicable for any object. However, we noted that it is often essential to refine the behavior of this built-in destructor in order to "take care" of subobjects that are exclusively associated with the deleted superobject. Otherwise, we run the risk that the object base becomes "trashed" with inaccessible objects—which can only be detected by a costly garbage collection.

A *destructor* operation has the same name as the type whose destructor it is. To distinguish it from the initializer, it is preceded by a tilde ("~"). As for the initializer, the declaration and implementation of a destructor is done the same way as for any other type-associated operation.

Let us now refine the behavior of the delete operation associated with *Cuboid* objects by a destructor ~*Cuboid*. If we assume that no two *Cuboids* share a common *Vertex*, it is certainly useful to (implicitly) destroy the eight vertices referenced by the *Cuboid* that is to be deleted. This is achieved by the following specification:

```
type Cuboid is
  public ...
  body ...
  operations
    declare ~Cuboid: → void code refinedCuboidDelete;
      ...
  implementation
    define refinedCuboidDelete is
      begin
        self.v1.delete;
          ...
        self.v8.delete;
      end define refinedCuboidDelete;
      ...
end type Cuboid;
```

Note that we have consciously omitted to (implicitly) delete the *Material* referenced by the attribute *mat*. Typically, *Materials* are highly shared subobjects; their deletion could have catastrophic side effects for other users.

8.7   Overloading

Reconsider the initializer specified for the object type *Cuboid* in Section 8.5. We can think of cases in which one might want to provide different initializers for the same type. For the *Cuboid* example, think of an additional initializer in which one may also specify the *scale* factors in x-, y-, and z-dimension in order to generate a *Cuboid* of arbitrary length, width, and height—instead of a unit cube. The concepts introduced thus far do not facilitate this functionality. However, *overloading*, which is introduced next, will allow this additional flexibility—not only for initializers but for other operations as well.

Overloading facilitates the reuse of operation names for semantically related yet different operations in the same name space. For the time being, we assume that each type has a separate name space, which will be somewhat revised when we introduce inheritance in Chapter 10. Overloaded operations in the same name space, i.e., in the same type definition frame, have to be distinguishable by at least one of the following criteria:

- The number of parameters

- The type of the parameters

Let us illustrate overloading on the example type *Cuboid* by providing two different, but identically named initializers:

     persistent type Cuboid **is**
       . . .
      operations
        declare Cuboid: Material $\rightarrow$ **void**
          code CuboidMatCode;
        declare Cuboid: Material, float, float, float $\rightarrow$ **void**
          code CuboidMatScaleCode;
       . . .
      implementation
        define CuboidMatCode(m) **is**
         . . .   !! as before
        define CuboidMatScaleCode(m, scaleLength, scaleWidth, scaleHeight) **is**
        begin
          self.mat := m;
          self.v1.create;   !! generate the *Cuboid* of specified dimensions
          self.v1.x := 0.0;
          self.v1.y := 0.0;   !! $v1 = (0,0,0)$
          self.v1.z := 0.0;
          self.v2.create;   !! $v2 = (scaleWidth, 0, 0)$
          . . .
          self.v8.create;
          self.v8.x := 0.0;   !! $v8 = (0, scaleLength, scaleHeight)$
          self.v8.y := scaleLength;
          self.v8.z := scaleHeight;
        end define CuboidMatScaleCode;
       . . .
     end type Cuboid;

Let us demonstrate the use of the two different (overloaded) initializers on the following program fragment:

     var myCuboid: Cuboid;
        yourCuboid: Cuboid;
        gold: Material;
        . . .
(1)   myCuboid.create(gold);   !! golden unit cube
(2)   yourCuboid.create(gold, 2.0, 3.0, 5.0); !! golden *Cuboid* of dimension $2 \times 3 \times 5$
        . . .

In statement (1) the first initializer with the implementation (code) *CuboidMatCode* is implicitly invoked; in statement (2) the initializer with coding *CuboidMatScaleCode* is implicitly invoked on the newly instantiated *Cuboid* object referred to by the variable *yourCuboid*. Which one of the initializers is to be invoked is determined—in this case—by the different number of parameters.

But the reader should not be misled that overloading is useful only for initialization. It can be very beneficial for other operations as well. Consider the operation *translate* of the type *Cuboid*: The flexibility would be greatly increased if we could invoke translate either with a *Vertex* as parameter (for specifying the translation offsets) or with three *float* parameters. This is achieved as follows:

```
persistent type Cuboid is
   ...
   operations
      ...
      declare translate: Vertex → void
         code translateVertexCode;
      declare translate: float, float, float → void
         code translateXYZcode;
         ...
   implementation
      define translateVertexCode(t) is
         begin
            self.v1.translate(t);   !! delegate translate to the Vertex instance v1
            ...
            self.v8.translate(t);   !! delegate translate to the Vertex instance v8
         end define translateVertexCode;
      define translateXYZcode(X, Y, Z) is
         begin
            ... !! code for translating the vertices
         end define translateXYZcode;
      ...
   end type Cuboid;
```

Consider the following program fragment that illustrates the use of the overloaded *translate* operations:

```
var myCuboid: Cuboid;
    translateVertex: Vertex;
    ...     !! create Cuboid and Vertex
myCuboid.translate(translateVertex);
myCuboid.translate(30.0, 2.0, 1.75);
    ...
```

Obviously, different *translate* operations—which are, however, semantically related—are invoked with different parameter sets. Again, the compiler can easily determine which one of the translate operations has to be bound (executed) from the number and types of parameters being passed in the two invocations.

8.8   Value-Receiving and Value-Returning Operations

So far, we have only considered operations that return a result upon invocation—the nonexistent result, denoted by **void**, is a special case. Aside from these so-called value-returning operations (VTOs), GOM also facilitates the definition of **value-receiving operations** (VCOs).

8.8.1   Built-In VCOs and VTOs

The VCOs and VTOs were already briefly introduced in Section 7.3 in the form of the built-in operations to write attributes in tuple structured types. Let us recall the built-in VCOs and VTOs for writing and reading attribute values. Consider the type *Cuboid* with the attribute *value*, which is constrained to *float*. This attribute has an associated pair of VCO and VTO operations:

> **declare** value: → float;
> **declare** value: ← float;

These operations can be invoked as demonstrated by the following program fragment:

> myCuboid.value←(yourCuboid.value→);

As "syntactic sugar," we also allow the "ordinary" assignment statement, that may be more convenient for people used to other programming languages, e.g.:

> myCuboid.value := yourCuboid.value;

In this example the value-returning operation—denoted as *value*→ —is implicitly invoked on the right-hand side of the ":="-sign. On the left-hand side of the ":="-sign, the value-receiving operation *value*← is implicitly invoked. Both operations are identically named in the program; the compiler determines whether the VCO or the VTO has to be invoked from the relative position of the invocation with respect to the ":="-sign.

8.8.2   User-Defined VCOs

In GOM, it is also possible that the user defines application-specific value-receiving operations—aside from the ones that are built in for attribute assignment. As an example consider the definition of a *Circle*:

```
persistent type Circle is
   public radius→, radius←, circumference→, circumference←
   body [radius: float;]
   operations
      declare circumference: → float
         code GetCircumferenceCode;
      declare circumference: ← float
         code SetCircumferenceCode;
   implementation
      define GetCircumferenceCode is
         return self.radius * 2.0 * 3.14;
      define SetCircumferenceCode is
         var c: float;
         begin
            receive c;
            self.radius := (c/3.14)/2.0;
         end define SetCircumferenceCode;
end type Circle;
```

Note, that *circumference* is not an attribute in the structural representation of *Circle*. Therefore, the value-receiving operation *circumference←* has to be coded such that the *radius* is modified according to the received *circumference* value.

This pair of VTO/VCO operations to read and modify the *circumference* of a *Circle* is utilized in the following program fragment:

```
var k, l: Circle;
   k.create; l.create;
   ...
   k.circumference := 10.0;
   l.radius := k.radius;
```

It is necessary that we can distinguish the identically named value-receiving (VCO) and value-returning (VTO) operations in the **public** clause in order to be able to selectively include the VCO or the VTO. For this purpose, we identify the VCO operation with a left arrow and the VTO operation with a right arrow—as the following **public** clause of the type *Cuboid* demonstrates:

```
persistent type Cuboid supertype ANY is
   public length, width, height, volume, weight, translate,
      scale, rotate, mat→, value
   body ...
```

This **public** clause specifies that the *mat* attribute can be read, but not written. This is determined by including *mat→* in the operation list. The inclusion of *value* in the **public** clause is equivalent to including both, *value*→ and *value*←, i.e., providing read and write access to the attribute to clients of this type.

8.9 Free Operations and Operators

Sometimes it appears awkward to associate certain operations with a distinguished receiver types, i.e., to invoke the operation on a distinguished receiver object. We included one such operation already in our examples: the operation *distance* which computes the distance between two vertices. The way distance was defined in Figure 8.1 it requires a receiver object—of type *Vertex*—and an additional argument object, also of type *Vertex*. Thus, invoking *distance* on a *Vertex* looks as follows:

```
var p1, p2: Vertex;
    d: float;
  . . .
  d := p1.distance(p2);
  . . .
```

Distance is a symmetric operation; i.e., the distance from *p1* to *p2* is the same as the distance from *p2* to *p1*. However, the type association obscures this symmetry by requiring a distinguished receiver object—in our case the *Vertex* referred to by *p1*—and an additional argument *Vertex*—the one referred to by *p2*. For many programmers, the following invocation would appear much more natural:

```
  . . .
  d := distance(p1, p2);
  . . .
```

To facilitate this, GOM allows the definition of so-called *free operations*, i.e., operations that are not associated with any receiver type.

8.9.1 Declaration and Definition of Free Operations

A free operation is declared and implemented similarly to type-associated operations, except that they are defined outside any type definition frame.

For our example operation *distance*, we have to provide the following declaration:

```
persistent declare distance: Vertex, Vertex → float
   code FreeDistanceCode;
```

The qualification **persistent** preceding the declaration specifies that this operation declaration should survive the termination of the application program. That is, the declaration is to be maintained by the schema manager for subsequent use by other programs. Type-associated operations that are defined in the type definition frame of a persistent type are automatically made persistent.

The implementation of the free operation *distance* can be founded on the existing type-associated operation *distance* (cf. Figure 8.1):

> **define** FreeDistanceCode(p, q) **is**
>   **return** p.distance(q);   !! invocation of the type-associated *distance* operation

Please note that this definition of *distance* is not overloading the *distance* operation associated with *Vertex*, since the type *Vertex* provides its own, separate name space. However, any other free operation named *distance*—e.g., one that computes the distance between two *Cuboid* objects—constitutes an overloading of this free operation.

8.9.2   Declaration and Definition of Operators

For certain operations, we are used to the so-called *infix* notation. Examples are: $+, -, /, *, <, \le, =$, etc. These operations are typically called *operators*. In GOM, we facilitate the definition of new operators—aside from the large set of built-in operators. Usually, new operators are defined as overloaded operators because they typically reuse some operator symbol that has been used before; either for a built-in or for a user-defined operator. Let us demonstrate this concept on an example operator "<" for two objects of type *Vertex* which compares their distance from the origin. That is, *v1* < *v2* holds, if the *Vertex* referenced by variable *v1* is closer to the origin than the *Vertex* referred to by variable *v2*.

The operator is declared as follows:

> **operator** <: Vertex, Vertex → bool
>   **code** vertexLessCode;

> **define** vertexLessCode(v1, v2) **is**
>   **begin**
>   **return** (v1.distance(0.0,0.0,0.0) < v2.distance(0.0,0.0,0.0));
>   **end define** vertexLessCode;

In the implementation of *vertexLessCode*, we assumed the existence of an overloaded operation *distance*, which takes three coordinates instead of a *Vertex* as a parameter. The declaration overloading is analogous to the one for *translate* in Section 8.7.

The now specified operator "$<$" can be used in *prefix* as well as in *infix* notation. For a prefix invocation, the keyword **operator** has to precede the operator symbol "$<$":

    var a, b: Vertex;
        flag: bool;
        . . .
        flag := **operator**$<$(a, b);   !! prefix notation
        flag := a $<$ b;   !! infix notation

8.10   Defining Operations Outside the Type Definition Frame

So far, the type-associated operations—i.e., those operations that require a receiver object—were always defined inside the receiver's type definition frame. There is, however, a reason that makes it desirable to define new operations outside the type definition frame: If a less experienced database user wants to define a new operation on an existing type one may want to exclude this user from accessing the inner signature of the type. Rather, the inexperienced user should implement the additional operation solely on the basis of the outer signature which ensures that a certain degree of consistency is always maintained.

We will illustrate the outside definition on an example operation called *mirror*, which is used to mirror the *Cuboid* object on which it is invoked about the plane determined by the three parameter vertices. Then *mirror* is declared as follows:

persistent declare mirror: Cuboid $\parallel$ Vertex, Vertex, Vertex $\rightarrow$ **void**;

The receiver type is separated from the parameter types by the symbol "$\parallel$". The implementation of *mirror* is then specified as

    define mirror (v1, v2, v3) **is**
        begin
            . . .
            self.rotate(. . .);
            . . .
        end define mirror;

It is important to observe that this implementation is now considered as being a client of the *Cuboid* object. Therefore, only operations included in the **public** clause of the type definition frame—i.e., only operations contained in the outer signature of *Cuboid*—are allowed to be invoked on **self**.

8.11   Parameter Passing

In the preceding examples, we described the means for declaring and defining operations. Furthermore, we already illustrated the invocation of operations—on an intuitive level. Here, we want to describe the parameter passing more precisely.

Parameters of operations are used to "channel" information from the location of the invocation of the operation to the operation's body, i.e., the execution of the operation. This information consists either of values, i.e., elements of sorts, or object identifiers; both of which are determined by evaluating expressions at run time.

In the most general case, an expression consists of several nested invocations of operations. In the invocation of an operation that is declared with parameters, we have to supply correspondingly typed expressions. These expressions are called the *actual* parameters of the invocations. The outcome of the evaluation of the actual parameters is assigned to the *formal* parameter, i.e., the parameter that was declared in the signature and the definition of the operation.

In the case of type-associated operations, we have one distinguished parameter, the receiver of the operation. The corresponding formal parameter is of the same name for all type-associated operations: **self**. However, from the point of parameter passing, the receiver is handled like any other parameter of the operation. Thus, an invocation of the type-associated distance operations looks as follows:

> myCuboid.v1.distance(yourCuboid.v7);

The receiver of the operation is determined by evaluating the expression *myCuboid.v1*.

In order to be precise, two cases have to be distinguished: *call by value* and *call by reference*. At run time, the expression corresponding to an actual parameter of the operation is evaluated. The result of evaluating the expression is substituted in place of the formal parameter of the operation definition. Thus, the parameter passing is equivalent with assignment of the actual parameters to the formal parameters. This form of parameter passing is termed *call by value* in the programming language literature—because the value of the actual parameter is passed, but not the parameter itself. In general, this applies to values, i.e., "instances" of sorts.

In case of parameters whose type is an object type—as opposed to a sort—the object identifier determined by evaluating the actual parameter is passed. Let us illustrate the effects on the basis of an operation *incAge* associated with the object type *Person*:

> **declare** incAge: Person $\|$ → **void**;
>
> **define** incAge **is**
>   **self**.age := **self**.age + 1;

Now reconsider the *Person* object base state shown in Figure 7.11. The invocation of

(1)   cityOfLA.mayor.incAge;
(2)   mickey.incAge;

both have the same effect. In both cases, the receiver parameter evaluates to the *Person* instance with OID id_{188}, i.e., the *Person* named "Mickey Mouse." The effect of the operation invocation is to increase the *age* of the object whose identifier is passed as a receiver parameter to the operation. Therefore, this kind of parameter passing is sometimes denoted *call by reference* because it establishes a reference from the formal parameter to the object determined by the actual parameter—thereby making the object a shared object. The effects of the operation execution become— of course—visible via all possible references leading to the object. This was already described in detail in Section 7.6.

8.12   Operations as Parameters

8.12.1   Motivation

Assume we need an operation that selects from a set of *Cuboids* those elements that satisfy certain search criteria. Let us, for example, find all those *Cuboid* elements of a *CuboidSet* whose *value* exceeds 500.00 and insert these qualifying *Cuboids* into a result set (of type *CuboidSet*). This can be achieved by the following free operation *selectExpensiveCuboids*:

```
declare selectExpensiveCuboids: CuboidSet, CuboidSet → void;

define selectExpensiveCuboids(allCuboids, selectedCuboids) is
   var candidate: Cuboid;
   begin
      foreach(candidate in allCuboids)
         if (candidate.value > 500.0)      !! selection predicate
            selectedCuboids.insert(candidate);
   end;
```

For selecting those *Cuboids* that have a *volume* exceeding 100.00, we would have to define yet another select operation, now called *selectLargeCuboids*:

```
declare selectLargeCuboids: CuboidSet, CuboidSet → void;

define selectLargeCuboids(allCuboids, selectedCuboids) is
   var candidate: Cuboid;
   begin
      foreach(candidate in allCuboids)
```

```
        if (candidate.volume > 100.0)     !! selection predicate
            selectedCuboids.insert(candidate);
end;
```

It is remarkable that the implementation of the two operations is almost identical. The only difference concerns the selection predicate, which selects *Cuboids* according to their

- *value* in *selectExpensiveCuboids*

- *volume* in *selectLargeCuboids*

The idea is to abstract from this (minor) difference and provide only one such *selectCuboids* operation. The necessary "tool" for doing so is given by passing operations as parameters.

8.12.2   Passing Selection Predicates as Parameter

Let us now define the *one* operation *selectCuboids*, which facilitates the selection of *large* and *expensive Cuboids*:

```
declare selectCuboids: CuboidSet, [Cuboid || → bool], CuboidSet → void;

define selectCuboids(allCuboids, selPred, selectedCuboids) is
    var candidate: Cuboid;
    begin
        selectedCuboids := {};
        foreach(candidate in allCuboids)
            if (candidate.selPred)     !! selection predicate
                selectedCuboids.insert(candidate);
    end;
```

In the declaration of *selectCuboids*, the parameter operation is delimited by square brackets [...]. Within these, we have to specify the signature of the parameter operation—in this case [*Cuboid* || → *bool*]—but not the name. This, later on, facilitates to pass arbitrary functions that select *Cuboids* based on different criteria to an invocation of *selectCuboids*. In the invocation it is, of course, required that the actual operation parameter satisfies the signature [*Cuboid*|| → *bool*].

In order to utilize *selectCuboids*, we could define the two selection predicates *isExpensiveCuboid* and *isLargeCuboid*—in the form of Boolean functions—as follows:

```
declare isExpensiveCuboid: Cuboid || → bool;
declare isLargeCuboid: Cuboid || → bool;
```

define isExpensiveCuboid **is**
   return self.value > 500.0;

define isLargeCuboid **is**
   return self.volume > 100.0;

These Boolean functions can then be passed to an invocation of *selectCuboids*, e.g.:

var   myCuboids: CuboidSet;
     largeOrExpensive: CuboidSet;
   ...
   largeOrExpensive.create;   !! generate the empty result set
   selectCuboids(myCuboids, isLargeCuboid, largeOrExpensive);
   selectCuboids(myCuboids, isExpensiveCuboid, largeOrExpensive);
   ... !! further processing of the result set *largeOrExpensive*

In the above program fragment, the *Cuboids* that are large or expensive are retrieved from the set *myCuboids* and inserted into the set referred to by *largeOrExpensive*. Of course, it is possible to define further selection predicates in the form of Boolean functions—e.g., *hasLargeSurface*, *liesInOrigin*, *isMadeOfGold*, etc.—and use them in invocations of *selectCuboids*. In Chapter 12 and in Chapter 14, we will augment the flexibility of such selection operations even further.

8.13   Exercises

8.1   Augment the type definition of *Vertex* by the following operations:

- *distanceToOrigin*, a function that computes the distance of the receiver *Vertex* object to the origin of the coordinate system

- *rotate*, the mutator that rotates the receiver *Vertex* about the specified axis by an angle that is passed as a parameter

- *onStraightLine*, a Boolean function that checks whether the receiver *Vertex* lies on a straight line determined by two parameter objects of type *Vertex*

- *onPlane*, a Boolean function that checks whether the receiver *Vertex* lies on the plane formed by the parameter *Vertex* instances

You should specify the three declarations as well as the implementations of these operations.

8.2   In Section 8.9.2, we used for the implementation of the $<$ operator the overloaded *distance* operation taking three coordinates instead of a *Vertex* as an
argument. Nevertheless, this operation does—hitherto—not exist. Provide
the declaration and implementation of this operation.

8.3   Define the object type *Tetraeder*. Make sure that you provide a satisfactorily
complete operational interface. In particular, you should include the analogous
operations that are (verbally) specified in Exercise 8.8.

8.4   In Exercise 7.6 the structural representation of a boundary representation
schema was developed in GOM. Augment the skeleton schema by operations
for geometric transformations.

8.5   Devise the kernel types of the constructive solid geometry representation of
solid objects. Sketch the combinational operations union, difference, intersection, etc.

8.6   In Section 3.1.2, the concept of *homogeneous coordinates* was introduced. Homogeneous coordinates require that a *Vertex* is represented by four, instead
of three, coordinates. Define the type *Vertex*, which implements the homogeneous coordinates. Implement the three geometric transformations—i.e.,
translate, rotate, and scale—as matrix multiplications. Of course, you have
to define the type *Matrix* first. Hint: Use lists of lists to implement the type
Matrix.

8.7   Realize the most important gate types, e.g.,

- AND

- NOR

- OR

- XOR

- NAND gate

as object types. What operations are meaningful for VLSI design applications?
Specify and implement them in GOMpl.

8.8   Augment the type definition of *Cuboid*. In particular, add the following operations:

- *surface*, yielding the surface value of the *Cuboid*

- *scale*, which scales the size of the *Cuboid*—be careful that the *scale* operation is properly implemented for *Cuboids* that have an arbitrary orientation in the coordinate system

- *center*, which determines the *Vertex* representing the center of the *Cuboid*

- *diagonal*, which computes the length of the diagonal of the *Cuboid*

- *minDistance*, a function that determines the minimum distance of a parameter *Vertex* from the receiver *Cuboid*—e.g., for assembly planning in order to avoid collisions

- *minDistance*, the overloaded function that determines the minimum distance of any boundary points of two *Cuboid* objects—one being passed as a parameter, the other one being the receiver object. Additionally, provide an alternative free operation for *minDistance*. You may reuse the type-associated operation *minDistance* for implementing the free operation *minDistance*.

8.9 The *call by value* or, in the case of object parameters, *call by reference* parameter passing sometimes leads to surprising effects. Consider the following operation *wrongSwap*:

declare wrongSwap: Cuboid, Cuboid → **void**;

```
define wrongSwap(v1, v2) is
   var h : Cuboid;
   begin
      h := v1;
      v1 := v2;
      v2 := h;
   end define wrongSwap;
```

What happens when *wrongSwap* is invoked in the following program fragment (*smallCube* and *largeCube* being variables referring to *Cuboid* objects):

```
   ...
   if (smallCube.volume > largeCube.volume)
      wrongSwap(smallCube, largeCube);
```

Illustrate your example on a sample object base. In particular, consider the two different cases:

1. *smallCube* refers to a Cuboid smaller than the one referred to by *largeCube*.

2. *smallCube* refers to a Cuboid larger than the one referred to by *largeCube*.

8.10   Implement an operation *can-cut* for the different cutting devices of a Swiss knife (cf. Exercise 7.11). Take into account the material and length of blades and the material and thickness of the things to be cut.

8.11   Add behavior to the model of GOM in GOM (cf. Exercise 7.12). Typical behavior would be the addition or the deletion of a new type or sort, modifying an operation's signature. Take especial care of the constraints imposed by the GOM data model. Examples of these constraints are

1. Type names are unique.
2. The operation of an operation call must exist and the parameters must be of the specified type.

Complete this list with the knowledge of the preceding chapters. Compare also to Exercise 8.14.

8.12   Implement the behavior for the object types given in Exercise 7.13. Take into account operations such as opening and closing a door, switching a light on and off, moving furniture and possibly installations around. Especially switching the lights and heating on and off can be important for two purposes:

1. Monitoring the usage of energy
2. Maintenance, since one has to know when to buy a new lightbulb

8.13   Implement the behavior for the object types given in Exercise 7.14. Typical operations might be correcting a paragraph, inserting a paragraph, changing a title, modifying a figure and its caption, etc.

8.14   Implement the behavior for the object types given in Exercise 7.15. Consider operations like modifying the signature of an operation, e.g., adding a new parameter, deleting an operation call, inserting a new operation for a frequently occurring piece of code and the like. Compare also to Exercise 8.11.

8.14   Annotated Bibliography

The language features of GOM discussed in this chapter heavily rely on the object-oriented programming languages Simula 67 [DN66] and its descendents, e.g., Eiffel [Mey88], Smalltalk-80 [GR83], C++ [Str90b], etc. Most of the relevant literature concerning object-oriented programming languages has already been cited at the end of the preceding chapter.

Liskov and Guttag [LG86] provide a very good treatment of program design using object-oriented features, such as information hiding, bottom-up design, etc. Their work is—at least partially—founded on the programming language Clu [LAB$^+$81]. Meyer [Mey87] discusses the advantages of the object-oriented languages with respect to program reusability. Another contribution on this topic was published by Micallef [Mic88].

Stefik and Bobrow [SB86] survey object-oriented languages and modeling in artificial intelligence applications.

Liskov and Snyder [LS79] discuss exception handling in the context of the language Clu.

Overloading of operations has been introduced in a variety of programming languages, in particular Ada [Boo83, ANSI83]. Ada also provides means for exception handling.

Derret, Kent, and Lyngbaek [DKL85] discuss the behavioral aspects of the object model Iris—which is heavily influenced by the functional data model DAPLEX [Shi81].

9

Implementing Relationships

In Chapter 2, relationships and relationship types were introduced as one of the main modeling concepts of the entity-relationship model—besides entities and attributes. While the previous chapters can be seen as a directive to implement entities within an object-oriented data model, the implementation of relationships and relationship types is still missing. Filling this gap is the goal of this chapter.

The set of all relationship types can be categorized along several dimensions. The first dimension is their corresponding arity. Here, we distinguish binary and n-ary relationships for $n > 2$. The former class is the most important and can be further divided according to the functionality that is obeyed by the relationship; i.e., $1 : 1$, $1 : N$, and $N : M$ relationships have been distinguished.

Another dimension that is important for the implementation of relationships and relationship types in an object-oriented data model is the number of attributes a relationship carries. More specifically, we are interested in the existence of attributes.

This chapter is organized along the above dimensions where the implementation of relationships is first discussed in general and subsequently demonstrated by concrete examples. Again, we will blur the distinction between relationship type and relationship.

9.1 Binary Relationships without Attributes

As is often the case, the implementation of a modeling concept depends on the requirements posed upon it. That is, the representation should take the foreseen usage in application programs into account. For implementing binary relationships this means that the object type representation should depend on the most frequent traversal direction. Consider the general binary relationship R_{bin} shown in Figure 9.1. The two object types T_{left} and T_{right} are related by the binary relationship R_{bin}—which

Figure 9.1: A Generic Binary Relationship Type

does not have any attributes on its own. The reference direction can be from T_{left} to T_{right} or from T_{right} to T_{left}—depending on the so-called *entry point*. A third possibility is a direct entrance through the relationship. The different entry points can be characterized by queries that are verbally phrased as follows:

1. For a given T_{left} object: Which T_{right} objects are related to it through R_{bin}? In this case, the entry point of the query is T_{left}.

2. For a given T_{right} object: Which T_{left} objects are related to it through R_{bin}? Here, the entry point is T_{right}.

3. Which T_{left} and T_{right} objects are related through R_{bin}? In this query the entry point is R_{bin}.

Let us now describe the alternative implementations of binary relationship types without any attribute.

9.1.1 $1:1$ Relationship Types

First, we consider a simple abstract $1:1$ relationship type in more detail, i.e.:

There exist three (meaningful) alternatives for representing this abstract relationship type. They are shown below:

type T_{left} **is**	**type** T_{right} **is**	**type** T_R
body	**body**	**with extension is**
[...	[...	**body**
R: T_{right};	R^{-1}: T_{left};	[left: T_{left};
...]	...]	right: T_{right};]
end type T_{left};	**end type** T_{right};	**end type** T_R;

On the instance level the different representations of the general $1:1$ relationship can be visualized as follows:

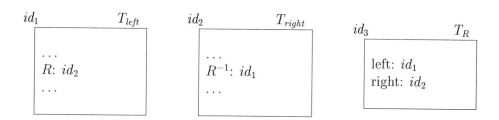

Either one of the three representations is sufficient to implement the relationship R. If the entry point of most applications is T_{left}, then the left-most representation is preferable because it allows to navigate from an object of type T_{left} directly to the associated object of type T_{right} via the attribute R. Thus, if o_{left} is a variable referring to an object of type T_{left}, then the associated T_{right} instance is retrieved simply by accessing $o_{left}.R$.

On the other hand, if the applications demand T_{right} to be an entry point, the representation shown in the middle is superior. In this representation, we can directly access the object of type T_{left} with which a given object of type T_{right} is associated. The attribute R^{-1} facilitates this access. If, for example, the variable o_{right} refers to an instance of type T_{right}, then $o_{right}.R^{-1}$ refers to the associated T_{left} instance.

The representation shown on the right-hand side separates the relationship from the entity (type) definitions. Here, the relationship is represented as a distinguished type, named T_R, with the attributes *left* and *right*, which are constrained to T_{left} and T_{right}, respectively.

The clause "**with extension**" specifies that the instances of T_R are to be collected in an implicitly maintained set which can be accessed as $ext(T_R)$—recall Section 7.8.2. This separate representation of the relationship type R is useful when application programs require the entry point R.

The following application scenario should illustrate the use of the entry point R. Assume we want to print the attribute pairs (*someAttr*, *someOtherAttr*) of object pairs (o_{left}, o_{right}) of type T_{left} and T_{right}, respectively, under the condition that the two objects are related via the relationship R. Then the following program fragment will do the job if we implement the relationship R as a separate type T_R:

```
foreach (r in ext(T_R))
    print(r.left.someAttr, r.right.someOtherAttr);
```

Of course, this task could also have been implemented under the representation shown on the left-hand side or in the middle—that is, with entry points T_{left} and T_{right}. Consider the representation of R in the type definition T_{left} (i.e., the representation with entry point T_{left} shown on the left-hand side):

```
    . . .
    foreach (o_left in ext(T_left))
        if (o_left.R ≠ NULL)
            print(o_left.someAttr, o_left.R.someOtherAttr);
    . . .
```

Note that this program requires that an extension of the type T_{left} exists—which was actually not specified in the above type definitions.

If only a very small fraction of the objects of type T_{left} participate in the relationship R then this last program is rather inefficient since it has to access *all* objects of type T_{left}—in order to inspect the value of the attribute R.

In relational modeling, one of the principal goals in implementation design is that redundancy should be avoided. In principle, this remains true for the object-oriented model as well. However, in object-oriented modeling, we can better control the schema redundancy by application-specific operations. This makes it possible to include *controlled redundancy* in the schema in order to increase the performance of the application programs. *Controlled redundancy* requires that the application-specific operations have to ensure the consistency of the object base, i.e., rule out the occurrence of update anomalies that manifest only in a part of the redundant object base state. Information hiding and encapsulation are very beneficial in accomplishing this task.

It may sometimes be worthwhile to consider a combination of the three alternative representations of the binary 1 : 1 relationship. Let us, for example, consider implementing R in the object type T_{left} (via the attribute R) as well as in the object type T_{right} (by the attribute R^{-1}). This has the advantage that two entry points for the relationship are provided: T_{left} and T_{right}. However, the cost is that this type definition contains redundancy, which may lead to update anomalies—if not properly controlled. In order to control the redundancy we could, for example, refine the value receiving operation $R\leftarrow$ associated with the type T_{left}:

```
    refine R: ← T_right code setR;

    define setR(o_right) is
        begin
            o_right.R^{-1} := self;
            self.R := o_right;
        end define setR;
```

In this implementation, we make sure that the attribute R^{-1} in the object o_{right} is updated in the appropriate way when the attribute R of a T_{left} instance is modified to refer to o_{right}. It should be noted that we cannot, at the same time, modify the value-receiving operation $R^{-1}\leftarrow$ of type T_{right} in the analogous way, because this would lead to an infinite chain of recursive operation invocations.[1]

   This is one reason, why these 1 : 1 binary relations, which constitute one case of *inverse relationships*, are supported explicitly by many systems. Inverse relationships will be discussed in Section 9.5. Nevertheless, also in the general case this problem can be avoided. Special insert and delete operations that take care of the controlled redundancy have to be defined, and encapsulation must prohibit the use of the value-receiving operations manipulating attribute values concerned with the implementation of a relationship.

   Redundancy is not only useful for providing an entry point for both types involved in a binary relationship. In some cases, it makes perfect sense to have an explicit type for the relationship and additional attributes in either T_{left} or T_{right} or in both. In these cases, the attributes do not necessarily point to the elements they are related to. Instead, they point to a tuple of the relationship type T_R. The modified type definitions then look as follows:

type T_{left} is	**type T_R**	**type T_{right} is**
body	**with extension is**	**body**
[...	**body**	[...
R: T_R;	[left: T_{left};	R^{-1}: T_R;
...]	right: T_{right};]	...]
end type T_{left};	**end type T_R;**	**end type T_{right};**

   Obviously, special update operations for manipulating the relationship should be provided and encapsulated. This implementation is useful especially where the relationship carries attributes (see Section 9.2 and the exercises).

9.1.2   1 : N Relationship Types

Let us now consider the case of a general binary one-to-many (1 : N) relationship type, e.g.:

This class of binary relationships can, again, be implemented in three alternative ways, which are very similar to the representation of the general 1 : 1 relationship types:

---

[1]This thought also prohibits the refinement altogether if T_{left} and T_{right} are actually the same type. Think of the attribute *spouse* modeling the symmetric relationship *marriage*.

type T_{left} **is**	type T_{right} **is**	type T_R
body	**body**	**with extension is**
$[\ldots$	$[\ldots$	**body**
$R: \{T_{right}\};$	$R^{-1}: T_{left};$	$[\text{left}: T_{left};$
$\ldots]$	$\ldots]$	$\text{right}: T_{right};]$
end type $T_{left};$	**end type** $T_{right};$	**end type** $T_R;$

In the above three alternative representations the one shown in the middle is the only one enforcing the $1 : N$ functionality of the relationship R on the structural level. It ensures that an instance of type T_{right} is related with at most one object of type T_{left}, since the attribute R^{-1} is single-valued. On the other hand, implementing the relationship via the set-valued attribute R in type T_{left} does not preclude the possibility of inserting the same object o_{right} of type T_{right} into the sets associated with two (or more) different objects of type T_{left}—which violates the consistency constraint imposed by the $1 : N$ functionality in the conceptual ER schema. Therefore, the enforcement of this consistency constraint would have to be realized at the behavioral level by refining the corresponding *insert* operation associated with the attribute R of the type T_{left}. The analogous holds for the separate representation of the relationship in a distinguished type T_R with the associated type extension.

9.1.3 $N : M$ Relationship Types

The most general binary relationship type obeys the functionality many-to-many ($N : M$), e.g.:

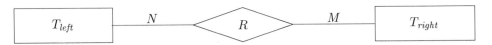

This most general class of binary relationships can, once again, be implemented in three alternative ways:

type T_{left} **is**	type T_{right} **is**	type T_R
body	**body**	**with extension is**
$[\ldots$	$[\ldots$	**body**
$R: \{T_{right}\};$	$R^{-1}: \{T_{left}\};$	$[\text{left}: T_{left};$
$\ldots]$	$\ldots]$	$\text{right}: T_{right};]$
end type $T_{left};$	**end type** $T_{right};$	**end type** $T_R;$

In implementing the binary $N : M$ relationship type, we can either include a set-valued attribute in the types T_{left} or T_{right} or, again, implement it as a separate type T_R. The best representation depends on the access behavior of the applications. That is, we should select the representation that provides the entry point for the most frequent and/or the most time-critical applications. Again, it is possible to include controlled redundancy by representing the relationship in more than one way and, thereby, providing more than one entry point.

9.2   Binary Relationships with Attributes

In Figure 9.2, a general binary relationship type with attributes is depicted. Here, the relationship type has k attributes $a_1, \ldots, a_k$, which are of type $T_1^R, \ldots, T_k^R$, respectively.

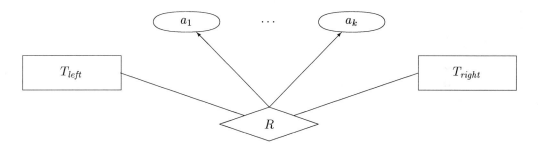

Figure 9.2: A General Binary Relationship Type with Attributes

The main question when implementing a relationship with attributes is where to place the attributes. There are three obvious possibilities: Store the attributes with T_{left}, T_{right}, or T_R, the new type representing the relationship. Storing the attributes of the relationship with one of the two entity types bears the problem that attributes of the relationship mash with the attributes of the entities. This might be okay for a single relationship, but the more relationships with attributes the type participates in, the worse it becomes.

It should also be observed that the inclusion of attributes of the relationship in the type of the participating entity types—i.e., T_{left} and T_{right}—works only in the case of $1:1$ and $1:N$ relationships. In the general case of an $N:M$ relationship, we cannot predict how many objects are related with a given instance of either T_{left} or T_{right}. Thus, we would have to include a varying number of relationship attributes—which may be possible by set- or list-valued attributes, but it is too cumbersome.

Therefore, we advise to implement a relationship R with attributes as a separate type T_R. This new type then has $2 + k$ attributes, if it is (1) a binary relationship and (2) has k attributes. Thus, the type definition for the generic relationship of Figure 9.2 is

type T_R
   with extension is
   body
      [left: T_{left};
      right: T_{right};
      $a_1 : T_1^R$;
      $\ldots$
      $a_k : T_k^R$;]
   operations
      declare T_R: T_{left}, T_{right}, T_1^R, $\ldots$, $T_k^R \rightarrow T_R$ **code** initT_R;
     $\ldots$
   implementation
      define init$T_R(o_{left}, o_{right}, o_1, \ldots, o_k)$ **is**
         begin
            self.left := o_{left};
            self.right := o_{right};
            self.a_1 := o_1; $\ldots$; **self**.a_k := o_k;
         end define initT_R;
     $\ldots$
   end type T_R;

In this type definition, we have included an initializer operation (cf. Section 8.5), which is a refinement of the built-in *create* operation. This initializer is merely used to set the attributes to the corresponding values passed as parameters to the *create* operation.

Note that implementing the relationship explicitly always necessitates searching the whole extension. Controlled redundancy, especially additional references to the tuples implementing the relationship within the two types involved as already discussed in Subsection 9.1.1, is of great value. Refer also to the exercises on this point.

9.3   *n*-ary Relationships

Figure 9.3 represents a general *n*-ary relationship type with associated attributes. Whenever encountering an *n*-ary relationship with or without attributes, it should be implemented by a separate object type. Following the previous sections, it is easy to comprehend the type definition for T_R:

type T_R
   with extension is
   body
      [entity$_1$: T_1;

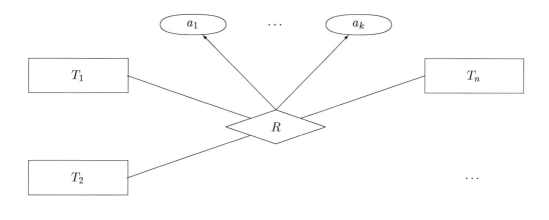

Figure 9.3: General *n*-ary Relationship

$$\ldots$$
$$\text{entity}_n:\ T_n;$$
$$a_1:\ T_1^R;$$
$$\ldots$$
$$a_k:\ T_k^R;]$$
operations
   declare T_R: $T_1, \ldots, T_n, T_1^R, \ldots, T_k^R \ \to\ T_R;$
   $\ldots$
implementation
   $\ldots$
end type $T_R;$

Again, we have included an initializer so that we can pass all relevant attribute values as parameters to the *create* operation.

9.4   Examples from Our *Company* Database

We will now illustrate the abstract discussion of relationship representations on some examples taken from our *Company* database schema introduced in Chapters 2 and 4.

9.4.1   1 : 1 Relationship

Since our original *Company* database does not include any 1 : 1 relationship, let us augment the schema by the relationship *residesIn* among *Engineers* and *Offices*. This relationship is one to one under the assumption that each *Office* is occupied by at most one *Engineer*. The relationship is graphically depicted as follows:

This relationship type can easily be implemented by an attribute *residesIn* that is constrained to the type *Office*, i.e.:

```
type Engineer is
   body
      [...
      residesIn: Office;
      ...]
end type Engineer;
```

Alternatively, we could have included an attribute *isOccupiedBy* in the structural representation of the type *Office*:

```
type Office is
   body
      [...
      isOccupiedBy: Engineer;
      ...]
end type Office;
```

The entry points determine which one of the two alternative implementations is preferable. If one accesses the *Offices* mostly by first accessing an *Engineer* instance and then finding out in what *Office* he/she resides, the first representation is surely superior. On the other hand, if the application frequently demands to find out who occupies a particular *Office*, the second representation is preferable.

It is, of course, also possible that both types include the respective attribute for implementing the relationship. However, this requires that the database user takes care of keeping the object base in a consistent state. Consider, for example the case that an *Engineer*, say *leonardo*, moves to a different *Office*, say to *thePenthouse*:

(1) **if** (thePenthouse.isOccupiedBy $\neq$ NULL)
         thePenthouse.isOccupiedBy.residesIn := NULL;
(2) **if** (leonardo.residesIn $\neq$ NULL)
         leonardo.residesIn.isOccupiedBy := NULL;
(3) leonardo.residesIn := thePenthouse;
(4) thePenthouse.isOccupiedBy := leonardo;

Statement (1) is used to "kick out" the previous tenant of *thePenthouse*—if it is still occupied. Then, in statement (2) *leonardo* "moves out" of his previous office—if he still resided in one. Finally, in statements (3) and (4) leonardo "moves into" *the Penthouse*.

It would, of course, be much more convenient to specify an operation *moveToOffice* that takes care of this tedious and error-prone procedure. Thereby, the redundancy inherent in the type definitions could be thoroughly controlled.

```
declare moveToOffice: Engineer ‖ Office → void;
define moveToOffice(newResidence) is
   begin
     if (newResidence.isOccupiedBy ≠ NULL)
        newResidence.isOccupiedBy.residesIn := NULL;
     if (self.residesIn ≠ NULL)
        self.residesIn.isOccupiedBy := NULL;
     self.residesIn := newResidence;
     newResidence.isOccupiedBy := self;
   end define moveToOffice;
```

The statements in this operation definition correspond to the above program fragment, where *leonardo* moved to *thePenthouse*. This move could now easily be achieved as

leonardo.moveToOffice(thePenthouse);

without knowing too much about the object base schema and the redundancy inherent in the schema.

9.4.2   1 : N Relationships

Recall the binary 1 : N relationship *worksFor* defined between the entity types *Engineers* and *Divisions*, i.e.:

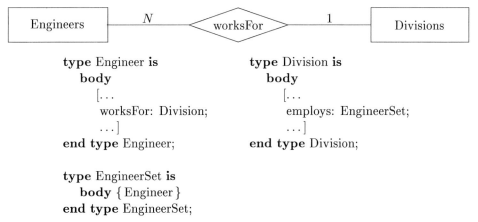

```
type Engineer is                      type Division is
   body                                  body
      [...                                  [...
      worksFor: Division;                  employs: EngineerSet;
      ...]                                 ...]
   end type Engineer;                    end type Division;

type EngineerSet is
   body { Engineer }
end type EngineerSet;
```

Again, it is possible to maintain both attributes, *worksFor* and *employs*, in the types *Engineer* and *Division*, respectively. Then, however, it is again the user's responsibility to keep the object base in a consistent state. The best way to achieve this is, of course, to provide corresponding operations to maintain the relationship. In our case, the two operations *hire* and *fire* could control the redundancy of the type schema:

> **declare** hire: Division ∥ Engineer → **void**;
>
> **declare** fire: Division ∥ Engineer → **void**;
>
> **define** hire(newEng) **is**
>   **begin**
>   **self**.employs.insert(newEng);
>   newEng.worksFor := **self**;
>   **end define** hire;
>
> **define** fire(badEng) **is**
>   **begin**
>   **self**.employs.remove(badEng);
>   badEng.worksFor := NULL;
>   **end define** fire;

The user can then apply the operations *hire* and *fire* and need not be concerned about maintaining the consistency of the redundant relationship representation—induced by the two semantically related attributes *employs* and *worksFor*..

9.4.3 Recursive $1 : N$ Relationship

Let us review the relationship *composition*, which relates *Products* as sub- and super-parts. The relationship is graphically depicted as follows:

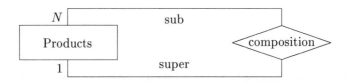

This relationship models a so-called parts explosion. This recursive $1 : N$ relationship type can be naturally represented in two variations:

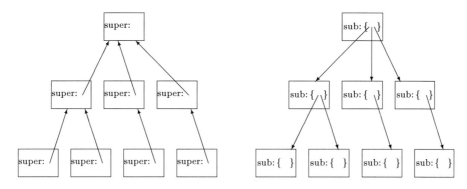

Figure 9.4: Alternative Representation of *composition*

type Product **is**
  body
    [...
      super: Product;
    ...]
  end type Product;

type Product **is**
  body
    [...
      sub: ProductSet;
    ...]
  end type Product;

type ProductSet **is**
  body { Product }
  end type ProductSet;

The left-hand representation of the type *Product* with the associated attribute *super* is useful when the application frequently requires to navigate from a *Product* to the super-*Products* in which it is used as a subpart.

The implementation on the right-hand side is useful to determine the constituent parts of a given *Product*. It appears that this kind of access is more frequent than the other.

In Figure 9.4 the two representations of the relationship *composition* are visualized at the object level. Each box represents a *Product* instance; the arrows represent references. On the left-hand side, we depict a sample object base state in which the type *Product* contains the attribute *super*. On the right-hand side, the same object base state is shown for the case that the type *Product* contains the set-valued attribute *sub*. Actually, this graphical representation blurs the fact that *sub* refers to a set object from which the references to the sub-*Products* emanate—for simplicity of the graphical representation the set objects are nested within the *Product* objects.[2]

---

[2]This corresponds to the representation if we had used set-structured sorts instead of set objects. Recall the intrinsic differences.

Let us now consider the implications of the two different representations on some
typical application programs. Assume that we need the following two operations:

1. *partList*, which returns the set of *Product* instances of which the *Product*, on
   which *partList* is invoked, is composed

2. *isPartOf*, a Boolean function that determines whether the *Product*, on which
   the operation is invoked, is a subpart of the *"higher-level" Product* passed as a
   parameter

Let us first implement these two operations under the assumption that the references are directed from sub- to super-*Product*, i.e., the scenario shown on the left-hand
side of Figure 9.4:

```
type Product
   with extension is
   body
      [...
       super: Product;
       ...]
   operations
      declare partList: → ProductSet;
      declare isPartOf: Product → bool;
      ...
   implementation
      define partList is
         var resultSet: ProductSet;
         begin
            resultSet.create;     !! create an empty set
            resultSet.insert(self);
            foreach (part in ext(Product))    !! iterate over the type extension
               if (part.super = self)
                  resultSet.setUnion(part.partList);     !! recursive call of partList
            return resultSet;
         end define partList;

      define isPartOf(theSuperPart) is     !! is self a subpart of theSuperPart?
         var part: Product
         begin
            part := self;
            while (part ≠ NULL)
               if (part = theSuperPart)
                  return true;
               else
```

```
                part := part.super;
            return false;
        end define isPartOf;

    . . .
  end type Product;
```

It turns out that this realization of *partList* is very inefficient because it involves iterations over the extension of the type *Product*. On the other hand, the implementation of *isPartOf* is very efficient, since we can easily navigate from the given *Product*—referred to by **self**—toward the root of the part hierarchy by following the attribute *super*, i.e., following the upward arrows on the left-hand side of Figure 9.4.

Let us now consider the case in which the relationship *composition* is implemented by including the set-valued attribute *sub* in Product:

```
type Product is
  body
    [. . .
    sub: ProductSet;
    . . .]
  operations
    declare partList: → ProductSet;
    declare isPartOf: Product → bool;

    . . .
  implementation
    define partList is
      var resultSet: ProductSet;
      begin
        resultSet.create;
        resultSet.insert(self);
        foreach (part in self.sub)
          resultSet.setUnion(part.partList);    !! recursive call of partList
        return resultSet;
      end define partList;

    define isPartOf(theSuperPart) is
      var isUsed: bool := false;
      begin
        foreach (part in theSuperPart.partList)
          if (part = self)
            isUsed := true;    !! we could already exit the loop here
        return isUsed;
      end define isPartOf;

    . . .
  end type Product;
```

Now, the realization of *partList* is very efficient, since we exploit the references leading from a *Product* to its sub-*Products*; i.e., we traverse the downward-directed arrows on the right-hand side of Figure 9.4. However, the realization of *isPartOf* suffers in comparison to the former realization: It now involves the computation of the *partList* of the *Product* passed as the parameter—called *theSuperPart*—and then checking whether **self** is contained in this set. Actually, this implementation of *isPartOf* could still be made somewhat more efficient (see Exercise 9.2).

In summary, the discussion indicates that this may be a case in which it is beneficial to maintain (controlled) redundancy in the schema by incorporating the attribute *sub* as well as *super* in the type *Product*, i.e.:

```
type Product is
   body
      [...
       super: Product;
       sub: ProductSet;
       ...]
   ...
end type Product;
```

Now we can realize the two operations *partList* and *isPartOf* in their most efficient way by exploiting the references from the super- to the sub-*Product* **and** from the sub- to the super-*Product*, respectively. Hence, this is a case for explicitly maintained inverse relationships as will be introduced in Section 9.5. Another possibility to express the semantics of a part-of hierarchy even better are *composite objects* as will be introduced in Section 9.6.

9.4.4 $N : M$ **Relationship**

This relationship type is exemplified by the relationship *designs* between *Engineers* and *Products*. Typically, the design of a *Product* is done in a team of *Engineers*, and one *Engineer* is involved in the design of several *Products*—which accounts for the $N : M$ functionality. The relationship *designs* is depicted as follows:

The representation in the generic object model (GOM) can be done in several ways. Let us first consider the implementation via attributes included in the types *Engineer* and *Product*, respectively.

```
type Engineer is                    type Product is
   body                                body
      [...                                [...
         designs: ProductSet;               designedBy: EngineerSet;
      ...]                                ...]
   end type Engineer;                  end type Product;


   type EngineerSet is                 type ProductSet is
      body { Engineer }                   body { Product }
   end type EngineerSet;               end type ProductSet;
```

In this example, it should also be considered to implement the relationship as a separate object type. In that case, we would define the object type *Designs* as follows:

```
type Designs
   with extension is
   body
      [theParticipatingEngineer: Engineer;
      theProduct: Product;]
end type Designs
```

An application in which the entry point *Designs* is exploited is as follows: Determine all pairs of *Engineers* who have participated at least once in the same design team.

9.4.5   Ternary Relationship

In our *Company* database, we encountered the ternary relationship *assembles*, which looks as follows:

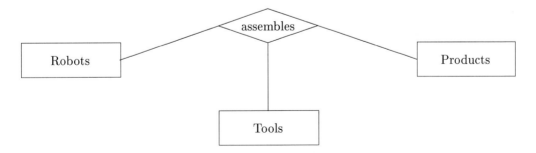

In this case, the "cleanest" representation is to define a separate type implementing the relationship. The definition of the type *Assembly* has the following form:

```
type Assembly is
   body
      [theBuilder: Robot;
       theToolUsed: Tool;
       theManufacturedProduct: Product;]
end type Assembly;
```

9.5   Implementing Inverse Relationships

9.5.1   Single-Valued Attributes Implementing 1 : 1 Relationships

Let us start our description of *inverse* relationships on a very easy yet illustrative example. Assume our object base includes types for *Man* and *Woman*. Then a relationship *Marriage* can be implemented by including an attribute *wife* in *Man* and, analogously, an attribute *husband* in *Woman*. For law-abiding people *Marriage* is a symmetrical 1 : 1 relationship because a *Man* can be married to at most one *Woman*, and vice versa. The 1 : 1 restriction of the relationship is easily ensured in the following schema:

```
type Man is                          type Woman is
   body                                  body
      [                                     [
            . . .                                 . . .
         wife: Woman;                          husband: Man;
            . . .                                 . . .
      ];                                    ];
end type Man;                         end type Woman;
```

This schema takes care of the constraint that every person is married to at most one person of the opposite gender. It does not, however, constrain the relationship *Marriage* to be symmetrical—also required by law. For example, the following assignments would (and could) not be rejected by the system:

```
   . . .
mickeyMouse.wife := miniMouse;
miniMouse.husband := donaldDuck;
```

We assume that *mickeyMouse* and *donaldDuck* are (appropriately declared) variables referring to instances of type *Man*; *miniMouse* is a variable of type *Woman*. The resulting state of the database violates the condition that *husband* and *wife* are really *inverse* to each other. That is, for all *Man* instances m and *Woman* instances w the following must hold:

$$w.husband = m \quad \text{if and only if} \quad m.wife = w$$

In general, it is rather difficult to keep a database in a consistent state when symmetrical relationships, such as *Marriage*, are implemented by two attributes that are inverse of each other. Each time one of the inverse attributes is modified the corresponding update on the other attribute has to be carried out by the database user. Here we want to show another solution: In preceding sections we showed that this can be taken care of by customized modification operations and encapsulation of direct attribute manipulation. Since this is not only tedious but also exhibits the disadvantage that the database system cannot make use of this special knowledge for optimization purposes, a special mechanism, called *inverse attributes*, is provided and delegates the consistency control of inverse properties to the system. We could implement our example as follows:

> **type** Man **is**
>   ...
>   wife: Woman **inverse** Woman$husband;
>   ...
> **end type** Man;
>
> **type** Woman **is**
>   ...
>   husband: Man **inverse** Man$wife;
>   ...
> **end type** Woman;

Let us now reconsider the effect of the two assignments stated before:

>   ...
> mickeyMouse.wife := miniMouse;
> miniMouse.husband := donaldDuck;

The first assignment sets the attribute *wife* of *mickeyMouse* to refer to *miniMouse*. At the same time, the specification of *husband* and *wife* as being **inverse** of each other causes the system to automatically set the attribute *husband* of *miniMouse* to refer to *mickeyMouse*. The second assignment first cancels this effect: *miniMouse.husband* as well as *mickeyMouse.wife* are reset to *NULL*. Subsequently, the settings of the *husband* and *wife* attributes for *miniMouse* and *donaldDuck*, respectively, are carried out in analogy to the description of the first assignment.

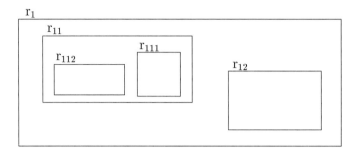

Figure 9.5: Containment of Rectangles

9.5.2   Multivalued Inverse Attributes

A multivalued attribute consists of a (possibly empty) set of references to objects of the specified type. As for single-valued attributes, the **inverse** concept is also useful in implementing inverse multivalued attributes that occur when redundantly representing $1:N$ or $N:M$ relationships.

We illustrate this concept on the example *Rectangle*. Suppose the type *Rectangle* is used to design a *window system* for a computer user interface. In this context we might be interested in recording all the *Rectangles* that are fully contained in a given *Rectangle*. Such a scenario is sketched in Figure 9.5.

In this figure the Rectangle r_1 *directly* and fully contains r_{11} and r_{12}. Let us assume that in this context the *Contains* relation always means direct containment; that is, a *Rectangle* such as r_1 contains another *Rectangle* such as r_{11} if and only if it is the smallest *Rectangle* to encompass r_{11} and r_{11} is smaller than r_1. In this sense r_1 does not (directly) contain r_{111} and r_{112}, because they are contained in the smaller *Rectangle* r_{11} (which itself is contained in r_1). Let us further assume that rectangles do not overlap. Then, the relationship *contains*, which thus relates a *Rectangle* to a set of other *Rectangle* instances, can be implemented as a set-valued attribute with element type *Rectangle*:

```
type Rectangle is
  body
    [
       height, length: float;
       contains: {Rectangle};
    ];
    ...
  end type Rectangle;
```

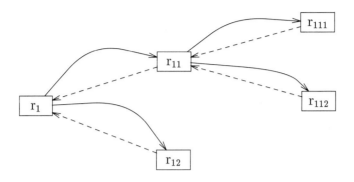

Figure 9.6: Database State of the Contained Rectangles Example

Let us now return to our *contains* example in the type definition of *Rectangle*. The *contains* attribute implements a relationship that has a natural inverse, which could be called *containedIn*. Thus,

$$\text{containedIn}(r') = \{r | r' \in \text{contains}(r)\}$$

Note that there is at most one *Rectangle r*, which (directly!) contains r'. Therefore, we can respecify the type *Rectangle* as follows:

```
type Rectangle is
  body
    [
        height, length: float;
        contains: {Rectangle}
            inverse Rectangle$containedIn;
        containedIn: Rectangle inverse Rectangle$contains;
    ];
    ...
  end type Rectangle;
```

A possible database state for our example rectangles of Figure 9.5 is shown in Figure 9.6.

Solid arrows represent the references emanating from the set-valued attribute *contains*; dashed arrows stand for references emanating from the attribute *containedIn*.

9.6   Composite Object Support

So far, our model supports only one (generic) kind of reference—relating an object o_1 to another object o_2 by storing o_1's object identifier (OID) in the object o_2.[3] In this respect, the two objects o_1 and o_2 are fully autonomous entities at the same semantic level. In particular, either one of the two objects may have any number of other outgoing and incoming references—there are no semantic restrictions imposed by the reference from o_2 to o_1. Therefore, we call this a *weak reference*.

It turns out that in many applications, especially in technical areas, it is advantageous to associate more semantics with some of the references. In particular, more semantics for supporting *composite objects* made up of several objects connected by a hierarchical *part-of* relationship is desirable. For this purpose, we distinguish between the weak (generic) references and the *composite object references*. A composite object reference relates the (super)object with its part object, i.e., the composite object with its constituent (part object).

9.6.1   Classification of Composite Object References

It is even meaningful to distinguish between several different composite object reference kinds according to the various semantics they carry. Let us now introduce the two dimensions along which composite object references can be classified:

- *Dependent/independent*
  This dimension differentiates between those part objects that can only exist, if their composite object—of which they are a constituent part—is "alive" and those whose existence is independent of the existence of their composite object.

- *Exclusive/shared*
  This category distinguishes part objects that can be associated with only *one* composite object, i.e., an exclusive association, and those that can be a constituent part of several composite objects.

Since the two dimensions can be freely combined, we obtain four kinds of composite object references, as summarized in the following table:

| Classification of Composite Object References ||
Dependent/Independent	Exclusive/Shared
dependent	exclusive
	shared
independent	exclusive
	shared

---

[3]This can be in the form of an attribute value—if o_2 is tuple structured—or as an element of a set or list—if o_2 is set or list structured, respectively.

Let us discuss these four reference types in more detail—and illustrate the discussion by examples—in the next four subsections.

9.6.2 Dependent Exclusive Composite Object Reference

This is the most stringent kind of composite object association. It ensures that the part object can be a constituent part of only one composite object. This, however, does not preclude that an arbitrary number of weak (generic) references refer to the part object. Furthermore, the part object depends on the existence of the composite object; that is, if the composite object is deleted the deletion of the part object is implicitly initiated as well.

An example of a dependent, exclusive part object is the following:

```
type Computer is
    body
      [cpu: dependent exclusive Processor;
       . . .]
    end type Computer;
```

In this example, we assume that the *cpu* (central processing unit) is never transferred from one *Computer* to another. Furthermore, *cpus* are not shared and are deleted when the *Computer*, to which they belong, is deleted.[4] Under these assumptions it makes sense to establish the *dependent exclusive* association between *Computer* and *Processor* via the attribute *cpu*.

9.6.3 Independent Exclusive Composite Object Reference

Under this composite object reference type, the part object is still exclusively associated to only a single composite object. However, it can autonomously exist in the object base even if the *one* composite object, of which it is a part, is deleted. An example is given as follows:

```
type Automobile is
    body
      [engine: independent exclusive Motor;
       . . .]
    end type Automobile;
```

The *engine* of an *Automobile* is, of course, exclusively built into a single car. However, the *engine* may be removed from the car and, possibly, be reused in a different *Automobile*—consider junk yards, which salvage spare parts.

---

[4]This example was chosen in analogy to a human's brain, which—as of the current state of medical technique—cannot be transplanted. Therefore, it is certainly an exclusive, dependent part of the human being.

9.6.4   Dependent Shared Composite Object Reference

A shared composite object reference facilitates the association of a part object to several different composite objects. This is particularly useful in modeling so-called *logical part-of* relationships—as opposed to a *physical part-of* relationship as in the automobile example above. As an example of a *logical part-of* association consider a *Chapter* which may be enclosed in several different *Documents*; e.g., in a particular *Book* and also in some related *UserGuide*. Thus, the *Chapter* should be modeled as a shared part object. Furthermore, the existence of the particular *Chapter* depends on the existence of at least one *Document*, of which it is a *part-of*. That is, as soon as the last *Document* including the particular *Chapter* ceases to exist, the *Chapter* should be deleted as well. This is implemented as follows:

```
type Document is
   body
      [chapter: dependent shared Chapter;
       ...]
   end type Document;
```

9.6.5   Independent Shared Composite Object Reference

The *Chapter* in the preceding example was a *dependent* shared part object. There are also *independent* shared part objects whose existence does not depend on their association with a composite object. An example is the *MotorDesign*, which is a *part-of* a *CarDesign*:

```
type CarDesign is
   body
      [engineDesign: independent shared MotorDesign;
       ...]
   end type CarDesign;
```

Since the *MotorDesign* is specified as a *shared* part object, it is possible that the same design of a motor is used in different car designs—which is common practice in the automobile industry.

It may become necessary—for reasons of a decreasing automobile market—to stop particular *CarDesign* efforts and even delete the design objects. In this case, it may still be worthwhile to save the (possibly finished) *MotorDesign*—even though no *CarDesign* is currently employing this particular *MotorDesign*. This is achieved by the shared independent association of *CarDesign* and *MotorDesign*.

9.7   Exercises

9.1   Reconsider the representation proposed for implementing binary $1 : N$ relationship types as a separate type, i.e.:

> **type** T_R
>   **with extension is**
>   **body**
>   [left: T_{left};
>   right: T_{right};]
>   **end type** T_R;

It is also possible to implement the relationship with a set-valued attribute *right* as follows:

> **type** T_R
>   **with extension is**
>   **body**
>   [left: T_{left};
>   right: $\{ T_{right} \}$;]
>   **end type** T_R;

- Investigate the pros and cons of these representations in comparison with the one proposed in the text.

- Outline the operations that are needed to enforce the consistency constraints imposed by the $1 : N$ functionality of the relationship type for both representation alternatives.

- Sketch sample applications with entry point R. Compare the two alternative representations with respect to the ease of coding these applications and the efficiency of the operations.

9.2   Reconsider the part explosion example of Section 9.4.3. Implement a more efficient version of *isPartOf* for the representation of the relationship *composition* based on the attribute *sub* in *Product*.

9.3   The relationship *composition* has to obey the functionality $1 : N$. Specify the operations that are needed to enforce this functionality. Also, devise the operations that control the redundancy inherent in the schema for the case that the relationship is represented with both attributes, *sub* and *super*, in the type *Product*.

9.4   Find some examples of n-ary relationship types with and without attributes—
for $n = 3, 4, \ldots$. Sketch the object representation of these relationships.

9.5   Compare the following two representations of a general $N : M$ relationships
type R between the entity types T_{left} and T_{right}:

> (1)   **type** T_R and   (2)   **sort** T_R **is**
>   **with extension is** $[\text{left}{:}T_{left};$
>   **body** $\text{right}{:} \ T_{right};]$
> $[\text{left}{:} \ T_{left};$
> $\text{right}{:} \ T_{right};]$ **var** $R : \{T_R\};$
>   **end type** $T_R;$

What is the major distinction between the object type representation (1) and
the value based representation (2)?

Hint: Consider the generation of duplicate entries; i.e., reestablishing a new
relationship among (o_1, o_2) that exists already.

9.6   [KBG89] Define the following sets for a (part) object o_p:

$$IX(o_p) \ = \ \{o_c | o_c \text{ has an independent exclusive composite reference to } o_p\}$$

$$DX(o_p) \ = \ \{o_c | o_c \text{ has a dependent exclusive composite reference to } o_p\}$$

$$IS(o_p) \ = \ \{o_c | o_c \text{ has an independent shared composite reference to } o_p\}$$

$$DS(o_p) \ = \ \{o_c | o_c \text{ has a dependent shared composite reference to } o_p\}$$

Discuss the following claims for any object o:

- $\text{card}(IX(o)) \leq 1$
- $\text{card}(DX(o)) \leq 1$
- $\text{card}(IX(o)) = 1 \quad \Rightarrow \quad \text{card}(DX(o)) = 0$
- $\text{card}(DX(o)) = 1 \quad \Rightarrow \quad \text{card}(IX(o)) = 0$
- $(\text{card}(IX(o)) = 1 \text{ or } \text{card}(DX(o)) = 1)$
  $\quad \Rightarrow \quad (\text{card}(IS(o)) = 0 \text{ and } \text{card}(DS(o)) = 0)$
- $(\text{card}(IS(o)) > 0 \text{ or } \text{card}(DS(o)) > 0)$
  $\quad \Rightarrow \quad (\text{card}(IX(o) = 0 \text{ and } \text{card}(DX(o)) = 0)$

9.7   Discuss a possible implementation that enforces that exclusive part objects
are associated with a single composite object and that dependent objects are

deleted as soon as the (last) composite object with which they are associated ceases to exist.

Hint: Consider particular counters in the part objects.

9.8   Reimplement the part explosion application, i.e., the *Composition* relationships described in the text using the *inverse* concept.

9.9   Estimate the costs induced by maintaining the consistency of *inverse* attributes for representing application scenarios of your area of expertise.

9.10   At the end of subsection 9.1.1, we provided part of the type definition frame for implementing a binary relationship without attributes redundantly by keeping additional references to the tuple objects representing the binary relationship within the objects involved in the relationship. We provided these type definition frame for the case of binary 1 : 1 relationships without attributes.

1. Give the type definition frames for binary 1 : 1 relationships with attributes.

2. Provide special insert and delete operations for manipulating the relationship.

3. Can you think of useful query operations on a binary relationship? If yes, give their implementation.

4. Perform the above three points for more complex relationships, e.g., an *n*-ary relationship with attributes.

9.11   What relationships do exists between the different tools and the handle of a Swiss Knife. Compare to Exercise 7.11 and Exercise 8.10.

9.12   In Exercises 7.13 and 8.12, we have been concerned with implementing the entity types necessary for a CAAD application. Now implement the relationships. Some of these relationships can be found in Exercise 2.12.

9.13   In Exercises 7.14 and 8.13, we have been concerned with implementing the entity types necessary for an office automation application. Now implement the relationships. Some of these relationships can be found in Exercise 2.13.

9.14   In Exercises 7.15 and 8.14, we have been concerned with implementing the entity types necessary for a software engineering application. Now implement the relationships. Some of these relationships can be found in Exercise 2.14.

9.8   Annotated Bibliography

The literature describing data models for conceptual modeling, in particular the entity-relationship model, was already cited in Chapter 2. Many of the ideas concerning the object-oriented representation of relationship types that were presented in this chapter can be found in a paper by Rumbaugh [Rum87]. The object model Iris—described by Fishman et al. [FBC+87]—incorporates explicit relationship types. This is similar to the separate representation of relationships discussed as one alternative in this chapter. Inverse relationships are automatically maintained by many systems. Among them *SIM* [JGF+88], O_2 [BDK92], *ObjectStore* [LLOW91], *Ontos* [AHS91].

The composite object support described in this chapter is based on the work of the Orion project group. In [KBC+87] a single kind of composite object reference was introduced. This was extended to the four kinds of composite object references in [KBG89].

Early concepts for composite object support can also be traced back to the work on XSQL [LP83, LKM+85], the design of the NF^2 model [SS86], and the work on aggregation by Smith and Smith [SS77b, SS77a]. Also, the molecular objects provide a classification of composite objects [BB84].

10

Inheritance

So far we have described means to implement entities and relationships. Especially, entities or objects with similar characteristics are grouped together and described by a single object type. Similar characteristics here means that all objects have the same structure and the same behavior. If objects differ only slightly in their structure or behavior, they nevertheless have to belong to different object types. Further, so far there is no relationship between object types hosting only slightly different objects. That this kind of relationship is useful—especially for factoring out commonalities between object types—is the subject of this chapter. In this respect inheritance supports the (dual) abstraction concepts generalization and specialization—which were described in Chapter 2.

10.1   Motivation

Similar objects are grouped together by describing their common characteristics in an object type, in which the data representation and operations for all member objects are supplied. Thus, an object type forms a template from which individual objects, called instances of the type, can be created.

The type concept of the generic object model (GOM)—as described so far—allows to model only objects that have identical attributes and operations to represent the individual instances' properties and behavior. The mechanism does not, however, account for representing slight differences, i.e., extensions or variations, of the properties or behavior that apply only for a subset of the instances of a type. For example, we may have a database type *Person*, which models the most basic properties and operations of a person such as *name*, *age*, and *spouse*.

The corresponding type definition frame for the object type *Person* looks as follows:

```
persistent type Person is
  public name, age, spouse, marry
  body [name: string;
        age: int;
        spouse: Person;]
  operations
    declare marry: Person → void;
      ...
  implementation
    define marry(victim) is
      self.spouse := victim;
      ...
end type Person;
```

Note that we consciously left out one-half of the implementation of the *marry* operation. The reasons for this will become clear pretty soon and are further discussed in Exercise 10.1.

Assuming that only some of the stored *Person* objects are at the same time *Employee* instances, which have additional features such as Social Security number (*ss#*), *salary*, and *boss* (an attribute referring to another *Employee*), one would need to include a separate type *Employee*. This type *Employee* contains all the features of *Person* plus the additional properties and operations that apply only to the *Employees*. The corresponding type definition could be given as follows:

```
persistent type Employee is
  public name, age, spouse, marry, ss#, salary, boss, isRetired
  body [name: string;
        age: int;
        spouse: Person;
        ss#: int;
        salary: float;
        boss: Employee;]
  operations
    declare marry: Person → void;
    declare isRetired: → bool;
      ...
  implementation
    define marry(victim) is
      self.spouse := victim;
    define isRetired is
      return (self.age > 64);
      ...
end type Employee;
```

Two severe problems can be identified from these two example object type definitions:

1. *Lacking reusability*

   The object type *Person* contains many concepts—e.g., the attributes *name*, *age*, and *spouse* and the operation *marry*—that had to be redeclared (and reimplemented) in the subsequently defined object type *Employee*. It would have been advantageous if the definition of type *Employee* could have been based on the specification of type *Person*—thus avoiding redefining already existing functionality.

2. *Lacking flexibility*

   There is another, even more severe problem concerning the lacking flexibility induced by introducing the two unrelated object types *Person* and *Employee*. Consider the attribute *spouse* which is constrained to reference objects of type *Person*. This implies that—in our current object base schema—nobody could possibly marry an *Employee*. The following program fragment highlights the problems:

   ```
   var mickeyMouse: Employee;
       miniMouse: Person;
   . . .
   mickeyMouse.marry(miniMouse);   !! okay, miniMouse is a Person
   miniMouse.marry(mickeyMouse);   !! illegal, mickeyMouse is not a Person
   ```

   The problem is caused by the lack of *subtyping*. That is, the *classification hierarchy*, that every *Employee* is also a *Person*—which is observed in the real world—is not accounted for in our object base schema. In our type schema, the two types *Person* and *Employee* are totally unrelated. However, reality requires the type *Employee* to be modeled as a *subtype* of *Person*.

In the following, we will introduce *inheritance* and *subtyping*, which will remedy both problems identified above.

10.2 The General Idea of Inheritance and Subtyping

The general idea behind inheritance is the implementation support of so-called **is-a** relationships between object types. Graphically, **is-a** relationships are represented as shown in Figure 10.1. The figure depicts three alternative representations for OT_{sub}

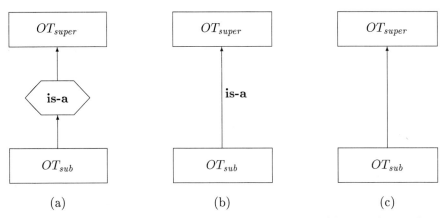

Figure 10.1: Alternative Graphical Representations of **is-a** Relationships

being a subtype of OT_{super}. The representation labeled (a) is the most precise one; however, when the meaning of the arrows is clear from the context it is common usage to choose the simpler notation shown in (b) or (c).

If two object types OT_{super} and OT_{sub} are related by an **is-a** relationship, all the instances of one of the types, OT_{sub}, can also be viewed as belonging to OT_{super}. Hence, the **is-a** relationship implies an explicit *subset* relationship between the extension of the supertype and the extension of the subtype. This is one of the aspects of the **is-a** relationship as often employed in the object-oriented context. There exist two other aspects, typically associated with the **is-a** relationship: *inheritance*, which is soon to come, and *substitutability*, to which a whole subsection is devoted.

Having established an **is-a** relationship between OT_{super} and OT_{sub}, OT_{super} is called the *supertype* of OT_{sub}; and, analogously, OT_{sub} is called the *subtype* of OT_{super}. Consequently, the relationship **is-a** is directed from OT_{sub} to OT_{super}, i.e., "OT_{sub} **is-a** OT_{super}". This viewpoint emphasizes that for any object o for which $o \in \text{ext}(OT_{sub})$ holds also $o \in \text{ext}(OT_{super})$ holds. In this sense, the super-/subtype relationship represents the concept of generalization and specialization: OT_{super} is a generalization of OT_{sub} and OT_{sub} is a specialization of OT_{super}.

In almost all object models—and also in GOM—*inheritance* is associated with the **is-a** relationships between two types. Hence, besides the *subset* relationship between the extensions, *inheritance* is the second aspect of the **is-a** relationship. An instance of the subclass OT_{sub} inherits all features (structural properties and operations) of the supertype OT_{super}—including those that OT_{super} may have inherited from its own ancestors. In addition, an instance o_{sub} of type OT_{sub} is supplied with all the features that the subtype OT_{sub} may have defined beyond those that were inherited from OT_{super}. Thus, the subtype—somewhat contradictory to the name—is an *augmentation* of the supertype. Consequently, inheritance leads to the fact that the subtype

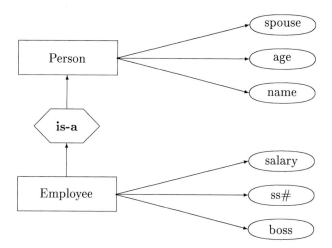

Figure 10.2: Relationship between the Types *Person* and *Employee*

has at least all the components the supertype possesses. The latter fact can also be discussed without inheritance. By accident or consciously, one type might possess all the structural and behavioral features another type exhibits. Then, one also speaks of the *subtype* relationship between these two types.

The concept of inheritance in combination with subtyping solves the problems outlined above. It allows to relate database types that model objects that are similar in one respect but different in another semantic context. This is very illustrative in our *Person/Employee* example. The **is-a** relationship between the types *Employee* and *Person* is graphically represented in Figure 10.2.

In the context of family life, an instance of type *Employee*, say, *mickeyMouse*, has a *name*, an *age*, and—optionally—a *spouse*. When the same database object, *mickeyMouse*, is dealt with in the context of a taxpayer (i.e., an *Employee*), the Social Security number (*ss#*), the *salary*, and the *boss* are of interest. At the same time, some other *Person*, say, *miniMouse*, could be stored in the database as a *"direct"* *Person* instance but not an *Employee* instance. Then this instance still provides the features *name*, *age*, and *spouse*; but not the features that are specific to *Employees*, i.e., *ss#*, *salary*, and *boss*.

Let us demonstrate the implementation of the *Person/Employee* conceptual schema in GOM under the use of inheritance.

```
persistent type Employee supertype Person is
    public ss#, salary, boss, isRetired     !! public clause is also inherited
    body [ss#: int;
          salary: float;
          boss: Employee;]
```

operations
   declare isRetired: → bool;
    . . .
implementation
   define isRetired **is**
     return (self.age > 64);
    . . .
end type Employee;

On the basis of these two types, we can now create the two database instances referred to by the variables *miniMouse* and *mickeyMouse*:

   var mickeyMouse: Employee;
     miniMouse: Person;

The two variables can be initialized by newly created database instances of the appropriate type—we assume that a corresponding initializer is defined to initialize the *name* and the *age* attributes:

   mickeyMouse := Employee\$create("Mickey Mouse", 60);
   miniMouse := Person\$create("Mini Mouse", 50);

So far only the structural issues of inheritance, such as the acquisition of attributes from a parent type by the subtype, have been outlined. But inheritance also has a behavioral component: Operations are inherited as well.

10.3   Substitutability

In addition to the reusability of type properties, we also promised to solve the problem incurred by the lacking flexibility, which was identified at the beginning of this chapter. In this context, we observed that the problem was caused by constraining attributes and formal parameters too rigidly to a particular type—in our case the attribute *spouse* and the formal parameter of *marry* were constrained to *Person* instances. The solution of the problem is to relax the type constraints in such a way that generally a subtype instance may occur everywhere an instance of a supertype is required. This principle is called *substitutability* of subtype instances for supertype instances. Typically, substitutability is associated with the **is-a** relationship and constitutes the third aspect intermingled for deriving the **is-a** relationship and its power.

In terms of our example this means: An *Employee* instance can always be substituted in places where a *Person* is required. Let us revisit our marriage problem:

var mickeyMouse: Employee;
    miniMouse: Person;

. . .

mickeyMouse.marry(miniMouse); !! okay, *miniMouse* is a *Person*
miniMouse.marry(mickeyMouse); !! now okay, *mickeyMouse* is also a *Person*

The first operation invocation remains, of course, valid. Note that we can no longer guarantee that the variable *miniMouse* actually refers to a *Person* object. By substitutability, the variable may very well reference an *Employee* object. However, the operation invocation still remains valid, since the substitutability principle also applies to receiver objects of operation invocations. Furthermore, the second invocation now becomes valid because the operation *marry* requires a *Person* object as an argument. The object referred to by *mickeyMouse* is an *Employee*, which, by substitutability of subtype instances in place of supertype instances, can always be substituted where a *Person* is required. Thus, the second invocation of *marry* is type consistent.

Let us briefly explain the validity of substitutability on an intuitive level—a more formal explanation will be provided in Section 10.8. Our inheritance concept requires that *all* features, i.e., structural representation and operations, of a supertype are inherited by the subtype. Therefore, all operation invocations—in particular attribute accesses and attribute assignments—that are valid on a supertype instance are also applicable on subtype instances. With respect to our example this implies:

- Every *Employee* object obtains *all* attributes of *Person*, i.e., *name*, *age*, and *spouse*. In addition, *Employee* objects have the attributes *ss#*, *salary*, and *boss*.

- Every *Employee* instance "understands" *all* the operations that a *Person* instance can respond to, i.e., *marry* and the attribute access and assignment operations. In addition, *Employee* objects provide further subtype-specific operations, e.g., *isRetired*.

Schematically, the situation is depicted in Figure 10.3. On the left-hand side of the boxes representing the type definitions, the features provided by an object o_{Person} of type *Person* are depicted. On the right-hand side we visualize the features that are associated with an object $o_{Employee}$ of type *Employee*. It is indicated that the *Person* object possesses a (true) subset of the features an *Employee* object owns. This implies that a client cannot distinguish between a *Person* and an *Employee* object as long as features owned only by *Person* objects are accessed. Therefore, type consistency is always maintained when we substitute an *Employee* object in place of a *Person* object—for example, in case of marriage.

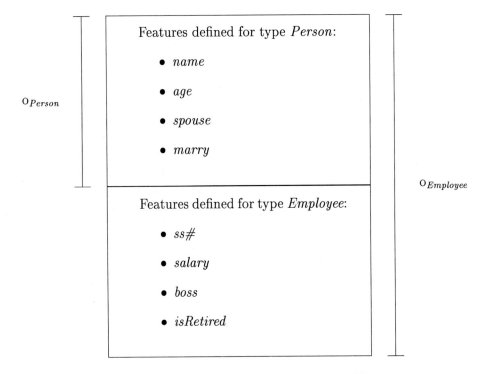

Figure 10.3: Schematical Illustration of Inheritance

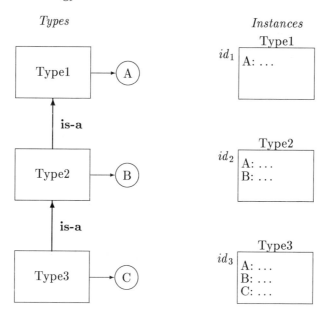

Figure 10.4: Schematical Visualization of an Abstract Type Hierarchy

10.4   Terminology

In this section, we want to summarize the terminology associated with inheritance and subtyping. Let us base our discussion on the abstract type hierarchy shown in Figure 10.4. We identify three types: *Type1* with the attribute A, *Type2* with an additional attribute B, and *Type3* with yet one more attribute C.

This type hierarchy is modeled in GOM as follows:

```
type Type1 is              type Type2                 type Type3
    public A→, A←              supertype Type1 is         supertype Type2 is
    body                       public B→, B←              public C→, C←
        [A: ...;];             body                       body
        ⋮                          [B: ...;];                 [C: ...;];
end type Type1;                    ⋮                          ⋮
                           end type Type2;            end type Type3;
```

We call *Type1* the direct supertype of *Type2*. *Type1* is also a supertype of *Type3*. However, it is not *Type3*'s direct supertype—which is, of course, *Type2*. Consequently, we call *Type3* the direct subtype of *Type2* and a subtype of *Type1*.

In Figure 10.4, we visualize the instances of the corresponding types on the right-hand side. As explained before, the subtype instances inherit all the features defined in the supertypes. Therefore, instances of *Type3*, as, e.g., the object identified id_3

has attributes A, B, and C. Instances of *Type2* have attributes A and B, whereas instances of type *Type1* have a structural representation of merely one attribute, namely, A, as shown for the object id_1. The operations associated with the types are analogously inherited—but this is not shown in our diagram.

In addition to the features, i.e., attributes and operations, also the **public** clause is inherited from the supertype to the subtype. Thus, in our abstract example, the instances of *Type3* can respond to invocations of the operations $A \rightarrow$, $A \leftarrow$, $B \rightarrow$, $B \leftarrow$, $C \rightarrow$, and $C \leftarrow$. It is important to observe that it would violate the principle of substitutability if a subtype had excluded some operation from its **public** clause that was included in the **public** clause of one of its supertypes.

All the instances of a particular type may be collected in an implicitly maintained set, called the *extension* of the type. For a particular type, say, *Type1*, the extension is denoted as *ext(Type1)*. As we have motivated before, the subtyping concept demands that all instances of a subtype can also be considered as instances of the supertype. Therefore, the extension of a subtype is contained in the extension of the supertype. For our abstract example, this is visualized in Figure 10.5. The extension of *Type3* is a (true) subset of the extension of *Type2*, which itself is a subset of the extension of *Type1*. The circles represent the objects of the respective type—the different sizes should visualize the extended set of features provided by instances of a subtype with respect to the objects of a supertype.

In some object models the extensions are also called *classes*, which is often used to refer to the type, as well. The word *class* is derived from *classification* in order to indicate that the extension of a type constitutes a classification of the objects. However, we prefer to clearly distinguish between the two different concepts *type* and *extension*.

For the time being, we consider only *single inheritance*, which is characterized by the fact that an object type has at most one supertype. Some object models also facilitate *multiple inheritance*—which we explain in Chapter 13.

10.5   ANY: The Common Root of All Type Hierarchies

There are several features that all object type instances, i.e. objects, provide in the same way. A few of those were already introduced in the preceding discussion, e.g.:

- *Object identity*
  Every instance in the database has a unique identifier (OID), which is automatically generated at instantiation time, i.e., the object's creation time.

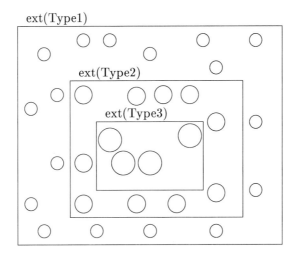

ext(Type1)

ext(Type2)

ext(Type3)

Figure 10.5: Visualization of Subtyping

- *Test for identity*
  Any two instances in the database may be tested for object identity, which results in *true* if and only if the OIDs are equal.

Of course, there are many further operations that must be provided for any object type and are, however, not visible to the user. Examples thereof are storage structure transformations, concurrency control, recovery mechanisms, etc.

One way to provide these features is by augmenting every user-supplied type definition by appropriate operations and attributes. In an object-oriented model with inheritance there is a much more elegant way: One common root type is provided and incorporates all these features that are common to all object types.

In GOM this root object type is called *ANY*. In other systems, e.g., Smalltalk-80, this type is called *Object*. The type *ANY* is implicitly assumed as the supertype of any type definition for which the **supertype** clause is left unspecified. Thus, the following two type definitions are equivalent.

type OT **is**	**type** OT **supertype** ANY **is**
public ...	**public** ...
...	...
end type OT;	**end type** OT;

The functionality of *ANY* is thus inherited—either implicitly or explicitly—by all object types defined in the database application.

An abstract type hierarchy that illustrates the preceding discussion—of *ANY* being the root of the type hierarchy—is visualized in Figure 10.6.

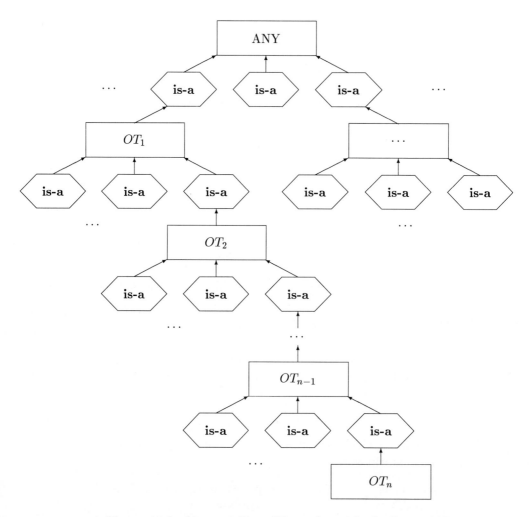

Figure 10.6: Abstract Type Hierarchy with the Root *ANY*

10.6 A Comprehensive Example of Inheritance and Specialization

Let us demonstrate the expressiveness and flexibility of the inheritance concept in combination with subtyping in an extensive example taken from the computer geometry application domain. Consider the type hierarchy shown in Figure 10.7. In addition to the subtype hierarchy, we also indicated attributes constituting the structural representation of the respective types. The type *Cylinder*—at the top of this subtype hierarchy—is structurally represented by three attributes:

- *center1* and *center2*, both being constrained to objects of type *Vertex*

- *radius* of type *float*

Instances of type *Pipe*, the (direct) subtype of *Cylinder*, have an additional attribute *innerRadius* of type *float*. The *innerRadius* constitutes the radius of the hollow part of the respective *Pipe* instance.

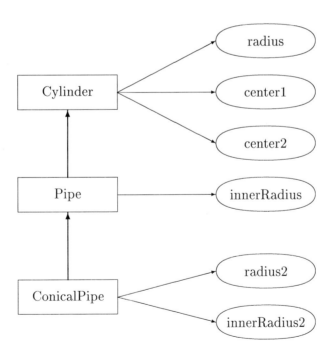

Figure 10.7: Type Hierarchy for Modeling *Cylinder*, *Pipe*, and *ConicalPipe* Objects

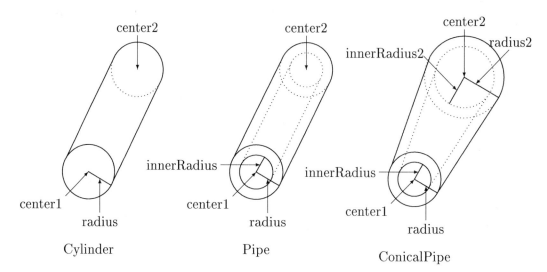

center2 — innerRadius — center1 — radius

Cylinder

Pipe

ConicalPipe

Figure 10.8: Structural Representation of *Cylinder*, *Pipe*, and *ConicalPipe* Objects

There is a third object type, called *ConicalPipe*, which models pipes that gradually decrease in diameter from one side to the other. For modeling *ConicalPipe* instances we have to store the *radius* and *innerRadius* on one side and—in addition, the *radius2* and *innerRadius2* on the other side.

In Figure 10.8 the meaning of the attributes is graphically indicated on a *Cylinder*, a *Pipe*, and a *ConicalPipe* instance.

Let us now specify the type definitions of these three objects types. Due to space limitations—and in order to reduce the complexity of the presentation—we will restrict the set of operations associated with these types to a minimum. Let us start with the type *Cylinder*. For the time being (cf. Section 11.1), we will assume that *Cylinder* has no explicit supertype; that is, it has implicitly the supertype ANY, which is the root of all type hierarchics. Our (rudimentary) type definition looks as follows:

```
persistent type Cylinder is
    public center1, center2, radius, length, volume
    body [center1: Vertex;
          center2: Vertex;
          radius: float;]
    operations
      declare length: → float;
      declare volume: → float;

        . . .
    implementation
```

```
      define length is
          return self.center1.distance(self.center2)     !! cf. Chapter 8
      define volume is
          return self.radius * * 2.0 * 3.14 * self.length;
          . . .
  end type Cylinder;
```

The only operations—aside from the built-in attribute access and assignment operations—that we associated with *Cylinder* are *length* and *volume*. Both operations are invoked on *Cylinder* instances without any further parameter and return a *float* value as a result. In the implementation of *length*, we relied on functionality that was defined before in the type definition of *Vertex* (cf. Figure 8.1). The *length* of a *Cylinder* instance is determined as the distance between the two center vertices, i.e., the distance between the *Vertex* instances referenced by *center1* and *center2*, respectively. Thus, the computation of the *distance* can be delegated to one of the *Vertex* instances, say, *center1*, with the other *Vertex*, *center2* as a parameter. In this respect, *Cylinder* is a client of the type *Vertex*, since it delegates the computation to the *Vertex* instances. The *volume* of a *Cylinder* instance is computed according to the well-known formula.

Next we introduce the type *Pipe* as a direct subtype of *Cylinder*—which is specified in the **supertype** clause of *Pipe*.

```
  persistent type Pipe supertype Cylinder is
      public innerRadius
      body [innerRadius: float;]
      operations
          declare hollowBodyVolume: → float;
          refine volume: → float;
          . . .
      implementation
          define hollowBodyVolume is
              return self.innerRadius * * 2.0 * 3.14 * self.length;
          define volume is
              return super.volume-self.hollowBodyVolume;
          . . .
  end type Pipe;
```

The difference between a *Pipe* and a *Cylinder* instance with respect to the structural representation is only the additional attribute *innerRadius*, which is introduced in *Pipe*'s **body** clause. Remember that the other attributes are implicitly inherited from *Cylinder*. We define one additional operation, called *hollowVolume*, in the type definition frame of *Pipe*. This operation determines the volume of the hollow part of the *Pipe*.

One major difference between *Pipe* and *Cylinder* instances is caused by the different formulas for computing the *volume*. The volume of a *Cylinder* is given as

$$radius ** 2.0 * \pi * length$$

whereas the volume of a *Pipe* instance is determined as

$$(radius ** 2.0 * \pi * length) - (innerRadius ** 2.0 * \pi * length)$$

Therefore, the inherited *volume* operation would yield a wrong result if it were applied to an instance of type *Pipe*.

To solve this problem object-oriented models like GOM facilitate the *refinement* of inherited operations. This allows the adaptation of an inherited operation's signature and implementation to the particular requirements of the subtype instances. In GOM, we can refine operations as shown in the type definition frame of *Pipe* for the operation *volume*. The refined operation is marked **refine** in the **operations** clause. It is even possible to specify a modified signature with respect to the originally defined signature in the supertype—however, strong typing restricts the allowable modifications (cf. Section 10.10).

In the **implementation** part of the subtype, in this case *Pipe*, a new coding of the refined operation is provided. In our case, we even utilized the original implementation of *volume* as provided in *Cylinder*. This version is invoked by **super**.*volume*, which specifies that the *volume* operation that is associated with the supertype *Cylinder* should be executed. Note that this is not a recursive implementation of *volume*—which would certainly be incorrect—but an implementation in which a different code version of *volume* is utilized. From the result obtained by **super**.*volume*, we subtract **self**.*hollowBodyVolume*, which yields the result specified by the above formula.

Finally, we introduce the third type in our type hierarchy, *ConicalPipe*. This type possesses the following rudimentary type definition frame:

```
persistent type ConicalPipe supertype Pipe is
    public radius2, innerRadius2
    body [radius2: float;
          innerRadius2: float;]
    operations
        refine hollowBodyVolume: → float;
        refine volume: → float;

        . . .
    implementation
        define hollowBodyVolume is
            return ((3.14 * self.length/3) *
                (self.innerRadius ** 2 + self.innerRadius * self.innerRadius2 +
                    self.innerRadius2 ** 2));
```

> **define** volume **is**
>   **return** $((((3.14 * \textbf{self}.\text{length}/3) *$
> $(\textbf{self}.\text{radius} ** 2 + \textbf{self}.\text{radius} * \textbf{self}.\text{radius}2 +$
> $\textbf{self}.\text{radius}2 ** 2)) - \textbf{self}.\text{hollowBodyVolume});$

      . . .
  end type ConicalPipe;

In this type definition two operations were refined:

- *hollowBodyVolume*, in order to account for the two different inner radius values stored in attributes *innerRadius* and *innerRadius2*

- *volume*, in order to take care of the two different outer radius values given by *radius* and *radius2*

10.7 Dynamic (Late) Binding of Refined Operations

It is not sufficient to merely refine inherited operations to account for the particular requirements of the subtype instances. It is absolutely necessary that the most specialized version of such a refinement hierarchy is executed—depending on the type of the receiver object. For example, when invoking *volume* on a *ConicalPipe* instance the system has to ensure that the version of *volume* that was implemented in *ConicalPipe* is executed—and not the original version provided by *Cylinder* or the (first) refined version defined in *Pipe*. This requirement appears trivial and straightforward to ensure. However, one should bear in mind that the principle of substitutability facilitates the substitution of a subtype instance wherever a supertype instance is required. Therefore, it is in general impossible to determine the direct types of objects referred to by variables at compile time.

An example should illustrate the need for dynamic binding (at run time): Consider the object base extension shown in Figure 10.9. In this object base, we identify one set object of type *CylinderSet* with OID id_0.

The set contains (references to) three elements:

- The object identified id_1 of type *Cylinder*

- The object id_2 of type *Pipe*

- The *ConicalPipe* instance id_3

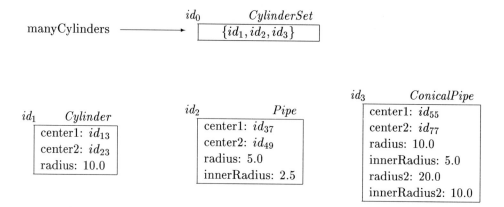

Figure 10.9: Object Base State with a *CylinderSet* Object

Even though the elements of *CylinderSet* instances are constrained to *Cylinder*, this is a perfectly legal database state—due to substitutability it is type consistent to insert *Pipe* and *ConicalPipe* instances into sets that "expect" *Cylinder* elements.

Let us now consider the following program fragment that is intended to total the *volume* of all geometric objects that are elements of the set referred to by variable *manyCylinders*:

```
var c: Cylinder;
    totalVolume: float;
    ...
foreach (c in manyCylinders)
    totalVolume := totalVolume + c.volume;
```

In the **foreach** loop c successively refers to the three elements of the *CylinderSet* object referenced by the variable *manyCylinders*. Even though this is not ensured by the system, let us—for simplifying the discussion—assume that the elements in the set object are "visited" in the sequence shown in Figure 10.9, i.e.:

1. In the first iteration c refers to the object identified id_1, which is a *direct Cylinder* instance.

2. Then, in the second iteration, c refers to the *Pipe* instance id_2.

3. In the third iteration c refers to the object with OID id_3, which is a *ConicalPipe* instance.

In all three iterations of the loop the volume of the object to which c refers is to be computed according to a different formula, i.e.:

1. $id_1.volume = \pi * r ** 2 * l$

2. $id_2.volume = \pi * r ** 2 * l - \pi * i ** 2 * l$

3. $id_3.volume = (\pi * l/3 * (r ** 2 + r * r_2 + r_2 ** 2)) - (\pi * l/3 * (i ** 2 + i * i_2 + i_2 ** 2))$

In these formulas, we have used the following abbreviations: r for **self**.*radius*, l for **self**.*length*, i for **self**.*innerRadius*, r_2 for **self**.*radius2*, and i_2 for **self**.*innerRadius2*. By close observation, we can easily verify that these formulas correspond exactly to the three different implementations of *volume* in the types *Cylinder*, *Pipe*, and *ConicalPipe*.

Dynamic (or late) binding achieves—with respect to our particular example application—the following:

1. When c is bound to object id_1, i.e., a *Cylinder* instance, the *volume* implementation defined in *Cylinder* is executed.

2. When c refers to the *Pipe* instance identified id_2, the version of the *volume* implementation specified in the type definition frame of *Pipe* is executed.

3. In the third iteration of the **foreach** loop, i.e., when c refers to the *ConicalPipe* object identified id_3, the version of volume refined in *ConicalPipe* is executed.

Incidentally, in our example application all types in the type hierarchy possess their own specialized *volume* implementation. This, in general, is not the case. The general procedure of dynamic binding starts with the direct type of the receiver object and follows the path in the type hierarchy toward the root *ANY*. The first implementation of the operation that is encountered on this path from the receiver type toward the root has to be executed. This guarantees that the most specialized version of the operation with respect to the receiver type is executed.

It now becomes clear that this concept is called *dynamic* or *late* binding: The binding of the refined operations has to be done at execution time. It is not possible to determine statically—with respect to our example—the direct type of the objects to which c refers in iterations of the loop. This depends on the state of the set object *manyCylinders*, which, of course, may change over time. Thus, at run time the system has to "ask" the object to what type it belongs and then, depending on the direct type of the receiver object, the correct, i.e., the most specialized, version of the refined operation has to be bound.

10.8   Typing Rules under Subtyping

In the design of GOM, two principal goals were pursued:

- *Flexibility of the model*
  The use of inheritance and the associated subtyping and substitutability provide a high degree of flexibility of the type scheme. This was illustrated on the example types *Person* and *Employee*. Furthermore, the type scheme introduced to model the *Cylinder*, *Pipe*, and *ConicalPipe* type hierarchy demonstrated the expressiveness of operation refinement (specialization) in conjunction with dynamic binding.

- *Type safety*
  The static type checking ensures that no run-time errors due to type inconsistencies can occur. That is, the type checker verifies at compile time that this class of errors cannot occur in a GOM program.

Unfortunately, the two goals *flexibility* and *type safety* appear to be contradictory. That is, in the design of the model one has to apply caution not to extend the flexibility and, in the process of gaining expressiveness, not to give up on type safety. In this section, we will detail the typing rules of GOM that ensure complete type safety.

10.8.1   Typing Rules under Substitutability

The substitutability of subtype instances in which supertype instances are required drastically increases the flexibility of the object model. However, we have to obey rigid rules in order to retain type safety, i.e., the verifiability of type consistency at compile time—in order to avoid run-time errors caused by type mismatches. The principal rule that has to be enforced by the compiler is twofold:

1. Ensure that the static type constraints imposed by the database designer are enforced. Whenever the static type constraint demands an object of type T make sure that it always—no matter what control path of the program is followed, or what state the persistent object base is in—gets an object of type T or a subtype instance of T.

2. Verify that all features of a type (operations or attributes) are available.

That is, in order to guarantee type safety the type checker—which is part of the compiler—has to statically verify the type consistency of a program on the basis of the type constraints provided for variables, attributes, set and list elements, and operation signatures.

The consequence can be demonstrated on a small example:

```
      var somePerson: Person;
          someEmployee: Employee;
      ...
(1)   somePerson := someEmployee;
(2)   ...
(3)   someEmployee := somePerson;        !! ILLEGAL
```

The assignment (1) is certainly type consistent: The variable *somePerson* demands—according to the type constraint in the **var** declaration—an object of type *Person* or a subtype thereof, e.g., an *Employee*. The variable *someEmployee* is guaranteed to refer to an *Employee* or a subtype of *Employee*. Therefore, the assignment is type safe. However, the assignment (3) is not type safe, since the variable *somePerson* may refer to an *Employee* due to the assignment (1), for example. But this cannot be verified statically—the compiler cannot keep track of what happened, for example, in line (2). This is especially true for persistent variables that may even have been assigned by different programs and/or users.

In general, an assignment is verified type consistent if the type that is determined statically for the expression on the left-hand side coincides with the type determined at the right-hand side of the ":=" sign—or if the left-hand side type is a supertype of the right-hand side type.

For the assignment statement (3), the compiler determines the following types:

$$\underbrace{\text{someEmployee}}_{Employee} := \underbrace{\text{somePerson}}_{Person};$$

Since the type of the left-hand side expression is not a supertype of the right-hand side type, the compiler signals an error.

The second part of the type verification process demands that the features accessed are verified available. Consider the following expressions:

```
(1)   somePerson.name;      !! okay
(2)   somePerson.salary;    !! ILLEGAL
(3)   someEmployee.salary;  !! okay
```

For a *Person* there is no feature *salary*; therefore, expression (2) has to be rejected. Of course, *somePerson* may actually refer to an *Employee*; but the compiler doesn't know. It has to assume the "worst case," which is: *somePerson* refers to a *Person* instance, which doesn't have an attribute *salary*. On the other hand, expression (3) is type safe, since the worst case for *someEmployee* to refer to is an *Employee*, which has a *salary* attribute.

10.8.2   Examples Illustrating the Typing Rules

Let us demonstrate the rules that have to be obeyed in order to facilitate the static
type consistency verification on the basis of an extended object base with *Person* and
Employee instances. Consider the following program fragment:

> **var** miniMouse: Person;
>   mickeyMouse: Employee;
>   chief: Employee;
>   i: integer;
>   . . .

(1)	mickeyMouse.spouse := miniMouse;	!! okay
(2)	miniMouse.spouse := mickeyMouse;	!! okay
(3)	mickeyMouse.boss := chief;	!! okay
(4)	miniMouse.spouse.boss := chief;	!! ILLEGAL
(5)	i := mickeyMouse.boss.ss#;	!! okay
(6)	i := miniMouse.spouse.boss.ss#;	!! ILLEGAL
(7)	i := miniMouse.spouse.spouse.age;	!! okay
(8)	i := mickeyMouse.spouse.boss.ss#;	!! ILLEGAL
(9)	mickeyMouse.boss.spouse.marry(chief);	!! type consistent, but legal problems

In Figure 10.10, a sample object base state is shown that may have been produced
by the first three statements of the above program fagment—assuming corresponding
initializations of the three sample objects. Let us now consider statement (4):

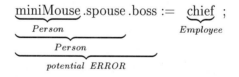

The underbraces of the assignment statement indicate the type of the (sub-)ex-
pressions that can be verified statically by the type checker. Note that the type
checker cannot rely on the dynamic object base state as shown in Figure 10.10. It
has to derive all its typing information from the type constraints provided by the user.
Therefore, the type checker determines that *miniMouse.spouse* is an expression yielding an object of type *Person*. It is pure coincidence that the actual object base state
yields an expression of type *Employee*, since *mickeyMouse*, the husband of *miniMouse*
happens to be an *Employee*, which could legally be substituted in place of a required
Person—because of substitutability. However, the compiler cannot rely on the object
base state, since this is due to change—for example, *miniMouse* could possibly get
divorced and remarry a *Person*. Therefore, the expression *miniMouse.spouse.boss* is
determined as potentially type inconsistent, since *Person* instances do not possess

a *boss* attribute. Consequently, the GOM compiler will reject statement (4). This appears odd because semantically statement (4) is equivalent to statement (3)—but only if one knows the state of the object base, which is, because of its dynamic nature, unknown to the compiler.

Among the other statements of the program fragment let us concentrate on statement (7), which is verified type consistent, as documented by the following representation:

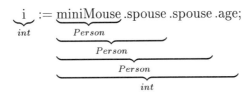

Since the type on the left-hand side of the assignment statement coincides with the statically determined type on the right-hand side, the statement is verified as being type safe. In general, due to substitutability the type derived for the left-hand side has to be a supertype of the type obtained for the right-hand side. Remember, it is okay to assign a subtype instance anywhere a supertype instance is required.

10.9   Anchor Type Constraints

In some cases, the static type constraints hinder the flexibility more than necessary. To illustrate the problem, consider the following two operation signatures (the implementation should be obvious):

> **declare** incAge: Person $\|$ $\rightarrow$ Person;
> **declare** incSalary: Employee $\|$ float $\rightarrow$ Employee;

These operations could be invoked on objects of type *Person* and *Employee*:

> **var** somePerson: Person;
>   someEmp: Employee;

(1)   somePerson.incAge.incAge;
(2)   someEmp.incAge.incAge;
(3)   someEmp.incSalary(1000.00).incAge;
(4)   someEmp.incAge.incSalary(1000.00);   !! REJECTED by the type checker

Statements (1), (2), and (3) are verified type consistent by the type checker. But statement (4) is rejected by the type checker because it determines *Person* as the result type of the invocation *someEmp.incAge*. Unfortunately, *incSalary* is not

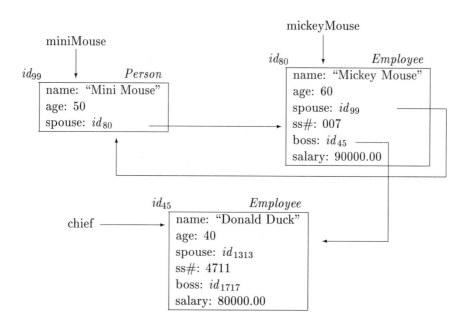

Figure 10.10: Sample Object Base State with *Person* and *Employee* Objects

applicable on *Person* instances—therefore, the compiler is forced to reject the entire statement (4).

What is the deeper cause of this problem? The compiler does not (and cannot) keep track of the object types that "float through" such an invocation chain. The type of a subexpression, e.g., *someEmp.incAge*, is determined from the signature of the operation. In our example, the signature specifies that the result object is a *Person*—which is the most general object type that can possibly returned by an *incAge* invocation. However, in reality the result type always coincides with the type of the receiver object. That is, an invocation of *incAge* on a *Person* returns a *Person* and the invocation on an *Employee* returns an *Employee*.

This can be specified in GOM using so-called *anchor* type constraints. Let us illustrate this on our example operations:

> **declare** incAge: Person ‖ → **like self**;
> **declare** incSalary: Employee ‖ float → **like self**;

The clause **like self** specifies that the return object's type coincides with the type of the receiver, i.e., **self**. Then in expression (4) the compiler verifies that *incAge* is invoked on an *Employee*—determined by the type constraint of the variable *someEmp*—and, according to the new signature, returns an *Employee*. Consequently, the invocation of *incSalary* on the object returned by *someEmp.incAge* is admissible.

Anchors can also be based on other parameters in the operation signature. For this purpose, the parameter has to be named in order to refer to it in the **like** clause. An example of an anchor in a free operation *meeting* is

> **declare** meeting: me: Person, he: Person $\rightarrow$ **like** he;

This signature states that the result object's type equals the type that can be derived for the second parameter, called *he* in the signature.

10.10 Legal Operation Refinement

We introduced the concept of operation specialization or refinement. This facilitates the adaptation of inherited operations to the particular requirements of the subtype instances. Note that an operation may be refined more than once—remember the refinement of *volume* in the types *Pipe* and *ConicalPipe*. However, a type inherits but one operation, e.g., *ConicalPipe* inherited the *volume* operation refined in *Pipe* and not the original version specified in *Cylinder*.

10.10.1 Typing Requirements

In this section we describe the rules that have to be obeyed in operation refinement in order to be able to verify type consistency at compile time. The refinement rules are specified in terms of the inherited operation; therefore, it does not make a difference whether the inherited operation constitutes a refinement itself or is the original version. In general, an operation refinement in GOM requires to obey two principal conditions:

- The originally specified name of the operation remains invariant. Otherwise, the system would treat the newly defined operation as being unrelated to the one that should be refined.

- The number of additional parameters has to be identical to the inherited version.

These two conditions are meaningful because we want to ensure that existing programs remain valid. In fact, type extensibility without affecting existing applications is one of the main goals of object orientation. In particular, the addition of a new type at the bottom of a type hierarchy should not affect any existing programs. In GOM, this is ensured by the compiler, which rejects any subtype definitions that

- Potentially violate type safety

- Invalidate existing applications

Therefore, in the refinement of inherited operations one has only limited freedom with respect to retyping the input parameters of operations and retyping the result type. In order to guarantee type safety, the following conditions have to be satisfied:

receiver type: The receiver type of a refined operation is always a subtype of the originally specified receiver type. Remember, a refinement is always specified in the subtype that inherits the operation.

additional argument types: The argument types of the refined operation have to be supertypes of the argument types of the the inherited operation.

result type: The result type specified for the refined operation has to be a subtype of the result type determined for the inherited operation.

In summary, the refined version of the operation has to accept more general (less specialized) input parameters and return a more specialized result.

The best way is to illustrate these refinement conditions—which may appear strange to the inexperienced user of object-oriented languages—on an example that violates the legal refinement criteria. The following is an example of *illegal* refinement of the operation *marry*:

> **declare** marry: Person ‖ Person → **void code** PersonMarriage;

> **refine** marry: Employee ‖ Employee → **void code** EmployeeMarriage;

> **define** EmployeeMarriage(victim) **is**
>   **begin**
>   **self**.spouse := victim;
>   **self**.salary := **self**.salary * 1.1;   !! married *Employee*s need more money
>   **end**;

According to the refinement rules the violation consists in retyping the input parameter of *marry* from *Person*—the originally specified type constraint in *Person*—to *Employee*. Intuitively, this leads to a "hole" in the applicability of *marry*, as demonstrated by the following program fragment:

> **var** anEmp: Employee;
>   aPerson: Person;
>   anotherPerson: Person;
>   · · ·

(1) aPerson.marry(anEmp);   !! okay, the original version of *marry* is applicable
(2) anEmp.marry(aPerson);   !! ILLEGAL, refined *marry* requires an *Employee*
(3) anotherPerson := anEmp;   !! okay because of substitutability
(4) anotherPerson.marry(aPerson); !! not applicable, because of dynamic binding

In particular, the principle of substitutability is violated. The invocation of *marry* with the receiver object of type *Person* is okay—as long as we assume that the variable *aPerson* refers to a *direct Person* instance. However, the invocation of *marry* in statement (2) is impossible because the refined version of *marry* that applies to the receiver object *anEmp* of type Employee requires an argument of type *Employee*. But *aPerson* may only be an "ordinary" *Person*; not necessarily an *Employee*.

Even the invocation shown in statement (4) is impossible: Here, the variable *anotherPerson* refers to an *Employee* because of the assignment in (3). Consequently, the run-time system should dynamically bind the version of *marry* as specified in *Employee*. But for this the argument is wrongly typed.

All these problems are caused by the illegal refinement of *marry*, which would, of course, be rejected by the GOM compiler. Let us now indicate a legal refinement of *marry*:

declare marry: Person || Person → **void code** PersonMarriage;

refine marry: Employee || Person → **void code** EmployeeMarriage;

define EmployeeMarriage(victim) **is**
  begin
    self.spouse := victim;
    self.salary := **self**.salary * 1.1;
  end;

We should note that the refinement conditions concern only the syntactically specified type constraints. It is not ensured that the refinement of an operation is semantically compatible. For example, the system cannot ensure that the refined *volume* operation in *Pipe* actually computes the (correct) volume and not, for example, the weight instead. In this respect, one has to rely on the considerate use of refinement by the database type implementor.

10.10.2 An Anecdotical Example to Illustrate the Refinement Conditions

Let us, once more, illustrate the refinement conditions—which may still appear strange to the inexperienced user—on an anecdotical application. Consider the type

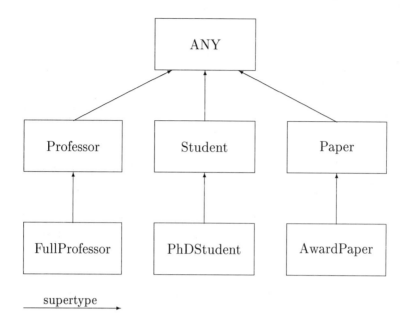

Figure 10.11: The University Type Hierarchy

hierarchy shown in Figure 10.11, which consists of a small fraction of the types contained in a university object base schema. There are three distinct subtype hierarchies below *ANY*:

- The object type *Professor* and its specialization (subtype) *FullProfessor*

- The type *Student* and its specialized subtype *PhDStudent*

- The object type *Paper* and its subtype *AwardPaper*

We omit the detailed type definition frames of these types—their specification should be obvious.

Intuitively, subtyping means that the subtype is, in some respects more specialized, i.e., "better," than the supertype. In this sense, a subtype instance can do more and/or better things than a supertype instance. This is very illustrative in our example type schema: A *FullProfessor* has more experience than an "ordinary" *Professor*. And a *PhDStudent* has more advanced knowledge than a general *Student*.

Let us illustrate this on one particular operation, called *writePaper*, which is associated with the type *Professor*:

    declare writePaper: Professor || PhDStudent → Paper
      code ProfessorWritesPaper;

For obvious reasons, we omit the implementation of this operation. However, from the signature it can be observed that a *Professor* can write a *Paper* if he or she has a *PhDStudent* as a coauthor. That is, the operation *writePaper* can be invoked on a receiver object of type *Professor*, with an additional argument of type *PhDStudent* and returns a result of type *Paper*.

Now let us consider the type *FullProfessor*. According to our intuitive understanding of subtypes as being "better" than supertypes, a *FullProfessor* should be able to author with a less qualified coauthor, e.g., *Student* instead of *PhDStudent*, and still yield a better result, e.g., an *AwardPaper* instead of an ordinary *Paper*.

This leads to the following refined signature of the operation *writePaper*:

> **refine** writePaper: FullProfessor || Student → AwardPaper
> **code** FullProfessorWritesPaper;

Of course, it is possible—again via substitutability—that a *FullProfessor* writes an *AwardPaper* together with a *PhDStudent*. In any case, whether the coauthor is a *Student* or a *PhDStudent* the invocation of *writePaper* on a *FullProfessor* is guaranteed to yield an *AwardPaper* as opposed to an *ordinary Paper*. Also, the result of coauthorship of a *Professor* with a *PhDStudent* may actually be an *AwardPaper*—however, it is only guaranteed by the signature that it is a *Paper*, not necessarily an *AwardPaper*.

10.11   Restrictions to Preserve Strong Typing

10.11.1   Retyping of Attributes Is Illegal

Retyping an inherited attribute is illegal in the context of strong typing even if the refined type is a subtype of the originally specified type. To illustrate this surprising fact consider the two types *Person* and *Employee*:

```
type Person                        type Employee
   supertype ANY is                   supertype Person is
   body                               body
      [name: string;                     [ss#: int;
      age: int;                          salary: float;
      spouse: Person;                    refine boss: Employee;];    !! ILLEGAL
      boss: Person;];                    :
      :                               end type Employee;
   end type Person;
```

In this schema the type *Employee* is a subtype of *Person*. The *boss* attribute that was originally introduced in *Person* and constrained to type *Person* is refined to type *Employee*, a direct subtype of *Person*. Now let us illustrate an example in which this refinement violates strong typing. Consider a database extension with two objects, one *Person* instance called *miniMouse* and one *Employee* instance called *mickeyMouse*.

> **var** miniMouse: Person;
>   mickeyMouse: Employee;
>   ...   !! create instances

(1)   miniMouse.boss := mickeyMouse;
(2)   mickeyMouse.boss := miniMouse;   !! *ILLEGAL*
(3)   miniMouse.boss.boss := miniMouse;   !! should be legal

Now the assignment (1) is clearly legal because it assigns an instance of type *Employee*, *mickeyMouse*, to the database component *miniMouse.boss*, which formally requires an instance of type *Person*, or a descendant thereof. The second assignment (attempt) is certainly illegal because *mickeyMouse.boss* requires an instance of type *Employee*—or a subtype thereof.

The third assignment, on the other hand, should be legal because by the static type definitions *miniMouse.boss* is constrained to type *Person* in the schema definition. Therefore, *miniMouse.boss.boss* also requires an instance of type *Person*, e.g., *miniMouse*. However, if we were to execute the assignments (1) and (3) in sequence, we would obtain the situation that *mickeyMouse.boss* is associated with a direct instance of type *Person*, which is a type violation!

Even though the example schema contains recursive type definitions, it is easy to show that the described conflicts could arise in nonrecursive type definitions in the same fashion. Even encapsulation of attributes would not solve this problem: The problem would only reappear in the form of conflicting argument types of operations that update and read the respective attribute states.

The solution to this problem under single inheritance is quite simple (and drastic): Outlaw all retyping of inherited attributes. However, in the case of multiple inheritance the situation is much more intrinsic because attributes of identical name but different type may be inherited from different supertypes. This is discussed further in Chapter 13.

10.11.2 Refinement on Sets and Lists Violates Strong Typing

It is also not possible to refine the element type of a set- or list-structured object types without losing the ability to guarantee type safety.

Let us illustrate this effect on our well-known *Person* and *Employee* example. Consider the two types *EmployeeSet* and *PersonSet* defined as

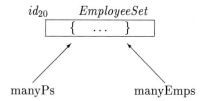

Figure 10.12: Sharing of a Set Object

type PersonSet **is**
  body
    { Person }
    . . .
end type PersonSet;

type EmployeeSet **is**
  body
    { Employee }
    . . .
end type EmployeeSet;

where *Employee* is a subtype of *Person*, both being tuple structured as defined before.

The tempting attempt to treat the type *EmployeeSet* as a subtype of *PersonSet*—and thus inherit all operations of *PersonSet*—will fail under strong typing. This is due to operations that cause side effects on the argument objects.

Suppose we have the following (persistent) variable declarations:

  var manyPs: PersonSet;
     manyEmps: EmployeeSet;

If *EmployeeSet* were a subtype of *PersonSet* the assignment

  manyPs := manyEmps;

would be valid and result in the sharing of the set object referred to by *manyEmps*, leading to a database state as pictured in Figure 10.12.

The dilemma is that via the variable *manyPs* one could insert *Person* objects into the set denoted by the identifier id_{20}, e.g.:

  manyPs.insert(somePerson);

Here, we assume that the variable *somePerson* refers to a direct *Person* instance. Then, the above program fragment has the catastrophic side effect that the variable *manyEmps* is no longer properly typed.

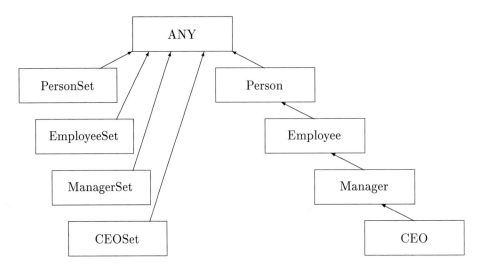

Figure 10.13: The Type Hierarchy with Set Types

The consequence is that *EmployeeSet* is not a subtype of *PersonSet*. Thus, the resulting type graph looks like the one represented in Figure 10.13. All set- and list-structured types are located directly below *ANY*, even if their element types form a type hierarchy.

The only modifications that are allowed in a subtype of a set- or list-structured supertype are *legal* operation refinements or the provision of additional operations not included in the supertype. For example, we could define *MyPersonalEmployeeSet* as a subtype of *EmployeeSet* as follows:

```
type MyPersonalEmployeeSet
   supertype EmployeeSet is
   body      !! must be empty
   operations
      declare avgSalary: → float;
      declare avgAge:  → float;
      ...
   implementation
      define avgSalary is
         var total: float := 0.0;
             card: int := 0;
         begin
            foreach (emp in self)
               begin
                  total := total + emp.salary;
                  card := card + 1;
```

> **end**
>   **return** (total / card);
>   **end define** avgSalary;
>   . . .
>   **end type** MyPersonalEmployeeSet;

Of course, we could have added this operation just as well to the type *EmployeeSet* as an "outer" operation, e.g.:

declare avgSalary: EmployeeSet $\parallel$ $\rightarrow$ float;

define avgSalary **is**
   . . .

10.12   Glossary of Inheritance Terminology

Those readers who have not yet had an extensive experience with the object-oriented paradigm might easily be overwhelmed by the vast number of new concepts and their terminology that were introduced in this chapter. In order to help them to cope with the various notations, we provide a glossary of the most important terms (directly or indirectly) related to inheritance.

Ancestor types: All direct and indirect supertypes of a type are called its *ancestors*. For convenience of notation, the type itself belongs to the set of its ancestors.

Class: The set of all *direct* and *indirect* instances of a type are referred to as the type's *class*. We chose to call this the *extension* or *extent* of the type—because the term *class* is often ambiguously used in the literature to refer to the type definition, too.

Descendant types: The dual concept of ancestor. The direct and indirect subtypes and the type itself belong to the set of *descendants* of a type.

Direct type: According to the typing rules every database instance has a unique *direct type*; this is the type of which it was instantiated using the *create* function. The direct type is the most specific type to which an instance belongs. Note that the instance also belongs to the extension (class) of all ancestors of the direct type; but not directly.

Dynamic binding: An operation may be refined in a subtype of the type that originally introduced the feature. The version that is actually executed when the operation is invoked is the one that is most specific with respect to the direct type of the receiver (first argument) of the invocation. The selection of the most specific version cannot—unfortunately—be done at compile time; rather, it takes place at run time. This process is called *dynamic binding*—sometimes it is also called *late binding*.

Extending a type: A type can be extended by additional features by introducing a subtype and including the additional features in it. This leaves the original type invariant; in particular, any existing, old applications based on the original type remain valid.

Extension: See *class*.

Inheritance: The acquisition of the features of a supertype by the subtype is called *inheritance*. Two different approaches to inheritance are distinguished:

- *Single inheritance*: A subtype has one unique supertype from which it inherits its features. In this case, the type structure corresponds to a hierarchy, i.e., a tree.

- *Multiple inheritance*: A subtype may have more than one supertype all of whose features it acquires through inheritance. In this case, the type structure represents a *directed acyclic graph* (cf. Chapter 13).

Instance: Any object in the database belongs to at least one type: the direct type (remember, through subtyping it may belong to more than one type). Then we say: The object o is an instance of type OT. If OT is the direct type of o, we say: o is a direct instance of OT.

Instantiation: Creating a new object of direct type OT by invoking the *create* function is called instantiation of OT.

Polymorphism: A database component is constrained to a particular type in the type specification that has to be obeyed throughout its existence. However, the typing rules under inheritance allow the component to be associated with a direct instance of the specified type or any descendant type thereof (see *substitutability*). Thus, the database component can still be associated with objects of different type (which have to be descendants of the specified type, though). This is referred to as *polymorphism*, which means that something can take up different forms. Polymorphism and dynamic binding are dual concepts that complement each other in providing the flexibility needed for object-oriented modeling. Some authors call this *inclusion polymorphism* to indicate that only subtype objects can replace supertype objects. A more general and more powerful kind of polymorphism will be introduced in Chapter 12.

A more restricted form of polymorphism, namely *overloading*, has also been introduced. Sometimes overloading is referred to as *adhoc polymorphism*.

Public features: The *public features* of a type include all *operations* that are included in the type's **public** clause—or in any of the inherited **public** clauses. Only the public features can be invoked by clients of the type.

Refinement: An operation defined in a type OT can be refined in a proper descendant OT' of OT. This process overrides the original version of the operation for all direct or indirect instances of OT'. It is also possible to refine a refinement—with the analogous semantics.

Specialization/generalization: This is the analogue to subtype/supertype from a more conceptual viewpoint. There exists an **is-a** relationship between a specialization and its generalization.

Strong typing: A *strongly typed model*, such as GOM ensures that no run-time errors occur due to incompatible types. A system guaranteeing strong typing is sometimes called *type safe*.

Subtype/supertype: If type OT_2 specifies "**supertype** OT_1", then OT_1 is called the *supertype* of OT_2 and OT_2 is the *subtype* of OT_1. The subtype OT_2 inherits all features of OT_1. OT_2 constitutes a specialization of OT_1. From the opposite viewpoint, OT_1 is a generalization of OT_2.

Substitutability: Objects of a subtype are *substitutable* at all places where a supertype object is required (see also *polymorphism*).

10.13 Exercises

10.1 In Section 10.1, only one-half of the *marry* operation was implemented. The spouse attribute of **self** was set to the *victim*, but the *victim*'s spouse attribute was not set to **self**. Give the full implementation taking care of both attributes. Provide this implementation for both types, *Person* and *Employee*. Which of the definitions is legal, which is not? Why?

10.2 Illustrate the problems that may occur if the **public** clause were not inherited by the subtype.

Hint: In this case, the outer signature of the subtype is not a superset of the outer signature of the supertype.

10.3   Inheritance and subtyping are commonly associated with one another—as in our object model GOM. However, this mash of two (slightly) different concepts may lead to problems. As an example consider modeling of *Cubes* and *Cuboids*. Assume that a *Cube* is modeled by one attribute *length* and a *Cuboid* by three attributes *length*, *width*, and *height*. It is common usage to represent the type *Cuboid* as a subtype of *Cube*, which inherits the one attribute *length* and augments the structural representation by two further attributes *width* and *height*.

- Illustrate the problems arising from this type structure. You should illustrate the discussion on a sample program fragment.

  Hint: The problems occur because the "real" **is-a** relationship between *Cubes* and *Cuboids* was inverted in the type definition in order to exploit inheritance.

- Consider the alternative modeling that a *Cube* is a subtype of *Cuboid*. What problems occur now?

  Hint: The type *Cube* now has more attributes than necessary and meaningful.

Reconsider the last point in more detail. Assume that *local* attributes are available and cannot be made public. These are *not* inherited. Further, it is allowed to implement two value-receiving and value-returning operations, which cover attribute accesses. These can be used used to "refine" attributes. How do these assumptions help? Aside: Why can local attributes not be public?

10.4   Extend the *Person/Employee* type hierarchy by additional types *Manager*, *CEO* (chief executive officer), *Student*, etc.

Augment the behavioral specifications of the types—in particular make use of operation inheritance and refinement.

10.5   Consider the following two operations in which the second is an attempted—though invalid—refinement of the first:

> **declare** connect: Pipe ‖ Pipe → **void**;
> **refine** connect: ConicalPipe ‖ ConicalPipe → **void**;

Why is this refinement invalid? Illustrate the problems that can arise due to the refined signature ("hole" in the applicability of the *connect* operation).

10.6 In the text, we outruled refinement of the type constraint of inherited attributes. Illustrate that a specialization (i.e., the new type constraint is a subtype of the original one) as well as a generalization (i.e., the new type constraint being a supertype of the original one) cause typing problems.

10.7 For attributes that are only read by clients—i.e., those for which only the VCO is made **public**—a specialization of the type constraint is legitimate. Verify this claim.

10.8 The typing rules can all be unified to valid operation refinement rules. For this purpose, an attribute is modeled as the pair of VCO/VTO operations and set- and list-structured objects are modeled via their operations provided for clients. Illustrate this idea.

10.9 Reconsider the Swiss knife example introduced in Exercise 2.10. Do you see any use for inheritance? Consider in detail the three constituents of the *supertype* relationship as introduced in this chapter.

10.10 Add the inheritance relationship to your meta-model of GOM. Compare to Exercises 7.12, 8.11, and 9.12.

10.11 Use inheritance to capture the similarities of different rooms (kitchen, living room, etc.). Are there any other useful applications of inheritance? The CAAD domain was introduced in Exercise 2.12.

10.14 Annotated Bibliography

The super-/subtype concept has its origin in the forerunner of all object-oriented systems: Simula-67 [DMN70, ND81]. Simula was the first language to support subtype substitutability; though the terminology of object-oriented programming had not been established at the time Simula was conceived.

In parallel to the object-oriented languages, inheritance concepts were developed in knowledge representation languages, e.g., in KRL by Bobrow and Winograd [BW77, BW79].

The inheritance of Simula was restricted to single inheritance. Only later developments, especially in Lisp-based languages, such as Loops [BKK+90], Flavors [Moo86], CLOS [Kee89], extended the single inheritance mechanism to multiple inheritance.

The typing rules, presented in this chapter, that reconcile strong typing with polymorphism by constraining a database component to refer to descendents of its specified type is adopted from Simula. Similar rules are incorporated in Eiffel [Mey88], a new object-oriented programming language that is similar to Simula (except that it provides multiple inheritance). In contrast, Smalltalk [GR83] is dynamically typed; a component of the system can refer to any object instance. The actual type of an object is determined at run time only. Two C-based object-oriented languages with inheritance are C++ [Str90b] and Objective C [Cox86].

Alan Snyder [Sny86] discusses the implications of inheritance on the encapsulation and information hiding of objects. Stein, Lieberman, and Ungar [SLU89] review the different aspects of behavior sharing among objects—inheritance is merely one approach to achieve this goal.

Bruce and Wegner try to formalize inheritance concepts in [BW90]. Another formalization, called *F-Logic*, of object-oriented features—including subtyping—was developed by Kifer and Lausen [KL89a]. In [KM90b] Kemper and Moerkotte discuss shortcomings of the inheritance concept, which are mainly due to exceptions that apply to subtypes.

There is some prior work on inheritance in the database context. The functional model DAPLEX [Shi81] incorporates inheritance as one of the earliest proposals. More recent database projects that are based on the functional approach of data representation adopted the DAPLEX inheritance concept. Among these are Iris [DFK$^+$86] and Probe [DMB$^+$87]. The database system Gemstone [BOS91] is founded on the Smalltalk object model and therefore provides only single inheritance; just like the object manager Vbase [Ont87] (for which—however—multiple inheritance is announced in future releases).

Type safety in the presence of subtyping is discussed by several authors. Among the first is Cardelli [Car88] and Cardelli and Wegner [CW85]. Danforth and Tomlinson [DT88] follow up this work and provide a thorough overview of typing issues. The typing rules that are employed in the GOM object model are formalized by the authors of this book in [KM89] and [KM92b]. A more thorough treatment can be found in a German monograph by Kemper [Kem92]. Type safety on the basis of type inference was studied in the context of the Machiavelli project [Oho88, OBBT89, BTBO89]. The seminal work on type inference was carried out by Milner [Mil78]. The *anchor* concept to increase the flexibility of otherwise purely static type constraints was borrowed from the Eiffel language [Mey88].

11

Virtual Types and Type Case

The type hierarchy facilitates generalization and specialization of object types. Thus, the supertype is used to factor out those features that all subtypes have in common. All the subtypes inherit these common features and may—within the limits specified in Section 10.10—refine the inherited operations. Despite the expressive power of the inheritance concept two problems sometimes occur:

1. The operation that one would like to factor out from the subtypes is not implementable in the supertype due to a lack of information; i.e., the supertype is not sufficiently specialized to implement the operation.

2. The operations that should logically be factored out exhibit different signatures; i.e., they cannot be formulated as *one* operation in the supertype.

In this section, we will describe two features to alleviate these problems. *Virtual types* are used to factor out operations that cannot be implemented in the supertype, and the *typecase* construct is used to specify programs whose control flow depends on the dynamically determined type of expressions.

11.1 Virtual Types

Not all types are really self-sufficient in the sense that they supply enough functionality to allow direct instances to be created. Rather, some types should be considered as the abstract framework for specifying the common functionality that all its subtypes have to implement. Thus, this kind of type is called a *virtual type* because it cannot be instantiated; i.e., no direct instance can be created—only instances of its subtypes are legitimate. *ANY*, the root of the type hierarchy, is one example of such a virtual, built-in type. But also user-defined virtual types are useful to factor out features that are common in a variety of subtypes.

Let us consider the type *GeometricPrimitive*, which is part of a geometric type hierarchy as shown in Figure 11.1. This type could contain the attributes *geoID*, *color*, and *mat* (with the obvious meanings).

```
persistent type GeometricPrimitive
    supertype ANY is
    public ...
    body [geoID: string;
          mat: Material;
          color: string;];
    operations
        declare specWeight: → float;
        ...
    implementation
        define specWeight is
            return self.mat.specWeight;
    ...
end type GeometricPrimitive;
```

This type cannot make any assumptions about the geometric properties of a potential instance. In particular, the functions *weight* and *volume* could not possibly be defined in *GeometricPrimitive*, because there is insufficient information for their implementation. They are defined only in the subtypes, when the geometric shape and its characteristic parameters are known. This is very unfortunate when we consider an application in which it is natural to include different subtype instances of *GeometricPrimitive* in one persistent set, like

```
var baseParts: GeometricPrimitiveSet;
```

We assume that *GeometricPrimitiveSet* has been defined as a set-structured type in the obvious way. *BaseParts* may store the constituent primitive parts of which a more complex geometric object is constructed. On the basis of this database schema, a query to obtain the total weight of the set *BaseParts* cannot be stated. The following program fragment, which we would like to express, is *illegal*:

```
foreach (b in baseParts)
    totalWeight := totalWeight + b.weight;
    totalVolume := totalVolume + b.volume;
```

The program is rejected by the compiler because the range variable b is of type *GeometricPrimitive*. This type, however, does not provide an operation *weight*. This is all the more disturbing when considering that the set *baseParts* contains only subtype instances of *GeometricPrimitive*, all of which actually do have an implementation for *weight*. Is this the penalty we have to pay in order to gain static type checking? Fortunately, the answer is *no*!

There are several misleading approaches that one could try to solve the problem:

1. Retyping the variable *baseParts*

   One could try to retype the variable *baseParts* to *CylinderSet*. Then the operations *volume* and *weight* are guaranteed applicable on all elements of the set. However, it is no longer possible to include *Polyeder* instances into the set object referenced by *baseParts*. This loss of flexibility is, of course, not tolerable, since we wanted to abstract from the geometric shape by introducing the types *GeometricPrimitive* and *GeometricPrimitiveSet*.

2. Introducing the operations in *GeometricPrimitive*

   This approach requires the definition of *volume* and *weight* in the type *GeometricPrimitive*. However, the structural representation of *GeometricPrimitive* instances does not provide sufficient information for computing these operations. Also, this structural representation cannot be factored out from the subtypes, since the geometric models of *Polyeder* and *Cylinder* are vastly different.

The solution is to include in the type *GeometricPrimitive* the *virtual* operations *volume*, *weight*, and *display* as follows:

```
persistent virtual type GeometricPrimitive supertype ANY is
    public geoID, color, specWeight, mat, weight, display
    body [geoID: string;
           mat: Material;
           color: string;];
    operations
        declare specWeight: → float;
        virtual declare display: → void;
        virtual declare volume: → float;
        virtual declare weight: → float;
        . . .
    implementation
        define specWeight is
            return self.mat.specWeight;
        . . .
end type GeometricPrimitive;
```

This specification makes *GeometricPrimitive* a *virtual type* of which no direct instances can be created. The object type *GeometricPrimitive* contains three operations marked **virtual**. For these operations only the abstract signature is specified; there is no implementation, yet. The implementation has to be provided in the subtypes of *GeometricPrimitive* by refining the inherited virtual operation (cf. Section 10.6). For that the usual refinement conditions—as specified in Section 10.10—have to be obeyed.

A virtual type cannot be used to instantiate new objects; i.e., it cannot respond to the operation *create*. Thus, there are no objects whose *direct* type is a virtual type. This has to be ruled out because virtual types contain operation signatures that cannot be executed; e.g.,

>   someGeoObject.display

could not be executed if *someGeoObject* referred to an instance whose *direct* type is actually *GeometricPrimitive*.

Every subclass of *GeometricPrimitive* has to either furnish an implementation of all virtual operations of the supertype such as *display*, *volume*, and *weight* or it is itself a virtual type. Let us first concentrate on the subtype *Cylinder*. In this subtype we refine all three inherited virtual operations and we do not introduce any additional virtual operations. Therefore, *Cylinder* becomes a "normal" instantiable type. The type definition is sketched as follows (we do not repeat the parts known from preceding discussions):

```
persistent type Cylinder supertype GeometricPrimitive is
    public ...  !! as before
    body ...    !! as before
    operations
        ...  !! as before
        refine display: → void;
        refine volume: → float;
        refine weight: → float;
    implementation
        define display is
            ...
        define volume is
            ...
        define weight is
            return self.volume * self.specWeight;
end type Cylinder;
```

On the basis of the new, virtual type definition of *GeometricPrimitive* our formerly invalid program for totaling the *weight* and the *volume* of all objects contained in *baseParts*

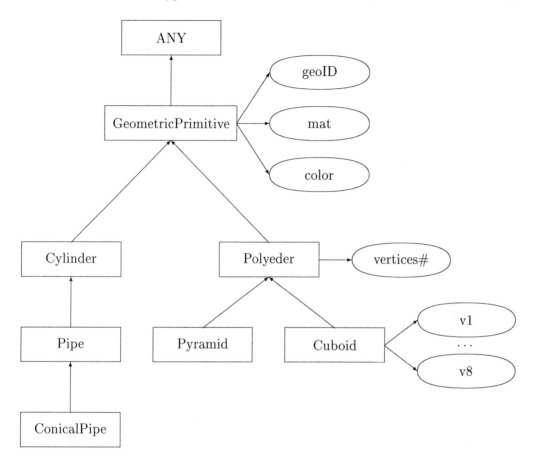

Figure 11.1: Geometric Type Hierarchy with Virtual Types

```
   ...
   foreach (b in baseParts)
      totalWeight := totalWeight + b.weight;
      totalVolume := totalVolume + b.volume;
   ...
```

becomes perfectly legitimate. Still, the variable *b* is of type *GeometricPrimitive*; but it is assured that any element of the set *baseParts* is a proper descendant of *GeometricPrimitive* for which an implementation of *volume* and *weight* exists. Through the dynamic operation binding mechanism, this version of *weight* is executed by the run-time system.

It is also possible that a virtual type has a virtual subtype. There are two possible causes for this to happen:

1. Not all inherited virtual operations are refined and implemented in the subtype.

2. The subtype itself introduces an additional virtual operation.

By cause 2 it is also possible that a "normal" instantiable type has a virtual subtype—this does not cause any problems.

The type *Polyeder* is now being defined as a virtual subtype of *GeometricPrimitive* due to cause 1. Of the three inherited virtual operations none is refined:

```
persistent virtual type Polyeder
   supertype GeometricPrimitive is
   body
      [vertices#: int;];
end type Polyeder;
```

Polyeder itself has the instantiable subtype *Cuboid*, which provides the implementations of the three virtual operations *display*, *volume*, and *weight*:

```
persistent type Cuboid
   supertype Polyeder is
   public ...
   body [v1, v2, v3, v4, v5, v6, v7, v8: Vertex;]
   operations
      declare length: → float;
      declare width: → float;
      declare height: → float;
      refine weight: → float;
      refine volume: → float;
      refine display: → void;
      ...
   implementation
      define length is
         return self.v1.distance(self.v2);
      define width is
         return self.v1.distance(self.v4);
      define height is
         return self.v1.distance(self.v5);
      define volume is
         return self.height * self.width * self.length;
      define weight is
         return self.volume * self.specWeight;
      ...
   end type Cuboid;
```

11.2   The typecase Construct

Virtual types—as introduced in the preceding discussion—are useful to factor out operations that are common to *all* instantiable subtypes. By dynamic binding the run-time system ensures that the most specialized implementation of a virtual operation is executed (with respect to the direct type of the receiver object). Virtual operations, thus, increase the flexibility of the object model. Nevertheless, the virtual operations have to obey the refinement condition; in particular, the number of parameters within all refined virtual operations has to be the same, and the type constraints of the parameters and the result have to be obeyed according to Section 10.10. In some applications this appears overly restrictive and may prohibit a "natural" modeling of the real world.

The **typecase** construct allows to specify code, which depends on the dynami-cally determined type of an expression, e.g., a variable. Let us illustrate this control construct on the basis of the type hierarchy shown in Figure 11.2.

The types are mostly self-explanatory. In the subsequent program fragment, we will focus on only very few attributes in the structural representation:

- *Student* instances possess an attribute *gpa* (grade point average) of type *float*

- An *Employee* object possesses the attribute *salary* of type *float*

- A *Manager* instance has an attribute *compCars* which refers to a set object of type *CarSet*, i.e., a set of *Cars* that the *Manager* is entitled to use.

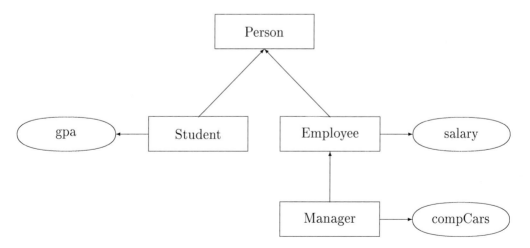

Figure 11.2: Type Hierarchy to Illustrate **typecase**

We will now sketch a program fragment in which a distinguished *Person* (remember: *Employees*, *Students*, and *Managers* are also *Persons*) referred to by the variable *laureate* receives a bonus. The bonus, however, is vastly different for the different subtypes of the *Person*, i.e.:

- A *Student* receives an upgrade of the grade point average (*gpa*).

- An *Employee* instance should receive a *salary* increase.

- A *Manager* is gratified by being entitled to use an additional company *Car*, i.e., the one referred to by the variable *jaguar*.

- An "ordinary" *Person* merely receives a "thank-you" note.

This gratification is achieved by the following program fragment:

```
var laureate: Person;
    jaguar: Car;
    gpaBonus: float := 1.1;        !! default gpa bonus
    salaryBonus: float := 1.1;     !! default salary bonus
typecase(p := laureate)
   begin
      case Student: p.gpa := p.gpa * gpaBonus;
      case Employee: begin
                        typecase (a := p)
                           begin
                              case Manager: a.compCars.insert(jaguar);
                              default: p.salary := p.salary * salaryBonus;
                           end typecase;
                     end;
      default: output("Nice job! Thank you.");
   end typecase;
```

In the statement "**typecase**(p := laureate)," the read-only variable *p* is assigned the value of *laureate*, i.e., the reference to the corresponding *Person* object. The variable *p* is implicitly declared as being constrained to the same type as *laureate*, i.e., *Person* in this example. Depending on the direct type of *p* one of the three cases is executed: *Student*, *Employee*, or **default**. The case *Employee* contains a nested *typecase* construct to distinguish between an ordinary *Employee* (who should receive a *salary* increase) and a *Manager* (who becomes entitled to drive the *jaguar*).

The **typecase** construct of GOM does not violate type safety. This is ensured by the fact that for any possible type derived by the **typecase** at least one branch has to be provided by the programmer (if more than one branch qualifies the textually first one that satisfies the type condition is chosen). Furthermore, we require a **default** entry that guarantees that a subsequently occurring extension of the type schema does not affect the validity of the **typecase** statement. For example, a new subtype *Secretary* of type *Employee* is handled by the **default** branch of the nested **typecase** construct. See Exercise 11.5 for a more elaborate discussion of the type safety of the **typecase** construct. Further, the **typecase** construct is not necessarily a sign of good programming style. In most cases, it can be avoided by utilizing overloading and/or refinement of type associated operations (see Exercise 11.6).

11.3   Exercises

11.1   The *virtual types* are only meaningful in strongly typed object models in which data components are constrained to a particular type. Why is the virtual type concept not needed in non-type safe models, i.e., those models where all data components can refer to any objects of any possible type? Illustrate your discussion on an example application.

11.2   Sketch the type definitions for the following types:

- *RealEstate* being a virtual type with the virtual operation *estimatedValue*.

- *Factory* being a subtype of *RealEstate* with attributes *profit*, *#employees*, and *sales*. The virtual operation *estimatedValue* should be refined and coded as a function of *profit*, *#employees*, and *sales*.

- *Housing* being a virtual subtype of *RealEstate* with an additional attribute *yearBuilt*.

- *Condominium* is a subtype of *Housing*. The operation *estimatedValue* is coded as a function of *yearBuilt* and *size*.

- *House* is also a subtype of *Housing* and contains also a refinement of *estimatedValue*—now as a function of *size* and *yearBuilt* and *yardSize*.

11.3   Illustrate the use of **typecase** on extracting particular typed objects from a heterogeneous set. Consider, for example, a given set *myGeoObjects* of type *GeometricPrimitiveSet* (see Figure 11.1). This set is suitable to store any kind of geometric objects. In an application however, we may have to manipulate the objects depending on their more specific type. Therefore, we want to separate the elements into *myCylinders*, *myPipes*, *myConicalPipes*, and *my-Polyeders*. Note that the set referenced by *myCylinders* should contain only direct instances of *Cylinder* and, e.g., no *Pipes*. Why is it advantageous to constrain *myCylinders* to the type *CylinderSet* as opposed to *GeometricPrimitiveSet*?

11.4 In Exercise 11.1 we stated that the virtual type concept is not needed in non-type safe models that impose no type constraints on data components. This is not true for the **typecase** construct. On the opposite, the **typecase** facility would be even more important in non-type safe models. Discuss and illustrate this statement on some example application.

11.5 In the text we indicated that the **typecase** construct as provided in the GOM language does not violate type safety. Elaborate this topic:

- Why is it required that at least one branch applies for each possible type encountered?

- Why is it necessary to include a **default** branch? Hint: Consider extensions of the type hierarchy, i.e., the definition of new types.

- Illustrate your discussion on an example type hierarchy of your choice (but it should not be the one used in the text).

11.6 Provide for each type in the hierarchy of Figure 11.2 a *laudat* operation without any parameter. These operations should behave like the specialized code to honor a *laureate* given in the **typecase** section. Do not use typecase! Give a second implementation of these operations in which the *gpaBonus* and the *salaryBonus* are given as parameters instead of being hard coded into the operation.

11.7 Enhance your model of GOM by modeling virtual types. Add the **typecase** statement to your type hierarchy of possible statements (cf. Exercises 7.12, 8.11, and 10.10).

11.4 Annotated Bibliography

Virtual types were already incorporated in the language Simula 67 [DN66, ND81, DMN70]. The virtual type concept is also used in Eiffel [Mey88]—there it is called *deferred class*, though. Also, C++ [Str90b] provides some form of virtual type. The *typecase* construct can be traced back to Smalltalk-80 [GR83], even though it is not an explicit syntactical construct there. Rather, the programmer can only inquire the type of an object and then, depending on the outcome of this message, channel the control flow appropriately.

12

Polymorphic Operations and Generic Types

In this chapter, we describe two concepts that are used to provide flexibility that goes beyond what can be achieved with inheritance and subtyping. It is noteworthy that the inheritance concept can be used only to pass functionality along the subtype hierarchy; that is, subtypes inherit only from supertypes that lie on the same branch leading to the root type *ANY*. In some cases, this turns out to be too restrictive in order to achieve a natural model of the real world. Sometimes it is convenient to define operations that apply to several different types that are not necessarily related by the super/subtype relationship. For this purpose, we incorporated the following two concepts in the generic object model (GOM):

1. *Polymorphic operations*

2. *Generic types*

Both constructs are based on *type parameterization*, by which some of the types on which the definition of the polymorphic operation (or the generic type) relies remains unspecified. This allows to specify the polymorphic operation (or the generic type) on an abstract and rather general level.

12.1 Polymorphic Operations

The word *polymorphism* has its roots in the ancient Greek language meaning *many forms* or *many structures*. The phrase *polymorphic operation* indicates that the operation is not only applicable on objects having all the same specified structure but is also able to deal with objects of different structure. Several forms of polymorphism are distinguished:

1. *Adhoc polymorphism*, which is nothing more than "ordinary" overloading of operations

2. *Inclusion polymorphism*, which allows the application of an operation not only to objects of a certain type but also to the instances of the subtypes

3. *Bounded polymorphism*

It is the latter kind of polymorphism this chapter treats in detail. The polymorphic operations introduced are able to deal with objects of different types that do not necessarily lie on the same branch of the super/subtype hierarchy. Instead, they only rely on the fact that these types provide at least some structure and/or behavior that is required as a minimum in order for the operation to work correctly. In order to specify these requirements, several constructs must be introduced or generalized:

- Type variables

- Type expressions

- Type boundaries

- Declarations

Type variables and *type boundaries* are the newly introduced constructs, whereas *type expressions* and *declarations* are generalized such that they can capture the structural and behavioral requirements imposed on the arguments.

12.1.1   Motivation for Polymorphism

This section motivates the polymorphic operations by means of several small examples. Especially, we argue that flexibility and reusability of software are increased by the usage of polymorphic operations. Note that we already achieved these goals to some extent by subtyping and inheritance. Increasing flexibility and reusability if subtyping and inheritance are *not* possible is the main motivation for the introduction of polymorphic operations. We clarify these points by examining two small examples, one for tuple-structured types and one for list-structured types.

Consider the following two tuple-structured types, called *Person* and *Wine*:

type Person **is**   **type** Wine **is**
   [name: string;   [name: string;
    age: int;   age: int;
    spouse: Person;];   bouquet: string;];

For objects of both types an operation *increaseAge* may be required. The two declarations look as follows:

(1)   **declare** increaseAge: Person ‖ int → Person
       code incAgeP;
(2)   **declare** increaseAge: Wine ‖ int → Wine
       code incAgeW;

Both operations have the same name; *increaseAge*. Operation (1) is the declaration of *increaseAge* associated with type *Person*, and operation (2) is associated with type *Wine*. It is important to note that the operation (2) is not related to (1)—remember that different unrelated types provide separate name spaces for operations. Nevertheless, their implementations look very similar:

(1)   **define** incAgeP(n) **is**
       self.age := **self**.age + n;
(2)   **define** incAgeW(n) **is**
       self.age := **self**.age + n;

In fact, the only difference is in the headline of the procedure; the code itself is identical. The question is whether we can do with only one declaration and one implementation of an operation that works on objects of both types, and possibly on all other types providing some *age* attribute of type *int*. Obviously, we cannot make *Person* a subtype of *Wine*, or vice versa. That is, subtyping does not solve the problem. As we will see, this is a typical case for bounded polymorphism by which the problems can easily be solved.

As another example consider the following two list-structured types called *CylinderList* and *WineList*:

        type CylinderList **is**   **type** WineList **is**
            <Cylinder>;   <Wine>;

For objects of either type a function *length*—which determines the cardinality of the respective list object—seems plausible:

(1)   **declare** length: CylinderList ‖ → int
       code lengthC;
(2)   **declare** length: WineList ‖ → int
       code lengthW;

Again, both operations have the same name. Operation (1) is associated with the object type *CylinderList*, operation (2) is associated with type *WineList*. This is the only difference concerning their signatures. Furthermore, their implementations look very similar:

```
(1)   define lengthC is                (2)   define lengthW is
          var cyl: Cylinder;                     var wine: Wine;
             i: int := 0;                            i: int := 0;
          begin                                  begin
             foreach (cyl in self)                  foreach (wine in self)
                i := i + 1;                             i := i + 1;
             return i;                              return i;
          end define lengthC;                    end define lengthW;
```

The implementation code of the two operations differs only in the headline and the
variable declarations where the iteration variable—*cyl* and *wine*, respectively—is constrained to the particular list element type.

Again, the problem is that writing many such *length* functions for any possible list-
structured type can be very tedious. Copying the operations using a text editor does
not really help, since the code must be individually modified for each type. Further,
if errors occur in one of the operations, then *all* existing copies of the operation have
to be modified. The goal is to provide a mechanism that is powerful enough such
that programming and maintenance of a single operation are sufficient.

12.1.2   Type Variables

In this section, we will unify the above two definitions for *increaseAge* and those for
length.

Let us consider *increaseAge* first. The only differences between the implementations are as follows:

1. Receiver types *Person* and *Wine*

2. Code names *incAgeP* and *incAgeW*

Since our goal is to maintain but one operation definition, i.e., only one piece of
code, the code names and local variable names will automatically become unique.
Thus, the only remaining problem consists in resolving the different type names.

In order to unify different type names, we introduce *type variables*. They provide
the necessary abstraction mechanism enabling to hide the differences of the declarations and definitions of operations. Since type variables are special syntactic constructs, a special notation is needed: They start with a backslash—i.e., "\"—followed
by an ordinary identifier. Thus, typical type variables might look like *AgeType*,
ListType, or *ElementType*. The semantics of a type variable is that every possible
type may be substituted for it.

Using the notion of type variables the unified version of the *increaseAge* operation
has the declaration

    polymorph declare increaseAge: \AgeType ‖ int → \AgeType
       code polyIncAge;

and the unmodified definition

    define polyIncAge(n) **is**
       self.age := **self**.age + n;

The keyword **polymorph** is used to indicate that the declared operation is polymorphic.

The role of the type variable $\backslash AgeType$ is to specify that the polymorphic operation *increaseAge* is valid for all types being *substitutable* for $\backslash AgeType$. Since there is, up to now, no restriction on the substitutability of types for type variables, all types may be substituted resulting in problems dealt with in subsequent sections. Basically, the problem is that the applicability of *incAge* is not at all constrained. Thus, even an (erroneous) invocation on, let's say, a *Cuboid* instance that does not have an *age* attribute is not yet ruled out.

Computing the length of a list-structured object requires not only the modification of the *length* declaration but also a modification of its definition:

    polymorph declare length: \ListType ‖ → int
       code polyLength;

    define polyLength **is**
       var item: \ElementType;
         i: int := 0;
      begin
        foreach (item **in self**)
          i := i + 1;
        return i;
      end define polyLength;

The local variable *item* within the above definition of *polyLength* is a *polymorphic variable*. It can hold object references to more than only a single type. Here, it is intended to hold any object that might occur as an item in some list-structured object.

There still exists a problem with the current polymorphic operation declarations and definitions as given above. Since a type variable that may stand for every possible type is used in the declaration of the operations and in the declaration of local variables no restriction is imposed on how to call the operation. For example, no mechanism prevents a programmer from accidentally invoking the operation *increaseAge* on some object not having the attribute *age*, which in turn is absolutely necessary for the implementation of *increaseAge*. Similarly, some user might invoke the operation *length* on some object that might not be list structured.

What is needed is a concept allowing us to specify requirements for types and relationships between types. For the above examples, the requirements we would like to express are the following:

- Only types having an attribute *age* of type *int* may be substituted for \AgeType

- Only list-structured types may be substituted for \ListType

- If \ListType is substituted by a list structured type whose element type is t, then only t may be substituted for \ElementType.

Providing support for expressing the above kinds of requirements is the issue of the next sections.

12.1.3   Type Expressions and Type Boundaries

In order to express requirements to constrain the substitutability of types for type variables, two basic mechanisms are introduced: *type expressions* and *type boundaries*.

Type expressions are built upon the expressions already utilized within the **body** clause of type definitions. There three different types of expressions are allowed:

- $[A_1 : T_1; \ldots, A_n : T_n;]$

- $< T >$

- $\{T\}$

where the A_i are distinct attribute symbols, and the T_i and T are type symbols. We generalize these simple expressions by allowing not only type symbols within the expressions but also type variables and type expressions. The latter results in the possibility of "nesting" type expressions. More formally, we define *type expressions* as follows:

1. Every type symbol is a type expression.

2. Every type variable is a type expression.

3. If $A_1, \ldots, A_n$ are attribute symbols and $e_1, \ldots, e_n$ are type expressions, then $[A_1 : e_1; \ldots; A_n : e_n;]$ is a type expression.

4. If e is a type expression, then $< e >$ is a type expression.

5. If e is a type expression, then $\{e\}$ is a type expression.

Examples of type expressions are as follows:

- [age: int;]

- [age: \AttrType;]

- < \ElementType>

- {\ElementType}

- [A: < \ElementType>;]

and so forth.

Recall the above example operations *increaseAge* and *length*. One constraint to be expressed was that for the type variable *AgeType* only those types that have at least an attribute *age* of type *int* are substitutable. The syntax for expressing exactly this is

$$\AgeType \le [age: int;].$$

We call this a *type boundary*. In general type boundaries are defined as follows:

- If \\$TV$ is a type variable and e is a type expression, then \\$TV \le e$ is a *type boundary*.

Another example for a type boundary is

$$\ListType \le < \ElementType>$$

It states that *ListType* can be substituted only by list-structured types—whose element type *ElementType* is not further constrained.

12.1.4 Polymorphic Operation Declarations

Summarizing the concepts developed so far, we are able to unify operation declarations and definitions by utilizing type variables and polymorphic variables whenever needed. Type boundaries enable us to represent constraints on the types that may be substituted for type variables. The missing step is to incorporate type boundaries within operation declarations. For this purpose, a list of type boundaries separated by commas and enclosed in parentheses can be specified in between the name of the defined polymorphic operation and the colon ("`:`")—which, in the nonpolymorphic case, follows the name immediately.

The complete polymorphic specification of *increaseAge* consists of the declaration

polymorph declare increaseAge (\AgeType $\le$ [age: int;]):
    \AgeType $\parallel$ int $\rightarrow$ \AgeType
        code polyIncAge;

and the unmodified definition is

> **define** polyIncAge(n) **is**
>   **self**.age := **self**.age + n;

For *length* we have

> **polymorph declare** length (\ListType $\leq$ < \ElementType>): \ListType $\|$ $\to$ int
>   **code** polyLength;

> **define** polyLength **is**
>   **var** item: \ElementType;
>   i: int := 0;
>   **begin**
>   **foreach** (item **in self**)
>   i := i + 1;
>   **return** i;
>   **end define** polyLength;

The type boundary $\backslash ListType \leq\ <\ \backslash ElementType>$ expresses that $\backslash ListType$ must be a list type having elements of type \ElementType. Since there is no further restriction on $\backslash ElementType$, any element type is allowed.

12.1.5   An Anecdotical Example

Consider the two types *Person* and *Swan*. The type hierarchy in which these two types participate is shown in Figure 12.1.

Their implementation could look like:

> **persistent type** Person
>   **supertype** LivingBeing **is**
>   **public** marry, ...
>   **body** [spouse: Person;
>   ...];
>   **operations**
>   **declare** marry: Person $\to$ **void**
>   **code** personWedding;
>   ...
>   **implementation**
>   **define** personWedding(victim) **is**
>   **begin**
>   **self**.spouse := victim;
>   victim.spouse := **self**;
>   **end define** personWedding;
>   ...
>   **end type** Person;

> **persistent type** Swan
>   **supertype** LivingBeing **is**
>   **public** marry, ...
>   **body** [spouse: Swan;
>   ...];
>   **operations**
>   **declare** marry: Swan $\to$ **void**
>   **code** swanWedding;
>   ...
>   **implementation**
>   **define** swanWedding(victim) **is**
>   **begin**
>   **self**.spouse := victim;
>   victim.spouse := **self**;
>   **end define** swanWedding;
>   ...
>   **end type** Swan;

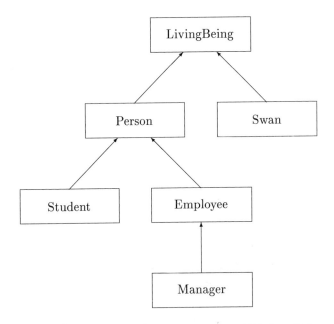

Figure 12.1: Type Hierarchy of Living Beings

Looking at the two definitions of the two operations *marry*, called *personWedding* and *swanWedding*, one easily recognizes that they are almost identical. Again, this is a case for a polymorphic operation. The development of the polymorphic operation *marry* differs a little bit from the other polymorphic operations as defined above. The requirement is that *Swans* may marry only *Swans* and *Persons* may marry only *Persons*. In general, any being's spouse must be of the same species as the being itself. This can easily be expressed by the following type boundary:

\WeddingType ≤ [spouse: \WeddingType;]

The polymorphic declaration of *marry* then is

polymorph declare marry (\WeddingType ≤ [spouse: \WeddingType;]):
    \WeddingType || \WeddingType → **void**
        code polyMarry;

The definition of *polyMarry* is just the same as the one of *personWedding* or the one of *swanWedding*.

We state some applications of the *marry* operation. Some of them are legal, some are not. Consider the following variable declarations and invocations of *marry* on the objects referenced by these variables:

var John: Employee;
    Paul, Maria: Person;
    Swen, Swenja: Swan;
      . . .

(1)	Paul.marry(Maria);	!! okay
(2)	John.marry(Maria);	!! okay
(3)	Maria.marry(John);	!! okay
(4)	Maria.marry(Paul);	!! okay
(5)	Swen.marry(Swenja);	!! okay
(6)	Maria.marry(Swen);	!! ILLEGAL
(7)	Swen.marry(Maria);	!! ILLEGAL
(8)	Paul.marry(Paul);	!! okay—as far as typing is concerned

We have complicated the situation of evaluating the correctness of the operation call by assuming the existence of a subtype *Employee* of the type *Person*. The question arises how to recognize whether a call of a polymorphic operation is legal or not. The general procedure works as follows. First, all type variables within the argument type specification have to be substituted by type symbols such that the arguments are of the indicated type or a subtype thereof. If this is not possible, the operation call is illegal. Second, if the first step succeeded, then the substitution is carried out on the type boundaries following the operation name. If they can all be verified, the invocation is legal; otherwise, it is illegal. We illuminate this process by considering some of the above example invocations and trying to validate them.

Consider (1): Since both *Paul* and *Maria* are *Person*s, we try to substitute the type variable \ *WeddingType* by the type symbol *Person*. This first substitution has to take place within the argument description of the declaration of *marry*, which is

$$\backslash \text{WeddingType} \parallel \backslash \text{WeddingType} \rightarrow \textbf{void}$$

The substitution of \ *WeddingType* by *Person* yields

$$\text{Person} \parallel \text{Person} \rightarrow \textbf{void}$$

Thus, this particular invocation of *marry* passes the first step, since both *Paul* and *Maria* are *Person*s.

In the second step, the type boundary

$$\backslash \text{WeddingType} \leq [\text{spouse: } \backslash \text{WeddingType;}]$$

in which *WeddingType* has to be substituted by *Person* has to be verified. The result of the substitution is

$$\text{Person} \leq [\text{spouse: Person;}]$$

This evaluates to true, since the type *Person* is defined such that it has an attribute *spouse* of type *Person*.

Consider (2): We know that *John* is an *Employee* which is a subtype of *Person*. Thus, *Person* is substituted for \WeddingType. The argument description of *marry* again yields

> Person ‖ Person → **void**

This expression passes the first test, since *John* is an *Employee* and since *Employee* is a subtype of *Person*, *John* is a *Person*. Note that the substitutability of subtype instances in places where supertype instances are required is, of course, also valid in evaluating polymorphic operation signatures. Further, *Maria* is a *Person*. The type boundary is the same as before and hence the invocation (2) is legal.

Consider (7): The invocation *Swen.marry(Maria)* of the operation *marry* is illegal. The reason is that we cannot find a type *T* that can successfully be substituted for the type variable \WeddingType. Remember that *Swen* is of type *Swan* and *Maria* is of type *Person*. In order to pass the first test, we need a common supertype. One common supertype of *Swan* and *Person* is *LivingBeing*. Thus, the first test can be passed with *LivingBeing* being substituted for \WeddingType, which yields

> LivingBeing ‖ LivingBeing → **void**

The same substitution applied to the type boundary results in

> LivingBeing ≤ [spouse: LivingBeing].

Since the attribute *spouse* has been introduced in the *Person* and *Swan* types—whereas *LivingBeing* does not provide a *spouse* attribute—this condition does not hold. Neither does this condition hold for any other common supertype of *Person* and *Swan*.

Statement (8) is type consistent—however, it may raise legal problems if someone wants to marry himself.

12.1.6   Behavioral Type Requirements

Up to now only type boundaries concerning structural requirements of the type parameters have been discussed. Sometimes it is necessary to incorporate behavioral requirements also. A typical scenario might be the following: A polymorphic operation *translateSet* has to be implemented that translates all elements—presumably geometric objects of various kinds—of a given set by a certain translation vector. Then the requirements of this operation are as follows:

1. The receiver type must be a set.

2. The members of the set must all provide an operation *translate*.

Further, the argument of the operation *translate* is a *Vertex* representing the argument vector for the translation process. The result of the *translateSet* operation is **void**.

While we are familiar with constraints of the first type, capturing behavioral requirements is new. The key idea to specify behavioral constraints is quite simple. We allow type expressions not only to contain structural descriptions but also behavioral descriptions in the form of operation declarations. Thus, a type expression representing a type having a *translate* operation with an argument of type *Vertex* and no return result looks as follows:

[translate: Vertex → **void**;].

Note that the specification of the receiver type has been dropped. Its specification is unnecessary, since the receiver type is assumed to be just the same type the type expression specifies.

From these generalized type expressions, we can again build type boundaries. The following boundaries express that the types being substitutable for the type variable \TraSet must be set types whose elements of type \TraElem provide at least an operation *translate* with an argument of type *Vertex*:

\TraSet ≤ {\TraElem},
\TraElem ≤ [translate: Vertex → **void**;]

Using these type boundaries, we get the definition of an operation *translateSet*, which translates all the elements of a given set by a certain amount provided by a *Vertex*:

```
polymorph declare translateSet
    (\TraSet ≤ {\TraElem}, \TraElem ≤ [translate: Vertex → void;]):
    \TraSet ‖ Vertex → void
    code translateSetCode;
define translateSetCode(t) is
    var geo: \TraElem;
    begin
        foreach (geo in self)
            geo.translate(t);
    end define translateSetCode;
```

Analogously, one could introduce polymorphic operations *scaleSet* and *rotateSet*.

The operation *translateSet* can now be used to translate a whole set of objects, e.g., a group of objects within a CAD application. The application of the operation *translateSet* is illustrated by the following example. Assume the variable declarations

> **var** myCylinderSet: CylinderSet;
>   myCuboidSet: CuboidSet;
>   translationVector: Vertex;

After the initialization of these variables the invocations

> (1)   myCylinderSet.translateSet(translationVector);
> (2)   myCuboidSet.translateSet(translationVector);

of *translateSet* are legal. The effect of (1) is the translation of all *Cylinders* contained in the set *myCylinderSet*, and the one of (2) is the translation of all *Cuboid* elements of the set referenced by *myCuboidSet*.

For validating the above calls to *translateSet*, the type variables \ *TraSet* and \ *TraElem* are substituted by

> (1)   *CylinderSet* and *Cylinder*, respectively
> (2)   *CuboidSet* and *Cuboid*, respectively

Since the types *Cylinder* and *Cuboid* both supply a *translate* operation, the invocations can both be verified type consistent.

12.1.7   Built-In Polymorphic Operations

While the reader may assume that the declaration and definition of polymorphic operations is not necessarily relevant for his or her work, it might be advantageous to be aware of many built-in polymorphic operations. Most of these operations concern list and set processing. In order to grasp the basic idea of these possibilities, this section introduces several declarations of useful operations.

Whereas the operation *first* retrieves the first element of a list, the operation *rest* returns the rest of the list, i.e., removes the first element of a list. Their polymorphic declarations look as follows:

> **polymorph declare** first (\ListType $\leq$ < \ElemType>):
>   \ListType $\parallel$ $\rightarrow$ \ElemType;

> **polymorph declare** rest (\ListType $\leq$ < \ElemType>):
>   \ListType $\parallel$ $\rightarrow$ \ListType;

Another useful operation for modifying a list is *append*. This operation appends to the receiver list the elements of the argument list. Its polymorphic declaration is

> **polymorph declare** append (\ListType $\leq$ < \ElemType>):
>   \ListType $\parallel$ \ListType $\rightarrow$ \ListType;

The most basic operations for set processing are *insert* and *delete*, which insert and delete a given element into/from a set, respectively. An operation implementable on the basis of the *delete* operation is *setminus*, which removes all elements of the argument set from the receiver set. Their respective polymorphic declarations are

polymorph declare insert (\SetType ≤ {\ElemType})
    \SetType ‖ \ElemType → \SetType;
polymorph declare delete (\SetType ≤ {\ElemType})
    \SetType ‖ \ElemType → \SetType;
polymorph declare setminus (\SetType ≤ {\ElemType})
    \SetType ‖ \SetType → \SetType;

There are many other possible and useful polymorphic operations for both list and set processing. The more operations there exist, the easier it is for the database programmer and designer to implement application specific tasks because it allows to build upon these building blocks. But even if the required functionality does not exist in the form of predefined operations, it is the main advantage of object-oriented databases, which provide bounded polymorphism, that these operations can easily be integrated into the system by the user.

12.2   Generic Types

Generic types facilitate the parameterization of type definitions with type parameters. We will first motivate the need for this concept and then illustrate generic types in the GOM model.

12.2.1   Motivation

Consider a simple (warehouse) stock that we would like to realize in GOM. It should be possible to check in goods of a particular kind into the stock and, subsequently, retrieve them from the stock. Furthermore, we would like to offer an operation that totals the value of all goods currently stored in the stock—e.g., for inventory purposes.

Already this high-level, verbal description of a stock indicates that the functionality of the *STOCK* is largely independent of the type of goods being stored. This observation is highlighted by the definition of two concrete *STOCK* types:

- *CuboidStock*, defined in Figure 12.2, is used to store *Cuboids*.

- *WineStock*, defined in Figure 12.3, can be utilized to model a wine cellar.

In both cases, we first created a corresponding *item* type, i.e., *CuboidItem* and *WineItem*, by defining it as a subtype of the respective objects' types that we want to store, i.e., *Cuboid* and *Wine*. Then the respective *STOCK* type is being defined such that it can hold items only of the respective type.

This way of defining the different *STOCK* types bears two severe disadvantages:

- If an additional *STOCK* type, e.g., a *CylinderStock*, is needed, we would have to define yet another *Stock* type whose type definition is—aside from the different element types—largely identical to the two existing.

- If one wants to extend the functionality of all *STOCK* types, one has to modify *all* the type definitions of all the different *STOCK* types. Consider, for example, an additional operation *avgValue*, which determines the average value of the stored goods.

These problems are caused by the fact that the type definitions of *CuboidStock* and *WineStock* did not realize the abstract idea of a *STOCK* but, rather, implemented two concrete *STOCK*s, one for *Cuboids* and the other for *Wines*.

This problem area is alleviated by the concept of *generic type*, which allows one to realize abstract functionality of a type that can then be instantiated to form a concrete *type instance*.

12.2.2   Sample Generic Type in GOM

Generic types provide the skeleton for the instantiation of concrete types. In Chapter 7, we used the metaphorical illustration that types are cookie forms that can be used to generate an arbitrary number of identically shaped cookies. Along these lines a generic type is a general mechanism from which these cookie forms—representing types—can be derived. Objects can only be instantiated from concrete types—generic types do not provide this functionality. Therefore, they first have to be instantiated themselves—yielding a concrete type—before one can create objects.

Generic Types can be parameterized by so-called type parameters. In the definition of the generic type, these type parameters are used like "ordinary" type constraints. The type parameters are typically further constrained (cf. Section 12.1). This is necessary whenever we make assumption about the structure and/or behavior of the type parameter—or, more precisely, about the instances of the type parameter. In an instantiation of a generic type to a concrete type, all type parameters have to be substituted by actual types.

Let us now illustrate generic types and their instantiation on our example type *STOCK*. In Figure 12.4 the generic type *STOCK* is defined and has just one type parameter, called *ItemType*. As before, type parameters generally start with a backslash (\) followed by any valid identifier. The type parameter is further constrained as follows:

$$\text{\textbackslash ItemType} \leq [\text{ itemNo: int; value: float; }]$$

This constraint denotes that—in an instantiation of *STOCK*—we can allow only types whose internal representation is tuple structured and contains the attributes

```
persistent type CuboidItem supertype Cuboid is
    [ itemNo: int; ];

persistent type CuboidStock supertype ANY is
    public store, get, totalValue
    body { CuboidItem }
    operations
        declare store: CuboidItem → void;
        declare get: int  → CuboidItem;
        declare totalValue:  → float;
    implementation
        define store(item) is
            self.insert(item);
        define get(no) is
            var item: CuboidItem := NULL;
            begin
                foreach (item in self)
                    if (item.itemNo = no)
                        begin
                            self.remove(item);
                            break;
                        end;
                return item;
            end define get;
        define totalValue is
            var val: float := 0.0;
                item: CuboidItem;
            begin
                foreach (item in self)
                    val := val + item.value;
                return val;
            end define totalValue;
    end type CuboidStock;
```

Figure 12.2: Definition of the *CuboidItem*-Stock

persistent type Wine **is**
  [name: string; age: int; value: float;];

persistent type WineItem **supertype** Wine **is**
  [itemNo: int;];

persistent type WineStock **supertype** ANY **is**
  public store, get, totalValue
  body { WineItem }
  operations
    declare store: WineItem → **void**;
    declare get: int → WineItem;
    declare totalValue: → float;
  implementation
    define store(item) **is**
      self.insert(item);
    define get(no) **is**
      var item: WineItem := NULL;
      begin
        foreach (item **in self**)
          if (item.item_no = no)
            begin
              self.remove(item);
              break;
            end;
        return item;
      end define get;
    define totalValue **is**
      var val: float := 0.0;
          item: WineItem;
      begin
        foreach (item **in self**)
          val := val + item.value;
        return val;
      end define totalValue;
end type WineStock;

Figure 12.3: Definition of the *WineItem*-Stock

```
persistent generic type STOCK ( \ItemType ≤ [ itemNo: int; value: float; ] ) is
   public store, get, totalValue
   body { \ItemType }
   operations
      declare store: \ItemType → void;
      declare get: int  → \ItemType;
      declare totalValue: → float;
   implementation
      define store(item) is
         self.insert(item);
      define get(no) is
         var item: \ItemType := NULL;
         begin
            foreach (item in self)
               if (item.itemNo = no)
                  begin
                     self.remove(item);
                     break;
                  end;
            return item;
         end define get;
      define totalValue is
         var val: float := 0.0;
             item: \ItemType;
         begin
            foreach (item in self)
               val := val + item.value;
            return val;
         end define totalValue;
end generic type STOCK;
```

Figure 12.4: Definition of a Generic *STOCK*

itemNo of type *int* and *value* of type *float*. Sample types that conform with this type constraint are *CuboidItem* and *WineItem*, as defined before. Therefore, these types can be used in an instantiation of *STOCK*:

> **persistent type** CuboidStock **is** STOCK(CuboidItem);
> **persistent type** WineStock **is** STOCK(WineItem);

12.2.3 Augmenting the Functionality of an Instantiated Generic Type

There exist two possibilities for augmenting the functionality of an instantiated generic type:

1. By adding operations to the type that was derived from the generic type by instantiation

2. By creating a subtype of the instantiated generic type

Let us illustrate the second possibility for augmenting the functionality of the type *WineStock*, which was generated as an instantiation of the generic type *STOCK*. Assume we want an additional function *averageAge*, which returns the mean age of all the wines stored in a particular *WineStock*. We add this functionality by defining the type *MySpecialWineStock* as a subtype of *WineStock*:

```
type MySpecialWineStock is
    supertype WineStock
    public averageAge
    body      !! empty, no additional attributes
    operations
        declare averageAge: → float;
    implementation
        define averageAge is
            var drink: WineItem;
                sum: float := 0.0;
                card: int := 0;
            begin
                foreach (drink in self)
                    begin
                        sum := sum + convertToFloat(self.age);   !! built-in conversion
                        card := card + 1;
                    end;
                return sum / card;
            end define averageAge;
    end type MySpecialWineStock;
```

This new subtype *MySpecialWineStock* can, of course, be used wherever a *Wine-Stock* instance is required.

12.3   Exercises

12.1 An inexperienced GOM programmer could think of the following solution to the marriage problem—that was used to motivate the need for polymorphic operations. Include the attribute *spouse* in the type *LivingBeing* and associate the operation *marry* with *LivingBeing*, i.e.:

> **declare** marry: LivingBeing ‖ LivingBeing → **void**;

Why is this representation problematic? What kind of semantics is missing with respect to the representation chosen in the text?

12.2 Give the definition of the polymorphic operation *setminus*.

12.3 Give the polymorphic definitions and declarations of the following useful list operations:

1. *member*, which returns *true* if the argument is contained in the receiver list, and *false* otherwise.

2. *nth*, which retrieves the *nth* element of the receiver list where *n* is given as an integer argument.

3. *substitute*, which substitutes a given element by another given element within the receiver list.

4. *sublist*, which produces a new list containing the *i*th to the *k*th elements of the receiver list. Assume that *i* and *k* are given as integer arguments.

12.4 Give the polymorphic definitions and declarations of your three favorite set operations.

12.5 Model the famous generic types *Stack* and *Queue* in GOM.

12.6 Consider a real warehouse stock, where items are as different as socks, refrigerators, wine, etc. Can a generic type be useful in defining a stock of items of different kind? Hint: Do virtual types help? Does the given generic type definition of *STOCK* have to be changed?

12.7 Continue modeling GOM within GOM. Add polymorphic operations and generic types to your model.

12.4 Annotated Bibliography

Type parameters in the context of database languages are surveyed by Cardelli [Car88]. Polymorphic operations are analyzed by Cardelli and Wegner [CW85]. They introduced the term *bounded polymorphism*. Also, Danforth and and Tomlinson [DT88] survey polymorphic operation concepts of various languages. The programming language ML—described by Milner [Mil78]—incorporates very general polymorphic operations. In ML the type consistency is verified by static type derivation, rather than type constraints specified by the user. A similar polymorphic operation capability forms the basis of the data model Machiavelli, designed by Ohori, Buneman, and Breazu-Tannen [OBBT89].

B. Meyer [Mey86] was one of the first authors to point out that inheritance and genericity are two dual concepts—none of which can be satisfactorily modeled by the other. Liskov and Guttag [LG86] wrote an excellent text on data modeling using generic type facilities. Their book is based on the type concepts of the Clu language [LAB+81]. The programming language Ada [ANSI83] provides a limited generic type concept.

13

Multiple Inheritance and Multiple Substitutability

In this chapter, we will describe two concepts that increase the expressiveness and flexibility of the object model beyond what can be achieved with single inheritance. These two data modeling constructs are the following:

1. *Multiple inheritance*

2. *Multiple substitutability*

The first concept, multiple inheritance, extends the super/subtype taxonomy in the way that types may have more than one supertype. The second modeling construct, multiple substitutability, allows to substitute objects even in those places where a different type is required.

13.1 Multiple Inheritance

Up to this point, when discussing inheritance we restricted ourselves to *single* inheritance; i.e., inheritance was restricted to the case where each type had at most one supertype. While this assumption is appropriate in most common cases, there are situations when it is useful to define types inheriting from several supertypes. As an example consider a factory and its products. These products play different roles within the factory. On the one hand, they are items for sale needing all the attributes a sales department likes to see. Let us call the type capturing the sales department's view of a product *CommercialProduct*. On the other hand, the manufacturing department views the same product from, e.g., a geometrical point of view. Call the type providing this view *GeometricObject*. In this case, the product should subsume the features provided by both types: *CommercialProduct* as well as *GeometricObject*. This can be modeled by a type inheriting features from both, *CommercialProduct* and *GeometricPrimitive*. As this mechanism generalizes single inheritance—where inheritance is allowed from only one supertype—it is called *multiple inheritance*. Of course, multiple inheritance also affects the substitutability in the way that an object is substitutable in all those places where any of its supertypes is required.

13.1.1   Introduction to Multiple Inheritance

Single inheritance is characterized by the requirement that each type directly inherits from *at most one* other type. The incorporation of the common root type *ANY* in the generic object model (GOM) actually leads to the situation that a type directly inherits from *exactly* one other type. This gives rise to a single hierarchical type structure (actually a tree), part of which is (abstractly) sketched in Figure 13.1. Under single inheritance there is always a unique, linear path from any type, such as OT_n, to the root (ANY) of the inheritance hierarchy, e.g.:

$$OT_n \rightarrow \boxed{\text{is-a}} \rightarrow \cdots \rightarrow \boxed{\text{is-a}} \rightarrow OT_1 \rightarrow \boxed{\text{is-a}} \rightarrow ANY$$

This characteristic guarantees the unique selection of an operation invocation or attribute access based on the direct type of the receiver. For example, upon invoking $o.op(\ldots)$ where o is a direct instance of, say, OT_n, the system searches—as described in Section 10.7—along the linear path from OT_n in direction to the root ANY until a matching feature *op* is found.

Multiple inheritance, on the other hand, relaxes the requirement that a type has at most one supertype. Rather, it allows a type to inherit the features of several supertypes. For example, a type OT_3 may inherit all the features of OT_1 and OT_2, which leads to a graphical representation of the **is-a** relationships as shown below:

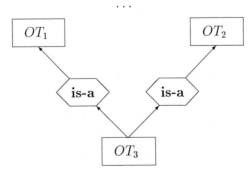

It is important to note that all three consequences—inheritance, subtyping, and substitutability—of single inheritance carry over to multiple inheritance. Thus:

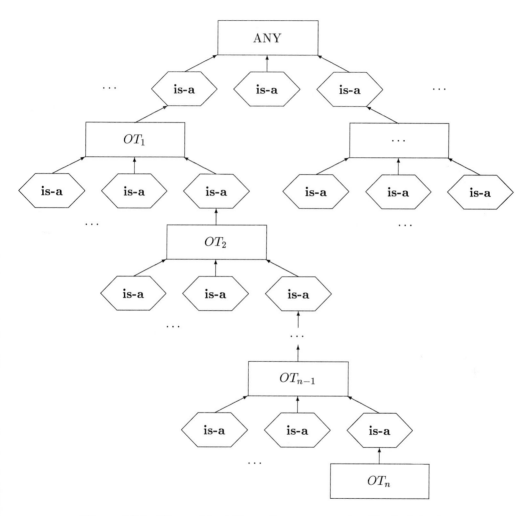

Figure 13.1: Hierarchical Type Structure under Single Inheritance

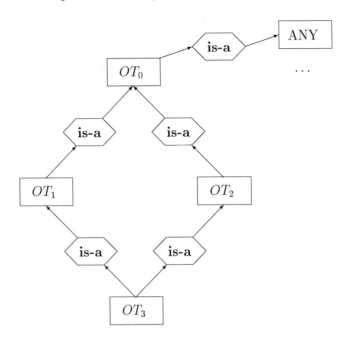

Figure 13.2: Multiple Inheritance Type Structure

- OT_3 inherits from OT_1 *and* from OT_2.

- An object of type OT_3 **is-a** OT_1 *and* OT_3 **is-a** OT_2. That is, the objects of type OT_3 belong to the type extensions of OT_1 and OT_2.

- An object of type OT_3 is substitutable for objects of type OT_1 *and* for objects of type OT_2.

Under multiple inheritance there may still be one common root, ANY, for all types. But it is no longer guaranteed that there is a unique linear path from a given type along the **is-a** relationship to the root ANY. In general, the type structure under multiple inheritance forms a directed acyclic graph (DAG)—that is, the type structure cannot contain any cycles.

It is even possible that a type, say, OT_3, inherits from a type OT_0 via two different paths. Consider, for example, the case in which OT_1 and OT_2 have a common supertype OT_0. This leads to the graphical representation of (a part of) the type structure as shown in Figure 13.2. Here, the features of OT_0 are propagated to OT_3 along two different inheritance paths: once via OT_1 and once more via OT_2.

The ability to uniquely determine the version of an operation op to be executed when invoked on an instance of a particular type can no longer be assured in a multiple-inheritance type system. Consider, for example, an operation op that is defined in both types, OT_1 and OT_2. Then we could invoke *op* on a direct instance of OT_3 to which the variable *v* may refer:

> **var** v: OT_3;
> v.create(. . .);
> . . .
> v.op(. . .);

This invocation of *v.op*(. . .) leads to ambiguity. It could mean either one of the following:

1. The operation *op* as defined in OT_1 is executed.

2. The operation *op* as defined in OT_2 is selected for execution.

Even features that are defined in only one subtype lead to problems. Consider—in our example type structure of Figure 13.2—an operation op' that is defined in OT_0. Then OT_3 inherits op' twice; via OT_1 and also via OT_2. This does not render any ambiguity as long as op' is not refined. But a refinement of op' in either OT_1 or OT_2 generates a conflict because no unique version of op' can be determined for an invocation of op' on an instance of type OT_3.

13.1.2   Illustration of Operation Ambiguity

Let us now demonstrate the ambiguity of operation invocations on a more intuitive example based on the type hierarchy shown in Figure 13.3.

This sample type hierarchy is mostly self-explaining. The type *TeachAsst* (teaching assistant) is defined as a supertype of both, *Student* as well as *Employee*. This accounts for the two separate contexts in which a *TeachAsst* can be viewed:

1. As a *Student* with all the features the *Student* type provides, such as *gpa* (grade point average)

2. As an *Employee* in his or her role as an *Employee* who receives a *salary*

In addition, a *TeachAsst* instance has its own, specialized features such as a set-valued attribute *tutoring*, which refers to the set of *Students* being tutored by the respective *TeachAsst*.

Let us now define two operations, both being identically named, for *Student* and *Employee*:

declare bonus: Student ‖ float → **void**	**declare** bonus: Employee ‖ float → **void**
code bonusForStudents;	**code** bonusForEmps;
define bonusForStudents(gpaInc) **is**	**define** bonusForEmps(salInc) **is**
self.gpa := **self**.gpa ∗ gpaInc;	**self**.salary := **self**.salary ∗ salInc;

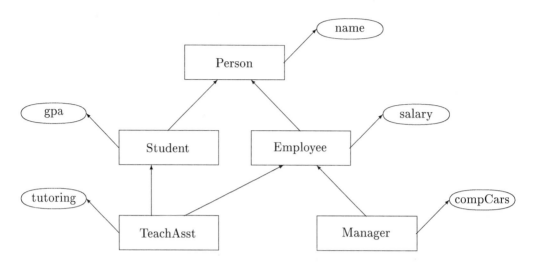

Figure 13.3: A Type Hierarchy with Multiple Inheritance

The above operation definitions render two different *bonus* operations:

1. The one associated with the type *Student* increases the *gpa* (grade point average) by the specified percentage *gpaInc*.

2. The operation *bonus* associated with the type *Employee* increases the *salary* by the specified percentage *salInc*.

An ambiguity problem occurs when the operation *bonus* is invoked on a *TeachAsst* instance as demonstrated in the following program fragment:

> **var** bestEmp: Employee;
>   bestStudent: Student;
>   myTA, your TA: TeachAsst;
>   boss: Manager;
>
> . . .

(1)	bestEmp := myTA;	
(2)	bestEmp.bonus(1.1);	!! supposedly the *salary* should be increased
(3)	bestStudent := your TA;	
(4)	bestStudent.bonus(1.05);	!! supposedly the *gpa* should be increased
(5)	boss.bonus (1.5);	!! no ambiguity can occur

In statement (1) the *TeachAsst* referred to by *myTA* is assigned to the variable *bestEmp*. When *bonus* is invoked on *bestEmp*, it is impossible by the system to determine whether *Student$bonus* or *Employee$bonus* is to be executed—supposedly

Employee$bonus is intended because *bonus* is invoked on a variable referring normally to an *Employee*.

The analogous situation occurs in statement (4) when bonus is invoked on the object referenced by the variable *bestStudent*, which, again, is a *TeachAsst* instance—as indicated by the assignment in statement (3).

However, no ambiguity can occur in the invocation of *bonus* in statement (5) because a *Manager* object inherits merely one *bonus* operation, i.e., the one from *Employee*.

Actually, the above program fragment leads to semantic ambiguity but not to a type conflict because the two *bonus* operations both have the same signature. In general, however, multiple inheritance may cause type conflicts that preclude static verification of type consistency (cf. Section 13.1.4).

13.1.3   Resolving Ambiguity under Multiple Inheritance

An object-oriented system that allows multiple inheritance has to provide a mechanism to deal with such binding conflicts. There are several possible approaches to resolve a binding ambiguity, some of which will be briefly discussed below.

User-Specified Priority

This approach relies on the user to define a priority among the types whose features are inherited. For example, for the type structure

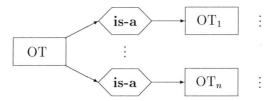

the object type designer could state the supertypes in the **supertypes** clause of *OT* in a specific order, like

> **type** OT
>   **supertypes** OT_1, ..., OT_n **is**
>   ...

This order indicates that upon invocation of

> o.op(...)

for an operation *op* and a direct instance *o* of type *OT* the system binds *op* to the version that is found first using the following search strategy:

(0) Determine whether *op* is defined in OT.

(1) Inspect the DAG—the directed acyclic type graph—starting at OT_1 whether it contains a definition for *op*; i.e., check whether OT_1 or any of OT_1's ancestors contains a definition (or refinement) for *op*. OT_1's ancestors are visited in the analogous order based on the order in the **supertypes** clause of OT_1.

. . .

(n) Inspect the DAG starting at OT_n whether it furnishes an implementation of *op*—proceed analogously to step (2).

The first implementation of *op* found by the inspection algorithm sketched above is the one that is selected by the run-time system for execution.

Explicit Renaming

This alternative avoids ambiguities altogether. It simply—and quite drastically—prohibits any ambiguities. It requires that all features inherited by OT from $OT_1, \ldots,$ OT_n have to have distinct names. However, if this is not fulfilled it is not necessary to modify the definitions of $OT_1, \ldots, OT_n$. This could have catastrophic effects on other, existing applications that are based on these types. Rather, it is possible to rename inherited features that generate potential ambiguities within OT. Renaming inherited features requires the modification of the **supertypes** clause. For each supertype, a list of renaming specifications for features inherited from that supertype is provided. The **supertypes** clause has now the following format for our abstract example:

    type OT
       supertypes OT$_1$ (**renames** op **to** op$_1$),

            . . .

            OT$_{m-1}$ (**renames** op **to** op$_{m-1}$),
            OT$_m$,

      is . . .

  . . .

    end type OT;

Here, we assumed that *op* is inherited by OT via m different supertypes $OT_1, \ldots,$ OT_m. Then *exactly* $m - 1$ versions have to be renamed in the type definition.

Now an invocation of *op* on a direct instance of OT will uniquely trigger the execution of the version of *op* that was inherited from OT_m—provided that OT did not itself refine this feature, in which case the renaming would have been unnecessary because all potentially inherited versions of *op* would have become invisible, anyway.

Refinement

Consider, again, the example type structure pictured in Figure 13.2, in which the type OT_3 has the supertypes OT_1 and OT_2, which in turn have the common supertype OT_0. There exist two possible cases of ambiguity for an operation op within type OT_3:

1. The operation op is defined in type OT_0—or, equivalently, in a supertype of OT_0—and at least one of the types OT_1 and OT_2 refines it.

2. The operation op is not defined in type OT_0, but both types OT_1 and OT_2 have a definition for op.

In both of the above cases the problem is which code to execute. This ambiguity can be resolved by explicitly stating which code to execute for the operation op. The natural way to do this is to refine op within the type definition frame of OT_3. Here, we have at least three possibilities:

1. The refined code of op within OT_3 is newly implemented.

2. The refined code of op within OT_3 is specified as being the same as the code for op within OT_1.

3. The refined code of op within OT_3 is specified as being the same as the code for op within OT_2.

For these possibilities the relevant part of the type definition frame is stated.

> **type** OT_3
>   **supertypes** OT_1, OT_2 **is**
>   . . .
>   **operations**
>   **refine** op . . .
>   **implementation**
>   **define** op . . .
>   . . . !! new implementation of op for OT_3
> **end type** OT_3;

By implementing OT_3 in the above way, calling op for an object of type OT_3 results in the execution of the code newly implemented.

In order to execute the code for op as defined in type OT_1, the following refinement is appropriate:

> **define type** OT_3
>   **supertypes** OT_1, OT_2 **is**
>   . . .

```
      operations
          refine op ...
      implementation
          define op ...
               self.OT₁$op(...);     !! execute OT₁'s implementation of op
      end type OT₃;
```

Here, the notation *type$operation* was utilized to directly indicate what type should form the start point for searching for the code to use for an operator invocation. Whenever an expression of this form is encountered, the search for the code to be executed does not start at the type of the receiver object but at the type *type* as specified preceding the "$" sign. Note that the **super** clause cannot be utilized in this case, since there exist multiple supertypes. Thus, the code referred to by **super**.*op* is not uniquely determined.

The resolution strategies for resolving ambiguities cannot deal with all possible cases. Especially in the context of strong typing, ambiguities can only be resolved if there does not exist a type conflict. This matter is discussed next.

13.1.4   Type Conflicts

The renaming of multiply inherited features and the inspection algorithm may achieve a unique selection of an invoked feature. However, these strategies cannot outrule type conflicts. For instance, if—in our example in Figure 13.2—the types OT_1 and OT_2 both provide an attribute A of incompatible types AT_1 and AT_2, there is no chance to resolve this type conflict without modifying the schema of either OT_1 or OT_2. This, however, may lead to the invalidation of existing applications that rely on the existing schema definitions. Note that single inheritance never requires modifications of a supertype's definition—thereby leaving all existing applications invariant.

Such a kind of conflict may happen if OT_1 and OT_2 furnish identically named operations, say, f, which have different result types. For example:

$$\text{declare } f: \ OT_1 \ \| \ldots \ \rightarrow T_1;$$

$$\text{declare } f: \ OT_2 \ \| \ldots \ \rightarrow T_2;$$

Then, for an instance o_2 of direct type OT_2 the invocation $o_2.f(\ldots)$ returns an object of type T_2. However, if o_2 referred to a direct instance of type OT_3—which is perfectly legal under our substitutability rules—this same operation invocation may yield a result of type T_1 under the condition that OT_1 is named first in the **supertypes** clause of OT_3. Since T_1 and T_2 may be of incompatible type, this could lead to a typing error, e.g., in the following nested invocation:

$o_2.f(\ldots).g(\ldots)$

if g were defined for instances of type T_2 but not for objects of type T_1.

Such a typing error could already manifest itself due to incompatible argument numbers or argument types in the operation f.

Under multiple inheritance, typing conflicts can, in general, only be avoided by outruling the inheritance of conflicting features. In such a system, only those types can serve as multiple supertypes, which provide distinct or compatible features.

In all other cases, the responsibility for avoiding run-time errors due to incompatible types is placed on the shoulders of the data type users. They have to—by thorough, yet error-prone inspection—verify that the invoked operations do not cause type mismatches of the kind described above.

Let us illustrate the occurrence of a type conflict on a more intuitive example based on our type structure shown in Figure 13.3. Assume we have two different operations *skill*:

- The *skill* operation associated with *Student* returning a *float* value, e.g., the *gpa*

- The *skill* operation associated with *Employee* returning a *string*, e.g., one of the values "low" or "high"

These operations could be defined as follows:

declare skill: Student ‖ → float
    code studentSkillCode;

define studentSkillCode **is**
    return self.gpa;

declare skill: Employee ‖ → string
    code employeeSkillCode;

define employeeSkillCode **is**
    if (**self**.salary > 100000.00)
        return "high"
    else
        return "low";

Now consider the following program fragment:

```
var myTA: TeachAsst;
    someStudent: Student;
    someEmp: Employee;
    empSkill: string;
    studentSkill: float;
    ...
(1)    someEmp := myTA; someStudent := myTA;
(2)    empSkill := someEmp.skill;
(3)    studentSkill := someStudent.skill;
```

One of the two assignment statement—either (2) or (3)—is bound to be type inconsistent. Since both variables, *someEmp* and *someStudent*, refer to the same object—the one referred to by *myTA*—the invocation of *skill* either returns a *float* value or a *string*. Thus, the program fragment leads to a type inconsistency that can be detected only at run time.

This type inconsistency is caused by the substitutability of subtype instances for any of their supertypes. Here, a *TeachAsst* can be substituted in places where an *Employee* is expected as well as in places where a *Student* is required.

It seems that a loss of type safety is the price one has to pay when using multiple inheritance—unless all the supertypes have nonconflicting features. It depends on the applications whether this price is worth the benefits gained by multiple inheritance. In most cases, a careful design of the applications will reveal that multiple inheritance is largely unnecessary to achieve a clean and natural modeling—some researchers even argue, that multiple inheritance is counterproductive in achieving a clean structuring of the types. This is especially true for the database context, in which the notion of extension is often used to support queries (cf. Chapter 14). Utilizing multiple inheritance just for code reusability can lead to unexpected results for queries.

13.1.5   Benefits of Multiple Inheritance

So far, we have only discussed the problems induced by multiple inheritance. Of course, multiple inheritance also provides benefits beyond those of single inheritance.

Multiple inheritance is advantageous whenever a single database instance should be treated under two or more rather unrelated semantic viewpoints. The functionality of each of the different semantic contexts could be inherited from a separate supertype under multiple inheritance. Let us illustrate this point on an example. The type hierarchy with the root *GeometricPrimitive* roughly speaking implements the geometric view of its instances. If some of the instances of this type hierarchy are also to be seen as, say, commercial products, one could acquire the associated features from a type *CommercialProduct*. This type can be outlined as follows:

```
type CommercialProduct supertypes ANY is
    body [price: float;
          qoh: float,           !! quantity on hand
          productionCost: float;
          distributor: Dealer;];
    operations
        declare profitMargin: → float;
        ...
    implementation
        define profitMargin is
            ...
end type CommercialProduct;
```

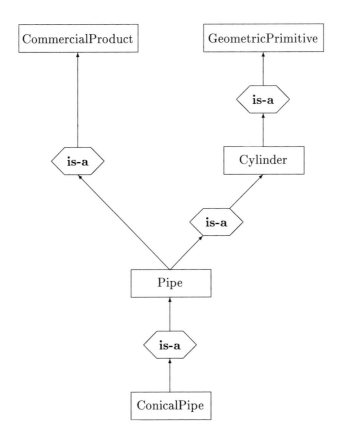

Figure 13.4: Conceptual Schema for *Pipes* and *CommercialProducts*

The *profitMargin* is computed based on the sales price (*price*) and the production costs (*productionCost*)—and possibly some other criteria, like storage costs, distribution costs, etc.

On the basis of the type definitions introduced so far, a company that manufactures and markets pipelines could define the type *Pipe* such that it inherits the features of *Cylinder* as well as those of *CommercialProduct*. Then the overall type hierarchy looks as shown in Figure 13.4.

In this type hierarchy, *Pipe* and *CommercialProduct* provide features under the two distinct semantic contexts:

1. The geometric modeling

2. The modeling of commercial aspects, such as sales price, production costs, etc.

13.2   Multiple Substitutability

One of the key concepts of object orientation that provides the main modeling power for complex applications is the possibility to build taxonomies utilizing the **is-a** relationship associated with subtyping and inheritance. One consequence of a type being a subtype of another is that objects of the subtype may be substituted for objects of the supertype. However, further substitution is not allowed. As turned out, this restriction inhibits the "true" modeling of certain situations common in many engineering applications. A mechanism is introduced to influence the substitutability property without affecting the taxonomy as represented by the supertype hierarchy. This mechanism also remedies some situations where multiple inheritance fails to correctly capture the semantics of the application domain.

13.2.1   Illustrating the Shortcomings of Multiple Inheritance

Consider modeling tools whose functionality is based solely on some constituent part. A typical example might be a knife consisting of a handle and a blade. The knife's cutting capability, i.e., its functionality, depends only on the blade. More specifically, the cutting capability of the knife depends on the length of the blade and its material, which must be harder than the material it is supposed to cut. Another cutting tool is a scissor, whose cutting capability depends on the material of its two (scissor) blades. Further, there exists a maximum thickness a scissor can cut—mainly depending on the blade's length.

   This situation has to be captured by a set of type definitions. Since we are only interested in the functionality of the tools and their respective components, we do not model the connections of the subparts to the whole. Nevertheless, we want to model the subparts itself. The type definitions for *Material, WorkPiece, Blade,* and *Knife* are as follows:

```
type Blade is                              type Knife is
   body                                       body
      [length: float;                            [handle: Handle;
       mat: Material;];                            blade: Blade;];
   operations                                 operations
      declare canCut: WorkPiece → bool;          declare canCut: WorkPiece → bool;
   implementation                             implementation
      define canCut(piece) is                    define canCut(piece)
         return piece.thickness ≤ self.length         return self.blade.canCut(piece);
            and piece.mat.hardness ≤     end type Knife;
               self.mat.hardness;
end type Blade;
```

                    type Material **is**   **type** WorkPiece **is**
                        body   **body**
                            [...   [thickness: float;
                              hardness: float;];   mat: Material;];
                        end type Material;   **end type** Piece;

The types *ScissorBlade* and *Scissor* are then specified as shown below:

type ScissorBlade **is**   **type** Scissor **is**
  body   **body**
    [bladeLength: float;   [blade1: ScissorBlade;
     mat: Material;];   blade2: ScissorBlade;
  operations   maxthickness: float;];
    declare canCut: WorkPiece $\rightarrow$ bool;   **operations**
  implementation   **declare** canCut: WorkPiece $\rightarrow$ bool;
    define canCut(piece)   **implementation**
      return piece.thickness $\leq$ **define** canCut(piece)
        self.bladeLength **and**   **return self**.blade1.canCut(piece) **and**
        piece.mat.hardness $\leq$ **self**.blade2.canCut(piece);
        self.mat.hardness;   **end type** Scissor;
end type ScissorBlade;

Assume the type *Handle* to be defined elsewhere. With each of the types *Blade*, *Knife*, *ScissorBlade*, and *Scissor* we have associated an operator *canCut*. Note that all the different declarations have the same argument and result types.

We now come to the crucial point of modeling a (simple) Swiss knife, which consists of a knife part and a scissor part. That is, the knife as well as the scissors are *part-of* the Swiss knife. It is common practice to model these situations by applying multiple inheritance as follows:

    type SwissKnife
      supertypes Knife, Scissor **is**;
      ...   !! see discussion below
    end type SwissKnife;

We start the analysis of the above model of a Swiss knife with a discussion of the appropriateness of the **is-a** relationship. We then turn our attention to the substitutability property. After that a brief look at the *canCut* operator is taken. A discussion of the inherited attributes completes our analysis. The latter includes a search for the subparts of a Swiss knife. During the analysis, we will talk about appropriateness rather than correctness, since there still exists a certain degree of freedom influenced by personal taste.

Most people will agree that a Swiss knife *is a* knife. Thus, the implication of *Knife* being a supertype of *SwissKnife* on the **is-a** relationship seems appropriate. Almost nobody will agree that a Swiss knife *is a* scissor. Thus, the implication of *Scissor* being a supertype of *SwissKnife* appears wrong. Consequently the use of multiple inheritance in this example results in an inappropriate manipulation of the taxonomy.

Assume a person having a Swiss knife is asked by another person for a knife. Surely, the former will hand his or her Swiss knife to the latter who in turn will *use* the Swiss knife as a knife. The same remains true if we replace *knife* by *scissor*. Consequently the use of multiple inheritance results in an appropriate manipulation of the substitutability relation in that *SwissKnife*s become substitutable for both, *Knifes* and *Scissors*.

Note that the operator *canCut* has to be refined in order to avoid ambiguities on the code selection if it is applied to an instance of type *SwissKnife*. The refinement can either mirror the *canCut* operator of *Knife* or *Scissor*, or some other implementation, e.g.:

> **define** canCut(piece)
>   **return self**.Knife$canCut(piece) **or self**.Scissor$canCut(piece)

Due to inheritance a *SwissKnife* has the structural representation:

> [handle: Handle;
> blade: Blade;
> blade1: ScissorBlade;
> blade2: ScissorBlade;
> maxthickness: float;]

This set of attributes is just the mash of all attributes of the supertypes *Knife* and *Scissor*. The only way to associate them to the appropriate subparts of a *SwissKnife* is to look at its supertypes. Moreover, the identity of the subparts is lost. There is no way to refer to the scissor of a *SwissKnife* as an object. Nevertheless, this might be very useful, e.g., if the scissor is broken and must be replaced. Writing an operation to replace the module for the above kind of modeling appears to be a cumbersome task. Further, any change to a subpart results in a modification of this procedure.

Moreover, considering a Swiss knife having more tools, like a corkscrew, a screw-driver, etc., the mash of attributes becomes even more severe and the probability of unrelated and unnecessary attribute/operator conflicts increases.

The conclusion we have to come up with is that the *part-of* relationship of a Swiss knife has been poorly modeled utilizing multiple inheritance.

13.2.2   Multiple Substitutability

Considering the above analysis an alternative modeling of a Swiss knife would be the following:

```
type SwissKnife
   supertype Knife is
   body [scissor: Scissor;];
      . . .
end type SwissKnife;
```

where *SwissKnife* has *Knife* as its only supertype. Here, we have avoided the mash of attributes. Also, the scissor remains identifiable as a subpart. This is not true for the knife. Later, this will be corrected. Further, a *SwissKnife* is-a *Knife* but *is not* a *Scissor*. We maybe content with this modeling alternative except for the effects on the substitutability relationship: A *SwissKnife* is substitutable for a *Knife*, but it is *not* substitutable for a *Scissor*.

In order to correct the latter deficiency, we introduce a new clause called **fashion**. It allows objects of a type A to be substitutable for objects of type B by explicitly stating these types and providing methods that allow the object to behave in a fashion as if it were an object of type B. We exemplify this with our Swiss knife example:

```
type SwissKnife
   supertype Knife is
   body [scissor: Scissor;];
   fashion Scissor via self.scissor;
      . . .
end type SwissKnife;
```

The intended semantics is that whenever a *SwissKnife* is used in the context of a *Scissor* all operation invocations are directed to the *scissor* component of the Swiss knife.

Consider the following piece of code:

```
var sk: SwissKnife;
    k: Knife;
    s: Scissor;
    p: WorkPiece;
p.create;
sk.create;
    . . .
k := sk; s := sk; sk.canCut(p); k.canCut(p); s.canCut(p);
```

The first assignment is valid, since *SwissKnife* is a subtype of *Knife*. The second assignment is valid, since the **fashion** clause allows the usage of a *SwissKnife* whenever a *Scissor* is demanded.

The first call of *canCut* invokes the *Knife$canCut*, since this is the only *canCut* operation inherited. The second call obviously invokes the same code. The third call of *canCut* results in an evaluation of the **fashion** clause. It indicates that all calls to a *SwissKnife* in the context of a *Scissor* are redirected to the result of the expression following the keyword **via**. On this result the actual operation, here *canCut*, is invoked. In our example, the whole procedure then resembles the evaluation of

> s.scissor.canCut(p)

In order to give a justification of the title of this section, we demonstrate the use of several (here two) **fashion** clauses.

```
type SwissKnife
    body [knife: Knife;
          scissor: Scissor];
    fashion Knife via self.knife;
    fashion Scissor via self.scissor;
    ...
end type SwissKnife;
```

The main difference to the first modeling alternative using the **fashion** clause is that a *SwissKnife* is no longer a subtype of *Knife*.

Let us characterize these first applications of the **fashion** clause in more general terms. An object has been modeled and consists of several subparts. Nevertheless, any of these subparts provides an object in its own right. Moreover, these subparts can be used ignoring the surrounding, i.e., the encompassing object. In fact, we have treated the whole object as if it were only a part (of itself). This became obvious when substituting the whole object where only a part of it was needed.

13.2.3   Representative Problems

The following scenario reflects a quite common situation in which a message is sent to an object that then chooses a representative who is responsible for taking care of it. A *Department* consists of a set of *Employees*. If a company meeting is scheduled, each *Department* receives an invitation and chooses a representative who has to attend the meeting. Usually the *manager* of the *Department* is chosen. Nevertheless, the invitations are often sent to the department as an institution and the secretary of the *Department* then redirects the incoming mail to the respective *Employee*, the *manager* in our case. Thus, the *Department* must be able to behave like an *Employee*. This can be modeled using the **fashion** clause:

```
type Department supertype ANY is
    body [ members: {Employee};
           manager: Manager;]];
    fashion Employee using self.representative;
    operations
```

```
        declare representative: → Employee;
        . . .
    implementation
        define representative is
            return self.manager;
        . . .
    end type Department;
```

We assume that by default the *representative* is set to the *manager* of a department. Whenever the manager is absent the *representative* is set to another employee of the department.

In this implementation, the **fashion** clause implements the secretary redirecting all incoming messages to the manager. The consequence is that each department can play the role of an employee. Especially, it may be inserted into a set of employees:

```
    var meetingParticipants: EmployeeSet;
        developmentDep: Department;
        bigBoss: Manager;
        someEmp: Employee;
    . . .
    meetingParticipants.insert(bigBoss);          !! no meeting without him
    meetingParticipants.insert(developmentDep); !! they will send a representative
```

A representative is selected via **self**.*representative* each time when we access *Employee* properties of a *Department*. If we print the name of all participants— for example, to get a mailing list—the representative is selected by evaluating the expression *emp.Name*:

```
        foreach (emp in meetingParticipants)
            print(emp.name);
```

Since the selection of the representative is hidden behind the interface of *Department*, it can easily be exchanged. For example, if the current representative is absent, we may point out another one without the need of updating the set *meetingParticipants*.

Another, very important problem in the class of representative problems is a *versioned object*. In engineering applications—no matter, whether electrical, mechanical, or software engineering—it is common practice to maintain several versions. As a default there often exists a *current version*. This situation can now be modeled by defining a versioned object to contain a set of objects that represent the different versions of the object to be designed. Each time an operator is invoked on the versioned object without specifying the version, this operator is applied to the current version. Thus, the versioned object, a collection of objects, must behave the same way as a single version. Again, the realization utilizes the **fashion** clause. Since modeling versioning and maintaining different versions is a very common and important problem in engineering applications, there exists a devoted chapter on it (cf. Chapter 17).

13.2.4   Treating Parts as a Whole

The problems of the last sections could be paraphrased as *substituting the whole where only a (sub-)part was actually required*. Now tables are turned: We demonstrate the possibility to simulate a whole object by "gluing" together some selected parts. Let us demonstrate this by means of an example. A house can be (partially) characterized by the following attributes and operations:

1. Number of floors

2. A list of tenants

3. The people moving in and out

This is reflected in the following type definition:

> **type** RealEstate **is** ... **end**;
>
> **type** House **supertype** RealEstate **is**
>   **public** noOfFloors, moveIn, moveOut
>   **body** [noOfFloors: int;
>   tenants: <Person>;
>   ...];
>   **operations**
>   **declare** moveIn: Person → **void**;
>   **declare** moveOut: Person → **void**;
>
>   ...
>
>   **implementation**
>   ...
>
> **end type** House;

With respect to these features a *HouseBoat* is very similar to a *House*: It possesses all of the features of a *House*. We admit that a slightly different terminology is used; e.g., the *floors* of the houseboat are called *decks*. Surely, nobody standing in front of a houseboat would classify it as a house since there is a significant difference between a house and a houseboat: A House **is-a** *Real Estate* and thus per definition not movable while a houseboat can easily change its location. Consequently, a *HouseBoat* cannot have *House* as a supertype.

Nevertheless, there are some cases, in which we want to treat the *HouseBoat* like a *House*, e.g., a move enterprise does not make any difference between a *House* and a *HouseBoat*.

> **type** MoveEnterprise **is**
>   ...
>   **operations**

> **declare** move: Person, House, House → float;   !! returning the cost
> . . .
> **implementation**
>   **define** move (a Person, from, to) is
>   . . .
> **end type** MoveEnterprise;

Assume the existence of some type definition for *HouseBoat*. To perform a *move* from a *House* to a *HouseBoat* some code along the lines of the following must be executed:

> **var** moveAndSonsInc : MoveEnterprise;
>   me   : Person;
>   oldResidence   : House;
>   newResidence   : HouseBoat;
> . . .
> moveAndSonsInc.move(me, oldResidence, newResidence);

In order to support the invocation of *move* the *HouseBoat newResidence* must be substitutable for *House* objects but, as mentioned before, the usage of inheritance again leads to a conflict with the real world's taxonomy.

A better solution to the problem is provided by using multiple substitutability:

> **type** HouseBoat **supertype** Boat **is**
>   **body** [noOfDecks: int;
>   tenants: PersonSet;
>   . . .];
>   **fashion** House
>   **where** noOfFloors: int **is self**.noOfDecks;
>   **where** moveIn: Person → **void is self**.moveIn(p);
>   **where** moveOut: Person → **void is self**.moveOut(p);
>   **operations**
>   **declare** moveIn: Person → **void**;
>   **declare** moveOut: Person → **void**;
>   . . .
>   **implementation**
>   . . .
> **end type** HouseBoat;

In the **fashion** clause, we map the properties of *HouseBoat* to the properties of *House*. Note that renaming of attributes—as we did for *noOfDecks*—is a special case of such a mapping.

It is important to notice that the simulated object (here, the *House*) does not have its own identifier. Instead, the identifier of the object providing the selected parts (here, the *HouseBoat*) is preserved. It is this fact that allows us to realize updatable views utilizing the **fashion** clause.

13.2.5 Discussion

There are three main reasons why (multiple) inheritance may be inappropriate:

1. Inheritance affects the taxonomy of types.

2. Inheritance clashes identities and mashes attributes.

3. Unexpected (and senseless) inheritance conflicts are possible.

In these cases, substitutability as provided by the **fashion** clause might be the right way to model the situation.

13.3 Exercises

13.1 Reconsider Figure 13.3. The call of the method *bonus* has been detected as ambiguous. Devise three solutions for resolving the ambiguity by applying

- User specified priorities
- Explicit renaming
- Refinement.

Try to specify (different) parameters for the different *bonus* calls, depending on the type the operation is associated to. Utilize overloading. Discuss the different possibilities.

13.2 Illustrate a type conflict resulting from incompatible operation parameters based on the type structure shown in Figure 13.3.

Hint: Revise the operation *bonus* such that it requires different parameter types for *Student* and *Employee*.

Illustrate that a refinement of the two different inherited *bonus* operations in *TeachAsst* cannot resolve the type conflict.

13.3 Why is it so bad to require the modification of the supertypes of a newly defined type in order to resolve typing conflicts and binding ambiguity? Illustrate your discussion on the basis of the type structure shown in Figure 13.3.

Hint: Consider existing applications that are based on the original type definitions. Illustrate the problems on examples.

13.4 Model the following types. A four-legged thing provides legs of a certain length and material. A platform has a geometric shape, consists of a material, and is able to hold things up to a certain weight. Model chairs and desks using multiple inheritance.

13.5 Assume a Swiss knife consists of a blade of a certain length that is able to cut different kinds of material. A small scissor that is able to cut paper and a screwdriver are also provided. Last but not least a corkscrew completes it. Every tool has a certain length and provides a certain capability. Model a Swiss knife using multiple inheritance. Try an alternative modeling with single inheritance only where the Swiss knife solely **is-a** knife, which additionally consists of the other tools. What is the difference?

13.6 From a modeling point of view, multiple inheritance has been used for different purposes within this chapter. In the example involving *CommercialProduct* and *GeometricPrimitive* (cf. Figure 13.4) multiple inheritance was utilized to simulate something similar to views as known from the relational area. Solely for software engineering purposes, e.g., code reusability, multiple inheritance was applied in Exercise 13.4. Modeling the Swiss knife in Exercise 13.5, multiple inheritance provides a tool for simplifying or supporting aggregation.

Discuss the different exploitations of multiple inheritance and point out the conceptual differences. In which cases is multiple inheritance appropriate? When is it not? Pay attention to the appropriateness of the **is-a**-relationship, inheritance (i.e., what attributes does each type provide?), and the substitutability.

13.7 Model the above introduced type *SwissKnife* consisting of a knife and a scissor without any inheritance. Provide an additional **fashion** clause that allows a *SwissKnife* to be substituted for *Knifes* as well.

13.4 Annotated Bibliography

Multiple inheritance was incorporated in many of the recent programming languages, such as C++ [Str90b]. Also, some Smalltalk-80 (dialects) support multiple inheritance. The semantics of multiple inheritance is studied by Cardelli [Car84].

The **fashion** concept was developed by Moerkotte and Zachmann [MZ92a]. The **fashion** concept resembles view mechanisms as discussed in, e.g., [AB91, SLT91].

14

Associative Object Access

In this chapter we describe the associative object access capabilities of the generic object model (GOM). The *associative* access can be characterized as a means to retrieve objects on the basis of describing the desired properties of the objects rather than specifying a (procedural) method for finding the sought objets. The associative access is based on selecting qualifying objects that are contained in collections. The possible collections, on which associative queries can be performed, are type extensions, user-defind set objects, and list objects.

In GOM there are two possibilities for associative access to objects:

- Within the object-oriented programming language, called GOMpl

- Via two declarative query languages, which are called GOMql and GOMsql

In GOMpl the operators for associative object access are predefined as polymorphic operations that are very flexible. The declarative query language GOMql is based on the relational language QUEL. It is mainly used for interactively accessing the object base, e.g., for adhoc queries or browsing. The other declarative query language GOMsql is based on the relational query language SQL, which was already described in Chapter 4.

14.1 Associative Selection of Objects

The polymorphic operations—as introduced in Chapter 12—are so expressive that they facilitate the realization of highly flexible search operations in an *orthogonal* way. In many other models these operations are constructed somewhat outside the paradigm of the language. In GOM we provide predefined *select* operations that determine the objects contained in a collection—i.e., in a set or list object or in a type extension—that satisfy the particular selection predicate. This selection predicate is

349

passed as an argument to the invocation of the select operation. However, we have to distinguish the *select* operations according to the number of additional arguments that this selection predicate expects. Therefore, in GOM we provide predefined polymorphic *select* operations with an arbitrary number (e.g., no, one, two) of additional parameters. By overloading, these operators all have the same name *select*.

14.1.1   Selection Predicate without Parameter

In this (easiest) case a single Boolean function is passed as the parameter to the *select* operation. The signature of the polymorphic *select* operation then looks as follows:

> **polymorph overload** select $(\backslash t_1 \leq \{\backslash t_2\}) : \ \backslash t_1 \ \| \ (\backslash t_2 \ \| \ \rightarrow \ \text{bool}) \ \rightarrow \ \backslash t_1$
> code selectNoParam;

The above signature specifies that *select* is an overloaded polymorphic operation—remember that we want to define further *select* operations, therefore the overloading—which has two arguments:

- The receiver object of type $\backslash t_1$ which is a set of elements of type $\backslash t_2$

- A Boolean function, which operates on objects of type $\backslash t_2$ and returns a result of type *bool*. Thus, the Boolean function has the signature $(\backslash t_2 \ \| \ \rightarrow \ bool)$.

We could easily define a similar *select* operation applicable for *list* objects—instead of *set* objects.

The principal implementation of this *select* operation, which is coded under the name *selectNoParamCode* as specified in the **code** clause of the signature, is semantically equivalent[1] to:

> **define** selectNoParam(selPred) **is**
>   **var** result: $\backslash t_1$;
>   candidate: $\backslash t_2$;
>   **begin**
>   result.create;   !! the empty result set is generated
>   **foreach** (candidate **in self**)
>   **if** candidate.selPred **then** result.insert(candidate);
>   **return** result;
>   **end define** selectNoParam;

In the implementation, we first initialize the variable *result* to refer to an empty set—this is implicit by the *create* invocation on *result*. The type of the *result* object coincides with the type of the type variable $\backslash t_1$, i.e., the type of the receiver object

---

[1] This is true only if *selPred* does not have side effects.

on which *select* is invoked. Then the variable *candidate* iterates through the set **self** and determines whether the selection predicate *selPred* is satisfied for the particular element of the set. If this is the case the respective element of the set—referred to by *candidate*—is inserted into the set *result*. Note that the *candidate* object nevertheless also remains in the set **self**. Finally, the *result* set is returned. A query returning a subset of the receiver set is sometimes called *selection, restriction,* or *subset query*. It is the most common form of query in object-oriented database management systems. The implementation of such a query is often not by a loop as indicated above, but by more efficient methods, which exploit index structures (cf. Part V of this book).

Let us now sketch a simple example application of the associative retrieval of objects using the *select* operation. For this purpose we define a new Boolean function *inOrigin* on the receiver type *Cuboid*. This operation determines whether one of the eight bounding vertices—referred to by the attributes *v1, v2, ..., v8*—lies in the origin of the coordinate system (since we deal with floating point arithmetic we have to account for computational imprecision, therefore the $\pm\varepsilon$):

> **declare** inOrigin: Cuboid $\|\ \rightarrow$ bool;

> **define** inOrigin **is**   !! check whether any boundary *Vertex* lies in the origin
> **return** ((**self**.v1.x $= 0.0 \pm \varepsilon$ **and**
> **self**.v1.y $= 0.0 \pm \varepsilon$ **and**
> **self**.v1.z $= 0.0 \pm \varepsilon$) **or**
>
> ...
>
> (**self**.v8.x $= 0.0 \pm \varepsilon$ **and**
> **self**.v8.y $= 0.0 \pm \varepsilon$ **and**
> **self**.v8.z $= 0.0 \pm \varepsilon$));

Based on this Boolean function—which serves as the selection predicate—we can now formulate the program fragment that retrieves the qualifying *Cuboids*:

> **var** myCuboids, theCuboidsInOrigin: CuboidSet;
>   ...
> theCuboidsInOrigin := myCuboids.select(inOrigin);

After executing the above program fragment the set object *theCuboidsInOrigin* contains all those *Cuboid* instances that are

- Contained in the set *myCuboids*

- Satisfy the selection predicate *inOrigin*

Note that the qualifying *Cuboids* are now elements of both sets; i.e., they remain elements of *myCuboids* and become elements of *theCuboidsInOrigin* set.

Note that the selection predicate *inOrigin* is only applicable on *Cuboid* instances— and instances of *Cuboid* subtypes, of course. On the other hand, the polymorphic *select* operation is applicable on any set type; not only on *CuboidSet* objects. It merely requires that a proper selection predicate—which is applicable on the elements of the receiver set object—is passed as a parameter.

For example, assume we have a type *Apple* with a Boolean function *isRed*:

> **declare** isRed: Apple ‖ → bool;
> **define** isRed **is**
>
>   ...

This function (somehow) determines whether or not the receiver *Apple* object is red. Then we can retrieve all red *Apples* from the *AppleSet* object *myApples* as follows:

> **var** redApples, myApples: AppleSet;
>
>   ...
>
> redApples := myApples.select(isRed);

After execution of the above program, the set *redApples* contains (references to) all the red *Apples* contained in the set *myApples*.

Especially when the selection predicate is a very simple Boolean function—as in our *Apple* example the function is Red—it appears rather tedious having to specify the signature and implementation of a type-associated function. To alleviate this problem it is also possible to merely pass the implementation of the operation in the form of the Lambda (λ) notation. For our *Apple* example, this can be done as follows—assuming an *Apple* has an attribute *color*:

> redApples := myApples.select(λ x: x.color = "red");

In this case, the selection predicate is passed as a so-called *anonymous* Boolean function. It is called anonymous because of the lacking function name; only the implementation is passed as a parameter. For computationally complex selection predicates, however, a *named* Boolean function should be specified in order to aid the readability and reusability of the programs. Note that anonymous functions cannot be reused; rather, they have to be respecified each time they are used.

Let us demonstrate the flexibility of the *select* operator on one more example. In Chapter 10, we defined the object type *Employee* with an associated Boolean operation *isRetired*, i.e.:

> **declare** isRetired: Employee ‖ → bool;

We can use this boolean function to retrieve all retired *Employees* from the extension *ext(Employee)*:

> **var** retiredEmps: EmployeeSet;
>
>   ...
>
> retiredEmps := ext(Employee).select(isRetired);

This example illustrates the invocation of *select* on a type extension. This, of course, demands that the *extension* is maintained for the respective type (which we assume here for *Employee*).

14.1.2   Selection Predicates with Parameters

We will now illustrate the principal use of parameterized selection predicates. Assume that the selection predicate requires one additional parameter. An example of such a selection predicate is

> **declare** bigCuboid: Cuboid ∥ float → bool;

This Boolean function returns *true* if the *Cuboid*, on which it is invoked, has a volume exceeding a certain threshold that is passed as the *float* parameter. That is, the operation *bigCuboid* is invoked on a *Cuboid*, say, *myCuboid*, as follows:

> myCuboid.bigCuboid(10.0);

and returns true if the volume of *myCuboid* exceeds 10.0.

Unfortunately, such a selection predicate cannot be utilized in the *select* operation as defined thus far. We, therefore, define another overloaded *select* operation that can handle this additional parameter:

> **poly overload** select $(\backslash t_1 \leq \{\backslash t_2\}) : \backslash t_1 \parallel (\backslash t_2 \parallel \backslash t_3 \rightarrow bool), \backslash t_3 \rightarrow \backslash t_1$
>   **code** selectOneParam;
>
> **define** selectOneParam(selPred, p1)
>   **var** result: $\backslash t_1$;
>   candidate: $\backslash t_2$;
>   **begin**
>   result.create;
>   **foreach** (candidate **in self**)
>   **if** candidate.selPred(p1) **then** result.insert(candidate);
>   **return** result;
>   **end define** selectOneParam;

In the above signature of *select*, the type parameter $\backslash t_3$ is not constrained in the type boundaries—the only condition that is implicitly imposed is $\backslash t_3 \leq$ ANY and the correspondence of the types of the third parameter to *select* and the first and only parameter of *selPred*. Thus, the additional parameter that is represented by $\backslash t_3$ can be of arbitrary type.

We should recall that the *select* operation is not only polymorphically defined; it is also overloaded because of the different numbers of parameters. The compiler can easily determine the "correct" version from the number of arguments passed in the invocation.

Let us now demonstrate the use of this additional *select* operation on the example *bigCuboid*, which was already motivated above. The Boolean function *bigCuboid* has the following (completed) specification:

> **declare** bigCuboid: Cuboid ‖ float → bool
>   **code** bigCuboidCode;
> **define** bigCuboidCode(threshold) **is**
>   **return** (**self**.volume > threshold);

This selection predicate is then used to determine those objects of the set *myCuboids*, whose volume exceeds 200.0:

> **var** myCuboids, myBigCuboids: CuboidSet;
>   . . .
> myBigCuboids := myCuboids.select(bigCuboid, 200.0);

Note that the parameter 200.0, which is the argument of the selection predicate *bigCuboid*, is passed as an ordinary parameter to the *select* invocation; in the implementation of *selectOneParam*, it is then "shifted" to become the argument of *selPred*.

Analogously, GOM provides built-in *select* operations for any number of additional parameters—and, due to polymorphism, for any types of the additional parameters. Let us demonstrate the invocation of the *select* operation with a predicate requiring two parameters. Consider the example Boolean function *volumeRange* that determines whether the *volume* of the receiver *Cuboid* lies in between the boundaries *low* and *high*. This function has the following form:

> **declare** volumeRange: Cuboid ‖ float, float → bool;
> **define** volumeRange(low,high) **is**
>   **return** (self.volume ≤ high **and** self.volume ≥ low);

Then we can specify any kind of *range query* on the *volume* of *Cuboids*, e.g.:

> ext(Cuboid).select(volumeRange, 10.0, 32.5);

This query determines all *Cuboids* in the object base whose *volume* ranges between 10.0 and 32.5.

14.1.3   Iterators

Iterators are useful for operating on large object sets. They can be viewed as a "filter" that selects all the relevant objects from the set and then passes these objects on for further processing. In GOM, we already introduced the **foreach** loop, which allows to sequentially iterate through a set of objects and process each one individually. We can now extend this concept with a so-called filter that first determines the qualifying objects; then the sequential processing is performed only on the subset that passed the filter. Such a filter can easily be modeled as a *select* operation. Let us demonstrate this on the following example application:

```
    declare bigCyl: Cylinder ‖ float → bool
        code bigCylCode;

  define bigCylCode(threshold)
        return (self.volume ≥ threshold);

    . . .
    var c: Cylinder;
        myCylinders: CylinderSet;
        bigCylindersTotalWeight: float := 0.0;
        . . .
    foreach (c in myCylinders.select(bigCyl, 20.0))
        bigCylindersTotalWeight := bigCylindersTotalWeight + c.weight;
```

In the above program fragment, we first specified a Boolean function *bigCyl* on the receiver type *Cylinder*—this function is analogous to the previously defined function *bigCuboid* on the type *Cuboid*. (Actually, we could have defined these two functions as one polymorphic operation—the polymorphic definition is left as Exercise 14.1 to the reader). In the **foreach** loop of the program the *weight* of those *Cylinders* that pass the *select* operation is totaled. That is, the *select* operation serves as a filter that lets only pass those *Cylinder* instances whose *volume* exceeds 20.0. Only those (big) *Cylinders* are processed in the **foreach** loop—therefore, only the *weights* of the big *Cylinders* are totaled.

14.2 GOMql: A Declarative Query Language

In this section, we introduce the declarative query language GOMql. This high-level language is based on the relational query language QUEL, which was developed for the relational database system INGRES.

14.2.1 The Example Object Base *DeptStore*

The subsequent discussion of the query language GOMql is based on a sample object base called *DeptStore*, which models some small fraction of a department store. The reader should not be misled to believe that object-oriented query languages are useful only for handling queries in such standard administrative application domains. We have chosen to start with a standard (administrative) application of which we trust most of the readers have a very intuitive understanding. Later on in this chapter we will illustrate the query language GOMsql on a technical application (boundary representation of geometric objects).

The definitions of the object types involved in the example are as follows:

type Emp **is**
    [name: string;
    worksIn: Dept;
    salary: int;];

type Manager
    supertype Emp **is**
    [cars: {Car};];

type Dept **is**
    [name: string;
    mgr: Manager;
    profit: int;];

type Car **is**
    [license: string;
    make: string;
    horsePower: int;];

We have shown only the structural representation of the involved types—the operations are handled in the query language like ordinary attributes (see below). Therefore, they are omitted in this type schema.

The above object base schema specifies that employees (*Emps*) work in at most one department (*Dept*)—this is modeled as a single-valued attribute *worksIn*. A department is managed by a *Manager*, which is represented via the single-valued attribute *mgr*. The type *Manager* is defined as a subtype of *Emp* with the additional attribute *cars*, which constitutes a set of company cars that the *Manager* is entitled to use. Thus, the attribute *cars* is set valued. The type *Car* contains merely atomic attributes; that is, no object references emanate from a *Car* object. A small extension of this schema is shown in Figure 14.1.

We want to emphasize a so-called *reference chain* that is present in this example schema. It corresponds to the *path expression P* specified as (a formal definition of path expressions will be given later in Chapter 20—for the time being an intuitive understanding suffices):

$$P \equiv Emp.worksIn.mgr.cars.make$$

This path expression is graphically visualized:

$$Emp \xrightarrow{\;worksIn\;} Dept \xrightarrow{\;mgr\;} Manager \xrightarrow{\;cars\;} Car \xrightarrow{\;make\;} string$$

The arrows are marked with the attribute names from which the object references emanate. The single-valued attributes are represented as single-headed arrows; set- or list-valued attributes (e.g., the attribute *cars*) are depicted as double-headed arrows.

Let us take a closer look at the attribute *cars*. It relates *Manager* instances with a set of *Car* objects (which may be used by the *Manager*). Even though our example object base in Figure 14.1 does not contain such an example, it is quite possible that the same *Car* is associated with two different *Managers*; e.g., if the company has *Car* sharing. This is achieved by inserting the respective *Car* into two different sets referred to by the different *Managers* via the *cars* attribute. Thus, the attribute *cars* models a general binary $N : M$ relationship.

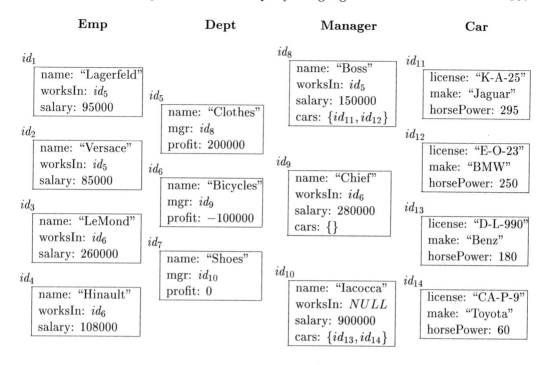

Figure 14.1: Sample State of the Object Base *DeptStore*

14.2.2   Base Concepts of GOMql

A (simple) query in GOMql has one of the following two syntactical structures:

range $r_1 : S_1, \ldots, r_m : S_m$	**range** $r_1 : S_1, \ldots, r_m : S_m$
retrieve r_i	**retrieve** $a_1 : r_{i_1}, \ldots, a_j : r_{i_j}$
where $P(r_1, \ldots, r_m)$	**where** $P(r_1, \ldots, r_m)$

The r_j $(1 \le j \le m)$ denote *range variables*, which are bound to collection-valued expressions in the *range* clause of the query. Thus, S_j $(1 \le j \le m)$ has to be one of the following:

- A type extension—which can just be denoted by the type name instead of $ext(\ldots)$

- A variable referring to a set-structured object

- A variable referring to a list-structured object

- A variable containing a set-structured value

- A variable containing a list-structured value

- An expression evaluating to a set-structured object

- An expression evaluating to a list-structured object

- An expression evaluating to a list-structured value

- An expression evaluating to a list-structured value

In all cases, the range variables r_j $(1 \leq j \leq m)$ are implicitly constrained to the element type of the respective collection S_j.

In the simplest case, which is shown on the left-hand side, the **retrieve** clause of the query contains merely one range variable. These queries are therefore called *single-target queries*. In general, however, a query may construct tuples in the **retrieve** clause by projecting onto several range variables—as shown on the right-hand side. Consequently, such a query is called *multi-target query*.

The *where* clause contains the selection predicate P. This predicate P is tested for any possible binding $(r_1, \ldots, r_m) \in S_1 \times \ldots \times S_m$. If the predicate $P(r_1, \ldots, r_m)$ is satisfied the actual binding of r_i is included in the result set—or, in the case of multi-target queries, the bindings of the $r_{i_1}, \ldots, r_{i_j}$.

14.2.3   Example Queries in GOMql

Instead of a detailed specification of the query language, we choose to give an intuitive explanation of the GOMql, constructs based on some example queries.

Simple Selection Predicate

In this first, very simple query we want to determine the *Emp*s whose *salary* exceeds 100000:

> **range**   *e*: Emp
> **retrieve** *e*
> **where**   *e*.salary > 100000.0

This query determines as a result—based on the object base state shown in Figure 14.1—the set $\{id_3, id_4, id_8, id_9, id_{10}\}$. Note that *Manager*s are also *Emp* objects because of the subtype relationship between *Manager* and *Emp*. Therefore, the qualifying *Manager* objects are contained in the result set.

Unfortunately, the result consists merely of a set of object identifiers. They may be useful for constructing temporary sets that are then used for further processing. But for a user interested in an interactive dialog with the object base system the (system-generated) identifiers are of little use. Therefore, the query is paraphrased to output the names of these *Emp*s:

range e: Emp
retrieve e.name
where e.salary > 100000.0

This query now yields the output

name
"LeMond"
"Hinault"
"Boss"
"Chief"
"Iacocca"

Path Expression in a Query

In this query, we will utilize the path expression that was already introduced above. This path expression

$$P \equiv Emp.worksIn.mgr.cars.make$$

will be part of the selection predicate in the **where** clause. The semantics of the query is verbally specified as:

"Find the *Emp*s whose *Manager* drives (among others) a Jaguar".

range e : Emp
retrieve e
where   "Jaguar" **in** e.worksIn.mgr.cars.make

For a given *Emp*-binding of the range variable e, the *string* values that can be reached via the path

$$e.worksIn.mgr.cars.make$$

are inspected. That is, the system determines—by traversing along the indicated reference chain—all strings and checks whether one of the strings equals "Jaguar". For example, for the binding $e = id_1$, we determine

$$e.worksIn.mgr.cars.name = \{\text{"Jaguar"}, \text{"BMW"}\}.$$

Thus, id_1 qualifies as a result of this query.

With respect to our example object base in Figure 14.1, this query yields the result set $\{id_1, id_2, id_8\}$. Again we could easily modify the query to return the names of the qualifying *Emps*. Note that the *Manager* instance identified by id_8 is contained in the result set—because of the substitutability of *Managers* for *Emps*.

The processing of this sample query causes tremendous costs in run time as our following analysis indicates. If the object base has no supplementary index structures that can be exploited for the evaluation of this query, the query processor has to inspect all objects of the types *Car*, *Manager*, *Dept*, and *Emp* at least once. Thus, the run time of processing the query is (at least) proportional to

$$\#(Emp) + \#(Dept) + \#(Manager) + \#(Car)$$

where $\#(t)$ denotes the cardinality of the type extension $ext(t)$.[2]

More Complex Selection Predicates

In this query we want to determine the *Managers* who lead a department (*Dept*) that generates a loss but, nevertheless, pay at least one of their *Emps* an exorbitant *salary* exceeding 200000. The query is formulated in GOMql as follows:

> **range** e : Emp, m : Manager
> **retrieve** m
> **where** $m = e$.worksIn.mgr **and**
> e.salary > 200000.0 **and**
> e.worksIn.profit < 0.0

The **where** clause contains several path expressions which are partially overlapping, i.e.:

- *e.worksIn.mgr* leading from a given *Employee* to his or her *Manager*

- *e.worksIn.profit* which leads from the same *Employee* e to the *profit* attribute of the department (*Dept*), in which e works

Even though this query is considerably more complex than the foregoing ones the high-level declarative formulation still enables an intuitive understanding. The result of the query—again, with respect to our sample object base in Figure 14.1—is the single *Manager* object with object identifier (OID) id_9.

A statement of the form $m = e$.*worksIn.mgr* is called a *functional join*. Note that the equal operator "=" on objects is implicitly interpreted as identity.

In the next query, we want to find those *Managers* who drive a *Car* that is too "noble" despite a low *profit* they achieve in their department. We interpret a "noble" *Car* as one that is manufactured by "Jaguar" or whose *horsePower* exceeds 150.

---

[2]Actually, in this analysis we assumed that each object of type *Car* is referenced by at least one object of type *Manager*, each *Manager* is referenced by at least one *Dept* via the attribute *mgr*, and each *Dept* is referenced by at least one *Emp* via the attribute *worksIn*. Otherwise, the processing time could be lower.

> **range** d : Dept, m : Manager, c : Car
> **retrieve** m
> **where** $m = d$.mgr **and**
> d.profit < 100000.0 **and**
> c **in** m.cars **and**
>   (c.horsePower > 150 **or**
> c.make $=$ "Jaguar")

This query yields the singleton set $\{id_{10}\}$ as a result with respect to our sample object base.

14.2.4 Quantifiers

For more complicated queries it is necessary or convenient to use quantifiers. GOMql provides both kinds of identifiers, i.e., *universal* (**forall**) as well as *existential* (**exists**) quantification.

Universal Quantification

Let us first illustrate the use of the universal quantifier **forall**. Assume we want to retrieve those *Emps* whose *Manager* drives only Jaguars. The following query would retrieve the set of qualifying *Emps*—which is empty for our example object base, though:

> **range** e: Emp, c: Car
> **retrieve** e
> **where forall** c **in** e.worksIn.mgr.cars
>   (c.make $=$ "BMW")

Existential Quantification

Existential quantification is already implicitly provided in GOMql by the **in** predicate. Nevertheless, GOMql also offers an explicit existential quantifier called **exists**. The use of the **exists** quantifier is demonstrated on an example query that was already stated before using the implicit existential quantification. Let us restate the query to retrieve those *Emps* whose *Manager* drives (among others) a *Car* that is manufactured by Jaguar:

> **range** e: Emp, c: Car
> **retrieve** e
> **where exists** c **in** e.worksIn.mgr.cars
>   (c.make $=$ "Jaguar")

14.2.5   Nested Queries

To enhance the power of the query language GOMql, *nested queries* are allowed. Any query expression is thought of as a regular set-valued expression. Hence, it can occur anywhere a set-valued expression is expected—especially in expressions within the **range**, **retrieve**, and **where** clauses. Nested queries are used for several purposes. In complicated queries it may be necessary or convenient to nest queries. If a nested result is desired, nested queries—where nesting takes place in the **retrieve** clause— might be necessary. An example of a nested subquery is given in the following query, which retrieves all those *Managers* who drive only *Cars* that are made by Jaguar or BMW or that have a *horsePower* exceeding 200.

```
range m: Manager, c: Car
retrieve m
where forall c in m.cars
    (c in
        (range v:  Car
         retrieve v
         where v.horsePower > 200 or
               v.make = "Jaguar" or
               v.make = "BMW"))
```

Actually, the above query could have been stated without nesting the subquery— see Exercise 14.3.

14.2.6   Invocation of Operations and Aggregate Functions

With the next two queries, we demonstrate the use of type-associated functions in the **where** clause and in the **retrieve** clause of GOMql queries. For this purpose, we return to our geometric object type hierarchy of Chapter 10. In the first query, we want to retrieve the *Cylinder*s whose *volume* exceeds 150.0:

```
range    c : Cylinder
retrieve c
where    c.volume > 150.0
```

Note that because of subtyping, this query also retrieves all *Pipe* and *ConicalPipe* instances whose *volume* exceeds 150.0. Therefore, the refined operation *volume* has to be bound dynamically in the course of processing this query—analogous to dynamic binding in GOMpl (cf. Section 10.7). Note that whenever an operation is used within a query, it has to be side-effect free. We will refer to a side-effect free operation as a *function*.

The next query is devised to illustrate the use of a type-associated function in the **retrieve** clause. Additionally, we will use the *aggregate* function *sum*. An *aggregate* *function* is a function returning a single value for a given set. Example aggregate functions are *count, sum, min, max, avg*, each of them returning a number.

In the example, the aggregate function *sum* is applied to total the *weight* values of the golden *GeometricPrimitive* objects. Note that *GeometricPrimitive* may still possess a type extension even though it is a virtual type—the type extension contains only subtype instances, e.g., *Cylinders* and *Cuboids*.

> **range**   *g* : GeometricPrimitive
> **retrieve** sum(*g*.weight)
> **where**   *g*.mat.name = "Gold"

Here, we have applied a type-associated function within the **retrieve** clause, and, on the result, an aggregate function.

14.3   GOMsql: An SQL-Based Object Query Language

As pointed out in Chapter 4 the standard query language in relational database systems is SQL. Therefore, several different SQL "dialects" were devised for object-oriented models. Here we describe one such SQL extension for our model, called GOMsql, which is a subset (abstracting from minor syntactical differences) of the query language O$_2$SQL. The objective of the GOMsql query language is to provide an SQL-based declarative, set-oriented interface to the object base, which also exploits the navigational links that are present in an object-oriented schema by tuple-, set-, or list-valued attributes, i.e., attributes that refer to one or more objects.

14.3.1   The Example Schema: Boundary Representation

Our discussion of the GOM query language will be based on an example schema to model geometric objects in boundary representation. The following type definitions form the schema of our example geometric (toy) database:

```
type BRep                         type Face is
   with extension is                 body
   body                                 [surface: float;
      [name: string;                     edges: EdgeSet;];
       weight: float;                  ...
       faces: FaceSet;];            end type Face;
   ...                             type FaceSet is
end type BRep;                        body
                                         { Face }
                                      ...
                                   end type FaceSet;
```

type Edge **is**
   body
      [length: float;
        vertices: VertexSet;];
      ...
  end type Edge;

type EdgeSet **is**
   body
     { Edge }
     ...
  end type EdgeSet;

type Vertex **is**
   body
      [x, y, z: float;];
     ...
  end type Vertex;

type VertexSet **is**
   body
     { Vertex }
     ...
  end type VertexSet;

The tuple-structured objects of type *BRep* constitute the *root* of each geometric object from which the subcomponents can be reached via attributes that reference sets of lower-level objects.

Let us assume that this example schema contains, among other geometric objects, a cuboid that we will call "cubo#5". This cuboid is modeled in the straightforward way: The corresponding *BRep* object has an attribute *name* with the value "cubo#5". Reachable from the *BRep* object via the attribute *faces* is the set of—in the case of a cuboid—six objects of type *Face*, each of which is a tuple with the atomic attribute *surface* and a set attribute *edges*. The *EdgeSet* referenced by *edges* contains references to four *Edge* objects. Note that *edges* models a general $N : M$ relationship, whereas *faces* models a $1 : N$ relationship (cf. Chapter 9). Each *Edge* object refers—via the attribute *vertices*—to a set of *Vertex* objects, which consist of only atomic-valued attributes.

We see that there is a reference chain modeled within our schema definition that can be represented as:

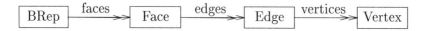

A small fraction of the database extension for our example object "cubo#5" is outlined in Figure 14.2. The object id_5 constitutes the tuple that models the root of our cuboid "cubo#5".

14.3.2   The Basic Query Constructs

The most basic building blocks for constructing queries in GOMsql are the

select ... **from** ... **where** ...

expressions—often abbreviated SFW. For example,

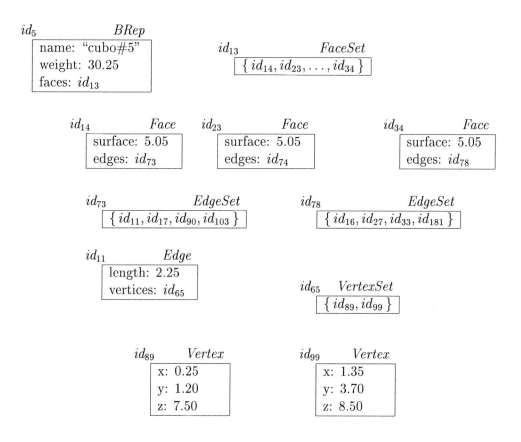

Figure 14.2: Object Base with a *BRep* Instance

select b.weight
from b **in** BRep
where b.name = "cubo#5"

retrieves the *weight* of all the *BRep* objects whose *name* attribute equals "cubo#5" (note that *name*—unlike the unique system-generated object identifier—is not necessarily unique; so there may be more than one object satisfying the condition).

This simple query exhibits already one difference to SQL concerning the specification of the range variables of a query. In GOMsql the explicit specification of range variables, such as b, is required whenever a query is supposed to "work on" a collection, i.e., a set or list of objects. On the other hand, SQL assumes an implicit range variable whenever none is specified. In this case, the relation (or view) name specified in the **from** clause serves as the implicit range variable. Range variables are therefore called *aliases* in SQL terminology because they rename the relation (or view) within the scope of the query. The specification of an explicit range variable is carried out in the **as** statement in SQL. Thus, an equivalent SQL query to the above GOMsql retrieval statement can be formulated in two ways: (1) without an explicit range variable, or (2) with an explicit range variable *b*.

(1) **select** weight
    from BRep
    where name = "cubo#5"

(2) **select** *b*.weight
    from BRep **as** *b*
    where *b*.name = "cubo#5"

A very major distinction between SQL and GOMsql concerns the object collections over which a query may range. In SQL a range variable—whether implicit or explicit—can only be bound to an existing, named set of tuple-valued objects, i.e., a relation or named view that has been created prior to stating the query. This, in particular, prohibits the creation of intermediate result relations on the basis of which further querying is done. Thus, the **from** clause of an SQL query can only contain the name of an existing base relation or of a previously defined view, but not a query itself.

GOMsql is much more orthogonal in that respect. A range variable can be bound to any object collection, whether the collection is associated with a named persistent set, a set- or list-valued attribute, or even a temporarily created object collection. In its full consequence, this orthogonality allows to bind a range variable to another SFW expression, since such an expression returns a collection of objects as a result. This leads to nesting of SFW constructs that is described in subsequent sections.

14.3.3 Range Variables

In GOMsql range variables are bound to object collections in the **from** clause of the SFW expression. The **from** clause has the following general form:

⟨from-clause⟩ ::= **from** ⟨rangeVar⟩ **in** ⟨objectCollection⟩
{ , ⟨rangeVar⟩ **in** ⟨objectCollection⟩ }

In this syntactical definition an *objectCollection* corresponds either to a stored or computed set or list, or it may itself be an SFW expression, i.e., a query, such that the range variable *rangeVar* is bound to the result of the query.

⟨objectCollection⟩ ::= (⟨SFW⟩) | ⟨Set⟩ | ⟨List⟩

14.3.4 Nested Query Expressions

Before turning to full-fledged realistic examples of nested queries, we want to illustrate the principle on a rather simple—yet contrived—query of nesting depth 2. For this purpose assume that our type definition of *Face* contained two operations *weight* and *surface*. On this basis, we could pose the query

> "Find the *Faces* that have a *surface* greater than 10 and belong to a *BRep* instance that has a *weight* greater than 1000."

This query can be easily formulated in two steps. In the first step we could select all *Faces* of the qualifying (i.e., heavy) *BRep* instances:

> **select** *b*.faces
> **from** *b* **in** BRep
> **where** *b*.weight > 1000

Based on this subquery, we can now restrict the set of *Faces* to those that satisfy the predicate "*f.surface > 10*".

> **select** *f*
> **from** *f* **in**
>   (**select** *b*.Faces
>   **from** *b* **in** BRep
>   **where** *b*.weight > 1000)
> **where** *f*.surface > 10

In this encompassing query, the range variable *f* is bound to the set of *Face* objects returned by the innermost SFW expression. In this binding of range variables, GOMsql does not distinguish between stored sets such as the extension of type *BRep* and intermediate, computed sets.

In GOMsql, a range variable has a unique type. In our example, the range variable *b* is of type *BRep* and *f* is of type *Face*. The type of a range variable can be deduced at compile time.

The foregoing example demonstrated the use of nested SFW expressions to traverse along set valued paths, e.g.:

In general, a range variable cannot only be bound to a set of objects that is retrieved via a path expression but to any object set. This is exemplified on the following rather contrived query, which retrieves the *name* of those *BRep* objects that have a *weight w* such that (100 < w < 1000). In a rather clumsy way, the query can be stated in two steps:

1. Create an intermediate result set of *BRep* instances whose *weight* is greater than 100.

2. On the basis of the intermediate result find those objects, whose *weight* is less than 1000.

The GOMsql query formulated in this way looks as follows:

select b_2.name
from b_2 **in**
  (**select** b_1
    from b_1 **in** BRep
    where b_1.weight > 100)
where b_2.weight < 1000

In this example the range variable b_1 is bound to the type extension of *BRep* as in the query in Section 14.3.2. The range variable b_2 is bound to the intermediate result set of *BRep* objects returned by the inner SFW expression, i.e., to those objects whose weight exceeds 100.

Before the reader of this book starts formulating queries in such a clumsy way, let us briefly state the retrieval in a much more elegant fashion:

select b.name
from b **in** BRep
where b.weight < 1000 **and** b.weight > 100

14.3.5   Binding Range Variables to Set-Valued Attributes

Let us once more illustrate the binding of range variables to set-valued attributes in GOMsql by way of a more complex example. In the following query, we want to retrieve the coordinate vectors of the eight bounding vertices of our cuboid "cubo#5":

(1)   **select** v.x, v.y, v.z
(2)   **from** v **in**
(3)   (**select** e.vertices
(4)   **from** e **in**
(5)   (**select** f.edges
(6)   **from** f **in**
(7)   (**select** b.faces
(8)   **from** b **in** BRep
(9)   **where** b.name $=$ "cubo#5")))

To explain this query, let us start with the innermost **select** ... **from** ... **where** ... (SFW) expression, i.e., lines (7)–(9). In this SFW statement, the *range variable b* is bound to the type extension *BRep*. The predicate $b.name =$ "cubo#5" restricts the range of b to only those *BRep* objects that satisfy this requirement. Let us assume that this predicate is true for only one object, the one that models our particular cuboid. Then the innermost SFW expression returns the objects of type *Face* that are contained in the set referred to by attribute *faces*; i.e., for the cuboid six tuple-structured *Face* objects are retrieved.

Analogously, in the SFW expression of lines (5)–(6) the tuple variable f ranges over all those objects returned by the innermost SFW expression, i.e., the six faces of the cuboid. The expression returns the set of all *Edge* objects that are referenced via the set-valued attribute *edges* of the *Face* objects. If we denote the *Faces* with $f_1, \ldots, f_6$ then this SFW expression returns the union

$$\bigcup_{f \in \{f_1, \ldots, f_6\}} f.edges$$

Thus, for our cuboid example 12 *different Edge* objects are returned by this nested SFW expression.

Similarly, the SFW statement (3)–(4) returns the set of (references to) the eight *Vertex* objects of the modeled cuboid, "cubo#5".

Finally, the outermost SFW expression retrieves the three attribute values $v.x$, $v.y$, and $v.z$ of each of the eight *Vertex* objects over which the tuple variable v ranges. The attributes x, y, and z are atomic-valued; hence, the values are literally representable, i.e., printable on the output device.

14.4 Query Nesting and Path Expressions

We now demonstrate the correspondence between a certain form of nested queries and path expressions. We do so only for GOMsql. It should be quite clear, that the same holds for GOMql as well. To illustrate this let us consider the following (abstract) schema:

type T_0 **is**	$\ldots$	**type** T_{n-1} **is**	**type** T_n **is**
body		body	body
$[\ldots;$		$[\ldots;$	$[\ldots]$
$A_1 : \{T_1\};$		$A_n : \{T_n\};$	$\ldots$
$\ldots];$		$\ldots];$	**end type** T_n
$\ldots$		$\ldots$	
end type T_0		**end type** T_{n-1}	

In this case, we have a reference from tuple type T_i to a set of tuples of type T_{i+1} via the attribute A_{i+1} for $(0 \leq i < n)$. In order to retrieve an attribute—say *someAttr*—of the tuple objects from the very end of this reference chain, i.e., tuples of type T_n for all root objects of type T_0 contained in the set *someSet*, we can formulate the following query:

select a_n.someAttr
from a_n **in**
    (**select** $a_{n-1}.A_n$
      from a_{n-1} **in**
            $\ddots$
               (**select** $a_1.A_2$
                  from a_1 **in**
                     (**select** $s.A_1$
                       from s **in** someSet) ...))

Utilizing a path expression, this query can be substantially shortened as follows:

select a_n.someAttr
from a_n **in** someSet.$A_1.A_2.\cdots.A_n$

In mathematical terms the resulting set corresponds to the union set U_n where U_i is recursively defined as

$$U_i := \begin{cases} someSet & \text{if } (i = 0) \\ \bigcup_{a_{i-1} \in U_{i-1}} a_{i-1}.A_i & \text{for } (1 \le i \le n) \end{cases}$$

14.5   Different Join Types

If we want to combine information of one object with properties from its associated objects, this is possible by formulating a *join query*, which is characterized by the fact that the **from** clause (GOMsql) or the **range** clause (GOMql) contains at least two range variables. We distinguish three types of join queries:

- Value-based join

- Identity join

- Functional join

These three different kinds may, however, be combined with each other; therefore, the three characteristics do not allow to strictly classify a query into either class. In the following, we demonstrate the different joins only for GOMsql. The same holds for GOMql as well.

Value-Based Join

This is the traditional join operation of relational query languages in which tuples are associated according to a comparison of the values of certain join attributes. A rather trivial example of a *value-based* join in the context of our geometric database is the following query:

> **select** b_1.name, b_2.name
> **from** b_1 **in** BRep,
> b_2 **in** BRep
> **where** b_1.weight $* 2.0 = b_2$.weight $\pm \varepsilon$

In this query, the pairs of *names* of two *BRep* objects are output for which the second one has twice the *weight* of the first one ($\pm \varepsilon$ to account for floating point arithmetic errors).

Identity Join

The *identity join* is based on object identity rather than on value equality, as in traditional relational join operations. For example, the predicate "$o_1 = o_2$" for two objects o_1 and o_2 is satisfied only if the two objects are identical; having equal values for all attributes is not sufficient. Analogously, the predicate "o $\in$ ⟨ObjSet⟩" is true only if o is an element of the set ⟨ObjSet⟩; it is not sufficient that ⟨ObjSet⟩ contains an element that has the exact same internal state as o. Such predicates are easily tested by comparing the object identifiers that are attached to each object.

To exemplify queries involving an identity join let us assume we have two object sets that share a common subentity type. In order to make the discussion a little more intuitive we call the two parent object types *Woman* and *Man*, whose instances could each contain (references to) associated entities of type *Child*. The (structural part of the) GOM schema definition for this database looks as follows:

> **type** Man **with extension is**
>   [name: string; ...; hasKids: { Child };];
> **type** Woman **with extension is**
>   [name: string; ...; hasKids: { Child };];
> **type** Child **with extension is**
>   [name: string; ...; father: Man; mother: Woman;];

We could then retrieve the pairs of children who have the same father:

> **select** c_1.name, c_2.name
> **from** c_1 **in** Child, c_2 **in** Child
> **where** c_1.father $= c_2$.father
>   **and** $c_1 \neq c_2$

This query returns the pairs of *Child*ren who have identical fathers. Contrast this with the query that retrieves *Child*ren pairs whose fathers have the same name, i.e., a value-based join:

> **select** c_1.name, c_2.name
> **from** c_1 **in** Child, c_2 **in** Child
> **where** c_1.father.name $=$ c_2.father.name
>   **and** $c_1 \neq c_2$

Let us now retrieve the *name*s of men and women who have "a common child" and, in addition, output the *name* of the *Child* that they have in common. This query makes use of the overloaded **in** operator, which here tests whether an object is contained in a collection of objects (and for binding range variables).

> **select** m.name, w.name, c.name
> **from** m **in** Man, w **in** Woman, c **in** Child
> **where** c **in** m.hasKids **and** c **in** w.hasKids

This query yields the same result as the following formulation—as long as the object base is in a consistent state:

> **select** m.name, w.name, c.name
> **from** m **in** Man, w **in** Woman, c **in** Child
> **where** c.father $= m$ **and** c.mother $= w$

Functional (or Implicit) Join

The *functional* (or *implicit*) *join* utilizes the reference chains in order to establish the association between objects from one set with objects from another. For example, if b is a range variable that is bound to the implicit set of *BRep* objects, then *b.faces* retrieves those objects of type *Face* that are associated with b via the *faces* attribute.

Let us illustrate such a join on the example query:

> retrieve the *name*s of those *BRep* objects together with the *surface* of their associated *Face* objects for which the value of the *surface* exceeds 20.

Navigating along path expressions this query can be formulated as follows:

> **select** b.name, f.name
> **from** b **in** BRep,
> f **in** b.faces
> **where** f.surface > 20

14.6   Exercises

14.1   As suggested in the text, specify a polymorphic Boolean function *bigThing* that can be invoked on any kind of object type that provides a function *volume*. The polymorphic function *bigThing* takes a *float* parameter and determines whether the receiver object's *volume* exceeds this parameter or not.

Illustrate the use of the Boolean function *bigThing* by specifying queries retrieving big *Cylinders* and big *Cuboids*.

14.2   Specify the signature of the *select* operation that takes a selection predicate with two parameters. Outline also the implementation of this operation.

14.3   Reconsider the query—specified in subsection 14.2.5—to retrieve those *Managers* who drive only *Cars* that are made by Jaguar or BMW or whose *horsePower* exceeds 200. Formulate the query in GOMql without nesting a subquery.

14.4   Give for all GOMql queries of this chapter their GOMsql equivalent, and vice versa.

All subsequent queries should be based on the extended *DeptStore* schema shown in Figure 14.3. For all subsequent exercises give the queries in both, GOMql and GOMsql.

14.5   Retrieve all *Managers* who share an *Office* with their *backUp*.

14.6   Retrieve all *Managers* whose *backUp* is also their chauffeur (i.e., the driver of one of their cars).

14.7   Retrieve the *Managers* all of whose *Projects*' budgets exceed 1 million.

14.8   Retrieve all *Emp*s who earn more than their *Manager*. But note that *Managers* are also *Emp*s and may work in their own *Dept*.

14.9   Retrieve all *Managers* of the R&D department(s), who supervise *Emp*s located in the *Building* called "E1".

14.10   Retrieve all *Managers* whose chauffeur is also a *Manager*—maybe someone who failed—and whose offices are located on the same floor.

14.11   Find the "MCPs" and "Traitors" in the *DeptStore*. An *MCP* is defined as a male *Manager* who pays one of his female *Emp*s less than one of his male *Emp*s even though the female *Emp*'s *skill* exceeds the male *Emp*'s *skill*. A *Traitor* is defined as a female *Manager* who does the same. Formulate the according queries in GOMql and GOMsql.

type Emp **is**
  body
    [name: string;
    worksIn: Dept;
    sex: char;
    salary: int;
    office: Office;
    projects: {Project};
    jobHistory: History;];   !! *History* being defined elsewhere
  operations
    declare skill: → int;
  implementation
    define skill **is**
      ... !! derived from attribute *jobHistory*
end type Emp;

type Dept **is**
  body
    [name: string;
    mgr: Manager;
    location: string;
    profit: int;
    travelBudget: int;];
  operations
    declare avgSkill: → int;
  implementation
    define avgSkill **is**
      ... !! average *skill* of all *Emp*s of the *Dept*
end type Dept;

type Manager **is**
  supertype Emp **is**
  body
    [backup: Emp;
    cars: {Car};];
  end type Manager;

type Car **is**
  body
    [license: string;
    make: string;
    horsePower: int;
    drivers: {Emp};];
  end type Car;

type Project **is**
  body
    [budget: int;
    no: string;];
  end type Project;

type Office **is**
  body
    [building: string;
    floor: int;
    roomNo: int;];
  end type Office;

Figure 14.3: Extended Schema of the *DeptStore* Object Base

14.12 Retrieve all the departments of *DeptStore* and the employees of the departments within a nested result. The resulting complex sort should be as follows:

$$[\text{depts: } \{[\text{dept: Dept, emps: } \{\text{Emp}\}]\}]$$

14.13 Retrieve the departments together with the average salary and the number of their employees.

14.14 List the name and salary of managers who manage more than 10 employees.

14.15 Find the names of those employees who earn more than any employee in the shoe department.

14.16 Find the departments whose employees cover all projects.

14.17 Find the departments whose employees have been involved in at least three projects.

14.18 Find the departments whose employees cover all projects of employees of departments in the second floor.

14.19 Give the number of departments having a travel budget below $1000 per employee.

14.20 Retrieve all locations together with the managers of departments in that location for which the number of cars in the fleet of the location (that is, all cars of managers of departments in that location) exceeds the number of managers within the location. Produce a nested result.

14.21 Restate the queries of Exercise 4.12 in GOMsql.

14.22 Restate the queries of Exercise 4.13 in GOMsql.

14.23 Restate the queries of Exercise 4.14 in GOMsql.

14.7   Annotated Bibliography

For a literature survey on polymorphic operations for associative object access we refer the reader to the bibliography at the end of Chapter 12.

The query language GOMql is based on the relational language QUEL, which was developed for the relational system INGRES [SWKH76]. An extension of QUEL to facilitate path expressions, i.e., the nested dot notation, was proposed in the context of the system GEM by Zaniolo [Zan83]. This proposal influenced the recent work by M. Stonebraker and his research group on the database language POSTQUEL [SAHR84], which is the query language of POSTGRES [SR86]. Our QUEL extension GOMql was also influenced by the language EXTRA, which was designed by M. Carey, D. DeWitt, and S. Vandenberg as the query language of the EXTRA [CDV88] data model. EXTRA is an experimental data model that was used to validate the virtues of the database system generator EXODUS.

The query language GOMsql—as the name indicates—is based on the relational database language SQL, which was originally designed for the relational DBMS System R [ABC+76]. The extensions to SQL were most notably influenced by the database language HDBL [PA86, PT86], an associative language to process nested relational databases. In [KLW87, LKD+88] a proposal for integrating abstract (user-defined) data types in HDBL is described. Another language that was developed for nested relational DBMS is SQL/NF [RKB87].

Several object-oriented database systems provide SQL dialects as query languages. Among these are Ontos (formerly Vbase) with its language ObjectSQL [HD91], O_2 with its language O_2SQL (formerly called O_2Query [BCD89], and OpenODB with its language OSQL. The extended relational system UniSQL also provides an SQL-based language, which is described in [KKS92]. GOMsql is a subset of O_2SQL with only slight syntactical differences. O_2SQL additionally provides a powerful grouping mechanism. For more literature on these systems, refer to Chapter 27.

Several object-oriented algebras have been proposed recently. Most of these algebras, however, are not to be viewed as query languages; they rather serve the purpose of internal query representation languages for query optimization and query processing. The most important algebras are described in [SO90], [SZ90], [VD91], [SS90], and [KM93]. A framework for optimization in object bases is presented in [KMP93].

F. Bancilhon discusses in [Ban89] the fundamental functionality required by query languages of object-oriented database systems. He concludes—in accordance with our experiences—that, even though declarative query languages play a less predominant role in advanced (engineering) applications in comparison to procedural object manipulation, they are still very important especially for interactive access to the object base.

Part IV

Control Concepts

15

Transaction Control

Typically, a database is a highly shared resource that is accessed concurrently by many different application programs and interactive users. This demands a control component that governs the access to shared data in order to avoid undesired interference between concurrently executing processes. The basic concept of this control component is the transaction. The properties of the classical transaction lead not only to the concurrency control concepts required in order to avoid the unwanted interference but also to the recovery concepts required to diminish the harm of failures. This chapter first reviews the classical notion of transaction, some concurrency control and recovery concepts. Then, some nonclassical notions of transactions better suited for advanced, e.g., technical design, transactions are outlined.

15.1   The Classical Notion of Transaction

Every application poses semantic restrictions on database states, and only those database states obeying these restrictions are legal or valid. Consider, for example, an accounting system. There, a possible restriction is that the total balance of all accounts remains invariant. Another example is an airline reservation system where the number of reserved seats must not exceed the number of totally available seats—plus some overbooking capacity. These restrictions capturing application-specific semantics are called *consistency constraints*. A database state fulfilling all imposed consistency constraints is called *consistent*.

During its lifetime, a database is regularly updated. Each update leads from one database state to another one. Thereby, an update often consists of several operations. Consider the accounting system again. An application which might want to transfer a certain amount of money from one account to another can do so by first withdrawing the amount (debit) from the first account and then crediting it to the second one.

Of course, it has to be ensured that the *debit* and *credit* operations are considered as "belonging together." This observation led to the introduction of the notion of *transaction*.

A *transaction* is a sequence of operations to be performed together. It starts with a **begin transaction** and ends with either

- **end transaction**: if the transaction terminates successfully

- **abort transaction**: if the transaction ended unsuccessfully, which may be due to, e.g., an explicit abort by the user, a programming error resulting in the transaction's termination, or a system failure

When a transaction ends successfully, all its changes are committed to the database and hence made visible for other users. Besides the elementary grouping property, a classical transaction is required to ensure some more properties.

Consider the accounting example again. The goal was to transfer money from one account to another by executing a debit followed by a credit operation. Performing only the debit without the subsequent credit clearly leads to a consistency violation, since the total balance of all accounts changed—money got lost. In order to avoid this, a transaction is required to execute in an *all or nothing* fashion. That is, either *all* of the transaction's operations execute or *none*. This property is called *atomicity*.

Also, an uncarefully chosen update operation may violate database consistency. If, for example, an airline reservation is executed despite the fact that the plane is already booked out, this results in a consistency violation. Thus, one required property of a database update is that it leads from one consistent database state to another (not necessarily different) consistent database state. This property is referred to as *consistency preservation*.

So far, we have not yet dealt with problems resulting from the concurrent execution of several transactions. Therefore, consider two transactions, TA_1 and TA_2, whose purpose is to increase a (persistent) counter x by 1. Both transactions perform this task by first reading the current value of x into a local variable *temp*, increasing it by 1, and then storing the value of *temp* back to x. Assume that the individual operations of the two transactions are executed in the following interleaved sequence:

Step	TA_1	TA_2
(1)	temp := x.read;	
(2)		temp := x.read;
(3)	temp := temp + 1;	
(4)		temp := temp + 1;
(5)	x.write(temp);	
(6)		x.write(temp);

First, transaction TA_1 reads the value of the global variable x into its local variable *temp*, then transaction TA_2 does the same for its local variable, and so on. Such a (possibly interleaved) execution of the operations of several transactions is called a *schedule*.

The above schedule obviously shows unwanted interference between the two transactions: The schedule increases x by 1 despite the fact that *serial* execution of TA_1 followed by TA_2 (or vice versa) would have increased x by 2. The reason for this is that in step (6) the effects of TA_1 on x are overwritten by TA_2. Avoiding this kind of interference between different transactions is called *isolation*—each transaction has to be isolated from the others. That is, each transaction logically perceives the database as if it were the only process accessing and modifying the data.

The fourth requirement imposed upon a transaction is a typical database requirement. The effect of executing a transaction successfully should be stored permanently; hence, *durability* is required for successfully terminating transactions.

To summarize, a transaction is characterized by the following so-called *ACID* properties:

Atomicity: The atomicity of transactions demands that either all the effects of the transaction become manifest in the database or none.

Consistency: The transaction transfers the database from one consistent state into another consistent state. This condition allows the database to assume an inconsistent state during the execution of the transaction; as long as consistency is guaranteed at the end of the transaction.

Isolation: Concurrent transactions are to be isolated from each other; i.e., there should be no interference among different transactions accessing the database concurrently.

Durability: The effects of a completed transaction remain persistent even if the database system encounters an error situation, e.g., a head crash, later on.

Classical transaction management was developed under the premise that no additional information about the applications is available. Consequently, there exists no explicit definition of database consistency. Further, the only semantic information available on transactions are their elementary accesses to data. These are the *read* and *write* operations a transaction performs on the database objects. Considering that the semantics of transactions are difficult to specify explicitly and every transaction can—from the database's point of view—eventually be broken down into these elementary operations, it seems reasonable to think of a transaction as a sequence of read and write operations.

Let us discuss briefly the consequences of this approach for the *ACID* paradigm. From the database's point of view a transaction consists of a set of read and write operations. These have to be executed atomically—that is, as one elementary, non-interruptable action. Since there is no way for the database system to check the consistency of a database state, consistency appears in a very restricted sense: The database system has to assume that a database state is consistent, if the transactions are executed isolated from each other. The notion of *isolation* is also influenced by this restriction. Since merely sets of reads and writes have to be treated, only three kinds of negative or unwanted interference have to be avoided:

1. *No lost update*
    This requirement demands that no updates will be lost due to two interfering transactions. An example of a lost update was already illustrated above in which a counter was increased by two parallel transactions.

2. *No dirty read*
    This ensures that no transaction will ever read an object state that—in the end—was not committed to the database. This could happen if, for example, in the accounting application a transaction reads the balance of an account after another transaction executed a debit operation. Then, the balance read is invalid if the latter transaction—i.e., the one that was supposed to do the debit/credit—aborts.

3. *Repeatable reads*
    This demands that a transaction can always perform two successive read operations on a data item and expect the same result—assuming that the transaction did not itself update the data item. This precludes a second transaction to perform an update on the data item in between the two read operations.

Since isolation implies consistency under the above-mentioned restrictions, three levels of consistency have been defined according to the avoidance of the above three types of isolation violations:

- *Level 1* consistency assures that no lost updates occur.

- *Level 2* consistency assures Level 1 consistency and that no dirty reads are possible.

- *Level 3* consistency assures Level 2 consistency and the guarantee of repeatable reads.

Note, that if the transactions execute *serially*, i.e., one after the other, Level 3 consistency is trivially guaranteed. In the following schedule the two transactions A and B updating accounts $account_1$ and $account_2$ execute serially (a and b are local variables of the transactions):

(1) **begin transaction** A

(2) a := $account_1$.readBalance;

(3) a := a + 100;

(4) $account_1$.writeBalance(a);

(5) b := $account_2$.readBalance;

(6) b := b − 100;

(7) $account_2$.writeBalance(b);

(8) **end transaction** A

(9) **begin transaction** B

(10) a := $account_1$.readBalance;

(11) a := a + 50;

(12) $account_1$.writeBalance(a);

(13) b := $account_2$.readBalance;

(14) b := b − 50;

(15) $account_2$.writeBalance(b);

(16) **end transaction** B

A schedule in which all transactions are executed serially is called a *serial schedule*. For the example, the other possible serial schedule would first execute transaction *B* and then transaction *A*.

The main correctness notion introduced in classical concurrency control is *serializability*, in which a schedule is called *serializable* if it is *equivalent* to a serial schedule, i.e., if the schedule produces the same output for each transaction and leaves the database in the same state.

Consider the following schedule:

(1) **begin transaction** A

(2)   **begin transaction** B

(3) a := $account_1$.readBalance;

(4)   a := $account_1$.readBalance;

(5) a := a + 100;

(6)   a := a + 50;

(7) $account_1$.writeBalance(a);

(8) $account_1$.writeBalance(a);

(9) b := account$_2$.readBalance;

(10)   b := account$_2$.readBalance;

(11) b := b − 100;

(12)   b := b − 50;

(13) account$_2$.writeBalance(b);

(14)   account$_2$.writeBalance(b);

(15) **end transaction** A

(16)   **end transaction** B

This schedule clearly is not serializable, since the database state after the execution of this schedule differs from the outcome of any of the two possible serial executions of *A* and *B*.

The main task of the concurrency control component—also called the *scheduler*—is to restrict the schedules of committed transactions to serializable schedules.

The next two sections present mechanisms guaranteeing serializability for a given set of concurrently executing transactions. Then the locking approach—which is the predominant method for guaranteeing serializability in existing systems—is refined in order to exploit *multiple granules of locking*. The assurance of atomicity and durability is treated in the section on *recovery*. A brief assessment of the *ACID* properties with respect to their appropriateness in advanced applications as well as an extensive requirements analysis for transaction models for advanced applications follow. Lastly, some recent proposals of transaction models for design applications are discussed.

15.2   Locking

An obvious solution to ensure serializability is the following. The concurrency control component—i.e., the scheduler—enforces a protocol based on *locking* objects among all transactions. Any transaction that wants to access an object—no matter whether the access is a read or a write—has to acquire a lock from the scheduler. After the lock has been granted, the transaction may access the object. All other transactions that want to access this object must wait until the lock-holding transaction terminates. Furthermore, the transaction has to keep the locks until no other locks (on different objects) are needed—see Section 15.2.2.

15.2.1   Lock Modes

The above locking protocol is a little bit too restrictive, since two reads on an object do not interfere. Thus, the locking protocol for objects is refined by introducing two different lock modes: *read locks (R)* and *write locks (W)*. Now, the following refined lock protocol also ensures Level 3 consistency, if all transactions T obey the following rules:

- T sets an R lock on any object it reads.

- T sets a W lock on any object it updates.

- T holds all locks until it terminates.

Locking an object in W mode ensures an exclusive access to the respective object; that is, no other transaction can read or update the object for the duration of the lock. A W lock can only be granted by the transaction manager if the object has not been locked by another transaction. The R mode is a shared lock; that is, several transactions can concurrently hold an R lock on the same object. However, no W lock can be granted on an object that has already an R lock. Likewise, no R lock can be granted on an object that has a W lock. We say that the R lock mode is incompatible with the W lock mode. Likewise, the W lock mode is incompatible with the W lock mode, whereas the R mode is compatible with the R mode. Thus, objects can be accessed simultaneously only by two reading transactions. This is the reason why the R lock is also called *share lock* (abbreviated S), since it allows sharing, and the W lock is called *exclusive lock* (abbreviated X), since it guarantees exclusive access for the transaction holding the lock.

The compatibility of lock modes is often summarized in a *compatibility matrix*. The left column indicates the lock currently associated with the object, and the upper row indicates the lock requested by another transaction. For the S and X locks the above discussion results in the following compatibility matrix:

	S	X
S	yes	no
X	no	no

The compatibility matrix is then used by the scheduler to decide if a lock can be granted or the lock-requesting transaction has to wait. If no lock is held for the object by any other transaction, or if the required lock is compatible with an already existing lock of some other transaction, then the transaction acquires the lock and may proceed. Otherwise, it has to wait until the current lock of the object has been released. Note that an object may be assigned an S lock by several transactions. The request of an X lock on this object can only be granted after *all* S locks have been released.

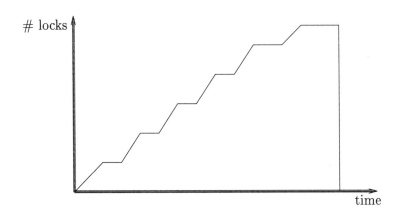

Figure 15.1: Graphical Representation of Strict Two-Phase Locking

15.2.2   Two-Phase Locking

Figure 15.1 illustrates one strategy of accumulating locks and releasing them all at once at the end of the transaction. The depicted strategy is called *strict two-phase locking* protocol because it consists of a *growing phase* and a *release phase*. The *strict* character of the strategy demands that all lock releases are directly prior to the end of the transaction; that is, the transaction releases its locks only after it has been assured that it will definitely terminate. This can only be determined after the last action has been performed.

The (*nonstrict*) *two-phase locking* protocol relaxes the requirement that all locks are to be released together at the end of a transaction and allows to release acquired locks as soon as the last lock has been requested for the particular transaction. Note that this may result in reading noncommitted data—i.e., so-called *dirty* data whose locks were released by a transaction that, in the end, aborted. To avoid this, in most commercial database management systems the classical transaction control with the strict two-phase locking protocol is applied.

For the application programmers, locking is invisible. The database system initiates a lock request prior to every read and write operation. If the lock is granted, the access operation is performed. If an S lock is already held by the transaction and a write is to be performed, the S lock is upgraded to an X lock if no other transaction has locked the object to be accessed. Let us consider the following partial schedule in which two transactions TA_1 and TA_2 access a counter object o with current value 1. Only the *read* and *write* operation calls are shown:

Step	TA$_1$	TA$_2$	lock handling
(1)	o.read		Since there is no lock held, TA$_1$ acquires an S lock on o.
(2)	o.write(2)		Since only TA$_1$ holds a lock, it acquires an X lock on o
(3)		*o.read*	Since TA$_1$ already holds an X lock which is incompatible with the required S lock, TA$_2$ has to wait. The intended operation invocation is written in italics to indicate that the operation cannot be executed now.
(4)	**end**		TA$_1$ terminates and releases all locks; TA$_2$ now acquires an S lock for o
(5)		o.read	TA$_2$ proceeds

Note that for consistency reasons TA$_2$ did not really have to wait, since TA$_1$ does not write o again. Thus, if TA$_1$ did not end in step (4) but proceed, e.g., executing many steps not involving o, then TA$_2$ would have to wait—according to the strict two-phase locking protocol—until TA$_1$ terminates despite the fact that this is unnecessary. The point that transactions have to wait for others to commit—necessarily or unnecessarily—is one obvious drawback of locking-based concurrency control.

15.2.3 Deadlocks

Another serious problem that might arise is a cycle in the wait relation of different transactions. Then, no transaction can proceed—a *deadlock* has occurred. These deadlocks have to be detected by the scheduler. In order to avoid *starvation* of the transactions, one transaction involved in the cycle of waiting transactions has to be selected and aborted by the scheduler. This results–due to the atomicity property of transactions—in its *rollback*; i.e., all its changes have to be undone such that the database state before the transaction started is restored.

Let us consider an example. Again, two applications try to increment counter o:

Step	TA$_1$	TA$_2$	lock handling
(1)	o.read		TA$_1$ acquires an S lock.
(2)		o.read	TA$_2$ acquires an S lock.
(3)	*o.write(2)*		TA$_1$ has to wait for TA$_2$ to terminate, since proceeding requires an X lock which cannot be granted. Again, the italics type face indicates that the operation cannot be executed.
(4)		*o.write(2)*	TA$_2$ has to wait for TA$_1$ to terminate, since proceeding requires an X lock that cannot be granted

This is a typical deadlock situation: TA_1 waits for TA_2 to terminate, and vice versa. One transaction has to be selected for aborting. Then, since the locks held by the aborted transaction are released, the other can proceed.

Summarizing, the locking-based concurrency control strategy induces the following overhead: First, there is the cost of maintaining the locks. Even if there are only read-only transactions, the locks must be maintained. Second, transactions often have to wait for required locks to be released. Third, deadlock detection procedures impose additional overhead. Fourth, rolling back a transaction in case of a deadlock produces costs that cannot be neglected. An alternative to locking-based protocols is described in the next section.

15.3   Optimistic Concurrency Control

Locking-based concurrency control can be termed *pessimistic* since a lock request is rejected if something might *possibly* go wrong. In order to avoid some of the disadvantages mentioned at the end of the previous section, *optimistic* concurrency control has been introduced. It is termed optimistic, since it allows all transactions to proceed until they try to commit their changes to the database. Then, and not earlier, a *validation* procedure is started and checks for conflicts that might have occurred. If a conflict is detected, the transaction is rolled back.

More precisely, for optimistic concurrency control a transaction is divided into three different phases: a *read phase*, a *validation phase*, and—possibly—a *write phase*, which is executed only if no conflict has been detected during the validation phase.

During the read phase, every object accessed by a transaction is copied into a private workspace of the transaction. All updates that are performed by the transaction are carried out on these local copies. In fact, the name *read phase* may appear as a misnomer, since updates do occur in this phase. But, since these are carried out on local copies, the transactions behave like read-only transactions from the database system's point of view. Further note that there is no blocking of transactions, even if other transactions acquire copies of the same object and possibly modify their local copies.

During the execution of the application's transaction in the read phase, the concurrency control component collects certain information that is then used in the validation phase for detecting conflicts. In order to support the validation phase's task, the concurrency control component maintains two sets for each transaction: the *read set*, containing all the objects read by the transaction, and the *write set* containing all objects updated by the transaction.

After the read phase, when the transaction enters the validation phase, a *time stamp* is attached to it. A *time stamp* is an element of a monotonically increas-

ing sequence such as the time point the transaction ended the read phase. These
time stamps together with the read and write sets of the transactions constitute the
information exploited by the validation procedure to detect conflicts.

During the validation phase of a transaction T, three criteria are evaluated for
each transaction T' whose time stamp precedes that of T—i.e., T' is "older" than T.
If either *one* of the criteria evaluates to true for *every* preceding T', serializability is
guaranteed. The criteria are as follows:

1. T' ended its write phase before T started the read phase.

2. The write set of T' does not intersect (overlap) with the read set of T, and T'
   ends its write phase before T starts its write phase.

3. The write set of T' does not intersect with the read set nor with the write set
   of T.

If one of the above conditions is fulfilled for every transaction T' preceding T according
to the time stamps, T may enter the write phase. Otherwise, T has to be aborted.
Note that the rollback of T is trivial, since all updates were performed on local copies.

During the write phase, all the objects changed by the transactions are written
back, i.e., committed, to the database. Hence, the updates performed by the transaction now become visible to other transactions. Note that this is the earliest point
in time changes are made visible to other transactions.

Let us consider the following example schedule of two transactions TA_1 and TA_2
operating on two objects x and y:

Step	TA_1	TA_2	read/write set of TA_1	read/write set of TA_2
(1)	x.read		$\{x\}$ / $\{\}$	$\{\}$ / $\{\}$
(2)		x.read	$\{x\}$ / $\{\}$	$\{x\}$ / $\{\}$
(3)	x.write($\dots$)		$\{x\}$ / $\{x\}$	$\{x\}$ / $\{\}$
(4)		y.write($\dots$)	$\{x\}$ / $\{x\}$	$\{x\}$ / $\{y\}$
(5)		**end**		
(6)	**end**			

Assume TA_2 starts its validation phase first. Thus, it receives a time stamp
preceding the time stamp of TA_1. Since no transaction precedes TA_2, nothing has
to be checked in the validation phase and TA_2 can commit. Now, for TA_1, the
validation procedure tries to find at least one of the above three criteria that evaluates
to true. The first criterion is obviously not fulfilled, since TA_1 did not start after TA_2
committed. The second criterion evaluates to true, since the intersection of the write
set of TA_2 and the read set of TA_1 is empty, and TA_2 committed before TA_1 started

its validation phase. Thus, TA_1 is allowed to enter its write phase during which the value of x is committed to the database.

Note that the above schedule would have been impossible by the two-phase locking protocol. That is, step (3) could not have been performed by transaction TA_1, since it would have had to wait for TA_2 to release its S lock on x.

Although optimistic concurrency control avoids some of the drawbacks of locking-based concurrency control, it exhibits its own disadvantages. First, there exists an overhead imposed by copying all data into the private workspaces of the transactions. Second, maintaining the read and write sets as well as performing the validation phase result in additional overhead. Note that rolling back a transaction is not expensive from the database management system's point of view, since the changes were carried out on local copies. Nevertheless, the work performed by the transaction to be rolled back is lost. This is a serious drawback, since the transaction is only aborted at its very end, i.e., when all its operations have already been performed—in vain, as it turns out during the validation phase.

15.4   Multiple Granularity Locking

For motivation, consider a query running against the extension of a certain type. Then, most likely, this query reads all objects contained in the extension. If locking is used, every single object has to be locked and, hence, the number of locks to be maintained is enormous. Thus, it might be worthwhile to lock the entire extension with a single lock and assuming the instances to be locked implicitly—that is, the instances are covered by the encompassing lock on the extension. An intuitive example where it is useful to lock a whole extension in exclusive mode is the yearly increase of the *salary* of all *Employees*.

For reducing the number of locks, *multiple granularity locking* has been introduced. Thereby, not every single object within the extension is locked; instead, only the extension is locked. Obviously, if a transaction only accesses a small part of an extension, then locking the whole extension unnecessarily diminishes concurrency in that it blocks all other transactions requesting access to objects in that extension.

On a smaller grain assume that a transaction only reads a single attribute of an object, while another transaction accesses another attribute of that same object. If the object is locked by the first transaction, the second transaction cannot access the object and, hence, the attribute. Applying a finer granule of locking—i.e., locking attributes instead of objects—allows for more concurrency at the expense of (many) more locks. This only pays off if the object is a so-called *hot spot*, i.e., an object frequently accessed by different transactions, and if the transactions are likely to access different attributes.

The above two scenarios give rise to organizing locks in a *granularity hierarchy*.

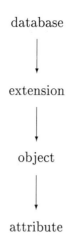

database

extension

object

attribute

Figure 15.2: A Granularity Hierarchy for Locking

The top level of the granularity hierarchy is the entire database. If it is locked, there is no more concurrency except for concurrent reads. The next finer granule is an extension followed by objects and attributes. The corresponding granularity hierarchy is shown in Figure 15.2.

For locking at different granularity levels the simple S and X locks are not sufficient. Consider a transaction TA_1 locking an object o of an extension E. Later, a transaction TA_2 requires a lock for the whole extension E. Obviously, the lock should not be granted if it is not compatible with the lock that TA_1 holds—e.g., if TA_1 holds an x lock. Nevertheless, browsing through all the objects of extension E and checking their locks for compatibility with the lock requested by TA_2 are prohibitively expensive. Therefore, in the multiple granularity locking scheme transaction TA_1 must indicate that it accesses an object of extension E by first acquiring an *intention lock* on the extension E. Without holding an intention lock for E, no object in E can be accessed. Thus, additional lock modes are needed.

There exist the following lock modes in multiple granularity locking for the two granules extension and object—for simplifying the discussion we restrict the granularity levels of Figure15.2 to the two middle levels:

S The *share* lock on objects implies that the object is locked in share mode as in the simple locking protocol already introduced. Locking an extension in share mode implicitly locks all instances of the extension in share mode.

X The *exclusive* lock can also be acquired for an extension or an object. Locking an object in exclusive mode has the same effect as in the simple locking protocol.

Locking an extension in exclusive mode results implicitly in locking all the instances of the extension in exclusive mode.

IS The *intention share* lock *IS* is used to lock an extension in order to indicate that the transaction will lock some instances of the extension in share mode *S*. The locks for the instances have to be required explicitly.

IX The *intention exclusive* lock *IX* is used analogously to the intention share lock.

SIX Acquiring a *share intention exclusive* lock for an extension results in locking the extension in share mode. Thus, all instances are implicitly locked in share mode. Additionally, it allows to acquire exclusive locks for some of the instances in the extension. The latter have to be acquired explicitly at the instance level.

Of course, we can replace *extension* and *object* in the above description by any two neighboring granules of the locking hierarchy—as exemplified in Figure 15.2. Thereby, the concept carries over to granularity hierarchies of arbitrary height.

To comprehend the motivation for introducing an additional *SIX* lock consider the following scenario. An update on only a few objects may often require that a large part of the extension has to be read. Then, without the *SIX* lock, one would have two possibilities. First, requesting an exclusive lock for the entire extension, which results in no concurrency, or second, requesting an intention exclusive lock for the extension and to lock each object in share mode and those to be updated in exclusive mode. The latter results in an increased locking overhead. Since the above-described situation is quite common, the *SIX* lock mode has been introduced for better performance. It allows to indicate that all objects of an extension are to be read and only a few are to be updated.

The compatibility matrix for the above locks is as follows (again, the leftmost column represents the lock currently associated with the considered granule and the top row represents the lock mode requested by a particular transaction):

	S	X	IS	IX	SIX
S	yes	no	yes	no	no
X	no	no	no	no	no
IS	yes	no	yes	yes	yes
IX	no	no	yes	yes	no
SIX	no	no	yes	no	no

In order to guarantee serializability a (strict) two-phase protocol for locking obeying the compatibility of lock modes and the following rules are applied:

1. If an *S* lock is required for a certain locking granule, all the ancestor granules in the granularity hierarchy have to be locked in *IS* mode.

2. If an X lock is required for a certain locking granule, all the ancestor granules in the granularity hierarchy have to be locked in IX or SIX mode.

3. All locks have to be set from the root of the granularity hierarchy down to the granule for which a lock is requested. Locks are released in the opposite direction (i.e., bottom-up).

4. For any lock required for an extension of a certain type, all extensions of subtypes have to acquire the same lock. This is actually a consequence of subtyping (and substitutability), since the extension of a subtype is implicitly contained in the extension of its supertype.

The last rule is necessary, since, for example, a query that is posed against a certain extension may also inspect the instances of the subtypes.

For the example hierarchy consisting of the database, extensions, objects, and attributes, we consider some examples of the locking protocol. First, we lock a certain object o:

1. Lock the database in IS mode.

2. Lock the extension E of the type of object o in IS mode.

3. Lock the object o in S mode.

Now, all attributes of o are implicitly locked in S mode.

Locking a whole extension E in exclusive mode, e.g., for updating all *Employees' salary*, proceeds as follows:

1. Lock the database in IX mode.

2. Lock the extension E in X mode.

Note that these lock requests conflict with those of the previous example—if o were of type *Employee* or a subtype thereof.

To lock an extension E for a complete read scan and occasional update, the following protocol is appropriate:

1. Lock the database in IX mode.

2. Lock the extension E in SIX mode.

Thereafter, those objects of E to be updated can be locked in exclusive mode. Notice that this locking is compatible with the one of the first example—that is, two parallel transactions could acquire the respective locks on the same database and the same extension.

15.5   Recovery

So far we have been concerned with avoiding unwanted interference between concurrent transactions. Recalling the *ACID* properties, *atomicity* and *durability* have not yet been discussed. Several circumstances may lead to a violation of these properties. All are of the same basic nature: A transaction is interrupted after having executed some (update) operations and before it commits. The reasons for these interrupts are manifold:

- A transaction may exhibit a run-time error, e.g., division by zero.

- A software failure of the database system or the underlying operating system occurs.

- A hardware failure occurs, e.g., a head crash, where portions of the data or all data may be lost.

- A power failure is experienced.

Failures due to erroneous software are called *soft failures* or *soft crashes*, while those due to hardware failure are called *hard failures* or *hard crashes*.

The task of the recovery scheme is to restore a consistent database state in case of a failure. For this purpose additional information is needed, e.g., a *back-up* copy of the entire database. In case of a failure, after the system recovered, the back-up copy is used to restore a consistent database state. A prerequisite is, of course, the consistency of the back-up copy itself. This is guaranteed most easily by not allowing any updating transactions to be executing while the back-up is copied. Then, whenever a crash occurs, one can use this back-up copy to restore a consistent database state.

Although this is a first attempt to recovery, additional care is needed, since the updates produced by committed transactions after the last back-up must not get lost—this is just the *durability* requirement. Since the production of back-ups is expensive in time and material, there is an upper limit on the frequency at which back-ups are cost effective. Thus, the loss may consist of the updates of days or weeks. In order to avoid these losses, more sophisticated recovery techniques have been devised. They all rely on the same idea, in that they store only the differences between database states. Before describing some simple recovery techniques in more detail, let us discuss the existing possibilities for transactions to be affected by a crash in between two back-up copies.

The possible situations are illustrated in Figure 15.3. All transactions are assumed to be updating transactions, since only these are of interest. The x-axis models the time, i.e., the time increases from left to right. The time where the last back-up copy prior to the failure was taken is indicated by the left vertical line. The time of

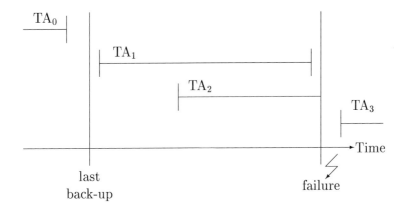

Figure 15.3: Transactions Relative to a Save Point and a Failure

the failure is indicated by the right vertical line. The transaction's execution times are those exhibited without a failure. Since no updating transaction is allowed to be executing during a back-up, all transactions start and end between two back-ups—or between a back-up and a failure. Thus, Figure 15.3 contains all possibilities for transactions to start and end relative to the back-up time and the failure time. Obviously, TA_0 is not affected by the failure, since its updates are already saved in the back-up copy. Additionally, TA_3 did not start before the failure and, hence, could not produce any updates by the time of the failure. Consequently, the only two transactions affected by the failure are TA_1 and TA_2.

Now, two cases have to be distinguished—depending on a possible loss of data:

no loss: Since transaction TA_1 already committed its data into the database,[1] and no loss of data occurred due to the failure, all the updates of TA_1 are still present in the database. A problem arises only for TA_2, which did not come to an end. Therefore, as required by the atomicity property, the database state valid before the start of the execution of TA_2 has to be restored. An obvious solution is to store for each object updated by TA_2 its so-called *before image*, i.e., the object's state before the update occurred. Then, in case of a failure, all *before images* produced by TA_2 have to be restored in order to regain a consistent database state. Note that if TA_2 does not produce any updates within the database, but instead—as in the context of optimistic concurrency control—performs its updates on local copies, no before images have to be stored explicitly.

loss: If the data on disk are lost, the only existing choice is to restore the last back-up copy. Then, the committed updates of transaction TA_1 are lost. This

---

[1]We assume that all modified data are written back to the secondary storage at commit time, i.e., at the end of a transactions.

can be avoided by two different techniques. First, the transactions that have been committed since the last back-up are memorized and in case of a failure reexecuted. This is called a *logical log*. The other possibility is to save the *after images* of all updated objects. An *after image* of an object is the object's state after the transaction committed. Following a failure and after restoring the back-up copy, only the after images of committed transactions have to be incorporated into the database to generate a consistent database state where no update of a committed transaction gets lost.

All the information necessary for recovery—before images, after images, and logical logs—are written to the *log file*. For each written item the identifier of the transaction responsible for this item has to be attached. Additionally, all transactions' start and end points are recorded in the log file.

15.6   Assessment of Classical Concurrency Control

The conventional concurrency control assumes transactions to be of short durations, i.e., in the order of milliseconds to seconds. This is reflected, for example, in the atomicity condition, which requires the complete rollback of an entire, not committed transaction in the case of a program error, system failure, or a conflict due to concurrency—e.g., in the case of a detected deadlock. This requirement is rather drastic in the context of design transactions, since one design transaction may reflect the designer's intellectual work of hours, days, or even weeks. Remember that in optimistic concurrency control protocols conflicts are detected only at the end of a transaction, i.e., when all work has already been performed and, hence, must be rolled back. This is unbearable for design transactions.

Another problem occurs especially if locking-based concurrency control is applied. A long-lasting transaction holds all acquired locks until it ends and, hence, blocks other transactions—that request (conflicting) access to the same data—as long as it lasts. Again, this is unacceptable, since it may prevent other designers from continuing their work for hours, days, or even weeks.

15.7   Requirements of Design Transactions

The requirements imposed upon transaction handling in design environments originate from observations under two different points of view of design:

1. The design process itself

2. The architecture of current design environments

Subsequently, these observations are discussed in turn and the requirements are derived. Before doing so, one remark may be necessary on the usage of the term *transaction* within the remainder of this chapter. Although, the term *transaction* originated in the area of classical transaction control where transactions obey the *ACID* paradigm, here we will use the word *transaction* also for more relaxed protocols. The idea behind this is that even protocols for advanced applications exert some concurrency control and since the notion of transaction always has the association of concurrency control this might be a better choice than introducing a new term. Also, we have to admit that we do not know of any better term. Whenever necessary, we attach the adjunct *classical* to avoid ambiguities.

We start out by summarizing the observations from the previous assessment. Two requirements can be deduced from the fact that long transactions have to be supported in design environments:

1. An entire rollback of a long-lasting design transaction has to be avoided; i.e., intermediate save points must be allowed even if the objects saved are not yet consistent.

2. Long-lasting blocking of transactions has to be avoided. Thus, more concurrency has to be allowed.

Another observation is that design is a *cooperative* task. Cooperation means sharing of information. This shared information is most probably inconsistent in some respect, since consistency is normally only reached at the end of the design process. Hence, it must be possible for designers to cooperate on a common task by exchanging inconsistent objects. In classical transaction control, this is prevented by the *isolation* property, which forces transactions to be *competitive* instead of cooperative. That is, in the classical transaction concept parallel transactions are isolated from each other, which really means that they compete for the access to the data items—at least in locking-based protocols.

Moreover, it might be necessary that several designers work on the same object(s) concurrently. This results in concurrent updates on the same object(s). Clearly, this is a very dangerous situation from the consistency preservation point of view. Nevertheless, this can often not be avoided. Thus, the design environment has to provide mechanisms to support concurrently designing engineers in avoiding inconsistencies and—if this is not possible—in detecting inconsistencies and notifying the involved designers.

The difficulties in maintaining consistency for the design objects have always been a concern in design processes. The solution applied is the organization of the teamwork to be carried out within a design process: The design process is organized into several tasks, and to each task a group of designers is assigned. This can be generalized to a hierarchy of tasks and subtasks: Tasks are broken down into subtasks, which in turn are broken down into subsubtasks, and so on. Ideally, the organizational structure of the personnel reflects this division. Cooperation is then strongest for designers within the same group. Since the degree of cooperation possible is an immediate consequence of the concurrency control protocol, the organizational structure of the design process should be reflected in it.

The concurrency control mechanism introduced in design environments should therefore support all the facets of cooperation mentioned:

3. It should allow to exchange inconsistent information among cooperating designers.

4. If necessary, it should support simultaneous or concurrent design by allowing concurrent updates on the same objects, and provide additional support for the detection of inconsistencies as well as advanced communication mechanisms such as notification of the involved designers in case of an inconsistency.

5. It ought to provide means to reflect the organizational structure of the design process.

After having turned some observations on the design process into requirements, we continue by taking a closer look at current design environments. Here, two important aspects have to be covered: the architecture and the interactive nature of design environments.

An important aspect in designing a transaction control mechanism for engineering applications stems from their typical *client/server architecture*—being introduced in the next subsection—within a network of workstations. Therefore, another requirement on the concurrency control mechanism employed in design environments is:

6. The transaction scheme should rovide support for *checkIn/checkOut* concepts in a client/server architecture.

During the design phase, conflicts cannot be avoided entirely. Resolving conflicts can either be performed in a predetermined way, i.e., by the standard roll back; but then no account on the current situation is possible. On the other hand, individual conflict resolution requires some intelligence and, hence, can only be provided by the user. Although this, in principle, is no problem due to the interactive nature of the design environments, the concurrency control mechanism has to allow for user manipulations affecting concurrency control resulting in the last listed requirement:

7. Allow specific user control.

After having posed so many requirements that lead to a relaxation of classical concurrency control mechanism, it might be necessary to warn the reader by mentioning that—despite its restrictive characteristics—in many cases standard serializability is still required, e.g., in a (system) catalog update. Thus, the proposed mechanisms should expand the classical transaction paradigm instead of replacing it.

The rest of the discussion on enhanced transaction concepts is organized as follows. After a short introduction of the client/server architecture, we start with an adaptation of classical transaction control to the client/server architecture. The adapted protocol is then stepwise refined in order to meet the other requirements.

15.8 Concurrency Control in Client/Server Architectures

In order to adapt the classical locking mechanism to the checkIn/checkOut paradigm as employed in client/server architectures, let us briefly introduce the client server architecture. In Figure 15.4 a typical client server architecture is sketched. There exists *one* server, which constitutes the *global* or *public* database, which may be accessed by all applications. The server constitutes a *logically* centralized database, which may, however, be physically distributed over several machines—which then form together the one (logical) server. In any case, whether physically centralized or distributed, the public database appears to the applications as a globally accessible data repository. Thus, the public database serves as an integration tool among the different applications. It constitutes their communication medium by which they can exchange objects in a controlled manner.

Each application is run in the form of a transaction that itself possesses a *private database*. This private database is created at the transaction's begin and deleted at the transaction's end. The private database constitutes an exclusively owned (private) workspace that can only be accessed by the one transaction. Therefore, it can be seen as a single-user database, which obviates any concurrency control mechanisms.

An application—or more precisely a transaction—retrieves objects for modification or for reading from the public DB by a *checkOut* operation. The *checkOut* of an object involves the physical transfer of the object (that is, a copy of the object) into the application's private database. This transferred object can then be accessed only by the *one* application—it is exclusively owned by the respective application. A modified object may be transferred back to the public database by a *checkIn* of the object.

Of course, the *checkOut* and *checkIn* of objects has to be performed in a controlled manner. An application may request the *checkOut* of an object:

- **for read**, in which case an S lock has to be maintained in the public database

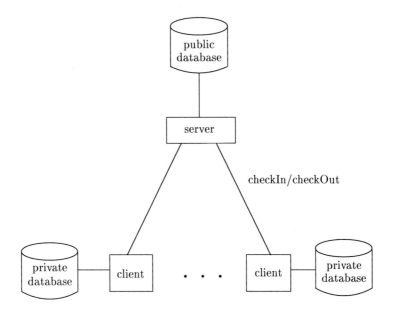

Figure 15.4: A Client/Server Architecture

for the respective objects. Other transactions may *checkOut* the same object,
but only for reading.

- **for update**, in which case an exclusive X lock has to be set. This object is
  now under exclusive access by the respective client until it checks it back into
  the public database.

The locks can be given to the *checkIn/checkOut* operations as an additional parameter; e.g., an object o can be checked out for read with *o.checkOut(S)*.

In order to prevent several clients to check in or out the same object concurrently,
which might cause inconsistencies, a short lock is used to guarantee atomicity and
isolation during the checkin or checkout process. However, this lock is released directly
after the end of the check in or check out. This is, of course, not true for the S and
X locks denoting the *checkIn/checkOut* status of the objects. These are *long-lasting
locks*. They are released only after a corresponding *checkIn* of the object or at the time
the application's transaction ends. It is the server's responsibility to lock checked-out
objects appropriately. These locks need to be maintained on secondary storage, since
they are of long duration and, therefore, must survive a failure of the server.

Let us illustrate this protocol on an example. Assume that a transaction T wants
to check out object o_1 for read and o_2 for update. Thus, T contains the two *checkOut*
statements:

      begin transaction T
      o_1.checkOut(S);
      o_2.checkOut(X);

      ⋮

These *checkOut* requests are sent to the server, which starts a short transaction—call it T_s. The transaction T_s determines whether the two *checkOut* requests can be granted; that is, it checks that there exists no long-duration X lock on o_1 and no lock at all on o_2. If this can be verified, T_s sets a long-duration S lock on o_1 and a long-duration X lock on o_2. Thereafter, the two objects are transferred to T's private database. Thus, the short transaction T_s is used to atomically set the long duration locks within the server.

At the end of transaction T the (modified) object o_2 is checked back into the server. Furthermore, the long duration X lock on o_2 and the long-duration S lock on o_1 have to be released. This is done analogously to the *checkOut* protocol by starting a short transaction to perform this task—atomically and in isolation of other concurrent activities within the server. Only after this has been successfully performed can T be terminated.

Thus far, not much has been gained with respect to the other requirements listed in the previous section. We will briefly indicate some straightforward ways to meet some of these requirements. The adaptations or extensions are in the cases discussed subsequently quite simple, since existing concepts are exploited in order to meet the requirements.

As already mentioned, design transactions are long in duration and, hence, they should not execute in an all or nothing manner (requirement 1). By supplying an additional *save* operation for objects that are checked out, it is possible to meet this requirement. Whenever an object is saved, it is stored permanently in the private database such that—in case of a failure—the latest saved version can be restored. This way additional save points—besides the beginning of the transactions—are provided. Also, for convenience, a general save operation could be supplied that saves all checked out objects. It is important to note that in either case these save operations do not check in the object(s) and, hence, the object(s) do not become visible to other transactions.

Requirement 3—allowing the exchange of information—can be met by expanding the set of possible locks, which allows reading an object in all cases, even if it is checked out from the public database. Since this lock meets the requirements imposed by a typical browsing operation it is called *browsing lock*, abbreviated by B. An example application in which browsing operations might be utilized is a report generator that generates approximate reports on the state or progress of a project. Since for this kind of operation the consistency of the examined data is not of major concern, they should always be executable. Thus, the compatibility matrix including the browsing lock looks as follows:

	S	X	B
S	yes	no	yes
X	no	no	yes
B	yes	yes	yes

Note that this allows for more concurrency, since transactions that "are content" with possibly outdated and inconsistent object states are neither hindered by nor hindering other transactions. In this sense, the B lock is actually equivalent to obtaining the object without setting any lock at all. Of course, objects that are checked out with a B lock cannot be checked in.

The classical protocol that regulates when locks may be acquired and released should also be relaxed. Whereas strict two-phase locking was necessary in order to guarantee Level 3 consistency (recall Section 15.1), it is inadequate for many design applications in which Level 3 consistency is not necessarily required—or achievable. Consider the following example in which a designer needs information about a certain part of the artifact to be designed. If this part has been checked out by a long transaction of another designer, then it might be reasonable for the latter to check in the needed parts if he or she has finished the partial design on these objects and no longer needs them. This need not necessarily be at the end of the transaction.

A typical transaction not adhering to the two-phase locking protocol might look as follows:

(1) **begin transaction**

(2) o_1.checkOut(S);

(3) o_2.checkOut(X);

(4) $\cdots$!! perform some work and finish the design for o_2

(5) o_2.checkIn;

(6) o_3.checkout(X);

(7) $\cdots$!! continue to adapt o_3 to the current design

By allowing the user to explicitly check out and check in objects whenever no conflict concerning the locks arises, more user control is exerted. The disadvantage of this approach is that the user—or the designer of application programs and tools like CAD systems—has to be aware of certain concurrency problems that might occur. The user also needs to think about the different levels of consistency desired at a

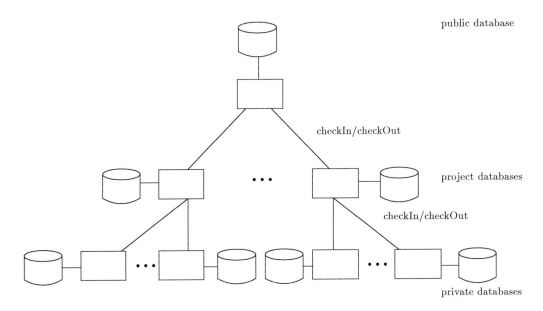

Figure 15.5: A Three-Layer CheckIn/CheckOut Model

certain stage of design and must know how they can be assured. In the above simple *checkIn/checkOut* environment this might become a hard problem, since conflicts can possibly occur not only between users working in the same project but between any two users of the public database. This is certainly due to the lack of mirroring the structure and organization of a design project—recall Requirement 7.

15.9 Group Transactions

A transaction model that especially supports the organization of a design process into (sub-)tasks is the *group transaction* model. Here, it is assumed that the design process is divided into several tasks. On each task a group of designers works cooperatively.

The basic idea to cover the hierarchical organization of the design process is to add an additional intermediate layer of *project databases*—also called *semipublic* or *group databases*—between the public database and the private databases. This results in the database hierarchy shown in Figure 15.5. Now, *checkIn/checkOut* operations can occur between the public database and the group databases as well as between the group databases and the private databases.

A special transaction model meeting the additional requirement has been developed: Incomplete designs can be made visible to other group members without making them visible to the public.

The new transaction model distinguishes two kinds of transactions: *group trans-actions* and *user transactions*. Each group transaction is associated with a group of designers that are supposed to cooperate on a certain task or project. At the creation time of a group transaction, a group database is created and associated to it. The group transaction may then *checkIn* and *checkOut* objects between the public database and its group database in the same way as described in the previous section; i.e., it copies objects from the public database into the group database and later checks them back into the public database. For group transactions serializability is assumed, since close cooperation takes place mostly within the groups and not between members of different groups.

For each group transaction, user transactions can be created that become *subtransactions* thereof. These user transactions totally depend on the group transaction in that they can only exist if the corresponding group transaction exists. For each user transaction a private database is created. After its creation at the user transaction's begin, it can start to check out objects from the group (project) database into its user database and check them back into the group database later on. The main restriction is that user transactions do not have access to the public database and can check out objects only from the group database, i.e., those objects that the corresponding group database has checked out from the public database and are therefore resident in the group database.

Besides the already-mentioned *checkIn/checkOut* operations, the group transaction model has to provide operations to create and modify new user groups, start and end group transactions, and start and end user transactions. The begin of a group transaction has to be parameterized by a group name and a group transaction name to which the user transactions can refer. Thus, a begin group transaction might look as follows:

begin group transaction A **for** $myGroup$

where A is the name of the group transaction and $myGroup$ is the name of a group of users working in this group.

A user transaction starts with

begin user transaction $myTA$ **within** A

where $myTA$ is the name of the user transaction and A is the name of the group transaction of which the user transaction becomes a subtransaction.

This model consists of only three levels of databases and two levels of transactions. A generalization of this model which allows arbitrary levels of nesting yields *nested transactions*.

15.10   Exploiting Versioning for Concurrency Control

Supporting a higher degree of concurrency and the exchange of partial designs are two of the major requirements for transaction models to be applied in design environments. Object versions are introduced in Chapter 17[2] and can be used to enhance concurrency and to facilitate the exchange of partially consistent design objects among cooperating transactions.

Two different versions of an object can be seen as two different—though related—objects. There is no reason not to allow concurrent modification of two versions, or to allow reading one version while another is updated. Thus, concurrent access to a single object can be supported by allowing concurrent access to different versions of the same object.

The most important operations on versions are as follows:

- Create a new version from an existing one.

- Update an existing version.

- Delete an existing version.

These operations on versions have to be synchronized. Among these operations deletion does not require any special treatment. We thus concentrate on the creation and modification of versions.

Obviously, for creating or modifying a version, the original version has to be read. Thus, the right to read the original version is a prerequisite for creating a successor version or modification. Note that a read lock on the original version does not prevent the concurrent derivation of two successor versions, although this might be unacceptable in some cases since merging the two new versions might be difficult or even impossible. In order to refine the locking protocol and provide more flexibility for restricting the derivation of new versions new locks have been introduced.

Remember that all locks have two different implications:

1. A transaction holding a lock is allowed to perform a certain operation.

2. Other transactions are limited to operations they can perform on the already locked object, i.e., only operations that do not conflict with an existing lock.

We list these two different implications for each lock utilized in the protocol for the concurrency control scheme incorporating versions:

---

[2]The reader not familiar with the basic concepts of versioning may want to first read the basics on versioning in that chapter and then return to this point.

R-lock: It allows the lock holder to read the locked version as well as parallel reads by other transactions. However, neither transaction—not the holder of the lock nor the parallel transaction—is entitled to derive a new version of the locked object.

RD-lock: The *RD* lock allows the lock holder to read the locked version as well as parallel reads by other transactions. Additionally, it allows other transactions—after obtaining a suitable lock—the parallel derivation of new versions from the locked object. However, the *RD* lock holder may not derive a new version.

DS-lock: This lock allows the lock holder to read the locked version and to derive new versions from it. Other parallel transactions may also read the locked object and derive new versions from it.

DX-lock: This allows the lock holder to read the locked version and to derive new versions. It allows other parallel transactions only to read the object but not to derive new versions.

X-lock: The exclusive *X* lock allows the lock holder to read and modify the locked version as well as to derive new versions. No parallel activities on the locked version are allowed.

These locks can then be used as parameters for the *checkIn* and *checkOut* operations. The compatibility matrix for these locks is as follows (again, the left column indicates the lock currently associated with the object and the upper row indicates the lock requested by a different transaction):

	RD	R	DS	DX	X
RD	yes	yes	yes	yes	no
R	yes	yes	no	no	no
DS	yes	no	yes	no	no
DX	yes	no	no	no	no
X	no	no	no	no	no

This compatibility matrix can best be understood when considering the so-called *strength* of a lock. We say that a lock mode L_1 *is stronger than* a lock mode L_2 if it grants its owner the same rights as L_2 or more rights than L_2 *and* L_1 grants other transactions (different from the holder) fewer rights than L_2 or the same rights as L_2. Under this definition of lock mode strength we can compare the strength of the five lock modes R, RD, DS, DX, and X as shown in Figure 15.6.

The graphic shows that the two modes R and DS are incomparable because DS grants the owner more rights—i.e., read and derive—than the R mode. At the same

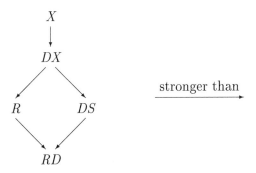

Figure 15.6: Comparison of Lock Mode Strengths

time, however, DS also grants other, parallel transactions more rights than an R mode would. It is this incomparability that has to be kept in mind when trying to understand the entries of the R and DS columns and rows in the compatibility matrix.

Opposed to traditional locking based concurrency control protocols additional user control on lock modes is supplied in order to meet Requirement 7 of Section 15.7. This control can be exerted not only by *checkIn* and *checkOut* of versions but also by explicit change of lock modes. For this purpose an operation *changeLock* is provided for each object that has the new lock mode as a parameter.

15.11   Exercises

15.1   Consider the following schedules. Which are serializable? Assume all *begin transaction* commands are executed before step 1 and all **end transaction** commands occur after step 6.

1.

Step	TA$_1$	TA$_2$	TA$_3$
1.	o1.read		
2.		o2.read	
3.			o3.read
4.	o1.write(...)		
5.		o2.write(...)	
6.			o3.write(...)

2.

Step	TA$_1$	TA$_2$	TA$_3$
1.	o1.read		
2.		o1.read	
3.			o1.read
4.	o1.write(...)		
5.		o1.write(...)	
6.			o1.write(...)

3.

Step	TA$_1$	TA$_2$	TA$_3$
1.	o1.read		
2.	o1.write(...)		
3.		o1.read	
4.		o1.write(...)	
5.			o1.read
6.			o1.write(...)

4.

Step	TA$_1$	TA$_2$	TA$_3$
1.	o1.read		
2.		o2.read	
3.			o3.read
4.	o2.write(...)		
5.		o3.write(...)	
6.			o1.write(...)

15.2   For the schedules above, give the locks to be acquired for each step and mark those that are granted. In which schedules can all transactions commit?

15.3   In which of the above schedules do deadlocks occur?

15.4   Assume two groups of designers are working on a car design. One group is responsible for the design of the chassis, while the other is responsible for the design of the engine. John belongs to the group of chassis designers, whereas Mary belongs to the group of engine designers. John wants to design a door, Mary wants to design the carburetor. Provide the necessary transactions for the group transaction model that allows each of them to read and modify the corresponding part of the car design.

15.5   Find illustrative examples of schedules in which optimistic concurrency control allows a higher degree of parallelism than the two-phase locking protocol would.

15.6 Show situations (i.e., sketch the parallel transactions) in which pessimistic locking-based concurrency control is far superior to optimistic synchronization.

15.7 Sketch a deadlock situation under two-phase locking that results from the cyclic dependency of three transactions T_1, T_2, and T_3. That is, T_1 is waiting for T_2, T_2 is waiting for T_3, and T_3 is waiting for T_1.

15.8 In the text we discussed a relaxation of the strict two-phase locking protocol for long transactions; i.e., transactions were allowed to check objects back into the public database before they terminated their work. Discuss potential consistency violations resulting from this relaxation. Illustrate the discussion by an example from your area of expertise.

15.12   Annotated Bibliography

The "traditional" approach to transaction control can be studied in [BG81]. The transaction concept of the relational DBMS System R was a driving force behind later works. These concepts are described in [ABC⁺76, EGLT76, GMB⁺81, Gra81]. The notion of transaction consistency, in particular the three-level classification, was introduced by J. Gray in [Gra78]. Optimistic concurrency control was introduced in [KR81]. Multiple granularity locking is discussed in [Gra78] and more recently in the context of the object-oriented database system Orion in [GK88]. An extension of multiple granularity locking for checking out and checking in nested relational (NF^2) objects is described in [DK88]. For further details on recovery see [GMB⁺81] and [Reu84]. A thorough treatment of conventional transaction control schemes is given in [BHG87] and in [Pap86].

Requirements for design transactions can be found in different papers, like [BK91, KW84, KKB88].

The cooperation between transactions via a semipublic database was developed in [KLMP84]. This approach is extended and formalized in [BKK85] and, more complete, in [KKB88]. For checkOut/checkIn of versions we followed the approach presented in [KSUW85]. Especially the concepts on checkOut/checkIn of versions presented in the text are based on the framework developed in [KSUW85]. The multilevel checkOut/checkIn concepts were heavily influenced by the notion of nested transactions, which was first worked out in [Mos81] and further generalized in [Wal84]. In the latter article, a framework was presented and covers different approaches to nested transactions. Besides the approach introduced in [Mos81] it also covers newer approaches like [BBG89]. Another source on concepts for nested transactions in advanced applications is [Wei88]. Recovery concepts for nested transactions are developed in [HR87].

Of course, we could not give an exhaustive overview of all existing approaches to concurrency control for design transactions. A good summary can be found in [BK91] where 21 different approaches are compared. Additionally, [Lan88] contains two evaluation criteria for concurrency control protocols and assesses five approaches. [YEEK87] contains a performance comparison of two different concurrency control schemes.

The exploitation of semantics of more complex type-associated operations has been totally neglected in this chapter. Many proposals exploiting semantics of type associated operations for more concurrency and for allowing more flexible protocols have been proposed, e.g., the work on synchronizing abstract data types [Wei89, Her90]. For a review see [SZ89b].

A framework that seems to succeed in capturing all approaches systematically has been proposed in [CR90]. Its underlying formal model is described in [CR91].

[Elm92] contains a collection of papers on advanced transaction models. [GR93] is a comprehensive volume on transaction processing.

16

Authorization

As already mentioned, a database is typically a highly shared resource. The data present in a database are accessed—read, modified, extended—by many users. Nevertheless, it is quite common that not every user is allowed to access each data item. On the contrary, some data should be accessible only by certain users or even a single user only. Some users may be allowed to read some attributes, while other attributes must be hidden for, e.g., privacy reasons. Some objects may be read by all users but be updated only by some particular user(s). Controlling these access rights to the database is called *authorization*.

16.1 Overview

Controlling access to a database has three different dimensions:

1. The *subject* (a user or a group of users) accessing information

2. The *object* which is to be accessed

3. The *mode* (an operation), in which the object is to be accessed by the subject

Thus, an *authorization* can be specified in terms of a triple (s, o, op) where s is the accessing *subject*, o is the *object* to be accessed, and op is the *operation* manifesting the access. All these authorizations are collected in an *authorization base*, called AB. Then, whenever a triple (s, o, op) exists within the authorization base AB, the subject s may invoke operation op on object o.

The above kind of authorization is called an *explicit positive* authorization. *Explicit* authorization means that the triple is stated explicitly; i.e., it is explicitly present in the authorization base.

To complement this, there exists *implicit authorization*. Consider a group of subjects (users), called *carDesigners*, who are granted the authorization to *read* an object *carDesign*. The explicit authorization is (*carDesigners, carDesign, read*). Further assume the existence of a set of *chiefDesigners*. It might be a good idea, to allow all the *chiefDesigners* to read everything the *carDesigners* are allowed to read. This is called an *authorization rule*. Now, if there is only the above triple in the authorization base *AB*, the *chiefDesigners* are not allowed to read anything. By explicit authorization, the only way to allow the *chiefDesigners* to read everything the *carDesigners* are allowed to read is to replicate the authorization triples with *carDesigners* as the subject with *chiefDesigners* as the subject. If there exist many of these triples, then this results in much work and a large authorization base that may lead to inefficiency. By *implicit* authorization, the above rule can be stated explicitly as part of the authorization model and the authorization triples for *chiefDesigners* are derived automatically. Hence, instead of giving a a very large of triples—which may even change over time due to newly created objects by *designers*—one rule suffices.

Implicit authorization can be present in all three dimensions—i.e., at the subject, object, and operation level—of authorization and is regulated by *implication rules*. The main advantage of implicit authorization lies in a vastly reduced number of explicit authorization triples in the authorization base. Thus, less effort is needed to build the authorization base and less storage is required to maintain it.

Next, *positive* and *negative authorizations* are distinguished. All the above authorization triples are positive, since they *allow* some subject to execute some operation on some object. On the opposite, *negative authorizations* disallow or forbid a subject to execute an operation on an object. As an example, take an accounting department whose members are not allowed to update the design of a car. The corresponding *explicit negative authorization* is the triple (*accountingClerks, carDesign, ¬write*). Here, the negation symbol "¬" indicates a negative authorization. As for positive authorizations, *explicit* and *implicit negative authorizations* are distinguished. In the above example, if *John* is an accounting clerk, belonging to the group *accountingClerks*, he is not allowed to update the object *carDesign* by the implications of the above explicit negative authorization. Again, the advantage is a reduced number of authorization triples.

The last distinction is drawn between *strong* and *weak authorizations*. *Weak authorizations* can be used to set up a default authorization. As an example, assume that all users are collected in a group *allUsers*. Then the *explicit weak authorization* [*allUsers, carDesign, read*] allows all users to read *carDesign* unless stated otherwise. Note, that the notation "[..., ..., ...]" is used for weak authorizations, whereas "(..., ..., ...)" is used for strong authorizations. The difference between strong and weak authorizations is that the latter can be overwritten, whereas the former cannot be overwritten. All weak authorizations are collected in the *weak authorization base*

WAB. As for strong authorizations, we can distinguish between *explicit/implicit* and *positive/negative* authorizations.

There exist two authorization bases:

- The authorization base *AB* containing all explicit strong authorizations—including both, positive and negative strong authorizations

- The weak authorization base *WAB* containing all explicit weak authorizations—including both positive and negative weak authorizations

They are maintained through the following four commands:

Grant strong authorization A strong authorization (s, o, op) is granted by the operation $grant(s, o, op)$. This operation adds the triple to the authorization base *AB*. The operation *op* may also be preceded by ¬ indicating a negative authorization.

Revoke strong authorization A strong authorization (s, o, op) is revoked by the operation $revoke(s, o, op)$. The operation removes the triple from the authorization base *AB*. The operation *op* may also be preceded by ¬ indicating a negative authorization.

Grant weak authorization A weak authorization $[s, o, op]$ is granted by the operation $grantWeak[s, o, op]$. This operation inserts the triple into the weak authorization base *WAB*. The operation *op* may also be preceded by ¬ indicating a negative authorization.

Revoke weak authorization A weak authorization $[s, o, op]$ is revoked by the operation $revokeWeak[s, o, op]$. This operation deletes the triple from the weak authorization base *WAB*. The operation *op* may also be preceded by ¬ indicating a negative authorization.

The remaining question is: Who is allowed to execute these operations? There exist at least the following three different approaches to the authorizations of the *grant* and *revoke* operations:

- Only a distinguished database administrator, often called DBA, is allowed to execute these operations.

- Only the owner or creator of an authorization object is allowed to grant and revoke authorizations concerning the object.

- Everybody who has a certain authorization to execute an operation on an object can grant and revoke this right to/from other users.

Obviously, the first possibility is the most restrictive one, whereas the last is the most relaxed paradigm. The disadvantage of the first approach is that in all cases the database administrator has to be asked to grant or revoke rights. This disadvantage is slightly diminished by implicit authorization. The disadvantage of the last alternative is the possible uncontrolled spreading of authorizations. In choosing the right paradigm for a particular application domain one must consider this tradeoff.

Up to now, we have not been concerned with the evaluation of the authorizations. Whenever a subject intends to execute an operation on a certain object, the following check is to be performed—let (s, o, op) be the authorization to be checked:

> **if** there exists an explicit or implicit strong authorization (s, o, op)
> > **then** allow the execution
>
> **if** there exists an explicit or implicit strong authorization $(s, o, \neg op)$
> > **then** forbid the execution
>
> **if** neither exists
> > **then**
> >
> > **if** there exists an explicit or implicit weak authorization $[s, o, op]$
> > > **then** allow the execution
> >
> > **if** there exists an explicit or implicit weak authorization $[s, o, \neg op]$
> > > **then** forbid the execution

Note that the algorithm assumes that there exists at least a weak authorization for each possible privilege. It is also assumed that there do not exist two conflicting strong or weak authorization triples. Further, the details on the implication of access rights are still missing. These are the subject of the next sections.

16.2   Implicit Authorization

As already mentioned, implicit authorization can take place along each of the three dimensions of authorization, namely, subjects, objects, and operations. To each dimension a subsection is devoted. Since the second dimension is the most difficult, the other two are treated first. Last, the possibilities for implicit authorization along type hierarchies are briefly discussed.

16.2.1   Authorization Subjects

Inheritance of authorizations from one subject to another subject has to be organized. For this purpose, a *role hierarchy* is introduced. This reflects the possibility that a subject can play different roles within a system with different associated authorizations. Further, not every role has to be occupied by a subject. Of course, the role hierarchy is application specific. Nevertheless, the following two roles must be present in all role hierarchies:

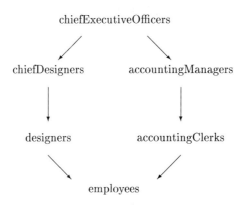

Figure 16.1: A Sample Role Hierarchy

- A distinguished *top* of the role hierarchy, e.g., the company manager, the database administrator, or a *superuser*.

- A distinguished *bottom* of the hierarchy, e.g., the *employees*.

All the other roles have to be organized in between the top and the bottom of the hierarchy. Arrows will point from an upper role to lower roles. As an example, consider the role hierarchy in Figure 16.1. A simple role hierarchy of a car manufacturer is depicted with the *chiefExecutiveOfficer* at the top and the *employees* at the bottom. In between, we have two roles for the design department and the accounting department. For each department, there exists a *manager role*. The members of the design department occupy the *designers' role*, those of the accounting department the *accountingClerks' role*.

For implicit authorization, we have to distinguish two different cases, one for positive and one for negative authorizations:

- Explicit positive authorizations for some role results in implicit positive authorizations for all higher-level (upper) roles.

- Explicit negative authorizations for some role results in implicit negative authorizations for all lower-level roles.

Intuitively, these two rules state that a higher-level role has at least the same privileges as the lower level role.

Thus, for a positive authorization, say (*designers, carDesign, read*), not only the *designers* are allowed to read the *carDesign* object, but also the *chiefDesigners* and the *chiefExecutiveOfficers*. On the other hand, if there exists a negative authorization, say (*accountingManagers, carDesign, ¬update*), then neither the *accountingManagers* nor the *accountingClerks* nor the *employees* are allowed to update the *carDesign*.

Figure 16.2: A Sample Authorization Operation Hierarchy

16.2.2   Authorization Operations

The treatment of authorization operations is quite similar. Here, the operations are organized into an *authorization operation hierarchy*. Considering only two types of operations, namely, those only reading and those also updating an object, a very simple authorization operation hierarchy containing just the two operations *read* and *write* is shown in Figure 16.2. Since writing an object without prior inspection via a read operation does not make much sense, an arrow is drawn from the *write* to the *read* operation type. This arrow indicates that the *write* authorization implies the *read* authorization. As for implicit authorization along the subject dimension, the two cases positive and negative authorization have to be distinguished:

- If a positive authorization is given for an operation in the authorization operation hierarchy, then this implies the authorization of all operations *below* it.

- If a negative authorization is given for a certain operation in the authorization operation hierarchy, then this also implies the negative authorization of all operations *above* this operation.

Thus, $(s, o, write)$ implies $(s, o, read)$ for all s and o, and $(s, o, \neg read)$ implicitly forbids the *write* operation, i.e., $(s, o, \neg write)$ is implied.

Of course, the shown authorization operation hierarchy is very simple. In general, more operation types, including *creation* and *deletion*, can be integrated. Moreover, object type specific authorization operation hierarchies are possible.

16.2.3   Authorization Objects

Implicit authorization along the authorization object dimension has a way larger potential for savings than implicit authorization along the other two dimensions. This stems from the simple fact that there normally exist far more objects than users or operations. Again, implicit authorization is organized along a *hierarchy*—the

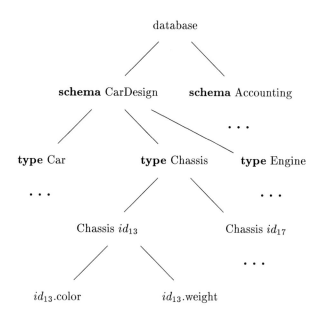

Figure 16.3: A Sample Database Granularity Hierarchy

authorization object hierarchy. Since it contains all objects, its explicit representation is too expensive. Therefore, a detour over the *authorization object schema* is taken. Then the authorization object hierarchy can be inferred from it.

A database can be organized into a *database granularity hierarchy*. An example of a database granularity hierarchy is shown in Figure 16.3. All the present granules belong to a certain granularity level. For example, all type definitions belong to the type definition level.

Abstraction from the instances of each granularity level is performed by the *authorization object schema*. An example is shown in Figure 16.4. The coarsest granule is the whole database; the finest granule is a single attribute. Note that in the authorization object schema, we distinguish between the type definition, the type extension, and the objects that are instances of a type. This is necessary, since an authorization for one operation type may have different impacts on each of these levels. For example, a *write* authorization for a type definition allows to change the type definition, whereas a *write* authorization for a type extension allows to update each instance of the type.

According to the authorization object schema we can now give an example *authorization object hierarchy*, which then serves as the basis for discussing implicit authorization along the authorization object dimension. Organizing the database

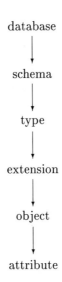

Figure 16.4: A Sample Authorization Object Schema

granularity of Figure 16.3 according to the authorization object schema of Figure 16.4 results in the authorization object hierarchy shown in Figure 16.5. It is important to note that the authorization object hierarchy is a virtual structure that is derived from the authorization object schema and the actual database contents. Thus, only the authorization object schema has to be explicitly stated. Nevertheless, implicit authorization along the object dimension is based on the authorization object hierarchy.

The direction of implicit authorizations along the object dimension is not as easy as for the other two directions, since the direction depends on the operation type considered. A read and a write operation authorization for a certain object requires to read the type definition of the object's type, as well. Thus, the direction of this implication is upward in the authorization object hierarchy. On the other hand, a read permission on the extension of a type implies a read permission on all the instances of that type. The direction of the implication of this permission is, therefore, downward. Furthermore, some permissions do not have any implications. An example is the definition (creation) of a new type. According to these cases all operations are organized into three classes:

1. *OpUp* contains those operations having upward implications.

2. *OpDown* contains those operations having downward implications.

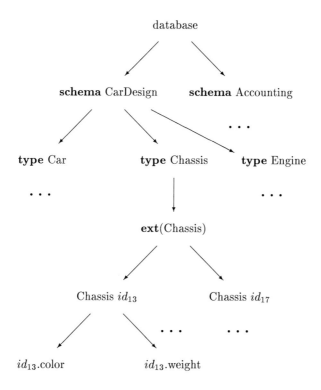

Figure 16.5: A Sample Authorization Object Hierarchy

3. *OpNil* contains those operations having no implication.

The operations considered here are *write*, *read*, *define*, and *readDefinition*. Their classification is as follows:

- OpUp = {readDefinition}

- OpDown = {write, read}

- OpNil = {define}

The authorization operation hierarchy for these operations is shown in Figure 16.6. We are now ready to define the implications of an authorization along the authorization object dimension separated for positive and negative authorizations:

1. For operations in *OpUp*:

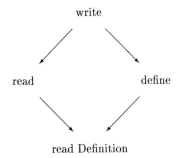

Figure 16.6: A Sample Authorization Operation Hierarchy

- For some authorization subject s and an authorization operation op in
  $OpUp$ let (s, o, op) be a given authorization for some authorization object o. Then (s, o, op) implies (s, o', op) for all objects o' above o in the
  authorization object hierarchy.

- For some authorization subject s and an authorization object type op in
  $OpUp$ let $(s, o, \neg op)$ be a given authorization for some authorization object
  o. Then $(s, o, \neg op)$ implies $(s, o', \neg op)$ for all objects o' below o in the
  authorization object hierarchy.

2. For operations in $OpDown$:

   - For some authorization subject s and an authorization operation op in
     $OpDown$ let (s, o, op) be a given authorization for some authorization ob-
     ject o. Then (s, o, op) implies (s, o', op) for all objects o' below o in the
     authorization object hierarchy.

   - For some authorization subject s and an authorization object type op in
     $OpDown$ let $(s, o, \neg op)$ be a given authorization for some authorization
     object o. Then $(s, o, \neg op)$ implies $(s, o', \neg op)$ for all objects o' above o in
     the authorization object hierarchy.

3. Operations in $OpNil$ do not have any implications along the authorization object
   hierarchy.

For example, $(designers, \mathbf{ext}(Chassis), write)$ implies

1. $(designers, id_{13}, write)$

2. $(designers, id_{17}, write)$

3. $(designers, id_{13}.color, write)$

4. ($designers, id_{13}.weight, write$)

for the authorization object hierarchy shown in Figure 16.5.

16.3 Type Hierarchies

For a given type hierarchy, there exist two different paradigms one can follow for authorization. Either, the authorizations given for a supertype are implied on its subtype, or no authorization is implied along the type hierarchy. Remember that authorizations on a subtype may imply authorizations on its instances. That is, whenever a user creates a subtype of a certain type and the system follows the first paradigm, then the instances he or she subsequently creates are accessible by all the subjects who can access instances of the supertype. This lack of privacy clearly discourages the creation of subtypes, and hence, the reusage of code.

On the other hand, following the second paradigm imposes problems when queries range over all instances of a type, since they naturally should also range over all instances of the type's subtypes. There is no way to get around this tradeoff by means of a general mechanism. The only possibility is to have weak implicit authorizations and then overwrite these by strong authorizations whenever necessary.

16.4 Exercises

16.1 Consider the role hierarchy of Figure 16.1. Specify the contents of the authorization base AB and the weak authorization base WAB such that the following three conditions hold:

1. The *designers* are allowed to read the object *carDesign*.
2. A *chiefDesigner* is allowed to read the object *carDesign*.
3. Nobody else is allowed to read the object *carDesign*.

16.2 Assume the problem of providing a fine-grained authorization control for the type-associated operations of a *Cuboid*. Give an authorization operation hierarchy such that the implicit authorization rules express the following:

1. The right to *write* implies the right to execute all other operations.
2. The right to *scale* implies the right to *rotate*, *translate*, and *read*.
3. The right to execute the *weight* operation implies the right to execute the *volume* operation and the *read* operation.
4. The right to execute *rotate* or *translate* implies the right to *read*.

    5. The right to execute *volume* implies the right to *read*.

    6. These are all implications of access rights.

16.3   Consider the authorization object hierarchy of Figure 16.5 together with the authorization operation hierarchy of Figure 16.6. Which are the implications of

    1. (*designers*, **ext**(*Chassis*), ¬*write*)

    2. (*designers*, **type** *Chassis*, ¬*define*)

    3. (*designers*, **type** *Chassis*, *define*)

    4. (*designers*, $id_{13}.color$, ¬*read*)

    5. (*designers*, $id_{13}.weight$, ¬*read*)

16.4   [Kim90] Extend the notion of implicit authorizations to versions—as discussed in Chapter 17. A user who has a certain type of authorization on a versioned object should be regarded as implicitly having this right on all versions. One possible solution consists of extending the *authorization object schema* by incorporating versions. Sketch the idea. Illustrate your solution by some examples.

16.5   Annotated Bibliography

This chapter followed largely [RBKW91], which is an extension of [RWK88]. The authorization triples can already be found in operating system literature like [GD72, Lam71]. Authorization in relational database systems can be found in [GW76]. Implicit authorization was first formalized in [FSC75, FSL75].

Authorization in the IRIS object-oriented DBMS is discussed in [ADG⁺92].

Authorization in the relational model is discussed in [Dat86] in the context of the DB2 relational database system of IBM.

Further work on authorization in object-oriented database systems can be found in [DHP89] and [GGF93].

17

Version Control

Object-oriented database systems have been developed in order to support advanced applications; e.g., engineering applications. One of the most difficult engineering tasks is the design of an artifact. No matter whether considering mechanical engineering, software engineering, very large scale integration (VLSI) design, architecture, etc., the design of an artifact evolves over time where the design process is iterative and tentative in nature and is by no means a straight linear sequence of steps undertaken by the designer. Instead, due to the complexity of design, the design process resembles more a search process comprising trial and error, the exploration of different alternatives and tracking back to previous design steps. During this process typically many *versions* of the same design object come into existence and must be managed. This chapter is devoted to *version management*, i.e., the organizational concepts introduced to support the handling of engineering design data evolving over time.

17.1 Motivation

One of the most salient features of engineering design artifacts is their intrinsic complexity. To illuminate this point consider the following examples of large amounts of data:

1. A million bank accounts

2. A million lines of code organized into hundreds of thousands of operations with their interfaces and implementations calling other operations, and hundreds of modules with a high degree of interrelationships

3. A million transistors organized into hundreds of thousands logic gates, their connections, and thousands of lines of description in a hardware description

language

Obviously, the first example is different from the other two in that the data is homogeneous with no obvious dependencies. Contrarily, the other two examples are typical for design: millions of objects with a high degree of interdependencies. Increasing complexity even further, a design object obtains many associated secondary objects, like documentation and test cases.

No wonder that the final artifact cannot be designed in a single step without any trial and error, exploration of alternatives, backtracking, etc. The design process is iterative and tentative in nature: The design of an artifact evolves over time through successive refinements—like the elimination of simple errors or enhancements of functionality—and exploration of *alternatives*. Furthermore, the same artifact may be used in different environments with slightly differing requirements giving rise to *variants*. In general, associated with each *design object* is a set of *versions* covering the *refinements*, *alternatives*, and *variants*. On first glance, one can think of the versions of a design object as snapshots of the design object taken during its design process. But on closer observation, it becomes clear that a version carries more semantics than just a time stamp; e.g., a version has predecessors and successors.

Moreover, to conquer the complexity of design, it typically exhibits a hierarchical structure like in software engineering and hardware design. Thus, a design object consists of several constituent design objects; e.g., a car consists of a chassis, an engine, etc. Each of these subobjects again has different versions giving rise to *configurations* for cars—for a specific version of a car, a version of a chassis and a version of an engine is necessary. Further, the creation of a new version of a chassis may result in a new version of a car. If the designer of the engine differs from the designer of the chassis, he or she should be *notified* when a new version of the chassis comes into existence.

The support of all the above mentioned facets of design is summarized under the term *version management* and is subsequently discussed in some more detail.

17.2   Organizing the Version Set

Our starting point in modeling the evolution of design objects over time is presented in Figure 17.1. It shows a design object with its associated (unstructured) set of versions. The versions are totally unrelated, which, in general, is not true. Instead, as mentioned above, the versions carry some semantics, and relationships exist between them. As a first step toward capturing more semantics, time stamps—called *version numbers*—are associated with the versions. In Figure 17.2 positive integers are used as version numbers. Lower numbers are created first. Thus, version 1 is created before version 2 and so forth. Sometimes, version numbers are composed of two

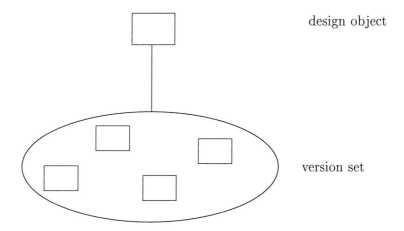

Figure 17.1: A Design Object and Its Version Set

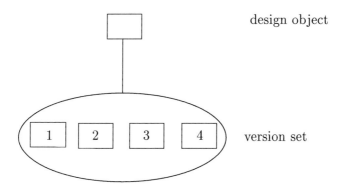

Figure 17.2: Numbering the Version Set

positive numbers separated by a dot. The first number is increased for major revisions, whereas the second number appearing after the dot is used for minor revisions. Hence, it can be assumed that the versions numbered 1.3 and 1.4 differ less from each other than the versions numbered 1.3 and 2.1. For the purpose of this chapter, simple positive numbers suffice as version numbers.

Now consider the following scenario. Version 2 is created from a copy of version 1 and version 3 is created from a copy of version 2 but version 4 is created from a copy of version 2 instead of version 3—giving rise to an *alternative*. Then the linear ordering of time stamps does not suffice to reflect the *ancestor/descendant relationship* between versions. Therefore, this relationship—being the most basic means to organize the version set—is maintained explicitly. Figure 17.3 shows the ancestor/descendant relationship for the scenario. Arrows point from the source version to the target ver-

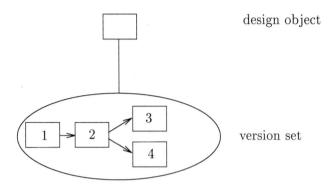

design object

version set

Figure 17.3: The Derivation History of Versions

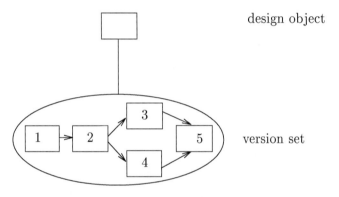

design object

version set

Figure 17.4: The Derivation History Forming a DAG

sion, which resulted from changes to the source version. Version 2 has two alternative successors: version 3 and version 4, which in turn do not have any successor versions. The graphical representation of the ancestor/descendant relationship is called *version graph*. The version graph of Figure 17.3 has a tree structure.

Assume that the versions 3 and 4 in the version graph of Figure 17.3 are not really design alternatives; rather, version 3 contains the next refinement step reflecting a major design decision and in version 4 a minor error in the design of version 2 is eliminated. Then in the final design, both changes should of course be incorporated. This gives rise to *merge* the two versions 3 and 4 into a version numbered 5. The resulting version graph, as shown in Figure 17.4, no longer exhibits a tree structure but is a directed acyclic graph (DAG). This is the most general form possible for a version graph.

Two other basic concepts aiding to organize the version set are the *current indicator* and the *main path*. The *current indicator* always refers to the current version from which the next version should be generated by default. In our example, the current indicator evolves as follows. Starting with the only version numbered 1, this clearly is the version the current indicator refers to. When creating version 2 it, in turn, becomes the current version. Then, version 3 should become the current version since it reflects a major design decision. However, after the creation of version 4—in which only a minor error of version 2 was eliminated—version 3 remains the current version. Finally, after its creation by merging versions 3 and 4, version 5 becomes the current version. If no explicit manipulation of the current indicator by the user arises, the *main path* contains all nodes from the root to the current version, which have been current versions. In our example, the main path is $1 \rightarrow 2 \rightarrow 3 \rightarrow 5$.

So far, the version set is organized by the ancestor/descendant relationship as well as a current indicator and a main path. In complex applications the need might arise to incorporate more relationships. Some of these relationships are useful in many applications and should therefore be supported as generic features by the system. Examples are the *equivalence* relationship, which expresses that two designs can be interchanged, the *is-upward-compatible-with* relationship, and the relationship *is-derived-by-tool-X*. Since the set of relationships between versions is open ended, the version management should be customizable by allowing to add new relationship types useful in specific applications.

While having covered relationships between two versions, properties of versions have not been treated. An example of an attribute attached to versions is a *valid/invalid* label. Since any version is either valid or invalid, this results in two (disjoint) *classes* of versions: one containing the valid versions and the other containing the invalid versions. Other classifications are necessary if the versions have to pass different tests, say, $test_A$ and $test_B$. Then the version set is partitioned into those versions having passed no test, either test or both tests. Since the passing of tests is not a rare situation, one could think of building the class handling for test classes into the version management system. Although very useful, it still does not suffice, since application-specific attributes may be necessary. Consider the version management for VLSI design. A partitioning of the version set of a full-adder into fast and slow alternatives as well as into low-power and small-area alternatives is useful for selecting versions fitting current requirements. Summarizing, new classes for versions provide a good means to further organize the version set according to application-specific criteria.

17.3 Reference Mechanisms and Configurations

As already mentioned, design objects are often hierarchically organized along the *part-of* relationship in order to scale down the complexity of design. In this case, a design object or a version of a design object refers to another design object or one

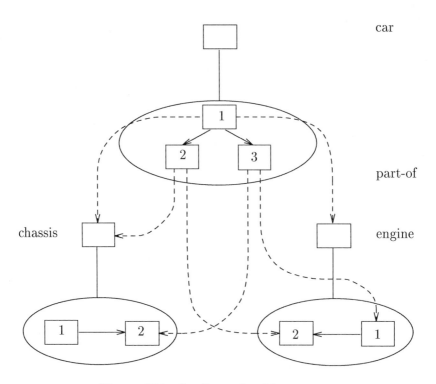

Figure 17.5: Configuration Management

of its versions. References pointing to a specific version are called *static references*, whereas those referring (abstractly) to a design object are called *generic* or *dynamic references*. The latter arise if the *part-of* relationship between design objects is known but not the specific versions that will be used as constituent parts.

Consider a car consisting of a chassis and an engine. Figure 17.5 shows this two-level hierarchical design. Here, the car has an associated version set consisting of three versions. For version 1 it is known only that it will consist of a chassis and an engine. The references from the car to the chassis design object and to the engine design object are generic. Version 2 of the car points to the chassis design object—this is a generic reference—and to a specific version (i.e., version 2) of the engine, e.g., one with 120 horsepower. The latter reference is a static reference. Version 3 of the car only consists of static references: one to version 2 of the chassis—say, one with a big trunk—and one to version 1 of the engine—the one with 75 horsepower, for example.

A version in which all *part-of* references point to explicit versions, whose references in turn also point to explicit versions, and so on, is called a *configuration*. Hence, version 3 of the car is a configuration. Often it is not possible to specify configurations statically, since they must be adapted to specific requirements. In software engineering, a program configuration may depend on the hardware and software platforms to be used and its parameters. In most disciplines, e.g., VLSI design or mechanical engineering, the configuration may depend on application-specific parameters such as cost, expected speed, reliability, total weight, etc. For example, the configurations of functionally equivalent robots will highly depend on the environment, e.g., a manufacturing cell or space, where they are supposed to be used. For this reason it is often convenient not to specify configurations but to use generic references and instantiate them by a reference to a specific version not before this is explicitly required by, e.g., the demand for a configuration meeting specific requirements. Another important aspect of using generic references is that it helps to keep the number of versions low on higher levels within the *part-of* hierarchy. In the car example, one version of the car with two generic references to the chassis and engine design object would suffice to reflect the possibility of four different configurations of a car: an economic car having a small trunk and 75 horsepower, a sporty car with a small trunk and 120 horsepower, a town car with a big trunk and 75 horsepower, and a travel car with a big trunk and 120 horsepower.

The problem to be solved next is the instantiation of the generic references at the time an actual configuration is required. The simplest instantiation mechanism relies on the current versions of each design object. This is useful for testing the latest designs within one configuration but it is, of course, unable to reflect requirements beyond this.

To provide a powerful configuration mechanism the notion of *environment* has been introduced. An environment is a table that specifies how to instantiate the generic references if the environment is activated. It consists of an environment name, and a version number entry for each referenced design object. For the car example, the two environments for specifying the configurations for the sporty and the economic car could look as follows:

Environment *EconomicCar*	
Design Object	Version Number
chassis	1
engine	1
. . .	. . .

Environment *SportyCar*	
Design Object	Version Number
chassis	1
engine	2
. . .	. . .

When the specific version is to be left unspecified for an environment, the version number can be replaced by *current*.

Having specified different environments then, whenever a configuration of a specific version is required under a certain activated environment, the generic references are instantiated by looking up the versions to be used in the particular environment. When requiring a configuration for version 1 of the car design object under the active

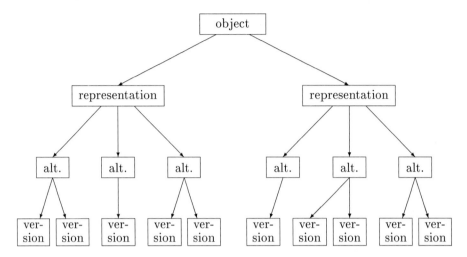

Figure 17.6: A Schema of Logical Version Clusters

environment *SportyCar*, the generic reference to the chassis is instantiated by using version 1 of the chassis design object, and the generic reference to the engine is instantiated by using version 1 of the engine design object.

17.4   Version Hierarchies

So far we have implicitly assumed a one-level version hierarchy; that is, an object has an associated flat version set. We have not yet discussed versions of versions.

In some design applications it is useful to extend the version hierarchy over several levels. Consider the schema shown in Figure 17.6. In this schema we assume several *representations* of a given design object. Let this, for example, be different representations at several levels of abstraction of the same engineering artifact, or different views on the artifact like a functional and a geometrical view. For each representation there are different *alternatives*, each of which may be composed of several *versions*. In this schema only the lowest-level objects, i.e., the versions, are self-contained in the sense that they contain all the necessary structure to model the entire design object. The intermediate levels, i.e., representations and alternatives, are only logical objects that are used to group the lower level versions into logically adjacent clusters.

In the schema of Figure 17.6 the version clusters are disjoint; that is, they form a hierarchy. In a more general setting also networks may be useful, in which case the version clusters form a directed acyclic graph. A further extension involves the semantics of the cluster hierarchies. Of course, it may be necessary for certain application domains to have a deeper nesting of the version clusters, e.g., *alternatives*, *representations*, again *alternatives*, and *versions*.

The identification of a version within the clusters involves specifying a path from the design object over the respective *representation* to the *alternative* to the *version*. Thus, an example version identifier looks like

$$i_1/R_2/A_3/V_2$$

This identifier specifies the version number V_2 within alternative A_3 within representation R_2 of the design object i_1.

It is useful to combine version clustering and dynamic binding of version references. For example, it may be desirable to specify the *current* version within a certain alternative or within a certain representation. This requires that *current* versions have to be labeled, i.e.,

- One for each design object

- One for each representation

- One for each alternative

Of course, some of these *current* versions may coincide.

Below, an example environment in which the two entries refer to dynamically evaluated versions within clusters is shown:

Environment E_1	
OID	Version ID
i_2	$i_1/R_1/current$
i_3	$i_2/R_2/A_3/current$
...	...

The entry $i_1/R_1/current$ refers to the current version within the dashed rectangular box shown in Figure 17.7.

An example of a *current* reference within a revision is the entry $i_1/R_2/A_3/current$, which refers to the *current* within alternative A_3 of representation R_2. In terms of our example in Figure 17.7, this generic reference points to the *current* version within the right-hand dotted box.

17.5   Change Propagation and Notification

The creation of a new version can give rise to a new version at an upper level of the *part-of* hierarchy. Consider the scenario depicted in Figure 17.8. Upon the creation of version 2 of design object B a new version of version 1 of design object A could be

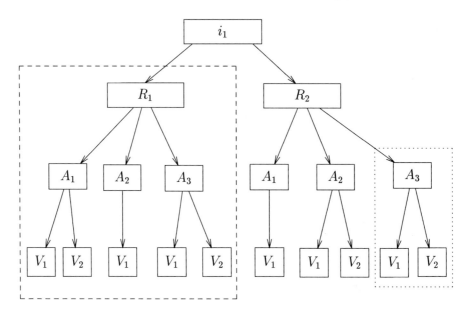

Figure 17.7: Dynamic Binding within Logical Clusters

generated automatically. This makes sense, if any version of the design objects B and C should be incorporated into at least one version of design object A. The automatic generation of new versions in upper levels of the hierarchy upon the creation of a new version in a lower level is called *change propagation*.

If, as in the scenario of Figure 17.8, two new versions of two different subobjects of a design object—here A—are created, then change propagation results in several possible new versions to be created. In our case, any new combination between the two versions of the design object B and those of C can lead to a new version of A. The result of change propagation is shown in Figure 17.9. Version 1 of design object A still points to version 1 of B and version 1 of C. The newly created versions 2, 3, and 4 of design object A cover the other possibilities. It is clear that the number of newly created versions explodes if several new versions are created at a lower level. The problem becomes even more severe if the hierarchy is deeper. Another problem appears if the different updates are semantically related. Assume design object A consists of a drawing (design object B) and a document (design object C) containing its verbal explanation. Then the two combinations reflected by versions 2 and 3 of design object A in Figure 17.9 are meaningless if version 2 of C reflects the necessary changes to the documentation resulting from updating version 1 of B. Hence, in this case change propagation should result in only a single new version of A reflecting both new versions of B and C.

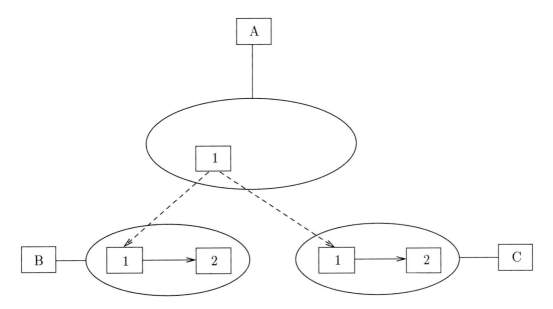

Figure 17.8: Deriving New Versions of Subobjects

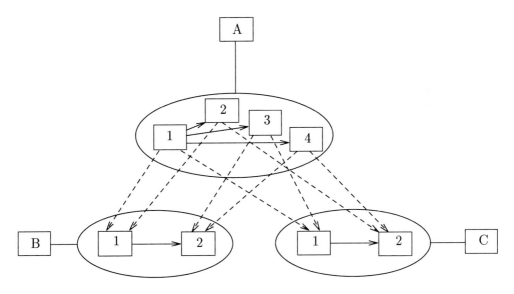

Figure 17.9: Change Propagation Generating all Possible Combinations

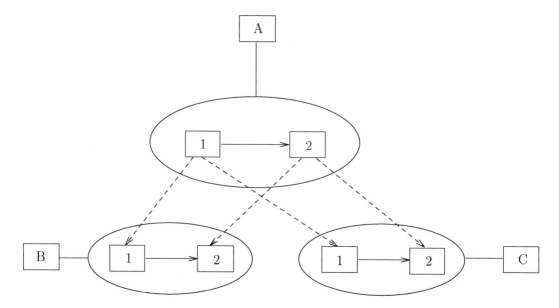

Figure 17.10: Change Propagation under *Group CheckOut*

To solve the above-mentioned problems the concept of *group check-out* of versions has been introduced. Within a design session it is possible to group several versions of different design objects into a single set. After updating all the versions within this version group, change propagation proceeds as follows. For every updated version all the versions on its path to the root are labeled. If a version has already been labeled, the second label is ignored. Then, after this labeling phase, for any labeled version a single successor version is created and connected to all the newly created versions of versions it pointed to before as well as those versions it pointed to that have not been updated. In our example, version 1 of A would be visited twice during the labeling phase, once for version 2 of B and once for version 2 of C. Following this labeling, a successor version 2 of A is created and connected to version 2 of B and version 2 of C. The result of this change propagation is shown in Figure 17.10.

Since not only one designer is responsible for the design of a complex artifact, but a whole team of designers, responsibility for different design subobjects is distributed among the members of the team. Nevertheless, changes in one design object often influence the work on other semantically related design objects. In such an environment it can easily happen that changes pass unnoticed by affected designers. To remedy this situation automatic *notification* mechanisms have been introduced. These take care of notifying other designers whose work may be affected by a specific change. Different *notification modes* are distinguished:

- *Message-based notification* sends a message to the affected designer whenever a certain design object—one in which the designer has some interest—is updated, deleted, or a new version is created. The message-based notification modes further fall into two categories:

- Within the *immediate notification mode* the designer is informed immediately when the change occurs.
- Within the *deferred notification mode* the designers are not immediately notified but only after the changes have been committed.

- *Flag-based notification* sets a flag only within a certain design object, which is affected by the change of another design object. Then, when the designer accesses the former design object, he or she is notified. Note that if the flagged design object is not accessed, no notification to the designer takes place.

17.6   Versioning in the Three-Level Database Hierarchy

The above version model assumes a global name server that is responsible to ensure uniqueness of all version names. It does not really account for a distributed environment in which different versions of the same object may be created in private workspaces—even those versions that are only transient and will never be visible to other users. Therefore, the version model has been adapted for the three-level database hierarchy consisting of one *public* database, several *project* databases, and several *private* databases, as described in Section 15.9. In this version model we distinguish three—semantically different—version types:

1. *Released version*: A released version resides in the public database and is protected against deletion and updates. Thus, a released version may safely be used (referenced) by other objects without danger of side effects due to updates of the shared version object.

2. *Working version*: Similarly to a released version a working version cannot be updated. However, it is still subject to deletion by its creator. A working version resides either in a private database or in a project database. A working version can be promoted to a released version by checking it into the public database.

3. *Transient version*: A transient version can be considered as an object version "under construction." It, therefore, resides in a private database that protects it from access (references) by other objects outside this private database. A transient version may be derived from either

- A working version
- A released version

A transient version is automatically upgraded to a working version by either

- Checking it into a project database
- Deriving a new transient version from it

The three type version model assumes a totally update free public database; that is, all released versions remain invariant. Therefore, the *checkIn* and *checkOut* operations are transformed into corresponding version creation primitives.

The *checkOut* of an object (version) from some database DB_1—which is either the public or a project database—into the private database DB_2 results in the creation of a new, transient version of this object in DB_2. This version may be arbitrarily modified because—initially—it is ensured that no other object references the respective version.

The *checkIn* of a version object from a database DB_2—where DB_2 is either a private or a project database—into a database DB_1—DB_1 being the project or the public database—involves installing a copy as a version to some specified object within DB_1. It is thus the user's responsibility to properly identify the object to which the checked-in version belongs. This provides flexibility in such a way that one can create a new version from scratch (or from another unrelated object) and later establish a version relationship to some existing object.

We can now summarize the possibilities for creating different types of versions.

Transient Versions   *Transient* versions are created by either

1. A *checkOut* of a released or a working version from some database, i.e., the public or a project database

2. An explicit derivation of a version from a working version

3. Creating one from scratch in the private database

Working Versions   *Working* versions can be obtained by either

1. Explicitly promoting a transient version in the private database to a working version

2. Checking in a transient version into a project database

Released Versions   A working version is promoted to status *released* by checking it into the public database. This is the only way of creating a released version.

17.7   A Simple Implementation

Since versioning is not necessarily an ingredient of an object-oriented database system, the design environment built around the database may be responsible for providing version management. Therefore, this section indicates how version management can be built on top of an object-oriented database. An example of a type definition for a design object and a version of this design object is presented. A more general implementation of versioning utilizing generic types and multiple substitutability is left to the exercises.

What follows is the code (skeleton) of a type definition for a *CarDesignObject*, a *VersionedCar* and a *Car* itself. The *CarDesignObject* is the most important data structure. It collects all versions in *versionSet*, maintains the *current* indicator and the counter *vCount* used to assign unique version numbers. Only a few example operations are given: The initializer for creating a new design object, the find operation to find a version with a certain number, and the creation of a new successor version. The predecessors and successors are maintained within each car version by the attributes *preds* and *succs*, respectively. Additionally, each car version carries its version number and possesses an attribute for the car containing the chassis and the engine.

```
    type CarDesignObject is
       body [ versionSet: {CarVersion};
              current: CarVersion;
              vCount: int; ]
       operations
          declare CarDesignObject: → CarDesignObject;
          declare findVersion: int → CarVersion;
          declare newVersion: int → CarVersion;
       implementation
          define CarDesignObject is      !! initializer
          begin
             self.vCount := 1;
             self.current.create;
             self.current.versionNumber := self.vCount;
             self.vCount := self.vCount + 1;
             self.versionSet.create;
             self.versionSet.insert(self.current);
             return self;
          end define CarDesignObject;

          define findVersion (vNumber) is     !! retrieve the version
          var cv: CarVersion;                 !! identified by vNumber
          begin                               !! from the versionSet
```

```
            foreach (cv in self.versionSet)
               if (cv.versionNumber = vNumber) return cv;
            return NULL;
         end define findVersion;

         define newVersion (predVersNumber) is      !! create a new
         var cvPred, cvNew: CarVersion;             !! version and
         begin                                      !! establish relationships
            cvPred := self.findVersion(predVersNumber);
            if (cvPred = NULL) return NULL;
            cvNew.create;
            cvNew.versionNumber := self.vCount;
            self.vCount := self.vCount + 1;
            self.versionSet.insert(cvNew);
            cvPred.succs.insert(cvNew);
            cvNew.preds.insert(cvPred);
            self.current := cvNew;      !! most recent becomes current
            return cvNew;
         end define newVersion;
      end type CarDesignObject;


   type CarVersion is
      [ versionNumber: int;          !! CarVersion contains
        preds: {CarVersion};         !! attributes for
        succs: {CarVersion};         !! version management
        car: Car; ];


   type Car is
      [ chassis: Chassis;       !! Car is the actual design object
        engine: Engine; ];      !! which may exist in several versions
```

The above type definitions constitute the skeleton for managing versios in a *Car* design process. Of course, ideally the object-oriented database system should facilitate generic version management capabilities that free the database designers from defining their own version handling. On the other hand, by designing their own version management the users can fully customize it to their particular needs.

17.8 Exercises

17.1 Provide simple type definitions for design objects for *Chassis* and *Engine*, and for versions of *Chassis* and *Engine*.

17.2   Provide for the design objects *Car*, *Chassis* and *Engine* at least the three useful operations *newVersion* (creating a new version directly from the current version; i.e., without any additional parameter), *mergeVersions* (creating a single successor version for two predecessor versions), and *deleteVersion* (deleting a version).

17.3   In the above modeling of versioning *Cars*, it is not possible to substitute a *CarDesignObject* for a *Car*, i.e., no generic references are possible. Further, it is not possible to substitute a *CarVersion* for a *Car*. Remedy this situation by applying multiple substitutability.

17.4   For every type whose instances are to be versioned, the type definitions for design objects and versioned objects have to be restated. To avoid this, provide generic types for design objects and versions.

17.5   We have seen that it is useful to capture even more semantics than incorporated so far in our example version model. Enhance the model by allowing for *version classes* and *relationships* between versions.

17.6   Implement a generic type *ENVIROMENT* that allows to define environment for specifying the resolution strategy for dynamic references. Be more general than we were in our presentation in that not only fixed versions or current indicators could be entries in the environment but furthermore operations can be used to retrieve the version by which a generic reference is resolved. Can you think of example applications or cases where this makes much sense?

17.7   In Section 15.10 we described a concurrency control concept for checking out and checking in of versioned objects. Illustrate this concurrency control scheme on the example *Car* design. In particular, come up with design scenarios in which the five lock modes R, RD, DS, DX, and X are useful.

17.8   Compare the concurrency control scheme of Section 15.10 with the version handling scheme within the three-level database hierarchy described in Section 17.6. In what respect can these two schemes be combined with each other? Sketch the system architecture.

17.9   Storing many different versions of the same design object induces high storage costs. Therefore, so-called *delta*-techniques have been developed. Thereby, not the entire version is stored; rather, the difference between the version and some reference version is stored—as the so-called *delta*.

Two different techniques have been devised:

1. *Forward deltas:* In this case, the first (i.e., the oldest) version is stored as a whole. All other derived versions are stored as a *delta* to (one of) the predecessors.

2. *Backward deltas:* The *current* version is stored in its entirety. The predecessors are stored as *deltas.*

Describe the two approaches—whose very basic idea is outlined above—in more detail. Discuss the pros and cons of the two strategies. Is it possible to combine the two approaches?

17.9 Annotated Bibliography

In the section on versioning we largely followed Katz's suggestion for a unified versioning framework presented in [Kat90]. Version graphs have been used in almost every versioning approach, e.g., [KSW86, DL88, Kat90, CK86, BM88, KCB86, Rum88, Tic88]. The same holds for the current indicator and the main path. The use of general relationships and classes has been proposed in [KSW86]. Automatic maintenance for the equivalence relationship has been discussed in [KCB86].

Also generic or dynamic references can be found in almost every approach, e.g., [KSW86, DL88, BM88]. Environments have been introduced in [DL88] and treated more generally as hierarchical environments. Mechanisms to build these environments can also be found there. A more general mechanism to instantiate generic references is proposed in [BM88] where a rule-based mechanism is applied.

Version clusters have been introduced in [DL88]. The special scheme of the hierarchy shown is adapted from [LAB+85]. The grouping of versions to limit change propagation is due to [KC88a]. In [Kat90] it is suggested to use the more general mechanism of propagation-controlling attributes of [Rum88] to limit change propagation.

The approach to adapt the versioning mechanism to the three-level hierarchy is taken from [CK86].

18

Schema Management

The definition of a set of types—including their structural specifications (i.e., **body** clause), operation declarations, and implementations—is a highly complex design task. Hence, additional structuring mechanisms are needed to allow breaking down this task into subtasks and to support its successive development over time. The additional structure imposed upon a collection of type definitions is the *schema*. Its adaptation to changing requirements over time is called *schema evolution*.

A *schema* is a set of type definitions—including their definitions of the structure and the behavior. A schema is not fixed once written but evolves over time in order to capture the ever-changing requirements. New types are introduced, old types deleted, operations modified, errors eliminated, and so on. So far, only the definition of new types has been discussed. The notion of schema as a structuring mechanism has not yet been introduced. We will do so in the first section of this chapter. Further, not every syntactically possible schema is legal; e.g., all types used must be defined. Therefore, the notions of schema consistency and schema/object consistency are introduced. Then, to account for evolution over time, a taxonomy of possible schema modifications is established and their possible impacts on schema consistency are discussed.

18.1 The Notion of Schema

Consider a company manufacturing mechanical artifacts. Typically, this company has at least a CAD, CAPP (computer-aided production planning), CAM, and a marketing department. Not all types required to represent the objects needed in one department are also necessary for all the other departments. For example, the CAD department utilizes types for representing geometric data. Dealing with 3-D models, boundary representations as well as constructive solid geometry representations are applied.

441

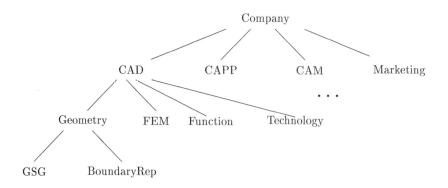

Figure 18.1: A Sample Schema Hierarchy

These types may not be needed, e.g., in the marketing department. Covering the whole spectrum of data to be processed within the company, often thousands of types are needed. Thus, it might be worthwhile not to have a large unstructured set containing all these types but, instead, to structure the set of all types, e.g., according to the varying requirements of different departments. Further, it is useful to structure the data for each department according to their own customized aspects. Therefore, a hierarchical structuring mechanism for managing the set of all available types is needed.

As an example, consider the hierarchical structure of the information as found in a typical company manufacturing mechanical artifacts. In Figure 18.1 the division of data processed in the whole company into four aspects corresponding to the departments *CAD*, *CAPP*, *CAM*, and *Marketing* is shown. On a second level, the data processed in the *CAD* department is further divided into data concerning geometry (*Geometry*), finite element models (*FEM*), functional aspects (*Function*), and technological aspects (*Technology*). On the lowest-level nesting of the structuring of the data, the object types necessary to capture geometry are classified into two partitions, one class comprising constructive solid geometry (CSG) models and the other comprising boundary representation models (*BoundaryRep*).

Note that, so far, not even a single type has been visible. Instead, only a division of all possible types into several sets has been sketched. By the mechanisms presented before in this book, it is impossible to express this partitioning of the company's information and make it known to the object base. Consequently, a new mechanism—the notion of *schema*—has to be introduced. As a first approximation, a schema can be thought of as a set of types. Then Figure 18.1 represents a *schema hierarchy*. The descendant schemata of a schema are called *subschemata*; e.g., *CAD*, *CAPP*, and

CAM are subschemata of *Company*. The schema *BoundaryRep* will then, e.g., contain the types *Cuboid*, *Surface*, *Edge*, and *Vertex*. Note that these types are of no relevance to the *Marketing* department. A typical example type of the *CAPP* subschema is the object type *Schedule*—representing processing schedules for manufacturing—which is fundamental for the *CAPP* department and of no interest for the *CAD* department.

Partitioning the set of all types into subsets—that is, providing a structuring mechanism—is not the only functionality a schema provides. In summary, there exist three different equally important aspects of schema management:

1. Structuring the set of all types and governing their persistence

2. Allowing high-level information hiding

3. Providing distinct name spaces

These points are discussed in more detail in subsequent sections.

18.2   The Schema Definition Frame

A schema with its subschemata can be specified by the *schema definition frame*. Part of the schema hierarchy of Figure 18.1 can be specified by the following three (preliminary) schema definition frames:

```
schema Company is                      schema CAD is
    subschema CAD;                         subschema Geometry;
    subschema CAPP;                        subschema FEM;
    subschema CAM;                         subschema Function;
    subschema Marketing;                   subschema Technology;
end schema Company;                    end schema CAD;

schema Geometry is                     schema BoundaryRep is
    subschema CSG;                         type Cuboid is ...;
    subschema BoundaryRep;                 type Surface is ...;
end schema Geometry;                       type Edge is ...;
                                           type Vertex is ...;
                                           var exampleCuboid: Cuboid;
                                       end schema BoundaryRep;
```

Here, only the schema *BoundaryRep* contains type definitions—i.e., the type specifications of *Cuboid*, *Surface*, *Edge*, and *Vertex*. However, in general every schema may contain actual type definitions. Note that we also defined a variable *exampleCuboid* within the schema *BoundaryRep*. Thus, a schema is not only used to structure the set of all types but also to group persistent variables. Types, variables, and (sub)schemata are generalized to *schema components*. Then a schema is a collection of schema components.

A schema also implicitly governs persistence:

> A schema is always *persistent*, and with it, all its schema components!

This means, that the keyword **persistent** is of no meaning within a schema definition frame. It might as well be dropped as we already did in the example schema definitions. Thus, all the schemata defined above and all the types and variables included are persistent.

18.3   Information Hiding for Schemata

Remember the type definition frame. There, the **public** clause controlled the possible operations to be executable on the instances of the type. Only those operations that were listed there were executable by clients of the type—or, more precisely, by clients of instances of the type. The same applies on a different level when considering schemata and contained types. Not all types have to be visible to all other schemata. In the above example, the type *Cuboid* in the schema *BoundaryRep* should be visible to other schemata, whereas the other types *Surface*, *Edge*, and *Vertex* should not be visible, since they are only used to implement the type *Cuboid*. This leads to information hiding on a higher level. Therefore, the **public** clause is also introduced for schemata:

```
schema BoundaryRep is
   public Cuboid
   interface
      type Cuboid is ...;
   implementation
      type Surface is ...;
      type Edge is ...;
      type Vertex is ...;
end schema BoundaryRep;
```

This schema realizes hiding the types *Surface*, *Edge*, and *Vertex*. Or, expressing it from a different viewpoint, only the type *Cuboid* may be used by other schemata. Further note that the type definitions of the types made public are contained in the **interface** section of the schema definition frame, whereas those only used for internal purposes are defined within the **implementation** section. Summarizing, a *schema definition frame* consists of three sections:

1. The **public** section, in which all the schema components made public to the superschema and—as we will see later—possibly other schemata are listed

2. The **interface** section, in which all the schema components made public are to be specified

3. The **implementation** section, in which all the schema components used for internal implementation purposes are specified

Remember that schema components can be types, variables, free operations, and (sub)schemata.

18.4   Name Spaces of Schemata

The third aspect of a schema is that it provides a *name space*. So far, without the notion of schema, all type names had to be distinct. Also, all names of global variables had to be distinct. There existed a single *global name space*. This is not necessarily easy to guarantee. In fact, in both of the schemata *CSG* and *BoundaryRep* a type representing cuboids may be appropriate. By allowing each schema to have its own *local name space*, it is possible to define a type *Cuboid* in the schema named *CSG* and another type *Cuboid* in the schema *BoundaryRep* without inducing a name conflict. Thus, the following two schema definitions are both valid, even in conjunction:

schema CSG **is**	**schema** BoundaryRep **is**
public Cuboid	**public** Cuboid
interface	**interface**
type Cuboid **is** . . . ;	**type** Cuboid **is** . . . ;
implementation	**implementation**
. . .	. . .
end schema CSG;	**end schema** BoundaryRep;

This becomes a very important feature when considering different designers developing the types for constructive solid geometry and boundary representation independently and in parallel. Thus, providing different name spaces for different schemata allows to design schemata independent of each other—thereby avoiding the necessity of a global consensus of naming.

Unfortunately, this advantage is not for free. As soon as the type *Cuboid*, which is a public type in both *CSG* and *BoundaryRep*, is referenced (i.e., used) in *Geometry*, this reference cannot be resolved uniquely. Both *Cuboid* types qualify. This problem is solved by explicitly renaming both types in the *Geometry* schema. Under the assumption that both *Cuboid* definitions are made public in *Geometry*, this is done by extending the *subschema* entries in the schema definition frame of *Geometry*:

    schema Geometry **is**
       public CSGCuboid, BRepCuboid
       interface

```
      subschema CSG with
          type Cuboid as CSGCuboid;
      end subschema CSG;
      subschema BoundaryRep with
          type Cuboid as BRepCuboid;
      end subschema BoundaryRep;
  end schema Geometry;
```

The **subschema** entry has been expanded as indicated by the keyword **with**. Following the keyword **with**, a list of schema components follows. Each entry in this list is preceded by the kind of the schema component to be renamed, followed by the "old" name, followed by **as**, followed by the "new" name. Thus, in the example, the type *Cuboid* of schema *CSG* is renamed to *CSGCuboid* and can be referred to by this new name within the schema *Geometry* and its superschema *CAD*. The latter holds because both new names for the *Cuboid* types have been made public. This **public** clause implies also that the renaming, that is the statement of the **subschema** entry, has to be in the **interface** section. If the types had not been made public, the subschema entry would have to be in the **implementation** section. For renaming imported types the qualifier **type** is utilized. If variables or operations are to be renamed, the qualifiers **var** and **operation** are used.

One might ask, whether the advantage of different name spaces for schemata becomes obsolete, since if name conflicts occur they have to be explicitly resolved by renaming and if they do not occur, there is no effect of different name spaces. Since a name conflict occurs only if two public schema components have the same name and both are used within a single schema, the number of name conflicts is limited. Further, independent development of schemata is not affected by the name conflicts. These have to be resolved within the single schema using the components whose names conflict.

18.5   Importing Schemata

Public types or other schema components implemented in a specific schema should be available to the implementor of other schemata. Consider, for example, the implementor of the schema *CAD*. He or she has free access to the public types specified in the schemata *Geometry*, *FEM*, *Function*, and *Technology* but to no other schema components. This is the—somewhat restrictive—default. That is, exactly those schema components of direct subschemata that are specified as being **public** are available to the implementor of the superschema.

There exist several reasons for this restrictive default. In general, a schema corresponds to a certain abstraction level of the application. The consequence of abstraction hierarchies often is that details at lower levels are of no relevance to levels higher than the direct upper level. Thus, the default provides for automatically hiding the details.

Of course, this is an idealized situation and there often exist good reasons to use schema components defined in a schema several levels down or even to use schema components of a schema that is not a direct or indirect subschema. If this is the case, the implementor can *import* other schemata. Since importing a schema implies a dependency, this mechanism should be handled with care. Therefore, it is required that any import of a schema has to be stated explicitly—hence the restriction on the default.

Yet another reason for the restrictive default is the name space. Since the imported schemata have most likely been developed independently, name conflicts may occur within the total set of locally defined schema components unioned with all imported schema components. Assume a type *Cuboid* has also been defined in the schema named *CAPP*. Then if the components of *CAPP* would automatically be visible to all other schemata, i.e., if there were an unconstrained import of all other schemata, another name conflict on *Cuboid* occurs within *Geometry*, although the type definition of *Cuboid* within *CAPP* is of no relevance to the schema *Geometry*. Resolving name conflicts for irrelevant schema components is a tedious task. The whole advantage of different name spaces would be nihilated.

Consider a tool that allows the automatic conversion of CSG into boundary representation. To integrate this tool into the schema hierarchy, a third subschema *CSG2BoundRep* of schema *Geometry* containing its implementation is introduced. Due to its functionality, it has to deal with both types of cuboids, the type *Cuboid* in *CSG* and the type *Cuboid* in *BoundaryRep*. Nevertheless, *CSG2BoundRep* has no implicit access to either of the *Cuboid* type definitions. What is needed is the explicit import of the schemata *CSG* and *BoundaryRep*. This is realized by utilizing the **import** clause within the schema definition frame:

```
schema CSG2BoundRep is
   public convert
   interface
      import /Company/CAD/Geometry/CSG with
         type Cuboid as CSGCuboid;
      end import schema CSG;
      import /Company/CAD/Geometry/BoundaryRep with
         type Cuboid as BRepCuboid;
      end import schema BoundaryRep;
      ...
end schema CSG2BoundRep;
```

The **import** clause is followed by a *schema path* specifying the schema to be imported. A schema path is a sequence of schemata separated by backslashes, i.e., "/". The first slash indicates that the specified path starts at the root. Thus, "/Company" refers to the root schema *Company*. "/Company/CAD" refers to its subschema *CAD*, and so on, until the needed schema is reached. A path starting at the root is called an *absolute schema path*. A relative schema path starts with one of the following:

- A schema name, in which case this schema name refers to the subschema of the enclosing schema and this is the start of the path

- A double dot, i.e., "..", in which case the superschema of the enclosing schema is the start of the path

Thus, in our example, the following paths are equivalent—remembering that the schema *CSG2BoundRep* is a subschema of *Geometry*:

- /Company/CAD/Geometry/CSG

- ../CSG

The double dot can also be iterated, e.g., "../.." refers to *Company*, if utilized directly within *Geometry*. It refers to *CAD*, if it occurs within *BoundaryRep* or *CSG*.

By starting with a schema name, only direct or indirect subschemata can be reached. In the example of Figure 18.1, the schema *CSG* can be imported into *CAD* by specifying the **import** clause utilizing the schema path "Geometry/CSG". The renaming within the **import** clause is analogous to renaming in the **subschema** clause. Clearly, in addition to the above definition of the schema *CSG2BoundRep*, we have to specify that *CSG2BoundRep* is a subschema of *Geometry* by adding the appropriate **subschema** entry to the schema definition of *Geometry*.

Let us briefly summarize the effects of importing a schema B by a schema A using the **import** clause: All the schema components—types, variables, and free operations—defined in B are readily available in A and can be accessed in the same manner as any other schema component directly defined in A. Also, all the schema components public in any subschema of A are available transparently in A. The only problem occurs in case of a name conflict. Name conflicts in a schema A can only occur if

1. The same name was used at least twice for the same kind of schema component, e.g., a type, within the set union of all schema component names of A, its subschemata and its imported schemata

2. This schema component was used within A, e.g., in an attribute definition

18.6   Schema Consistency

Up to now, we have discussed how to build schema hierarchies, import schemata and rename schema components. The issue of validity of a schema has not yet been touched. However, not all syntactically possible schemata are valid. There exists a whole set of possible errors that might occur, e.g., the usage of a type that is not defined, calling a type-associated operation not present in the type, misplaced arguments within an operation call, referencing variables not declared, and so on. The property of a schema being free of errors is referred to as *schema consistency*.

Schema consistency is specified by a set of *schema consistency constraints*. These constraints state that no errors, e.g., of the above kinds, occur. The set of schema consistency constraints can be classified into *uniqueness*, *existence*, and *subtyping* constraints. We give some examples of each class of constraints.

Uniqueness Constraints   These constraints require the uniqueness of names. Due to the schemata having different name spaces, the uniqueness is restricted to the names occurring within the same schema. Further, schema components of different kind may have the same name. For example, there might exist a type and a variable with the same name. The following are typical examples of uniqueness constraints:

- Within a schema, the type names—including the imported types—have to be unique. The same holds for all other kinds of schema components.

- Within a tuple-structured type, the attribute names have to be unique. This includes the attributes inherited from supertypes.

Existence Constraints   Existence constraints require the existence of something that is referred to explicitly. Examples are as follows:

- Every subschema has to exist.

- Every imported schema has to exist.

- Every type referenced—e.g., in an attribute declaration, a list or set specification, a parameter specification, a declaration of a (local) variable—has been defined in the schema or has been imported.

- Every operation called in the code of some operation has been declared—or is a built-in operation, e.g., to read and write attributes.

- For every declared operation there has to exist an implementation.

Subtyping

- The sub/supertype relationship has to be acyclic.

- If only single inheritance is allowed, only a single supertype must be specified.

- There must not exist any conflicts due to multiple inheritance (see Chapter 13).

- The refinement property must hold (see Section 10.10).

A schema satisfying all the schema consistency constraints is called *consistent*.

Besides a schema being inconsistent, i.e., violating one of the schema consistency constraints, another kind of problem can arise. Consider a tuple-structured type having two attributes *length* and *height* of type *float*. Then every instance, i.e., object, of this type should also have these two attributes with values both of type *float*. This results in a *schema/object consistency constraint*: All the (possibly inherited) attributes of a type have to be present in all its instances. This constraint is typically taken care of during the instantiation of the type, i.e., the creation of the object. That is, so far no violation can occur if the system works correctly. Nevertheless, assuming the possibility of changing type definitions—treated in the next section—the above schema/object constraint may be violated if an attribute *width* is added to the type and some objects have already been created. They lack the newly added attribute.

18.7   A Taxonomy of Schema Evolution Operations

A schema is a set of schema components, in particular type definitions—including the definitions of the structure and the behavior. Designing a schema hierarchy together with all its components included is a highly complex task. Hence, errors may occur and have to be eliminated. Further, the requirements may change and the schemata have to be adapted. New types are introduced, old types deleted, operations modified, and errors eliminated. Thus, a schema is subject to frequent changes. This section gives a taxonomy of possible changes to a schema. All these changes should be supported by an object-oriented database system.

According, to the constituent parts of a schema definition frame, the set of all possible changes to a schema can be classified as follows:

1. *Changes to the schema hierarchy*

   (a) Add a new **subschema** entry (possibly with renaming)

   (b) Delete a **subschema** entry

2. *Changes to the import of a schema*

   (a) Import a new schema with possible renamings

   (b) Drop the import of a schema

3. *Changes to the set of all types*

   (a) Add a new type

   (b) Delete an existing type

   (c) Change the name of an existing type

4. *Changes to parts of an existing type*

   (a) Changes to the body

      i. Exchange the total body

      ii. Changes to attributes

         • Add a new attribute

         • Delete an existing attribute

         • Change the name of an existing attribute

         • Change the type of an existing attribute

      iii. Change the list element type

      iv. Change the set element type

   (b) Changes to the behavior

      i. Add a new operation declaration

      ii. Delete an existing operation declaration

      iii. Change an operation's implementation

      iv. Refine an inherited operation

      v. Rename an existing operation

5. *Changes to the sub/supertype relationship*

   (a) Add a new supertype

   (b) Delete an existing supertype

   (c) Exchange a supertype

6. *Changes to the substitutability relationship*

   (a) Add a new **fashion** construct

   (b) Delete an existing **fashion** construct

(c) Change an existing **fashion** construct

Note that in this taxonomy, we did not mention all possible changes, e.g., changing the argument types of a type defined as an instance of a generic type or changing an existing type to a virtual type, or vice versa. Completing the taxonomy for these cases is left to the reader as an exercise.

An important point is that these schema evolution operations affect the schema consistency. For example, adding a new type may violate the required uniqueness of type names. The modifications to be performed to regain consistency may be enormous. Consider, for example, the change of an attribute's name. This requires a change of the name in all places the attribute has been used.

The verification (and maintenance) of schema consistency should be supported by the schema evolution component of the database system. The functionality of this component resembles that of a traditional compiler's semantic analysis.

18.8   Conversion and Masking

While the functionality of the component verifying schema consistency is similar to that of a traditional compiler's semantic analysis, the enforcement of schema/object consistency is a typical database problem resulting from the persistence of objects. Consider, once again, the example of an attribute being added to a type. Then, a brute force method to regain consistency is to delete all instances of this type. Since the information contained in the persistent objects of this type may represent an enormous value for the company, this clearly is an inappropriate measure. Therefore, two more appropriate mechanisms have been developed: *conversion* and *masking*. For the addition or deletion of an attribute, each of these techniques is discussed in turn.

18.8.1   Conversion

In this approach the already existing objects are modified in order to reflect the changes to the type they are instances of. In case of an addition of an attribute, this attribute is added to all the instances. The problem arising is the value to be used for initialization of the new attribute. There exist several possibilities:

1. A default value is used and initializes the attribute of all objects with the same value.

2. An operation is supplied and computes the value for the new attribute.

3. The user is asked for the values. This approach is, of course, limited, since the number of objects to be changed may be enormous.

In case an attribute of a type is deleted, no problems occur, since it can be deleted from the objects without requiring any further information.

Depending on the time when the conversion takes place, different modes of conversion are distinguished. *Immediate conversion* takes place as soon as the type has been changed by adding or deleting an attribute. *Lazy conversion* takes place either at times of low database load or—at the latest—when an affected object is accessed. The disadvantage of the former is that the system "stops" for some time in order to perform the conversions. The disadvantage of the latter is a performance penalty of all subsequent applications using the objects to be converted.

18.8.2   Masking

This approach takes advantage of the fact that the presence of an attribute has two foreseeable consequences: The definition of an attribute a establishes an operation $a\rightarrow$ to read the attribute and an operation $a\leftarrow$ to write the attribute. The idea is not to provide a piece of memory where the actual attribute value can be stored and read—as in case of conversion—but instead to provide specific operations mimicking the read and write of the new attribute. Then, whenever an object is accessed that does not provide the newly defined attribute, these two operations are used. The advantage of this approach is that the expensive conversion may be omitted. The disadvantage is that the mimicking imposes an overhead whenever the newly defined attribute is accessed. Note that in case of masking no additional action has to be provided by the user if an attribute is deleted.

18.9   Exercises

18.1   Consider the schema hierarchy in Figure 18.2. Give the schema definition frames including the subschema clauses.

18.2   Consider again Figure 18.2. Assume B has types S and T. Where are the type definitions of these types if S has to be public and T private?

18.3   For schema B of Figure 18.2, specify the following import clauses:

- Import G with an absolute path

- Import G with a relative path

- Import H with an absolute path

- Import H with a relative path

Figure 18.2: A Sample Schema Hierarchy

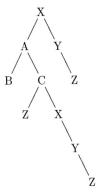

Figure 18.3: A Sample Schema Hierarchy

18.4 Consider the schema hierarchy in Figure 18.3. Which schema named Z is imported into schema C when each of the following paths is utilized?

- Z

- X/Y/Z

- /X/Y/Z

- ../../Y/Z

18.5 Consider the possible schema modification operations and the schema consistency constraints. Provide a matrix in which the columns correspond to the operations and the rows correspond to the constraints. Each entry should be an "X" if the modification affects the constraint, and a "−" otherwise.

18.6   Provide a taxonomy for possible changes of generic types and their instantiation.

18.7   Model a type. What are the entities of a type, what are their relationships? Draw an entity relationship diagram. Turn the diagram into a set of type definitions. How can the schema consistency constraints be assured? Can you provide operations that check the schema consistency constraints—one operation per constraint?

18.8   Discuss in detail the effects/problems that might occur on existing **fashion** clauses when changing the type hierarchy, and vice versa.

18.9   A typical schema evolution operation that occurs often in practice is that of subsequent splitting of a type. Here, a given type is split into two often disjunctive subtypes. Consider, for example, the car type, which splits as soon as the catalyst appeared. Then, cars with and without catalyst have to be distinguished, which wasn't the case before. Or a company splits into two companies with different objectives. Here, projects and employees have to be distributed over the two new companies. Model these examples before and after the change occurred. After that, implement the schema changes that manifest the necessary change.

Do not only consider schema consistency but also schema/object consistency. Furthermore, note that adjustments at the object level (adjustment of references) might be necessary even if schema/object consistency is already guaranteed. To see this, assume, for example, that a project leader moves to another part of the company than the project moves to but there is a constraint that does not allow a project leader to be of a different company. Write a short program that looks for such constraint violations and asks the user for a new project leader for every detected violation, and then adjusts the references accordingly.

18.10   In this exercise, we consider the opposite problem, i.e., two types that have been different types before are to be subsumed by a new type. Which problems do occur? How do the solutions look like? Which mechanisms do you need in order to cope with the situation? Consider two cases in which only single inheritance occurs:

   1. The two types are direct subtypes of a single type.
   2. The two types do not have a common direct supertype.

In which ways does multiple inheritance complicate the situation? Additionally, consider the presence of multiple substitutability.

18.10   Annotated Bibliography

The notion of schema as introduced in this chapter is due to Moerkotte and Zachmann [MZ93]. Schema consistency and schema/object consistency has been treated in almost every proposal for schema evolution—independent of the system the proposal was made for. The schema evolution mechanism for GemStone is described in [PS87]. The proposals for O_2 can be found in [Zic89b, Zic91, Zic91, DZ89, DZ91].These approaches follow the conversion approach to schema/object consistency. The proposal for schema evolution for Encore can be found in [SZ86, SZ87, Zdo89]. Here, the masking approach to schema/object consistency is taken. In Orion, schema consistency constraints are called invariants [BKKK87]. Later, this proposal was enhanced by a versioning mechanism for schemata [KC88b]. Schema versioning is also discussed in [CJ90, CJK91].

Schema evolution approaches are also discussed in [NR88], [NR89], and [TK89]. [Boc87] discusses several possible semantics for simple schema update operations like deleting a type. A formal approach to schema consistency can be found in [AKW90, Wal91].

Part V

Physical Object Base Design

19

Clustering

The *clustering* problem is the problem of placing objects onto pages such that for a given application the number of page faults becomes minimal. This problem is computationally very complex—in fact, the problem is NP-hard. Hence, several heuristics to compute approximations of the optimal placement have been developed. Some of these will be discussed in this chapter.

19.1 Introduction

19.1.1 Motivation for Clustering

As already mentioned, the *clustering* problem consists of finding an optimal or near optimal placement of objects onto pages such that the number of page faults becomes minimal. Consider the *Cylinder* example

> **type** Cylinder **is**   **type** Vertex **is**
>   [center1, center2: Vertex;   [x, y, z: float;];
>   radius: float;];

where a *Cylinder* is represented by the two center points *center1* and *center2* of type *Vertex*, and its *radius*. As usual, the type *Vertex* is described by its x, y, and z coordinates. Thus, a complete description of a *Cylinder* consists of three objects. As an instance of this schema consider a *Cylinder* with object identifier (OID) id_1 and identifiers id_2 and id_3 for its center vertices.

If the *Cylinder* object with id_1, and the vertices with identifiers id_2 and id_3 are scattered onto pages on disk—as depicted on the left-hand side of Figure 19.1—three disk accesses are needed to fetch the complete *Cylinder* representation into main memory. Assuming an average access time of 10 ms per page access, this fetch phase lasts 30 ms. The result after fetching these objects is shown on the right-hand side

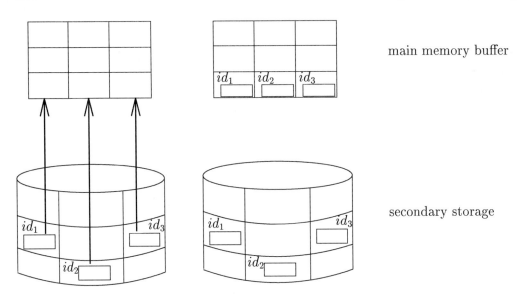

Figure 19.1: Distribution of a Cylinder onto Pages

of Figure 19.1. Since the involved objects are quite small, they could all easily fit on a single page. If all the objects reside on a single page, only one page access—taking approximately 10 ms—is needed to fetch the complete description of the *Cylinder* into main memory. A factor of three is saved. Obviously, the saving increases the more logically related objects fit onto a single page.

Besides this obvious saving, there exists another less obvious advantage of clustering several logically related objects onto a single page. We first observe that all pages fetched into main memory occupy buffer space. Further, buffer space is usually restricted. Hence, if too many pages are needed, some of them must be stored back onto disk despite the fact that during the continuation of the application certain objects they contain are again accessed. This results in more page faults and, hence, in decreased performance. Less buffer space is wasted if the percentage of objects on a page needed by an application is high. Clustering of those objects that are accessed together in an application increases this percentage and, hence, increases performance. From this point of view, filling a page totally with objects always needed together is the best clustering strategy possible.

The optimal clustering for the *Cylinder* example is very intuitive. To illustrate that this is not always the case consider the so-called *SIMPLE*-example [TN91] exhibiting an interesting pitfall when following the above, intuitively straightforward clustering strategy. There exist objects with identifiers S, I, M, P, L, and E. They reference each other in the way indicated in Figure 19.2. The application we consider is characterized by the following access pattern or *reference string*:

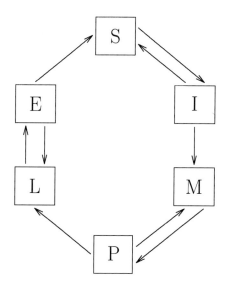

Figure 19.2: Referencing within the *SIMPLE*-Example

$$(SI)^{99}(MP)^{99}(LE)^{99}$$

where first object S is accessed, subsequently object I. Then 98 further accesses go from I to S and back ending at I. From here, M is accessed. Again, the application switches 99 times between M and P. Last, L is accessed and 99 mutual accesses between L and E take place. Consider the case that three objects fit onto a single page. Then consider the following placement of objects onto pages—the brackets [· · ·] indicate page boundaries:

- $[S, I, M], [P, L, E]$

is a reasonable clustering of the objects since

- The space occupied is minimized.

- The number of outgoing references, i.e., references between objects on different pages, is minimized.

Nevertheless, assuming a page buffer with a capacity of only one page, the above application leads to 198 page faults. This can be seen as follows. The application first accesses the object S. This leads to the first page fault. Switching between S and I does not produce any further page fault. This also holds for accessing M after $(SI)^{99}$. Accessing P leads to the next page fault where the page $[S, I, M]$ is replaced by the page $[P, L, E]$. Switching back to M again requires a page fault. The next page fault occurs when accessing P again. Hence, executing $(MP)^{99}$ after $(SI)^{99}$ has been executed leads to $2 * 99 - 1$ page faults. After the last access to P, no further page fault occurs while executing $(LE)^{99}$. Hence, in summary 198 page faults occurred.

A much better placement is

- $[S, I, -], [M, P, -], [L, E, -]$

where '$-$' indicates unoccupied space. For this placement, the above application induces only three page faults.

In subsequent sections, we will introduce a variety of different clustering strategies applied in object-oriented database management systems. More specifically, we will concentrate on *sequence-based clustering* strategies. They are termed sequence-based because they transform the object net into a (linear) sequence of objects.

19.1.2   Reclustering Possibilities

Before we describe the different sequence-based clustering strategies, we would like to stress the impact of physical versus logical object identifiers on clustering. Typically, optimal clustering requires to periodically transfer objects from an initially assigned page to another page. This is quite easy if logical object identifiers are used, since these are location independent. However, problems occur if the system is based on physical object identifiers, since these are location dependent. That is, whenever an object is moved, all the references pointing to it must be adjusted. This either requires to hold all backward references, which is quite expensive during the operational phase, or it requires an exhaustive database scan to search for the according references, which is prohibitively expensive. As a consequence, systems maintaining physical object identifiers usually do not permit any reclustering. These systems place all the objects created by a single transaction at its commit time. A placement for the newly created objects is selected and the objects are stored accordingly. Note that there is no guarantee that this really reflects an average access behavior of later transactions. Typically, reclustering results in better performance.

Object-oriented database systems that rely on physical object identifiers often apply a simple clustering strategy by exploiting *user hints*. In this clustering strategy, the user can indicate where a newly created object is to be placed. This is mostly done by allowing the user to specify another object in whose proximity the newly created object is to be stored. This approach exhibits two disadvantages. First, solving the complicated clustering problem is totally left to the user. In fact, every programmer implementing an operation that creates a persistent object must—in addition to the operation's semantics that are to be implemented—be aware of good clustering possibilities. A prerequisite is that the implementor knows a good deal about the other applications and their access behavior. Second, if the object in whose proximity the newly created object is to be placed is stored on a page on which the newly created object does not fit any more, there is no more control on the clustering possible. The result is a "random" placement of the object.

19.2   An Introduction to Sequence-Based Clustering

The basic idea of sequence-based clustering is quite easy and consists of two steps:

1. A sequence of the objects contained in the object base is obtained—the *clustering sequence*

2. The objects are stored onto pages according to this sequence.

While the second phase is obvious, there exists a wide variety of approaches to obtain a clustering sequence for a given object base. Nevertheless, all the known approaches can be described by a *generic algorithm* consisting of two steps. We denote this generic algorithm in the style of a UNIX[1] pipe[2] consisting of the two stages *PreSort* and *Traversal*:

   PreSort | Traversal

In the first stage, the objects to be clustered are sorted applying a method *PreSort*, resulting in the input sequence for the second stage. During the second stage, the object net is traversed using the method *Traversal*. *Traversal* is parameterized by the first nonvisited object of the input sequence and then traverses all objects reachable from this one. This is repeated until all objects have been visited. In the resulting clustering sequence, the objects occur in the order in which they have been visited by *Traversal*. Then the second step takes place and stores the objects onto pages according to the clustering sequence. Often, the objects are directly stored when they are accessed, resulting in a merge of the *Traversal* and the storing phase.

To obtain a good clustering sequence the *Traversal* component should exhibit an access behavior that closely resembles the actual applications. Let us illustrate this point by an example. Consider the database extension shown in Figure 19.3. It consists of a number of grid points with x, y, and z coordinates constituting a regular three-dimensional grid. The type definition of a *GridPoint* is

   type GridPoint **is**
      body [x, y, z: GridPoint;];
      operations

---

[1] UNIX is a registered trademark.

[2] A pipe gives the objects produced by its first component—the one before '|'—to its second component—the one after '|'—by obeying the order in which they were produced.

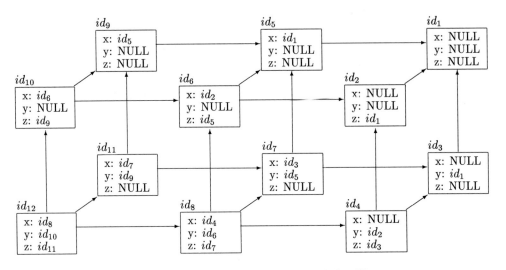

Figure 19.3: A Simple Extension of *GridPoint*

```
declare xmove: int → GridPoint;
declare x*move: → GridPoint;
declare y*move: → GridPoint;
implementation
define xmove(n) is
   var current: GridPoint;
   begin
     current := self;
     while (current.x ≠ NULL and n > 0)
     begin
       current := current.x;
       n := n − 1;
     end while;
     return current;
   end define xmove;

define x*move is
   ...      !! move as far right as possible
end type GridPoint;
```

The operation *xmove(n)* follows a trajectory of length n in x direction of the grid.
   To illustrate the impact of the traversal phase, we fix the following assumptions:

- The application considered is $id_9.xmove(2)$

And, hence, the application visits the objects with identifiers id_9, id_5, and id_1, in this order. Further, we assume:

- Three *GridPoint*s fit onto one page.

- The order produced by *PreSort* is $id_{12}, id_{11}, \ldots, id_1$.

Let us refer to this *PreSort* order by toc^{-1}. Assume that this is the reverse order in which the objects have been created; i.e., the objects have been created in the order $id_1, \ldots, id_{12}$. Here, *toc* stands for *time of creation* and toc^{-1} for the reverse of this.

We will compare the clustering performance of the following two different *Traversal* operations:

1. For a given *GridPoint* move as far as possible in y direction—denoted by y^*move.

2. For a given *GridPoint* move as far as possible in x direction—denoted by x^*move.

Applying the clustering strategy

$$toc^{-1} \mid y^*move$$

results—under the above assumptions—in the following placement of the *GridPoint* objects onto pages:

$$[id_{12}, id_{10}, id_{11}], \ [id_9, id_8, id_6], \ [id_7, id_5, id_4], \ [id_2, id_3, id_1]$$

For the operation $id_9.xmove(2)$, this object placement results in three page faults, whereas the clustering strategy

$$toc^{-1} \mid x^*move$$

results in

$$[id_{12}, id_8, id_4], \ [id_{11}, id_7, id_3], \ [id_{10}, id_6, id_2], \ [id_9, id_5, id_1]$$

and produces only one page fault for our particular operation invocation $id_9.xmove(2)$. As could be expected, the more similar the *Traversal* to the actual application, the smaller the number of page faults.

The impact of the *PreSort* order is less obvious but can also be highlighted by means of an example. Again, the goal is to cluster the objects of the *GridPoint* extension given in Figure 19.3 for the operation invocation $id_9.xmove(2)$. We apply x^*move as the *Traversal* and compare the two different *PreSort* orders *toc* and toc^{-1}. As already seen, clustering according to

$$toc^{-1} \mid x^*move$$

results in

$[id_{12}, id_8, id_4], [id_{11}, id_7, id_3], [id_{10}, id_6, id_2], [id_9, id_5, id_1]$

leading to one page fault only. Clustering according to

$toc \mid x^* move$

leads to the placement

$[id_1, id_2, id_3], [id_4, id_5, id_6], [id_7, id_8, id_9], [id_{10}, id_{11}, id_{12}]$

inducing three page faults.

Despite the fact that we have chosen a *Traversal* which could possibly produce a clustering inducing only one page fault, the "wrong" *PreSort* order results in a three times higher number of page faults as compared to the optimal *PreSort*. What happens is that the objects are already placed according to the *PreSort* operation, before the *Traversal* has a chance to really traverse the object net. More specifically, the trajectories of the traversals of $x^* move$ are trivial under the applied *PreSort* order *toc*; that is, they are of length one. The first objects in the *toc* order represent the *end* of the object net, preventing any real traversal. Consequently, one goal in finding a good *PreSort* method is to find the (frequently used) entry points in the object net and let these occur before the other objects.

19.3   PreSort Orders

From the database system's point of view, the simplest way to come to a *PreSort* operation is to ask the user. Very common are *type-based* orders in which the user specifies an order on types. Then all objects of a certain type t preceding type t' occur before the objects of type t'. Clearly, this results in a partial order only, since the objects of a single type cannot be ordered by this information. To remedy this situation, the user may specify an operation ordering the objects of a single type or specify a distinguished attribute by which objects of the same type are ordered.

Consider again the *Cylinder* example. If *Cylinder*s with approximately the same volume are often accessed together, a reasonable *PreSort* order is specified by

> **define cluster on** Cylinder, Vertex
>   **sort** Cylinder **by increasing** volume;
> **end cluster**

where we assume that *volume* is an operation associated with type *Cylinder* returning values of type *float*. The order of occurrence of the types in the first line of the clustering specification is the order chosen on the type—the *type order*. In the above example, all *Cylinder*s are sorted before the objects of type *Vertex*. Further, the

*Cylinder*s are sorted by increasing volume. It is also possible to specify **decreasing** instead of **increasing**, with the obvious meaning.

Note that this way we cannot specify any of the *PreSort* methods applied to the *GridPoint* example. The *PreSort* method *toc* is specified by

> **define cluster on** GridPoint
>   **sort** GridPoint **by increasing time-of-creation**
> **end cluster**

and toc^{-1} is specified by replacing **increasing** by **decreasing**. The use of the creation time as a clustering heuristic is useful if objects, which are created together, are often accessed together in the application programs.

As mentioned above, choosing the entry points first for traversal highly affects the clustering performance. Since the time of creation does not necessarily reflect entries in the resulting object order, the user may also specify the entry points explicitly. Let us assume that there exists a variable *gridEntries* of type *GridPointSet*, and it contains exactly the entry points of the grid—for the example extension in Figure 19.3 the objects with identifiers id_9, id_{10}, id_{11}, and id_{12}. Then

> **define cluster on** GridPoint
>   **entries** gridEntries;
> **end cluster**

causes the database system to sort all the objects contained in the set *gridEntries* before those not contained in it.

In general, it is possible to specify different sets of entries. Then the order of their occurrence in the **entries** clause is reflected in the resulting *PreSort* order. The restriction on the *PreSort* orders as imposed by the **entries** clause are obeyed also in case an additional **sort** clause is specified.

Besides time of creation, there exist two other orders that are often built into database management systems. These orders utilize *static* and *dynamic* reference counts.

The *static reference count* (src) of an object is the number of references to this object. In the *GridPoint* example of Figure 19.3, the object with identifier id_1 has a static reference count of 3 whereas id_{11} has the static reference count 1. Intensive benchmarking has shown that sorting objects by increasing static reference counts in the *PreSort* phase is often superior to sorting by decreasing static reference count values. This becomes plausible when considering that objects not referenced by other objects often constitute the entry points to the object net. Note that sorting by static reference counts may often result in a partial order, since many different objects will typically have the same *src* value.

The *dynamic reference count* (drc) of an object reflects the number of times this object has been visited, i.e., accessed, by the applications. Assume that the *GridPoint* operation *xmove* is called twice:

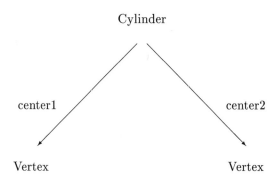

Figure 19.4: A Possible Placement Tree for *Cylinder*s

id$_9$.xmove(2); id$_5$.xmove(1);

Then the dynamic reference counts of the objects with identifier id_9 and id_5 are 1 and 2, respectively.

19.4   Traversal Strategies

The second stage of the generic clustering algorithm is the *Traversal*. It can be seen as an operation with one receiver argument. For each object in the outcome of *PreSort*, this operation is called—obeying the order of the objects. Roughly this operation works as follows. If the receiver object is not-yet-stored and it fits on the current page, it is stored there. If it is not yet stored but does not fit on the current page, it is stored on the next selected page. Then a new object to be processed is chosen. Most commonly, it is an object that is referenced by some object already visited during the current traversal. This object is then stored and a new object chosen. If no new object can be chosen, the traversal ends and a new one starts until all objects to be clustered are processed. Clearly, the step of choosing the new object to be processed requires some information upon which the decision can be based.

A special case of specifying the information needed by the traversal operation is the *placement tree*. A *placement tree* is a tree whose nodes are labeled by types and whose edges are labeled by attribute names. For the *Cylinder* example, Figure 19.4 represents a possible placement tree. The root is labeled by the type *Cylinder*. It has two successor nodes both labeled by the type *Vertex*. The edges are labeled by the attribute names *center1* and *center2* defined for type *Cylinder*. Of course, placement trees can be deeper. Consider the boundary representation in which we could specify a placement tree for the *BRep* type together with its *Faces*, *Edges*, and *Vertices*.

If a placement tree is applied, the actual *Traversal* operation is of minor importance, since the assumption is made that *all* objects visited during a *Traversal* fit on a *single* page. The *Traversal* itself works by starting with an object of the root's type. Then the attributes corresponding to the labels of the outgoing edges of the placement trees are traversed for the starting object. Then, the outgoing edges of not yet processed nodes in the placement tree are traversed. This proceeds until all edges have been traversed and, hence, all nodes have been visited. The result is a set of objects (of different types), which are then stored onto a single page. Another interpretation is that the placement tree is matched onto the current object net and all objects involved in the match are stored on a single page. In the case of the *Cylinder* example, the *Cylinder* object with identifier id_1 is the start of the traversal. Subsequently, its vertices with identifiers id_2 and id_3 are visited. Since all objects fit onto one page, it is of no importance in which order id_2 and id_3 are visited. Summarizing, for each starting object the traversal processes the entire placement tree and thereby collects the visited objects. The result is a set of objects to be stored together on one page.

This set of objects may contain objects already stored—for example, due to shared subobjects. These are then eliminated from the set of extracted objects and the remaining objects are stored together on a page. If they do not fit onto the current page, a new page is started. To demonstrate this, assume the existence of two *Cylinder* objects with identifiers id_1 and id_4. Let id_2 and id_3 (id_5 and id_6) be the identifiers of the vertices representing the center points of cylinder id_1 (id_4). Further assume that five objects of either type—*Cylinder* and *Vertex*—fit on a page and the order on the *Cylinder* objects produced by *PreSort* is id_1 followed by id_4, which could represent the time of creation order. Then the objects with identifiers id_1, id_2, and id_3 are stored onto the first page. Next, the objects with identifiers id_4, id_5, and id_6 have to be stored together on a page. Since only five objects fit on a page, a new page has to be started. Thus, the resulting placement of the cylinders and their center points is

$$[id_1, id_2, id_3, -, -], \ [id_4, id_5, id_6, -, -]$$

where '−' denotes unoccupied space.

Placement trees can be seen as a specification of weights 0 and 1 for attributes at the schema level. Only those attributes with weight 1 are represented in the placement tree and—on the object level—traversed. A more subtle weighting allows not only to specify weights 0 and 1 but, for example, all floating point numbers or positive integers. This generalization is absolutely necessary, if not all objects visited during the traversal of the placement tree fit onto one page. Then there is information needed on which objects are to be placed together onto a page or, the other way round, where to put the page boundaries.

The most common system-generated weights are *dynamic transition counts*. There

exist two fundamental differences between weights as presented by the placement tree
and dynamic transition counts:

1. Dynamic transition counts are positive integers including 0, whereas placement
   trees imply weights 0 and 1.

2. Dynamic transition counts are attached to references in the object base and not
   to attributes at the type level. As a consequence, they can differ for references
   between objects of the same type, even if the references correspond to the same
   attribute. This is not possible for the weights implied by placement trees.

   More specifically, for a given reference between two objects, its dynamic transition count represents the number of times this reference has been traversed by some
monitored applications. To demonstrate how dynamic transition counts are accumu-
lated, consider again the *GridPoint* example together with its extension as shown in
Figure 19.3 and the following application:

$$id_9.xmove(2);$$
$$id_5.xmove(1);$$

Before the application starts, the counts of all references are set to 0. Each time
a reference is traversed, its count is increased by one. Thus, after the execution of
$id_9.xmove(2)$, the count of the reference pointing from id_9 to id_5 is 1. The same holds
for the reference between id_5 and id_1. The execution of $id_5.xmove(1)$ only increases
the count of the reference between id_5 and id_1. The resulting weights are shown in
Figure 19.5.

All traversal strategies rely on the weights assigned to the references between the
objects in the object base. These weights need not necessarily be dynamic transition
counts. Hence, in the subsequent discussion of traversal strategies, we will abstract
from the method the weights are derived by.

Of the many existing *Traversal* strategies, we have chosen to discuss the three
most commonly applied. These are:

- Depth-first traversal

- Breadth-first traversal

- Best-first traversal

Each of these will now be discussed in turn.

The *depth-first* traversal works as follows. Starting from the root node, one child,
i.e., an object referenced by an outgoing reference, is selected—the one with the
highest weight. Then one child of the selected child is visited, and so on. Thus,
the depth-first traversal visits a branch from the root down to a leaf, i.e., an object

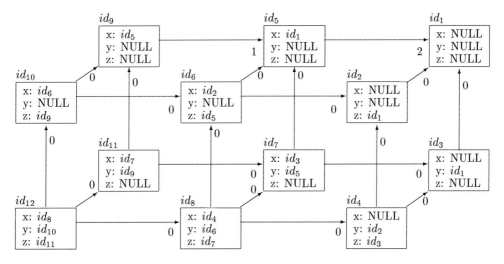

Figure 19.5: The *GridPoint* Extension with Weighted References

without any outgoing reference with weight greater than 0. When a leaf is reached, its siblings are visited. As soon as all siblings are visited, depth-first visits the not-yet-visited siblings of a node one level higher. The algorithm can be roughly described as follows:

```
define depthFirst (rootNode) is
    nodeQueue := <rootNode>;
    while (not empty(nodeQueue))
    begin
        next := pop(nodeQueue);
        visit(next);
        nodeQueue := concatenate(childrenSorted(next),nodeQueue);
    end;
end depthFirst;
```

In this pseudo-code

- *rootNode* is assumed to be the entry point where the traversal starts. It is the only argument to the operation *depthFirst*.

- *nodeQueue* is a list of nodes that is initialized to the argument of the traversal, i.e., *rootNode*.

- *empty* returns *true* if its argument list is empty; it returns *false* otherwise.

- *pop* returns the first element of a list and removes it from the list.

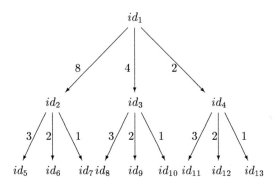

Figure 19.6: A Simple Object Tree with Weighted References

- *visit* visits its argument node.

- *childrenSorted* returns a list of all children of the argument node. If the argument is a leaf node, the empty list is returned. Further, the returned list is sorted by decreasing weights of the references pointing to the respective children.

- *concatenate* concatenates two lists. Note that the sorted child nodes are concatenated to the front of the *nodeQueue*.

We demonstrate the behavior of *depthFirst* by means of the tree shown in Figure 19.6. In the first step, the *nodeQueue* is initialized with the root node id_1. Then the first element of *nodeQueue*—here id_1—is assigned to the variable *next* and removed from *nodeQueue*. Next, id_1 is visited. The children of id_1 in sorted order are id_2, id_3, and id_4. After the assignment, *nodeQueue* contains exactly these nodes in the order of the previous step. Since the *nodeQueue* is not empty, the loop body is executed once more. The node id_2 is visited and its children constitute the first elements of the *nodeQueue*. Hence, after the execution of the loop body, *nodeQueue* contains the following elements: $< id_5, id_6, id_7, id_3, id_4 >$. Now, id_5, id_6, and id_7 are visited in the next three loop iterations. Since they do not have children, the only effect on *nodeQueue* is that these nodes are eliminated from it. Now, the algorithm steps back to visit node id_3 being the first element of *nodeQueue*. Its children are visited the same way as the children of id_2. Last, id_4 is visited and, subsequently, its children.

The order in which all nodes of the tree in Figure 19.6 are visited is

$$id_1, id_2, id_5, id_6, id_7, id_3, id_8, id_9, id_{10}, id_4, id_{11}, id_{12}, id_{13}.$$

Assuming that three nodes fit on a page, this yields the following placement:

$$[id_1, id_2, id_5], [id_6, id_7, id_3], [id_8, id_9, id_{10}], [id_4, id_{11}, id_{12}], [id_{13}, -, -].$$

The *breadth-first* traversal first visits all children of the latest visited node. For a tree, this results in visiting the nodes levelwise. First, all nodes at level 0—the root node only—are visited. Second, all nodes at level 1—the children of the root—are visited. Third, all nodes at level 3—the children of the children of the root—are visited. The algorithm implementing a *breadth-first* traversal is very similar to the one implementing *depth-first* traversal:

```
define breadthFirst (rootNode) is
    nodeQueue := <rootNode>;
    while (not empty(nodeQueue))
    begin
        next := pop(nodeQueue);
        visit(next);
        nodeQueue := concatenate(nodeQueue,childrenSorted(next));
    end;
end breadthFirst;
```

The only difference is that the children sorted by their weight are not placed at the beginning but at the end of *nodeQueue*. For the tree shown in Figure 19.6 the order in which its nodes are visited is

$$id_1, id_2, id_3, id_4, id_5, id_6, id_7, id_8, id_9, id_{10}, id_{11}, id_{12}, id_{13}.$$

Again assuming that three nodes fit on a page, this yields the following placement:

$$[id_1, id_2, id_3], [id_4, id_5, id_6], [id_7, id_8, id_9], [id_{10}, id_{11}, id_{12}], [id_{13}, -, -].$$

The *best-first* traversal cannot be described that easily, because its traversal is not independent of the actual placement of object onto pages. Whereas the decision of which node to visit next is local for the above two traversals—in that only the children of a node are sorted—this decision is dependent on all objects already stored on the current page under the best-first strategy. That is, all the objects that are referenced by objects stored on the current page are sorted according to the weight of the respective references pointing to them. The algorithm for the *best-first* traversal given in pseudo-code is as follows:

```
define bestFirst(node) is
    create a new page
    store node onto the new page;
    while (page not full)
    begin
        Select a not-yet-stored object fitting onto the page such that:
            a) it is referenced by an object stored on the page
            b) the accumulated weights of all references from objects on the page
               to this object are at a maximum
```

place this object onto the page.
    end
  end bestFirst;

To demonstrate how the algorithm works, consider again the example tree in Figure 19.6. Assume that the operation *bestFirst* is called with the root id_1 of the tree as the actual parameter and that three objects fit onto a page. Then the algorithm proceeds as follows:

1. A new page is created and id_1 is stored on it.

2. The not-yet-stored objects referenced by id_1—the only stored object—are id_2, id_3, and id_4 in order of decreasing weights of their references.

3. id_2 is selected and stored onto the page

4. In the next loop execution, the stored objects are those with identifiers id_1 and id_2. The referenced, not-yet-stored objects are id_3, id_5, id_4, id_6, and id_7 where they are ordered by decreasing reference weights. Note that the order between id_4 and id_6 is arbitrary, since they have the same reference weights.

5. id_3 is selected and stored on the page.

6. The page now contains three objects and, by assumption, is full. The traversal terminates.

The remaining nodes of the tree are stored during subsequent traversals that have to be initiated by successively invoking *bestFirst* until all nodes are stored. This obviously depends on the *PreSort* order.

Besides the traversals presented above, it is possible that the user specifies a traversal operation. For the *GridPoint* example, the following cluster specification chooses static reference counts for presorting and the x^*move operation for traversal:

    define cluster on GridPoint
      sort GridPoint **by increasing static reference count**;
      traversal is x^*move;
    end cluster

The result is that the objects are stored in the order they are visited by x^*move called on all *GridPoint* objects sorted by their increasing static reference counts.

19.5   Exercises

19.1   Consider the objects with identifiers $id_1, id_2, \ldots, id_9$. For the placements

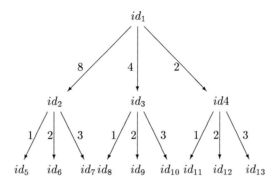

Figure 19.7: A Simple Object Tree with Weighted References

- $[id_1, id_2, id_3], [id_4, id_5, id_6], [id_7, id_8, id_9]$
- $[id_1, id_4, id_7], [id_2, id_5, id_8], [id_3, id_6, id_9]$

give the number of page faults produced by the following access profiles:

- $id_1, id_2, \ldots, id_9$
- $id_9, id_8, \ldots, id_1$
- $id_1, id_3, id_5, id_7, id_9, id_2, id_4, id_6, id_8$

Assume a buffer size of one page.

19.2 Consider the tree in Figure 19.7. Give the sequence in which its nodes are visited utilizing

- Depth-first traversal
- Breadth-first traversal

19.3 For the tree in Figure 19.7 give the placement of its nodes for the following clustering strategies:

1. toc | depth-first
2. toc | breadth-first
3. toc | best-first
4. toc^{-1} | depth-first
5. toc^{-1} | breadth-first
6. toc^{-1} | best-first

Assume different page sizes of 2, 3, and 4 objects; that is, 2, 3 or 4 objects fill a page.

19.4 For the placements produced by the previous exercise, give the number of page faults for the two applications traversing the tree in

- Depth-first order
- Breadth-first order

Assume a page buffer size of one page; that is, whenever an object not on the one page that is currently in the buffer is accessed, a page fault arizes.

19.6 Annotated Bibliography

The SIMPLE-example that was used to motivate sophisticated placement algorithms was adapted from [TN91].

In [BKKG88] *user* hints are exploited to select the starting points of the traversal—leading to a partial order on the objects in which selected objects are sorted before nonselected ones. A whole class of orders found in the literature is based on an order of types. Type-based *PreSort* orders have been examined by [Sta84, HZ87, BD90]. In [Sta84] the *PreSort* order *trace* has been proposed to obtain a near-optimal clustering sequence. *Trace* sorts objects according to their first appearance in the access trace of a specific application. This requires monitoring the trace of the application. The clustering strategy incorporated in the Cactis prototype [HK89] sorts objects by descending dynamic reference counts (abbreviated *drc*) before starting the traversal. The *drc* value of an object is given by the number of times the object is referenced in the trace of a former application. In [CH91] the Kruskal algorithm [Kru56] is utilized to compute the maximum weight spanning tree of the object net. This spanning tree is then transformed into an object sequence which is stored on consecutive pages. If an arbitrary *PreSort* order is applied, we denote this by λ.

For the traversal algorithm, many different alternatives like depth-first, breadth-first, children-depth-first, best-first, etc., can be found in the literature. In the Cactis prototype [HK89] a best-first traversal is utilized. We denote this traversal strategy by *best-first.dtc* (*dtc* abbreviates *dynamic transition count*).

Besides dynamic transition counts, reference counts for objects can be used to guide the traversal. Here, the successors of objects are sorted by increasing *src* or decreasing *drc* values [Sta84].

In [BD90] *placement trees* were introduced to describe the access patterns of applications.

The simplest traversal strategy is to copy the input sequence to the clustering sequence (denoted by *id*). This strategy has been applied by [Sta84, HZ87, CH91].

Sequence-based clustering algorithms:

[Sta84]	λ | depth-first	λ | breadth-first	λ | id
	λ | depth-first.src	λ | breadth-first.src	trace | id
	λ | depth-first.drc	λ | breadth-first.drc	type | id
[HZ87]	type | id		
[BKKG88]	user | depth-first		user | breadth-first
	user | children-depth-first		
[HK89]	drc | best-first.dtc		
[BD90]	type | placement-trees		
[CH91]	kruskal.dtc | id		

Partitioning-based clustering algorithms:

[TN91]	Kernighan & Lin graph partitioning [KL70]
[GKKM92b]	greedy graph partitioning

λ Arbitrary input sequence
src Stat. ref. counts (increasing)
drc Dyn. ref. counts (decreasing)
dtc Dyn. transition counts (decr.)

Table 19.1: Overview of Literature

Besides the sequence-based clustering algorithms as presented in this chapter, there exist so-called partitioning-based clustering algorithms. They build upon graph-partitioning algorithms. The first such clustering algorithm was introduced by Tsangaris and Naughton [TN91]. They use the Kernighan and Lin graph-partitioning algorithm [KL70]. The algorithm shows good performance but exhibits for practical use some major disadvantages [GKKM92b]. One of the most critical points is the run-time behavior of the algorithm, which is $O(n^{2.4})$. This led to the development of a greedy graph partitioning (GGP) algorithm for clustering problems [GKKM92b]. Although first benchmark results show good performance of the proposed algorithm, there does not exist any practical experience yet. The related literature comprising both, sequence-based as well as partioning-based clustering strategies, is summarized in Table 19.1.

20

Index Structures

In this chapter, we describe an indexing scheme, called *access support relation* (ASR), that goes beyond the common indexing as known from relational systems. Access support relations are tailered for the characteristics of associative access in object bases in which evaluating queries will often require "navigation" along object reference chains. The basic idea is to extract the most frequently traversed reference chains from the object base and maintain them redundantly in separate data structures that are called access support relations.

20.1   The Running Example

The discussion of this section is based on the example object base schema *DeptStore* that was already introduced in Chapter 14. For convenience, it is iterated here. The schema is as follows—only the structural representation is shown because the behavioral dimension is less significant in this discussion:

```
type Emp is                    type Manager
   [name: string;                 supertype Emp is
   worksIn: Dept;                 [cars: {Car};];
   salary: int;];

type Dept is                   type Car is
   [name: string;                 [license: String;
   mgr: Manager;                  make: String;
   profit: int;];                 horsePower: int;];
```

A sample object base extension of the object schema *DeptStore* is shown in Figure 20.1.

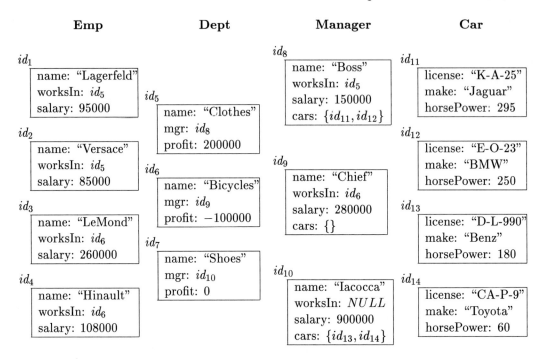

Figure 20.1: Sample State of the Object Base *DeptStore*

20.2   The Basics of Indexing

The most basic concepts of indexing in relational systems were already introduced in Chapter 4. We want to review this basic indexing in the context of object-oriented data modeling. Assume we frequently need to access *Emp* instances on the basis of their *salary*. For example, if we want to retrieve the *names* of those *Emps* whose *salary* ranges between 90000 and 100000, we can formulate the following GOMql query:

> **range** e: Emp
> **retrieve** e.name
> **where** e.salary > 90000 **and** e.salary < 100000

Assuming no further access support, the query has to be evaluated by locating every object in the extension of *Emp*—including, of course, all *Managers*—and inspecting whether the *salary* attribute satisfies the selection predicate of the **where** clause.

In order to expedite the associative access based on attribute values indexing is employed. An index could be created—just like in the relational systems—by the

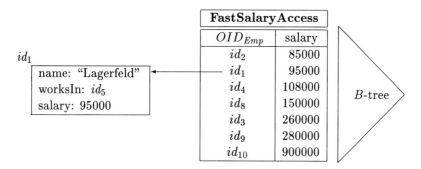

Figure 20.2: Illustration of the Index on the Attribute *salary* of *Emp*

following statement:

create index FastSalaryAccess **on** Emp (salary)

This statement initiates the creation—and subsequent maintenance—of an index that is illustrated schematically in Figure 20.2. Typically, such an index is maintained as a *B*-tree whose keys are the *salary* values of the objects of type *Emp* (and *Manager*). This *B*-tree is indicated by the triangle on the right-hand side of the diagram. The table *FastSalaryAccess* represents merely the logical contents of the index structure—the information is actually contained in the *B*-tree, i.e., the triangle.

From the diagram it becomes obvious that the index *FastSalaryAccess* is only useful for retrieving the object identifiers (OIDs) of objects whose *salary* attribute equals a certain value (i.e., an *exact match query*) or whose *salary* attribute falls within a specified range (i.e., a *range query*). However, the index is not adequate to obtain the *salary* of a given object, i.e., when the OID of the *Emp*loyee is known. The diagram in Figure 20.2 illustrates that we cannot efficiently search for a specific OID, say, id_8, and obtain the associated *salary* attribute. Even though this information is contained in the index structure, it takes too long to retrieve it because it requires an exhaustive search for the specified OID—id_8 in our example—within the entire *B*-tree. Of course, for a short path like the one considered here, the index does not make much sense, since a single object retrieval is much cheaper than an index access. But as soon as we consider longer paths (as we soon will), an index access will pay off.

Inspired by the above examples, we distinguish between *forward* and *backward* queries with respect to a certain attribute or, in general, a path expression. Consider the following two abstract queries involving the *salary* attribute of *Emp*:

range *e*: Emp	**range** *e*: Emp
retrieve *e*	**retrieve** *e*.salary
where $P(e.\text{salary})$	**where** $Q(e)$

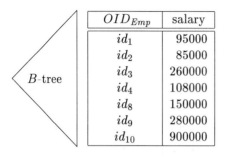

Figure 20.3: Illustration of the Binary ASR $[\![Emp.salary]\!]$

The query on the left-hand side is called a *backward* query on *salary* because its search predicate P is based on the *salary* attribute. That is, the so-called entry point of this query is *Emp.salary*. On the right-hand side a *forward* query is shown. The entry point of this query is an *Emp*—or a set of *Emp* instances. From those *Emp* instances for which the predicate Q is satisfied the corresponding *salary* attribute is retrieved. Note that the evaluation of such a query need not necessarily involve the access of the *Emp* objects for evaluating Q if Q were supported by some other index structure.

As noted before, the conventional indexing—such as the index on the *salary* attribute shown in Figure 20.2—supports only backward queries over atomic attribute values.

20.3   Binary Access Support Relations

Let us now extend the conventional indexing in three dimensions:

1. Provide support for backward as well as forward queries.

2. Allow indexing of attributes that refer to objects.

3. Allow the index to be defined over collection-valued attributes.

The concept of binary *access support relations* achieves all three extensions. Later on, we will augment the concepts to nonbinary access support relations that support arbitrarily long path expressions, i.e., path expressions that follow reference chains with arbitrary, though fixed number of intermediate objects.

20.3.1   Atomic Attributes

The binary access support relations over atomic attributes are very similar to conventional indexes except that they provide entry points from both directions. Let us illustrate this on our foregoing example: *salary* of *Emps*. The binary ASR is denoted as [Emp.salary]. The physical representation of this binary ASR [Emp.salary] is depicted in Figure 20.3. The diagram illustrates that the relation (table) relating OIDs with their *salary* is stored twice, i.e., redundantly in two different physical representations:

1. First, in the right-hand *B*-tree which is keyed on the *salary* values—indicated by ordering the tuples according to their *salary* value

2. Second, in the *B*-tree on the left-hand side whose keys represent the OIDs of *Emp* instances—again indicated by ordering the tuples according to the OIDs of the *Emp* instances

Utilizing this storage representation, it is straightforward to support the evaluation of both kinds of queries, backward as well as forward queries on the *salary* attribute, by exploiting the access support relation. For both kinds of queries a corresponding entry point is provided. Forward queries are supported by the left-hand *B*-tree; backward queries are supported by the right-hand *B*-tree.

20.3.2   Complex Attributes

The preceding binary access support relation [Emp.salary] relates objects with the value of one of their atomic attributes. It is also possible to relate objects with other objects that are referred to by an attribute constrained to an object type. An example is the binary access support relation [Emp.worksIn], which relates *Emp* instances with the *Dept* in which they work. The logical representation of this binary access support relation for our sample object base *DeptStore* is as follows:

[Emp.worksIn]	
OID_{Emp}	OID_{Dept}
id_1	id_5
id_2	id_5
id_3	id_6
id_4	id_6
id_8	id_5
...	...

In this representation, we omitted the illustration of the two index structures keyed on

1. OIDs of objects of type *Emp*

2. OIDs of objects of type *Dept*

which provide the entry points for forward and backward queries, respectively.

20.3.3   Collection-Valued Attributes

So far we have restricted the attribute on which the (binary) access support relation was defined to be single valued. However, the access support relation concept also allows to support collection-valued attributes. The only such attribute in our example object base schema *DeptStore* is the attribute *cars* in the object type *Manager*. This set attribute relates *Managers* to the company's *Cars* that they may use. Thus, the binary access support relation [[Manager.cars]] contains tuples with pairs of OIDs: an OID of a *Manager* and an OID of a *Car*, which may be used by the respective *Manager*. For our sample object base state of Figure 20.1, the binary ASR [[Manager.cars]] contains the following tuples:

[[Manager.cars]]	
$OID_{Manager}$	OID_{Car}
id_8	id_{11}
id_8	id_{12}
id_{10}	id_{13}
id_{10}	id_{14}

Again, we omitted the physical representation of the access support relations and showed only the logical information contents.

20.4   Path Expressions

So far we have considered only binary access support relations. Such a binary access support relation $[[t_{i-1}.A_i]]$ can be viewed as a materialization of elementary paths (i.e., references) from objects of type t_{i-1} to objects or values of type t_i via the attribute A_i—in the case of an atomic attribute A_i it is an elementary path from the object to the value. In the following sections we will generalize the concept of access support relations to paths of arbitrary length. For this purpose, we have to introduce some formalism first.

20.4.1   The Syntax

Let $t_0, t_1, \ldots, t_{n-1}$ be—not necessarily distinct—object types and let t_n be an object type or a sort. Furthermore, let $A_1, \ldots, A_n$ be—not necessarily distinct—attribute names. Then, the expression

$$t_0.A_1.\cdots.A_n$$

is a valid generic path expression if for all $1 \leq i \leq n$ either one of the following conditions is satisfied:

- The object type t_{i-1} is defined such that it contains an attribute A_i, which is constrained to type t_i. That is, t_{i-1} has the following form:

$$\textbf{type } t_{i-1} \textbf{ is } [\ldots; A_i : t_i; \ldots]$$

  In this case A_i is a single-valued attribute.

- The type t_{i-1} contains a set-valued attribute A_i whose elements are constrained to objects of type t_i. That is, t_{i-1} has the following structural representation:

$$\textbf{type } t_{i-1} \textbf{ is } [\ldots; A_i : \{t_i\}; \ldots];$$

  In this case the path $t_0.A_1.\cdots.A_n$ contains a set attribute at position i.

A path expression without any set-valued attributes is called *linear*—since it represents those paths that have no "forks."

The definition of a path expression $t_0.A_1.\cdots.A_n$ is given in such a way that the subpaths, which start at the anchor type, say, $t_0.A_1.\cdots.A_i$ for $1 \leq i < n$, lead to objects of type t_i. These objects are then required to possess an attribute A_{i+1} in order to "match" the next transition from t_i to t_{i+1} instances. Schematically the "chain-up" can be visualized as follows:

$$
\underbrace{\underbrace{\underbrace{\underbrace{t_0.A_1}_{t_1}.A_2}_{t_2}.\cdots.A_i}_{t_i}.A_{i+1}.\cdots.A_n}_{t_n}
$$

Let us now illustrate path expressions on a more intuitive example based on our *DeptStore* schema. The following is a valid path expression:

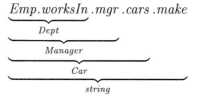

The underbraces, again, illustrate the domain—i.e., the result type—of the subpaths originating in the anchor type *Emp*.

20.4.2   The Semantics

So far we have only discussed the syntactical characteristics of (valid) path expressions. Let us now briefly describe the semantics of a path expression.

A generic path expression $t_0.A_1.\cdots.A_n$ becomes an evaluable path expression if an anchor for the path expression is specified. The anchor—as opposed to the anchor type—is a single object or a set of objects of the anchor type. In the easiest form, the anchor is a single object, say, o, of type t_0. Then, intuitively, the path expression $o.A_1.\cdots.A_n$ returns a set—or a single object if the path expression is linear—of objects of type t_n that can be reached from o via the specified attribute chain.

Alternatively, a set of objects of type t_0 can be used as an anchor—let s denote this set. Then the result of the path expression $s.A_1.\cdots.A_n$ is the set of all objects of type t_n that can be reached from any one of the objects contained in s via the specified attribute chain.

More formally the semantics can be specified recursively. Let us define the following sets $R_0, R_1, \ldots, R_n$:

$$R_0 \quad := \quad s$$

$$R_i \quad := \quad \bigcup_{v \in R_{i-1}} v.A_i \quad \text{for } 1 \leq i \leq n$$

Then the result of the path expression $s.A_1.\cdots.A_n$ is just the set R_n.

Let us illustrate this definition on our example path $Emp.worksIn.mgr.cars.make$ where we choose id_1 as the only starting point. Hence, s contains id_1 as its only element. By the first part of the definition, R_0 becomes $\{id_1\}$. R_1, by the second part of the definitions for the case $i = 1$, is assigned $\bigcup_{v \in \{id_1\}} v.worksIn = \{id_5\}$. For $i = 2$, $\bigcup_{v \in \{id_5\}} v.mgr = \{id_8\}$ is assigned to R_2. $\bigcup_{v \in \{id_8\}} v.cars$ now contains two elements, namely, id_{11} and id_{12}. These are both assigned to R_3. Hence, $R_3 = \{id_{11}, id_{12}\}$. The last step is to compute $R_4 = \bigcup_{v \in \{id_{11}, id_{12}\}} v.make = \{\text{"Jaguar"}, \text{"BMW"}\}$.

20.5   Join Operators

Before we can formally define the access support relations for arbitrarily long path expressions, we need to review different join types that were developed in the relational database theory—they will be used to derive the information contents of ASRs here. We distinguish four different join operators:

- $\bowtie$ represents the *equi-join*.

- $\bowtie$ represents the *outer join*.

- equi-join

L				R				Result				
A	B	C	⋈	C	D	E	=	A	B	C	D	E
a_1	b_1	c_1		c_1	d_1	e_1		a_1	b_1	c_1	d_1	e_1
a_2	b_2	c_2		c_3	d_2	e_2						

- left outer join

L				R				Result				
A	B	C	⟕	C	D	E	=	A	B	C	D	E
a_1	b_1	c_1		c_1	d_1	e_1		a_1	b_1	c_1	d_1	e_1
a_2	b_2	c_2		c_3	d_2	e_2		a_2	b_2	c_2	—	—

- right outer join

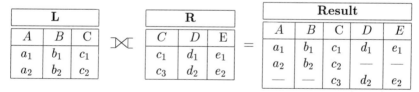

L				R				Result				
A	B	C	⟖	C	D	E	=	A	B	C	D	E
a_1	b_1	c_1		c_1	d_1	e_1		a_1	b_1	c_1	d_1	e_1
a_2	b_2	c_2		c_3	d_2	e_2		—	—	c_3	d_2	e_2

- outer join

L				R				Result				
A	B	C	⟗	C	D	E	=	A	B	C	D	E
a_1	b_1	c_1		c_1	d_1	e_1		a_1	b_1	c_1	d_1	e_1
a_2	b_2	c_2		c_3	d_2	e_2		a_2	b_2	c_2	—	—
								—	—	c_3	d_2	e_2

Figure 20.4: Example Applications of the Join Operators

- ⟕ denotes the *left outer join*.

- ⟖ is the sign for the *right outer join*.

Let us briefly illustrate the semantics of these four join operators. In Figure 20.4 the four join operators are applied to two relations L and R. For all four cases, we show the resulting relation—called *Result*—under the condition that the rightmost attribute C of the relation L and the leftmost attribute C of the relation R are the join attributes, respectively.

In the case of the *equi-join*, only those tuples of L and R are generated as a result which have a matching C attribute value. Thus, the *Result* relation contains only one tuple $(a_1, b_1, c_1, d_1, e_1)$ generated from the tuples (a_1, b_1, c_1) of L and (c_1, d_1, e_1) of R. On the other hand, the tuples (a_2, b_2, c_2) of L and (c_3, d_2, e_2) of R are "lost" in the *Result*, since they have no match-up in R and L, respectively. That is, (a_2, b_2, c_2) of L gets lost because R does not contain any tuple $(c_2, \dots, \dots)$ and (c_3, d_2, c_2) of relation R gets lost because L does not contain any tuple $(\dots, \dots, c_3)$ to match in the join attribute. Thus, the relation *Result* actually contains less information than the two relations L and R, in combination.

This observation led to the definition of so-called *outer join* operators. When the *left outer join* operator (denoted $⟕$) is applied, the information contents of the left argument relation is assured to be transferred to the *Result* relation. This is achieved by "filling up" tuples of relation L by *Null* values to the right if a corresponding matching tuple in R is missing. Thus, the tuple $(a_2, b_2, c_2, -, -)$ is contained in *Result* where "—" denotes the *NULL* value.

Analogously, the *right outer join* ensures that the information contained in the right argument relation—R in our example—is saved in the *Result* relation. If necessary, tuples of R that lack a matching tuple in L are filled up by *Null* values to the left. An example is the tuple $(-, -, c_3, d_2, e_2)$ in our scenario of Figure 20.4.

Finally, the *outer join* is used to ensure that the information of both argument relations, here L and R, is transferred to the *Result* relation—even if matching tuples in either argument relation are missing. If necessary, tuples are filled up to the right and to the left with NULL values. Such examples are $(a_2, b_2, c_2, -, -)$ and $(-, -, c_3, d_2, e_2)$.

20.6   Extensions of Access Support Relations

In the following, we will—if nothing else is explicitly stated—always assume the abstract path expression $t_0.A_1.\cdots.A_n$. An access support relation for such a (long) path expression is no longer a binary relation. Rather, the access support relation $[\![t_0.A_1.\cdots.A_n]\!]$ is a relation of arity $n+1$. The ASR $[\![t_0.A_1.\cdots.A_n]\!]$ contains (partial) paths relating objects of type t_{i-1} with objects of type t_j ($1 \leq i \leq j \leq n$), which can be reached via the attribute chain $A_i.\cdots.A_j$. However, the ASR does not merely contain the starting and ending points of these paths. Rather, it contains also all intermediate objects that lie on the paths from t_{i-1} to t_j instances. Furthermore, we distinguish four different extensions of access support relations—depending on the information contained in the ASR. We distinguish whether the ASR contains only those paths that originate in an object of type t_0 and go "all the way through" to an object of type t_n or whether it also contains partial paths with respect to the given path expression.

For the purpose of defining the four extensions of the access support relation that supports the abstract path expression $t_0.A_1.\cdots.A_n$, we assume that the binary access

[Emp.worksIn]	
OID_{Emp}	OID_{Dept}
...	...
id_2	id_5
id_1	id_5
id_3	id_6
id_4	id_6
id_8	id_5
...	...

[Dept.mgr]	
OID_{Dept}	$OID_{Manager}$
...	...
id_5	id_8
id_6	id_9
id_7	id_{10}
...	...

[Manager.cars]	
$OID_{Manager}$	OID_{Car}
id_8	id_{11}
id_8	id_{12}
id_{10}	id_{13}
id_{10}	id_{14}
...	...

[Car.make]	
OID_{Car}	$STRING$
id_{13}	"Benz"
id_{11}	"Jaguar"
id_{12}	"BMW"
id_{14}	"Toyota"
...	...

Figure 20.5: Binary ASRs of the Path Expression *Emp.worksIn.mgr.cars.make*

support relations $[t_0.A_1]$, $[t_1.A_2]$, ..., $[t_{n-1}.A_n]$ have all been computed. Note that these binary ASRs correspond to the elementary transitions in the path expression $t_0.A_1. \cdots .A_n$.

For our example path expression *Emp.worksIn.mgr.cars.make* the binary access support relations are the following:

1. [Emp.worksIn]

2. [Dept.mgr]

3. [Manager.cars]

4. [Car.make]

Their logical representation for our sample object base *DeptStore* is depicted in Figure 20.5.

20.6.1 The Canonical Extension

The canonical extension of the ASR is denoted $[t_0.A_1. \cdots .A_n]_{can}$. This extension contains only complete paths that originate in an object of type t_0 and lead to an object of type t_n via the attribute chain $A_1.A_2.\cdots.A_n$. Formally, the canonical extension is computed from the given binary ASRs by applying the equi-join:

$$[\![t_0.A_1.\cdots.A_n]\!]_{can} := [\![t_0.A_1]\!] \bowtie \cdots \bowtie [\![t_{n-1}.A_n]\!]$$

For our example path expression *Emp.worksIn.mgr.cars.make* the canonical extension of the access support relations—denoted $[\![Emp.worksIn.mgr.cars.make]\!]_{can}$—has the following contents (only part of the information is shown):[1]

$[\![\mathbf{EMP.worksIn.mgr.cars.make}]\!]_{\mathbf{can}}$				
OID_{Emp}	OID_{Dept}	$OID_{Manager}$	OID_{Car}	*string*
...	...	...	...	...
id_1	id_5	id_8	id_{11}	"Jaguar"
id_1	id_5	id_8	id_{12}	"BMW"
id_2	id_5	id_8	id_{11}	"Jaguar"
id_2	id_5	id_8	id_{12}	"BMW"
...	...	...	...	...

This example ASR $[\![Emp.worksIn.mgr.cars.make]\!]_{can}$ illustrates that the canonical extension contains only those tuples representing reference chains that originate in an *Emp* (or *Manager*) object, lead to a *Dept* object via the attribute *worksIn*, then lead to a *Manager* instance via the attribute *mgr*, then lead to a *Car* object via the set-valued attribute *cars*, and finally lead to the *string* representing the *make* of the *Car*. The canonical extension contains only those paths that start at the leftmost border of the specified path expression—here *Emp* objects—and lead all the way through to the rightmost border of the path expression—here *strings* representing the *make* of a *Car*.

20.6.2   The Full Extension

The canonical extension of an access support relation constitutes the minimal information contents useful for indexing on a path expression. We will now define the *full* extension—denoted $[\![t_0.A_1.\cdots.A_n]\!]_{full}$—which contains the maximum information with respect to the path expression $t_0.A_1.\cdots.A_n$. The full extension contains three kinds of paths:

1. The complete paths that are also contained in the canonical extension

---

[1]In particular, the *Manager* instances, which are—by substitutability—also *Emp* instances, are left out as entry points of the paths.

2. Those paths that lead to a NULL value because of an undefined attribute A_i or an empty set associated with attribute A_i

3. Those paths that originate in an object o_i of type $t_i \neq t_0$ because none of the objects of type t_{i-1} references o_i via the A_i attribute

In the formal definition of the full extension, we utilize the *outer join* operator, denoted $\bowtie$:

$$[\![t_0.A_1.\cdots.A_n]\!]_{full} := [\![t_0.A_1]\!] \bowtie \cdots \bowtie [\![t_{n-1}.A_n]\!]$$

For our example path expression *Emp.worksIn.mgr.cars.make* the full extension, which is denoted as $[\![\text{Emp.worksIn.mgr.cars.make}]\!]_{full}$, contains, among others, the following tuples:

\[EMP.worksIn.mgr.cars.make\]$_{full}$				
OID_{Emp}	OID_{Dept}	$OID_{Manager}$	OID_{Car}	*string*
...	...	...	...	...
id_1	id_5	id_8	id_{11}	"Jaguar"
id_1	id_5	id_8	id_{12}	"BMW"
id_2	id_5	id_8	id_{11}	"Jaguar"
id_2	id_5	id_8	id_{12}	"BMW"
id_3	id_6	id_9	—	—
id_4	id_6	id_9	—	—
—	id_7	id_{10}	id_{13}	"Benz"
—	id_7	id_{10}	id_{14}	"Toyota"
...	...	...	...	...

The first four tuples are examples of complete paths that were also contained in the canonical extension. The tuple $(id_3, id_6, id_9, —, —)$ represents a path leading into an empty set attribute—also, if the attribute *cars* of the *Manager* id_9 were NULL, the tuple would have had the same form. The tuple $(—, id_7, id_{10}, id_{13}, "Benz")$ is an example of a path originating at an intermediate position in the path expression—rather than at the left-hand border. The *Dept* id_7 does not have any *Employees*—thus, no *Emp* instance references id_7 via the *worksIn* attribute.

20.6.3   The Left-Complete Extension

The information contents of the other two extensions, called *left-complete* and *right-complete extension*, range somewhere between the canonical extension and the full extension.

Let us first introduce the *left-complete* extension, which contains all paths originating at the left-hand border of the path expression. The paths may lead "all the way" to the right-hand border—i.e., to an object of type t_n—or, alternatively, end in a NULL because of an undefined attribute or an empty set.

Formally, the left-complete extension for our path expression $t_0.A_1.\cdots.A_n$ is defined using the left outer join:

$$[\![t_0.A_1.\cdots.A_n]\!]_{left} := \left(\cdots\left([\![t_0.A_1]\!] \bowtie [\![t_1.A_2]\!]\right) \cdots \bowtie [\![t_{n-1}.A_n]\!]\right)$$

For our example path of the *DeptStore* object base the left-complete extension—denoted as $[\![Emp.worksIn.mgr.cars.make]\!]_{left}$—contains, among others, the following tuples:

$[\![\textbf{EMP.worksIn.mgr.cars.make}]\!]_{\textbf{left}}$				
OID_{Emp}	OID_{Dept}	$OID_{Manager}$	OID_{Car}	*string*
...	...	...	...	...
id_1	id_5	id_8	id_{11}	"Jaguar"
id_1	id_5	id_8	id_{12}	"BMW"
id_2	id_5	id_8	id_{11}	"Jaguar"
id_2	id_5	id_8	id_{12}	"BMW"
id_3	id_6	id_9	—	—
id_4	id_6	id_9	—	—
...	...	...	...	...

As mentioned before, the information contents of the left-complete extension range between the canonical extension (the minimal information) and the full extension (the maximal information). The left-complete extension contains only those paths originating in an *Emp* object. The paths may lead all the way to a string representing the *make* of a *Car* or end in a NULL—or an empty set. In the latter case, the paths are filled up with NULL values to the right.

20.6.4   The Right-Complete Extension

The right-complete extension is formally defined by applying the right outer join on the elementary binary ASRs of the underlying path expression:

$$[\![t_0.A_1.\cdots.A_n]\!]_{right} := \left([\![t_0.A_1]\!] \bowtie \cdots \left([\![t_{n-2}.A_{n-1}]\!] \bowtie [\![t_{n-1}.A_n]\!]\right)\cdots\right)$$

Thus, the right-complete extension contains all those paths—that is, the tuples representing these paths—that lead to the right-hand border of the specified path expression. These paths may originate at the left-hand border of the path expression—in which case they constitute complete paths—or may originate somewhere at an intermediate type within the path expression. This, however, can only occur if the object in which they originate, say, object o_i of type t_i, is not referenced by any object o_{i-1} of type t_{i-1} via its A_i attribute.

For our example path of the *DeptStore* database the right-complete extension, which is denoted $[\![\text{Emp.worksIn.mgr.cars.make}]\!]_{right}$, contains, among others, the following tuples (paths):

$[\![$EMP.worksIn.mgr.cars.make$]\!]$ $_{\textbf{right}}$				
OID_{Emp}	OID_{Dept}	$OID_{Manager}$	OID_{Car}	*string*
. . .	. . .	. . .	. . .	. . .
id_1	id_5	id_8	id_{11}	"Jaguar"
id_1	id_5	id_8	id_{12}	"BMW"
id_2	id_5	id_8	id_{11}	"Jaguar"
id_2	id_5	id_8	id_{12}	"BMW"
—	id_7	id_{10}	id_{13}	"Benz"
—	id_7	id_{10}	id_{14}	"Toyota"
. . .	. . .	. . .	. . .	. . .

Note that the *Dept* instance id_7 is not referenced by any *Emp* via the *worksIn* attribute. Thus, the tuples $(-, id_7, id_{10}, id_{13}, \text{"Benz"})$ and $(-, id_7, id_{10}, id_{14}, \text{"Toyota"})$ represent paths that are right-complete but not complete.

20.7 The Storage Model of ASRs

The storage model is analogous to the storage model for *binary* access support relations—as discussed in Section 20.3. Let us illustrate the storage model on the full extension of the access support relation $[\![\text{Emp.worksIn.mrg.cars.make}]\!]_{full}$, which is depicted in Figure 20.6.

The diagram illustrates that the access support relation is redundantly stored in two *B*-tree representations:

1. The triangle on the left-hand side represents the *B*-tree whose keys are the OIDs of *Emp* instances.

2. The triangle on the right-hand side represents the *B*-tree that is keyed on *string* values, i.e., the values of the *make* attribute of *Cars*.

Both *B*-trees contain records that constitute the (entire) tuples of the ASR.

This storage representation facilitates the efficient evaluation of forward as well as backward queries over the path expression *Emp.worksIn.mgr.cars.make*. Let us recall the following query from Section 14.2:

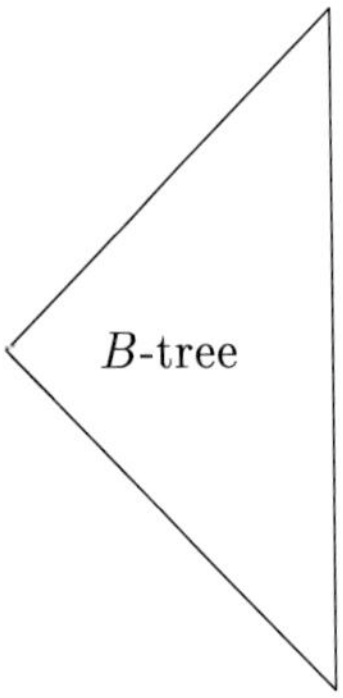

[EMP.worksIn.mgr.cars.make]$_{full}$				
OID_{Emp}	OID_{Dept}	$OID_{Manager}$	OID_{Car}	$STRING$
. . .	. . .	. . .	. . .	. . .
id_1	id_5	id_8	id_{11}	"Jaguar"
id_1	id_5	id_8	id_{12}	"BMW"
id_2	id_5	id_8	id_{11}	"Jaguar"
id_2	id_5	id_8	id_{12}	"BMW"
id_3	id_6	id_9	—	—
id_4	id_6	id_9	—	—
—	id_7	id_{10}	id_{13}	"Benz"
—	id_7	id_{10}	id_{14}	"Toyota"
. . .	. . .	. . .	. . .	. . .

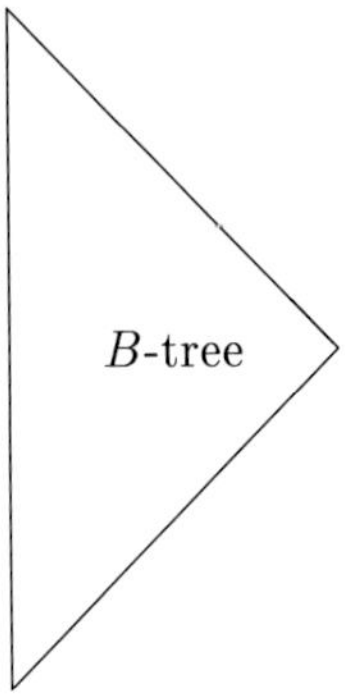

(left side: B-tree, right side: B-tree)

Figure 20.6: Storage Representation of a General (Nonbinary) ASR

range e: Emp
retrieve e
where "Jaguar" **in** e.worksIn.mgr.cars.make

This is a prototypical backward query. It can now efficiently be evaluated by utilizing the access support relation [Emp.worksIn.mgr.cars.make]$_{full}$. More precisely, the right-hand B-tree of Figure 20.6 is used as an entry point for retrieving those paths ending in a "Jaguar" value. Thus, the two *Emp* instances with OIDs id_1 and id_2 are determined as the result of the above query.

Analogously, the left-hand B-tree can be used as an entry point for evaluating a forward query, e.g., when we want to determine the *make* of the *Cars* the *Manager* of *Emp* id_2 drives. Thus, we want to evaluate the expression $id_2.worksIn.mgr.cars.make$, which yields the set { "Jaguar", "BMW"}—for our example database.

The above discussion shows that the storage representation of ASRs allows very efficient evaluation of both forward and backward queries—as long as they span the entire path expression. Unfortunately, the storage model is not well suited to evaluate forward or backward queries that start at some intermediate position within the path expression for which the ASR is maintained.

Consider the following two queries:

- Find the *make* of the *Cars* that *Manager* id_8 uses.

- Find the *Emps* that are directed by *Manager* id_8.

The first query is a forward query—with respect to the example generic path expression *Emp.worksIn.mgr.cars.make*—since it requires to evaluate the path expression $id_8.cars.make$. The second is a backward query, since it requires to start at a particular *Manager* instance (here id_8) and "go backward" to retrieve those *Emp* who work in a *Dept* whose *mgr* attribute refers to id_8.

Unfortunately, neither of the two queries is well supported by the storage structure shown in Figure 20.6. Neither query is supported by an explicit entry point in the form of a *B*-tree, in the ASR representation. In both cases an exhaustive search is needed—within one of the two *B*-trees—in order to find those tuples (paths) whose third column equals id_8. In case of a large extension, this exhaustive search might be prohibitively expensive. If this is to be expected, the path should be decomposed into several subpaths and for each subpath, an access support relation is to be defined. For the example, the path is split into *Emp.worksIn.mgr* and *Manager.cars.make*.

20.8 Applicability of ASR Extensions

The four extensions of access support relations contain different information for the same object base state. The information contents can be characterized as follows:

$$[\![t_0.A_1.\cdots.A_n]\!]_{can} \quad \leq \quad [\![t_0.A_1.\cdots.A_n]\!]_{left} \quad \leq \quad [\![t_0.A_1.\cdots.A_n]\!]_{full}$$

$$[\![t_0.A_1.\cdots.A_n]\!]_{can} \quad \leq \quad [\![t_0.A_1.\cdots.A_n]\!]_{right} \quad \leq \quad [\![t_0.A_1.\cdots.A_n]\!]_{full}$$

That is, the canonical extension contains less information than the left- or right-complete extensions. These, in turn, contain less information than the full extension. The left-complete and the right-complete extension are incomparable—meaning the left-complete extension contains paths that are not contained in the right-complete extension, and vice versa.

If NULL-values are present in the materialized paths, the different information contents induce different applicabilities of the four extensions in query evaluation. The canonical extension can be used to evaluate only forward or backward queries spanning the entire path expression. The left-complete extension can be exploited for evaluating a query—backward or forward—which originates (forward) or leads to (backward) objects of type t_0. Analogously, the right-complete extension can be used for backward queries starting in t_n or forward queries leading to t_n instances. Finally, the full extension supports any kind of queries over the underlying path expression.

The following summarizes the applicability of the four extensions for evaluating the path expression $t.A_i.\cdots.A_j$ under the assumption that the access support relation $[\![t_0.A_1.\cdots.A_n]\!]_X$ is maintained.

An access support relation $[\![t_0.A_1.\cdots.A_n]\!]_X$ under extension X is *applicable* for evaluating a path $t.A_i.\cdots.A_j$ with $s \leq t_{i-1}$ under the following condition, depending on the extension X:

$$Applicable(\llbracket t_0.A_1.\cdots.A_n \rrbracket_X, t.A_i.\cdots.A_j) = \begin{cases} X = \textit{full} & \wedge \quad 1 \leq i \leq j \leq n \\ X = \textit{left} & \wedge \quad 1 = i \leq j \leq n \\ X = \textit{right} & \wedge \quad 1 \leq i \leq j = n \\ X = \textit{can} & \wedge \quad 1 = i \leq j = n \end{cases}$$

Here $t \leq t_{i-1}$ denotes that type t has to be identical to type t_{i-1} or a subtype thereof.

Let us again illustrate this abstract formalism on our example access support relation $\llbracket \textit{Emp.worksIn.mgr.cars.make} \rrbracket$. Assume, we want to evaluate the following—verbally phrased—queries:

1. Find the *Emp*loyee whose *Manager* drives a "Jaguar".

2. Find the *Emp*loyees who are managed by the *Manager* id_8.

3. Find the *Manager* who drives a "Toyota".

The first query spans the entire path expression *Emp.worksIn.mgr.cars.make* and can, therefore, be evaluated using either of the four extensions of the corresponding ASR.

The second query involves the backward traversal of the path *Emp.worksIn.mgr*. This is a (true) prefix of the original path *Emp.worksIn.mgr.cars.make*. Thus, only the *full* and the left-complete extension of the ASR $\llbracket \textit{Emp.worksIn.mgr.cars.make} \rrbracket$ contain the necessary information—the reader may verify this on the tables shown in Sections 20.6.1 – 20.6.4.

Finally, the third query involves backward traversal of the (true) suffix path expression *Manager.cars.make*. The needed information is contained only in the right-complete and the full extensions of the ASR $\llbracket \textit{Emp.worksIn.mgr.cars.make} \rrbracket$.

20.9   Physical Database Design

It is the user's—or, more precisely, the database administrator's (DBA)—responsibility to create the access support relations that are best suited for the particular application profile. This is a very important task that should be very carefully carried out. In summary, the DBA has two dimensions along which one access support relation can be specified:

- The four extensions *can, full, left,* and *right*

- The many different decomposition choices of paths

Along the first dimension, the decision about which of the four extensions is to be chosen depends on the kinds of queries. If they mostly span the entire path then the canonical extension would be most appropriate. If any kind of subpath is to be supported, one should choose the full extension. If mostly prefix paths are traversed, the left-complete extension is most advantageous. If the majority of queries involves forward or backward traversal of suffix paths, then the right-complete extension ought to be selected.

Along the second dimension the DBA has the possibility to create entry points for intermediate positions within the materialized paths. Special care has to be taken for overlapping paths (see exercises).

Furthermore, the update behavior of the database applications has to be taken into account. In general, decomposition into small partitions and/or full or left-complete extensions are less costly under updates than larger partitions and/or canonical and right-complete extensions.

20.10   Exercises

20.1   Access support relations are based on generic path expressions of arbitrary, though fixed length. Consider the "part-explosion" example of Chapter 9 (Section 9.4) in which a *Product* is modeled as a tuple type containing a set-valued attribute *sub*.

The access support relation ⟦Product.sub⟧ contains tuples of the form (id_1, id_2), where id_1 is the OID of the superpart and id_2 the OID of the direct subpart.

Discuss the following issues:

- What kind of entry points are supported by this access support relation ⟦Product.sub⟧?

- How can the access support relation be exploited to evaluate the operations *partList* and *isPartOf*—which were specified in Chapter 9?

20.2   In Exercise 20.1 the access support relation ⟦Product.sub⟧ was described. For evaluating the operation *partList* it is still necessary to compute the transitive closure—starting at a particular *Product*—within the ASR. What kind of augmentation of the ASR concept is necessary to allow the materialization of the transitive closure in an ASR?

20.3   The path expressions of two different access support relations may possibly overlap. Examples are the two path expressions

$$P_1 \;=\; Emp.worksIn.mgr.cars.make$$

$$P_2 \;=\; Emp.worksIn.mgr.cars.horsePower$$

If two separate access support relations, e.g., $\llbracket Emp.worksIn.mgr.cars.make\rrbracket_{full}$, and, in addition, $\llbracket Emp.worksIn.mgr.cars.horsePower\rrbracket_{full}$ were maintained, a great deal of redundancy would occur. Illustrate this on the example object base *DeptStore*.

- Devise a solution to this problem based on decomposing the paths.

- Illustrate the problems occuring when the ASRs corresponding to the two overlapping paths are maintained in different extensions. Devise a solution.

- Illustrate your solution also on ASRs that overlap in some subpaths in the middle of the path expression—as opposed to overlapping prefix and suffix paths.

20.4   Devise the algorithms for (automatic) maintenance of access support relations upon object base modifications. Consider the following two operations—based on the path expression $t_0.A_1.\cdots.A_n$:

- Inserting an object o_{i+1} of type t_{i+1} into the set $o_i.A_i$

- Removing an object o_{i+1} from the set $o_i.A_{i+1}$

Illustrate the abstract algorithms on examples from our *DeptStore* database. In particular, discuss the different maintenance costs induced by the four extensions.

20.11   Annotated Bibliography

Access support relations were developed by the authors [KM90a]. A more detailed description can be found in [KM92a]. Furthermore, a discussion of access support relations is included in the book [Kem92].

The basic idea of access support relations was borrowed from the GemStone object-oriented database system [MS86]. However, GemStone supports—according to our terminology—only access support relations in full extension and binary decomposition. Furthermore, the indexing concept in GemStone prohibits set-valued attributes within the supported path expressions.

The concepts developed in GemStone were also incorporated in the Orion database project [BK89]. Keßler und Dadam [KD91] applied the ideas to the nested relational data model.

Similar optimization methods were also incorporated in the EXODUS database (generator) system. Unlike the separate data structures used for access support relations, the EXODUS developers investigated the augmentation of objects by redundant access information [SC89]. This can be seen as maintaining—for certain frequently used objects—the forward reference as well as the (redundant) backward reference.

The storage model we used for access support relations is adapted from the work on binary join indexes, carried out by P. Valduriez [Val87]. The binary join index work is based on a very early proposal for a general index structure in relational databases, called *links*, by T. Härder [Här78].

The storage representation of access support relations is based on B-trees, which were introduced in [BM72] and are thoroughly described in a survey paper by Comer [Com79]. Also, almost any text book on data structures—e.g., [HS76]—or on database systems—e.g., [KS91]—contains material on B-tree realizations.

21

Function Materialization

In this chapter, we introduce another optimization concept that can also be viewed as an index to objects. The basic idea is quite simple: Precompute those functions that are frequently used in selection predicates. However, the maintenance of this index is quite complex. Of course, the consistency of the stored precomputed function values has to be automatically assured by the object base management system. It is obvious that such an index favors retrieval operations—which are based on the precomputed function results. On the other hand, this index consisting of precomputed function results penalizes the update operations—since the index's consistency has to be maintained.

21.1   The Running Example

The discussion of this chapter will be based on our well-known *Cuboid* example. For convenience, the type definitions of *Vertex*, *Material*, and *Cuboid* are iterated in Figure 21.1.

We have intentionally made all parts of the structure of a *Cuboid* visible (**public**) to the clients of the type; e.g., all the boundary *Vertex* objects—represented by the attributes *v1*, ..., *v8*—are accessible and, therefore, directly modifiable. This is needed to demonstrate the function materialization approach in its full generality. Later on (in Section 21.7), we will refine the definition of *Cuboid* by hiding many of the structural details of the *Cuboid* representation—and, thus, drastically decrease the invalidation penalty of many update operations.

Note that the function *distance* depends on the object types *Cuboid* and *Robot*— the type definition of *Robot* is omitted here. In order to simplify the discussion, we assume that all operations have unique names, even across different type definitions. Therefore, we purposely called the function that computes the distance between two

type Vertex **is**
  public x→, x←, y→, y←, z→, z←,
          translate, scale, rotate, dist
  body [x, y, z: float;]
  operations
    declare translate: Vertex → **void**;
    declare scale: Vertex → **void**;
    declare rotate: float, char → **void**;
    declare dist: Vertex → float;
  implementation
     . . .
end type Vertex;

type Material **is**
  public specWeight→, specWeight←,
          name→, name←
  body [name: string;
          specWeight: float;]
end type Material;

  type Cuboid **supertype** ANY **is**
    public length, width, height, volume, weight, rotate, scale, translate, distance
            v1→, v1←, ..., v8→, v8←, value→, value←, mat→, mat←
    body [v1, v2, v3, v4, v5, v6, v7, v8: Vertex;
            mat: Material; value: decimal;]
    operations
      declare length: → float;   !! v1.dist(v2)
      declare width: → float;   !! v1.dist(v4)
      declare height: → float;   !! v1.dist(v5)
      declare volume: → float;
      declare weight: → float;
      declare translate: Vertex → **void**;
      declare scale: Vertex → **void**;
      declare rotate: char, float → **void**;
      declare distance: Robot → float;   !! *Robot* is defined elsewhere
    implementation
      define length **is**
        return **self**.v1.dist(**self**.v2); !! delegate the computation to *Vertex v1*
         . . .
      define volume **is**
        return **self**.length * **self**.width * **self**.height;
      define weight **is**
        return **self**.volume * **self**.mat.specWeight;
      define translate(t) **is**
         . . .
  end type Cuboid;

Figure 21.1: Type Definitions of *Vertex*, *Material*, and *Cuboid*

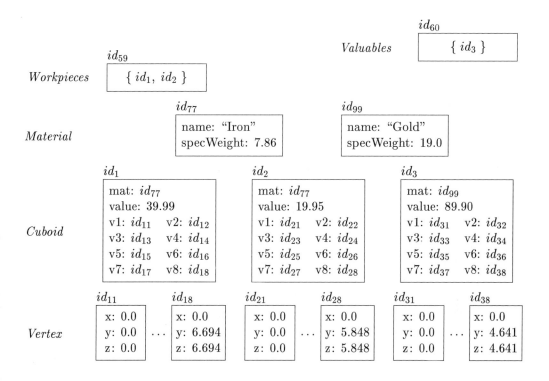

Figure 21.2: A Sample Object Base Extension

Vertex instances *dist* as opposed to the semantically related operation *distance* on *Cuboids*.

A sample database is shown in Figure 21.2. In this diagram, we also show two set-structured instances: id_{59} of type *Workpieces* and id_{60} of type *Valuables* whose types were not introduced. Sets of type *Workpieces* contain *Cuboids*, which are used in some manufacturing process, whereas sets of type *Valuables* contain *Cuboids*, which are interesting because of their high value, e.g., gold ingots. Interesting functions for *Workpieces* instances are *totalVolume* and *totalWeight*; set objects of type *Valuables* respond to the function *totalValue*.

21.2   Motivation for Function Materialization

Consider the above definition of the type *Cuboid* with the associated functions *volume* and *weight*. Assume that the following query, which is phrased in GOMql, is to be evaluated:

> **range** c: Cuboid
> **retrieve** c
> **where** c.volume > 20.0 **and** c.volume < 100.0 **and**
> c.weight > 40.0 **and** c.weight < 200.0

For the evaluation of this query, the selection predicate has to be evaluated by invoking the functions *volume* and *weight* for each *Cuboid* instance. It is obvious that this query evaluation imposes a tremendous complexity in run time—at least, if the extension of *Cuboid* contains a large number of instances.

To expedite the evaluation of this query the results of *volume* and *weight* can be precomputed: We call this the *materialization* of the functions *volume* and *weight*. In GOMql, the materialization of the functions *volume* and *weight* is initiated by the following statement:

> **range** c: Cuboid
> **materialize** c.volume, c.weight

In this statement the variable c is bound to the extension of type *Cuboid*, i.e., the set of instances of type *Cuboid*. Thus, the results of *volume* and *weight* are precomputed for each *Cuboid* instance. Of course, the variable c could also be bound to some set- or list-structured object with elements of type *Cuboid*. However, in the subsequent definitions we consider only variables bound to type extensions—but these definitions could easily be extended to include variables bound to sets or lists.

After the materialization of *volume* and *weight* the above query can use the precomputed results instead of invoking the functions *volume* and *weight* for each *Cuboid* instance. This query is called a *backward query*, as it uses a predicate on the function results to select the corresponding argument objects. As the evaluation of this query needs the *volume* and *weight* of all *Cuboid* instances, it would be advantageous to have these functions precomputed for all argument objects.

Another class of queries that can be supported by function materialization are *forward queries*, which are characterized by the retrieval of function results for particular objects. Consider, for example, the forward query

> **range** c: myValuableCuboids
> **retrieve** sum(c.weight)

where *myValuableCuboids* is a variable of (the set-structured) type *Valuables*. This query could exploit precomputed results of the function *weight*. However, in this case, it is not necessary to have the function *weight* precomputed for all *Cuboid* instances but only for those instances contained in the set *myValuableCuboids*.

Besides forward and backward queries there is a large variety of queries that are able to exploit precomputed function results. In Section 21.4, we discuss the support of these queries by materialized functions in more detail.

21.3   Generalized Materialization Relations

There are two obvious locations where materialized results could possibly be stored: in or near the argument objects of the materialized function or in a separate data structure. Storing the results near the argument objects means that the argument and the function result are stored within the same page such that the access from the argument to the appropriate result requires no additional page access. In general, storing results near the argument objects has several disadvantages:

- If the materialized function $f : t_1, \ldots, t_n \to r$ has more than one argument (i.e., $n > 1$), one of the argument types must be designated to hold the materialized result. But this argument has to maintain the results of all argument combinations—which, in general, may not fit on one page.

- Storing the results with the argument objects leads to a spreading of the function values throughout the object base. On the other hand, the clustering of function results would be beneficial to support selective queries on the results, e.g., a backward query as shown above. Then one could restrict the search for qualifying objects to the cluster of precomputed function results. But this is not possible if the location of the materialized results is determined by the location of the argument objects.

Therefore, it appears most advantageous to store materialized results in a separate data structure disassociated from the argument objects.

If several functions are materialized that share all argument types, the results of these functions may be stored within the same data structure. This provides for more efficiency when evaluating queries that access results of several of these functions and, further, avoids storing the arguments redundantly.

We formally define so-called *generalized materialization relations* (GMRs), which constitute the logical representation for materialized function results. Let $t_1, \ldots, t_n$ be object types and $r_1, \ldots, r_m$ be types—either sorts or object types. Further, assume $f_1, \ldots, f_m$ are functions of the following kind:

$$f_1 \quad : \quad t_1, \ldots, t_n \to r_1$$

$$\vdots$$

$$f_m \quad : \quad t_1, \ldots, t_n \to r_m$$

Thus, all m functions $f_1, \ldots, f_m$ have the same argument types—but (possibly) different result types. The materialized results of all m functions can then be maintained in one GMR—denoted $\langle\!\langle f_1, \ldots, f_m \rangle\!\rangle$—which has arity $n + 2 * m$. The GMR $\langle\!\langle f_1, \ldots, f_m \rangle\!\rangle$

has the following relational schema:

$$\langle\langle f_1, \ldots, f_n \rangle\rangle : \{[O_1 : t_1, \ldots, O_n : t_n, f_1 : r_1, V_1 : bool, \ldots, f_m : r_m, V_m : bool\,]\}$$

Intuitively, the attributes $O_1, \ldots, O_n$ store the arguments—or, more precisely, references to the argument objects, since the argument types are assumed to be complex structured. The attributes $f_1, \ldots, f_m$ store the results or—if the result is of complex type—references to the result objects of the invocations of the functions $f_1, \ldots, f_m$. Finally, the attributes $V_1, \ldots, V_m$ (standing for *validity*) indicate whether the stored results are currently valid.

Here, we will discuss only the materialization of functions having complex argument types. As can easily be seen it is not practical to materialize a function for all values of an atomic argument type, e.g., *float*. Therefore, we omit the discussion of materialized functions with atomic argument types in this presentation. However, it can be done using the concept of *restricted* GMRs, in which the function results are only precomputed for a restricted range of the argument combinations.

Let us now illustrate this abstract formalism of GMRs on our example object base schema of Figure 21.1. Based on the sample object base extension of Figure 21.2 the GMR $\langle\langle$volume, weight$\rangle\rangle$ can be defined for the functions

$$\text{volume} \quad : \quad \text{Cuboid} \rightarrow \text{float}$$

$$\text{weight} \quad : \quad \text{Cuboid} \rightarrow \text{float}$$

This GMR $\langle\langle$volume, weight$\rangle\rangle$ then has the following form—for our sample object base extension of Figure 21.2:

$\langle\langle volume, weight \rangle\rangle$				
$O_1 : Cuboid$	$volume : float$	$V_1 : bool$	$weight : float$	$V_2 : bool$
id_1	300.0	*true*	2358.0	*true*
id_2	200.0	*true*	1572.0	*true*
id_3	100.0	*true*	1900.0	*true*

As mentioned before, the GMRs contain so-called validity indicators—denoted $V_1, \ldots, V_m$ for the GMR $\langle\langle f_1, \ldots, f_m \rangle\rangle$—for each entry, i.e., for each precomputed function result. These validity indicators are set to

- *true*, if the corresponding entry is guaranteed correct

- *false*, if the corresponding entry may be invalid, e.g., due to some database update that occurred after the stored result was computed

Formally, the GMR $\langle\langle f_1, \ldots, f_m \rangle\rangle$ is *consistent* if for every tuple $\tau \in \langle\langle f_1, \ldots, f_m \rangle\rangle$ and for every j $(1 \le j \le m)$ the following holds:

$$\tau.V_j = true \quad \Rightarrow \quad \tau.f_j = f_j(\tau.O_1, \ldots, \tau.O_m)$$

This condition formally states that a true validity indicator V_j implies that the stored result $\tau.f_j$ equals the result of $f_j(\tau.O_1, \ldots, \tau.O_m)$ if computed in the *current* database state. Subsequently, we will always assume that the considered GMRs obey this condition; that is, we will only deal with *consistent* GMRs.

However, consistency is only a minimal requirement on GMR extensions. Further requirements like completeness and validity are introduced in the next subsection, where the retrieval of materialized results is discussed.

21.4 Retrieval of Materialized Results

All GMR extensions are maintained by the so-called *GMR manager*. The GMR manager offers retrieval operations to access argument objects and materialized results. Retrieval operations on GMRs can be represented in a tabular way—similarly to the relational query language *Query by Example (QBE)*.[1] The table below represents two abstract retrieval operations on a GMR $\langle\langle f_1, \ldots, f_m \rangle\rangle$ with complex argument types $t_1, \ldots, t_n$ and atomic result types:

$\langle\langle f_1, \ldots, f_m \rangle\rangle$							
$O_1 : t_1$	$O_2 : t_2$	$\ldots$	$O_n : t_n$	f_1	f_2	$\ldots$	f_m
id_1	id_2	$\ldots$	id_n	?	?	$\ldots$	?
?	?	$\ldots$	?	$[lb_1, ub_1]$	$[lb_2, ub_2]$	$\ldots$	$[lb_m, ub_m]$

The first retrieval operation is a *forward query*, in which all argument objects are specified and the corresponding function values are obtained from the GMR. The second retrieval denotes a *backward range query*, in which a range—$[lb, ub]$ standing for the range from **lower** bound to **upper** bound—is specified for each function value and the corresponding argument objects are obtained. In general, the columns of a GMR query can contain any combination of constants (or ranges), "?"-signs and "don't-care"-signs (denoted by "–") facilitating a large variety of possible retrieval operations on GMRs.

The GMR retrieval operations can be used to support queries containing invocations of materialized functions. For example, if all results of the materialized function f_j are requested by some query, e.g., to perform some aggregate operation on the results, all results currently being valid can be obtained from the GMR, and all invalid

---

[1] The tabular notation improves clarity of presentation in this subsection. However, in general, GOMql is used to denote queries.

results have to be (re)computed. These (re)computed results are also used by the
GMR manager to update the GMR. Further, if the GMR is not complete, i.e., does
not contain an entry for each argument combination, the results of missing argument
combinations have to be computed as well. These should be inserted into the GMR,
too.

The example query—retrieving *Cuboids* based on their *volume* and *weight*—formu-
lated at the beginning of this chapter can be evaluated on the GMR $\langle\!\langle volume,weight\rangle\!\rangle$
as follows:

$\langle\!\langle$volume,weight$\rangle\!\rangle$		
O_1: Cuboid	volume	weight
?	[20.0, 100.0]	[40.0, 200.0]

The question mark in the column O_1—i.e., the column that contains object identifiers (OIDs) of *Cuboid* instances—indicates that we want to retrieve the *Cuboid*
instances that satisfy the two range specifications, i.e., [20.0, 100.0] in the *volume*
column and [40.0, 200.0] in the *weight* column.

In order to evaluate the above-shown query it might be necessary to recompute
some invalidated GMR entries first. However, note that invalidated or missing results
need not necessarily all be (re)computed. For example, if, for the above query, a
Cuboid is encountered having a *weight* greater than 200.00, it is disqualified regardless
of its volume. Hence, if the *volume* entry is invalidated, this does not matter.

Subsequently, we formalize the notions of *valid* and *complete* GMR extensions. A
consistent extension of the GMR $\langle\!\langle f_1,\ldots,f_m\rangle\!\rangle$ is *valid* for the function f_j $(1 \leq j \leq m)$
if *all* V_j entries in the GMR table are set to *true*.

A consistent extension of the GMR $\langle\!\langle f_1,\ldots,f_m\rangle\!\rangle$ is *complete* if it contains the
function results for *all* argument combinations. Formally, if

$$(o_1,\ldots,o_n) \in ext(t_1) \times \cdots \times ext(t_n)$$

then there exist values f_i and v_i for all $(1 \leq i \leq m)$ such that the GMR contains a
tuple:

$$(id(o_1),\ldots,id(o_n), f_1, v_1,\ldots, f_m, v_m)$$

Upon the creation of a new GMR the database administrator can choose whether
the GMR extension has to be complete or whether the extension may be set up
incrementally (starting with an empty GMR extension). Incrementally set up GMR
extensions can be used as a cache for function results that were computed during the
evaluation of queries. If the number of entries is limited (due to space restrictions),
specialized replacement strategies for the GMR entries can be applied. Note that
GMRs must be set up incrementally if they contain at least one partial function.

21.4.1   Storage Representation of GMRs

The flexible retrieval operations on the GMRs require appropriate index structures to avoid the exhaustive search of GMR extensions. Obviously, we need to provide entry points for each argument type and for each column representing a precomputed function result. For that, well-known indexing techniques from relational database technology can be utilized. The easiest way to support the flexible and efficient access to any combination of GMR fields would be a single multidimensional index structure, denoted MDS, over the fields $O_1, \ldots, O_n, f_1, \ldots, f_m$:

$$\text{MDS} \quad \boxed{O_1} \; \cdots \; \boxed{O_n} \; \boxed{f_1} \; \cdots \; \boxed{f_m} \; \| \; \boxed{V_1} \; \cdots \; \boxed{V_m}$$

Here, the first $n+m$ columns constitute the $(n+m)$-dimensional keys of the multidimensional storage structure. The m validity bits $V_1, \ldots, V_m$ are additional attributes of the records being stored in the MDS.

Unfortunately, the (currently existing) multidimensional storage structures, such as the Grid-File [NHS84], are not well suited to support more than three or four dimensions. Therefore, the GMR manager should utilize more conventional indexing schemes to expedite the access to entries in GMRs of higher arity. The index structures are chosen according to the expected query mix, the number of argument fields in the GMR, and the number of functions in the GMR.

21.5   Invalidation and Rematerialization of Function Results

The incorporation of validity indicators in the GMRs—in conjunction with the above definition of consistency—provides for some tuning measure with respect to the invalidation and rematerialization of precomputed results. Upon an update to a database object that invalidates a materialized function result, we have two choices:

1. *Immediate rematerialization*: The invalidated function result is immediately recomputed as soon as the invalidation occurs.

2. *Lazy rematerialization*: The invalidated function result is only marked as being invalid by setting the corresponding V_i attribute to *false*. The rematerialization of invalidated results is carried out as soon as the load of the object base management system falls below a predetermined threshold or—at the latest—at the next time the function result is needed.

In Figure 21.3, the possible rematerialization strategies are sketched for the *volume* result of a *Cuboid* referred to by some variable c relative to an invalidation by the

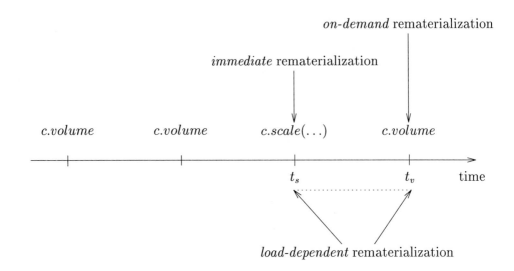

Figure 21.3: Illustration of the Different Rematerialization Strategies

transformation *c.scale*. The diagram graphically depicts three different rematerialization strategies for a sequence of operations on the *Cuboid* object *c*. *Immediate* rematerialization requires recomputing the value *c.volume* directly after the *Cuboid* *c* has been updated; that is, at time t_s. *Lazy* rematerialization is further divided into two substrategies: *load-dependent* and *on-demand* rematerialization. Under on-demand rematerialization, the recomputation is delayed until the next time the *volume* operation is invoked on the object *c* at time t_v. The *load-dependent* rematerialization can take place anywhere in the time interval $[t_s, t_v]$, that is, after the update of *c*. The latest possible time is the next invocation of *volume* at time t_v—corresponding to the on-demand rematerialization.

The following program fragment demonstrates the problems of *immediate* rematerialization:

```
for (i in {1, ..., CONST})
    c.scale(...);
```

In this example, the *immediate* rematerialization alternative would "blindly" recompute the function *volume* after each iteration. In our—admittedly contrived— example the immediate rematerialization takes place *CONST* times of which the first (*CONST* − 1) materializations are obviously wasted because no intermediate invocation of *volume* occurs in between the *CONST* *scale* operations.

21.6   Detection of Invalidated Function Results

21.6.1   Maintaining Reverse References

When the modification of an object o is reported to the GMR manager, the GMR manager must find all materialized results that become invalid. This task is equivalent to determining all materialized functions f and all argument combinations $o_1, \ldots, o_n$ such that the modified object o has been accessed during the materialization of $f(o_1, \ldots, o_n)$. Note that in GOM—as in most other object models—references are maintained only unidirectionally. That is, there is no efficient way to determine for a given object o the set of objects that reference o via a particular path. Therefore, the GMR manager maintains reverse references from all objects that have been used in some materialization to the appropriate argument objects in a relation called *reverse reference relation* (*RRR*). The RRR contains tuples of the following form (lists are denoted by the angle brackets $\langle \ldots \rangle$):

$$[id(o),\ f,\ \langle id(o_1), \ldots, id(o_n) \rangle]$$

Herein, $id(o)$ is the OID of an object o utilized during the materialization of the result $f(o_1, \ldots, o_n)$. Note that o may not be one of the arguments $o_1, \ldots, o_n$; it could be some object related (via attributes) to one of the arguments. Thus, each tuple of the RRR constitutes a reference from an object o influencing a materialized result to the tuple of the appropriate GMR in which the result is stored. We call this a *reverse reference*, as there exists a reference chain in the opposite direction in the object base.[2]

The reverse references are inserted into the RRR during the materialization process. Therefore, each materialized function f and all functions invoked by f are modified—the modified versions are extended by statements that inform the GMR manager about the set of accessed objects. During a (re)materialization of some result the modified versions of these functions are invoked.

Besides the already discussed GMR $\langle\langle volume,\ weight \rangle\rangle$, let us introduce a second GMR $\langle\langle distance \rangle\rangle$, which materializes for a given *Cuboid* and *Robot* their distance. This is important to know, whenever we have to retrieve a *Robot*, able to manipulate a certain *Cuboid*. Since this retrieval ought to be fast, we create the GMR $\langle\langle distance \rangle\rangle$. Note that now the *Cuboid*s occur in two GMRs.

Let the GMRs $\langle\langle volume,\ weight \rangle\rangle$ and $\langle\langle distance \rangle\rangle$ be defined (again, this example is based on the object base extension shown in Figure 21.2). The extensions of the RRR and the two GMRs are shown in Figure 21.4. Note that two *Robots* with the identifiers id_4 and id_5 are assumed to exist in the object base.

---

[2]This holds only if no global variables are used by the materialized function. Otherwise, the RRR contains reverse references not only to the argument objects but also to the accessed global variables.

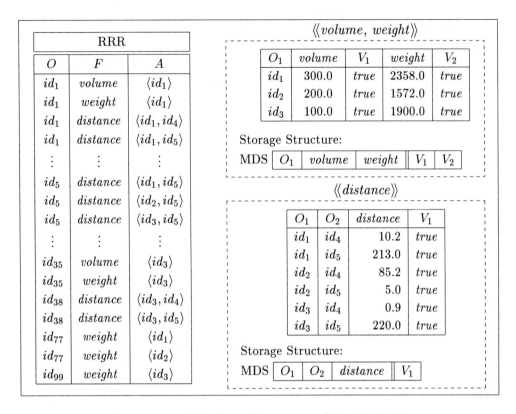

Figure 21.4: The Data Structures of the GMR Manager

Based on the RRR, we can outline the algorithms for invalidating or rematerializing a stored function result, i.e., the computations that have to be performed by the GMR manager when an object o has been updated. The GMR manager is notified about an update of some object o by the following statement:

GMR_Manager.invalidate(o);

The algorithms below reflect the two different possibilities of *lazy rematerialization* and *immediate rematerialization*.

The *lazy rematerialization* algorithm only invalidates the affected GMR entries:

lazy(o) $\equiv$ **foreach** triple $[id(o), f_i, \langle id(o_1), \ldots, id(o_n) \rangle]$ **in** *RRR* **do**
(1) **set** $V_i := $ *false* for the appropriate tuple **in** $\langle\!\langle f_1, \ldots, f_i, \ldots, f_m \rangle\!\rangle$
(2) **remove** $[id(o), f_i, \langle id(o_1), \ldots, id(o_n) \rangle]$ **from** *RRR*

Step (2) of the algorithm—i.e., the removal of the RRR entry—ensures that for the same, repeatedly performed object update the invalidation is done only once.

Subsequent invalidations due to updates of o will be blocked at the beginning of $lazy(o)$ by not finding the RRR entry that was removed upon the first invalidation—thus, the unnecessary penalty of accessing the tuple in the GMR to reinvalidate an already invalidated result is avoided. By the next rematerialization of $f(o_1, \ldots, o_n)$, all relevant RRR entries are (re)inserted into the RRR—analogously to the immediate rematerialization algorithm shown below.

Under the *immediate rematerialization* strategy, the GMR manager has to recompute the affected function results:

immediate(o) $\equiv$
  foreach triple $[id(o), f_i, \langle id(o_1), \ldots, id(o_n) \rangle]$ in RRR **do**
   (1) **remove** $[id(o), f_i, \langle id(o_1), \ldots, id(o_n) \rangle]$ **from** RRR
   (2) **recompute** $f_i(o_1, \ldots, o_n)$ and
      * **remember** all accessed objects $\{o'_1, \ldots, o'_p\}$
      * **replace** the old value of $f_i(o_1, \ldots, o_n)$ in $\langle\!\langle f_1, \ldots, f_i, \ldots, f_m \rangle\!\rangle$
   (3) **foreach** v in $\{o'_1, \ldots, o'_p\}$ **do**
      * **insert** the triple $[id(v), f_i, \langle id(o_1), \ldots, id(o_n) \rangle]$ **into** RRR (if not present)

We will explain step (1) of this algorithm last. In step (2), we recompute the function result $f_i(o_1, \ldots, o_n)$ and remember all objects visited in this process in order to insert them into the RRR in step (3). However, it cannot be guaranteed that the RRR does not contain any obsolete entries that constitute "leftovers" from the previous materialization(s) of $f_i(o_1, \ldots, o_n)$—this happens whenever two subsequent materializations of $f_i(o_1, \ldots, o_n)$ visit different sets of objects. Let $[id(w), f_i, \langle id(o_1), \ldots, id(o_n) \rangle]$ be such a leftover entry meaning that in an earlier materialization of $f_i(o_1, \ldots, o_n)$ the object w was visited; but the current materialized result of $f_i(o_1, \ldots, o_n)$ is not dependent on the state of w. Then the next (seemingly relevant) update on w will remove the triple $[id(w), f_i, \langle id(o_1), \ldots, id(o_n) \rangle]$ from the RRR by step (1) of the above outlined algorithm, while steps (2) and (3) do not inject any new information that is not already present in the GMR and the RRR.

In most cases, an object will be reused after an update—thus, the same RRR entry that has been removed in step (1) of the above algorithm will be reinserted into the RRR. This situation could be remedied by a *second chance algorithm*, which is based on marking RRR entries instead of removing them in step (1).

With respect to removing leftover entries the RRR maintenance algorithm can be termed *lazy* because leftover entries are removed only when the corresponding object is updated. An alternative to this strategy would be a periodic reorganization of the RRR.

The rematerialization of function results that constitute complex objects may lead to the creation of new objects. Invalidated result objects cannot be deleted by the GMR manager as they may be referenced in other contexts independently of the materialization of the function. Thus, to minimize the number of unreferenced but undeleted result objects, GMRs with complex result types should be maintained under lazy rematerialization. A garbage collection mechanism can be employed to remove unreferenced objects.

In addition to the invalidation upon an object update, the GMR manager also has to take proper action upon insertion of a new object into the database and deletion of an object from the database. For this purpose, the two functions *newObject* and *forgetObject* are provided by the GMR manager (see Exercise 21.3).

21.6.2   The Update Notification Mechanism

The operations of the GMR manager to keep the GMR extensions in a consistent state (i.e., *invalidate, newObject, forgetObject*) could be invoked either by the object manager or by the updating operation.

The object manager can inform the GMR manager about relevant object modifications when the updated object is stored in the object base. This approach makes the adaptation of the object manager necessary, which may be prohibitively difficult in existing systems. Further, the adaptation of the object manager has the following shortcomings:

- Every user of the object base will be penalized by the materialization of functions—even if only those parts of the object base are accessed that are not involved in any materialization.

- As applications may first modify some objects and then access materialized results being affected by the former updates, all updates must immediately be propagated to the object manager in order to keep the GMR extensions consistent. This need to store updated objects immediately prevents optimization strategies based on deferring the storage of modified objects.

Therefore, in the generic object model (GOM) we chose the schema rewrite approach that is based on analyzing the materialized functions and modifying the relevant parts of the object base schema, i.e., those update operations that affect materialized results. The state of the object base can be modified only by the elementary update operations $t.create$ and $t.delete$ for any type t, $t.A{\leftarrow}$ for any tuple-structured type t and any attribute A of t, and $t.insert$ and $t.remove$ for any set-structured type t. Every elementary update operation associated with some type t involved in the materialization of any function is modified and recompiled, such that each time the update operation is invoked, the invocation of one of the functions *invalidate*, *newObject*, and *forgetObject* will be triggered.

The approach of injecting the notification mechanism into the primitive updating operations has the advantage that no adaptation of the object manager is needed.

Instead, it requires a modification of the updating functions and their recompilation. Further, the schema rewrite approach guarantees that the GMR manager is immediately informed when an update occurs—by that, the extensions of the GMRs will remain consistent.

Figure 21.5 shows the modified versions of the update operation $A\leftarrow$ for a tuple-structured type t with attribute A and of the delete operation *delete* for type t. The operation $A\leftarrow$ is extended by the statement

> GMR_Manager.invalidate(**self**);

such that the invalidation occurs *after* the value of attribute A has been updated. If the materialized results are invalidated before the update, the immediate rematerialization strategy would lead to wrong results. The *delete* operation is extended by the statement

> GMR_Manager.forgetObject(**self**);

that is invoked *before* the object is deleted.

In the next section, we show how the set of update operations that have to be modified can be drastically reduced.

```
declare A←: t || ← t'                    declare delete: t || → void
   code set_A';                             code delete';
define set_A' (x) is                     define delete' is
   begin                                    begin
      self.A := x;                             GMR_Manager.forgetObject(self);
      GMR_Manager.invalidate(self);            self.system_delete;
   end define set_A';                       end define delete';
```

Figure 21.5: Modification of Primitive Update Operations (Version 1)

21.7   Strategies to Reduce the Invalidation Overhead

The invalidation mechanism described so far is (still) rather unsophisticated and, therefore, induces unnecessarily high update penalties upon object modifications. In the following, we will describe four dual techniques to reduce the update penalty— consisting of invalidation and rematerialization—by better exploiting the potential of the object-oriented paradigm. The techniques described in this section are based on the following ideas:

1. *Isolation of relevant object properties:* Materialized results typically depend on only a small fraction of the state of the objects visited in the course of materialization. For example, the materialized *volume* certainly does not depend on the *value* and *mat* attributes of a *Cuboid*.

2. *Reduction of RRR lookups:* The unsophisticated version of the invalidation process has to check the *RRR* each time any object o is being updated. This leads to many unnecessary table lookups that can be avoided by maintaining more information within the objects being involved in some materialization—and thus restricting the lookup penalty to only these objects.

3. *Exploitation of strict encapsulation:* By strictly encapsulating the representation of objects used by a materialized function, the number of update operations that need be modified can be reduced significantly. Since internal subobjects of a strictly encapsulated object cannot be updated separately—without invoking an outer-level operation of the strictly encapsulated object—we can drastically reduce the number of invalidations by triggering the invalidation only by the outer-level operation.

4. *Compensating updates:* Instead of invoking the materialized function to recompute an invalidated result, specialized compensating actions can be invoked that use the old result and the parameters of the update operation to recompute the result in a more efficient way.

21.7.1   Isolation of Relevant Object Properties

Suppose that *volume* and *weight* have been materialized. Then these two materialized functions surely do not depend on the attribute *value*. Nevertheless, under the unsophisticated invalidation strategy the operation invocation

$$id_1.\text{value}\leftarrow(123.50); \quad \text{!! equivalent to: } id_1.value := 123.50;$$

does lead to the invalidation of $id_1.volume$ and $id_1.weight$, both of which are unnecessary. Likewise, the operation invocation

$$id_1.\text{mat}\leftarrow(\text{copper}); \quad \text{!! } copper \text{ being a variable of type } Material$$

leads to the necessary invalidation of $id_1.weight$, but also to the unnecessary invalidation of $id_1.volume$. In order to avoid such unnecessary invalidations, the system has to separate the relevant properties of the objects visited during a particular materialization from the irrelevant ones. Then invalidations should only be initiated if a relevant property of an object is modified.

The relevant properties of a materialized function f can automatically be derived from the implementation of the function f—of course, also inspecting all functions invoked by f.

The relevant properties, called *RelAttr*, for the function *volume* are given below:

$$RelAttr(volume) = \{Cuboid.v1, Cuboid.v2, Cuboid.v4, Cuboid.v5, Vertex.x, Vertex.y, Vertex.z\}$$

From this it follows that the stored results of the function *volume* can be invalidated only by the update operations $v1\leftarrow$, $v2\leftarrow$, $v4\leftarrow$, and $v5\leftarrow$ associated with type *Cuboid*, and by the $x\leftarrow$, $y\leftarrow$ and $z\leftarrow$ operations of type *Vertex*.

For each primitive update operation, the schema manager maintains a set, called *SchemaDepFct*. This set contains those materialized function identifiers that may possibly be (partially) invalidated by the invocation of the primitive update operation. For example, *volume* is contained in the sets *SchemaDepFct(Cuboid.v1$\leftarrow$)*, *SchemaDepFct (Cuboid.v2$\leftarrow$)*, ..., *SchemaDepFct(Vertex.z$\leftarrow$)*.

21.7.2   Marking "Used" Objects to Reduce RRR Lookup

The improvement of the invalidation process developed in the preceding subsection ensures that no more unnecessary invalidations occur.[3] However, one problem still remains: The GMR manager is invoked more often than necessary to check within the RRR whether an invalidation really has to take place. Suppose object o of type t is updated by operation $o.A\leftarrow(\ldots)$ and all functions that have used o for materialization are *not* dependent on the attribute A of object o. In this case there cannot be a materialized value that must be invalidated due to the update $o.A\leftarrow (\ldots)$. Consider, for example, the update

$$id_{111}.x\leftarrow(2.5); \quad !! \ Vertex \ id_{111} \ \textbf{not} \ being \ a \ boundary \ Vertex \ of \ any \ Cuboid$$

of the *Vertex* instance id_{111} that is not referenced by any *Cuboid*. Since the functions *volume* and *weight* do depend on the x attribute of (some) *Vertex* objects, the GMR manager is being invoked—only to find out by a RRR-lookup that no invalidation has to be performed. This imposes a (terrible) penalty upon geometric transformations of "innocent" objects, e.g., *Cylinders* and *Pyramids*, if the *volume* of *Cuboid* has been materialized—due to the fact that all three types are clients of the same type *Vertex*.

The goal is to invoke *GMR_Manager.invalidate* only when an invalidation has to take place. Therefore, each object o is appended by the set-valued attribute *ObjDepFct* that contains the identifiers of all materialized functions that have used o during their materialization. Now, the GMR manager is informed only if the updated

---

[3]Actually, in the case that the same object type is utilized in the same materialization in different contexts there may still be an unnecessary invalidation.

object was actually used to compute some materialized function result. Thereby, the
update penalty of function materialization is drastically reduced.

Figure 21.6 shows the modified versions of the update operations that exploit the
attribute *ObjDepFct*.

declare $A\leftarrow$: t $||\leftarrow$ t$'$ **declare** delete: t $|| \rightarrow$ **void**
  code set_A$'$;   **code** delete$'$;
define set_A$'$ (x) **is**   **define** delete$'$ **is**
begin   **begin**
  self.A := x;   **if self**.ObjDepFct $\neq$ { } **then**
  RelevFct := **self**.ObjDepFct $\cap$ GMR_Manager.forgetObject(**self**);
   SchemaDepFct(t.$A\leftarrow$);   self.system_delete;
  if RelevFct $\neq$ { } **then**   **end define** delete$'$;
   GMR_Manager.invalidate(**self**,RelevFct);
end define set_A$'$;

Figure 21.6: Modification of Primitive Update Operations (Version 2)

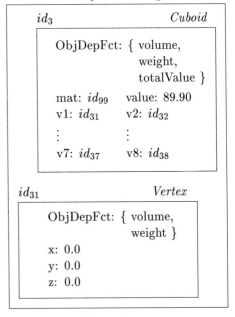

Schema Manager

declare $x\leftarrow$: Vertex $||\leftarrow$ float
  code set_x;

define set_x(newX) **is**
  self.x := newX;

schema rewrite

declare $x\leftarrow$: Vertex $||\leftarrow$ float
  code set_x$'$;

define set_x$'$(newX) **is**
begin
  self.x := newX;
  RelevFct := **self**.ObjDepFct
      $\cap$ SchemaDepFct(Vertex.$x\leftarrow$);
  if RelevFct $\neq$ { } **then**
  GMR_Manager.invalidate(**self**,RelevFct);
end define set_x$'$;

Object Manager

id_3 *Cuboid*

ObjDepFct: { volume,
          weight,
          totalValue }

mat: id_{99} value: 89.90
v1: id_{31} v2: id_{32}
 ⋮   ⋮
v7: id_{37} v8: id_{38}

id_{31} *Vertex*

ObjDepFct: { volume,
         weight }

x: 0.0
y: 0.0
z: 0.0

Figure 21.7: Interaction between Schema and Object Manager

Recall the database extension shown in Figure 21.2. Suppose that the following GMRs were introduced: $\langle\langle totalVolume, totalWeight\rangle\rangle$ for the type *Workpieces*, $\langle\langle totalValue\rangle\rangle$ for the type *Valuables*, and $\langle\langle volume, weight\rangle\rangle$ for the type *Cuboid*.

Consider the invocation $id_{31}.x\leftarrow(\ldots)$, which modifies the x coordinate of Vertex id_{31}. Figure 21.7 shows the modification of the update operation $Vertex.x\leftarrow$. The set of materialized functions that is dependent upon the update $id_{31}.x\leftarrow(\ldots)$ is then given by the intersection of the sets $SchemaDepFct(Vertex.x\leftarrow)$ and $id_{31}.ObjDepFct$.

$$SchemaDepFct(Vertex.x\leftarrow) = \{volume, weight, totalVolume, totalWeight\}$$
$$id_{31}.ObjDepFct = \{volume, weight\}$$

In this case, the intersection coincides with the set $id_{31}.ObjDepFct$. However, in general this is not the case, e.g., for the operation $Cuboid.v1 \leftarrow$ and the update $id_3.v1\leftarrow(\ldots)$.

21.7.3 Information Hiding

Despite the improvements of the invalidation mechanism outlined in the previous two subsections, three problems that can be avoided by exploiting information hiding remain.

First, the improvements incorporated so far do not totally prevent the penalization of operations on objects not involved in any materialization. For example, update operations defined on other geometric objects, e.g., *Pyramids*, are penalized to some extent by the materialization of *Cuboid.volume* if the type *Vertex* is utilized in the definition of both types. This is a consequence of modifying the update operations of the lower-level types, e.g., $Vertex.x\leftarrow$, which is then invoked on *every* update of attribute x of type *Vertex*.

Second, a single update operation consisting of a sequence of lower-level operations may trigger many subsequent rematerializations of the same precomputed result. For example, one single invocation of $id_1.scale(\ldots)$ triggers 12 invalidations of $id_1.volume$ initiated by the $x\leftarrow$, $y\leftarrow$, and $z\leftarrow$ operation invocations on the *Vertex* instances. Obviously, one invalidation should be enough.

Third, the algorithms detailed so far cannot detect the irrelevance of an update operation sequentially invoking lower-level operations that neutralize each other with respect to a precomputed result. For example, the invocation of $id_1.rotate(\ldots)$ performs 12 invalidations of $id_1.volume$ despite the fact that *no* invalidation is required, since the volume stays invariant under rotation.

We can exploit information hiding to avoid the unnecessary overhead incurred by the three above mentioned problems. Analogous to information hiding in traditional software design we call an object *strictly encapsulated* if the direct access to the representation of this object—including all its subobjects—is prohibited; manipulations are only be possible by invoking public operations defined on the type of that object. These operations constitute the *object interface*. In GOM strict encapsulation is realized (1) by discarding all access operations for attributes from the **public** clause, (2) by creating all subobjects of an encapsulated complex object during the initialization of that object, and (3) by enforcing that no public operation returns references to subobjects. Thus, no undesired access to subobjects via, e.g., object sharing is possible.

By enforcing strict encapsulation, only updating interface operations have to be modified to perform invalidations. Further, the number of invalidations due to the invocation of an update operation is reduced to one. Last but not least, update operations leaving the result of a materialized function invariant need not be modified. Thus, by specifying and exploiting a set of *invalidated functions* for each invalidating public operation the above-mentioned problems can easily be eliminated.

In terms of our *Cuboid* example we could specify—and thus "inform" the GMR manager—that a materialized *volume* can be invalidated only by a *scale* invocation; a materialized *weight* can be invalidated by scaling the Cuboid or changing the material (i.e., assigning a new *Material* instance to the *mat* attribute or changing the *specWeight* of the referenced *Material* instance).

Of course, it is the object type implementor's responsibility to ensure strict encapsulation and to determine the set of invalidated materialized functions for every update operation. In general, these cannot be deduced automatically. Consider, for example, the operations *rotate* and *translate* associated with the type *Cuboid*. It requires (human) intelligence to infer that these two operation—if correctly implemented—leave the materialized *volume* and *weight* invariant.

21.7.4   Compensating Actions

A materialized result that has been invalidated by an update can be recomputed either by an invocation of the materialized function or by a specialized function compensating the update. For example, consider the GMR $\langle\langle totalVolume \rangle\rangle$.[4] When a new *Cuboid* instance is inserted into a set of type *Workpieces* the result of *totalVolume* can be recomputed by adding the *volume* of the inserted *Cuboid* to the old result of *totalVolume*—instead of having to recompute *volume* for all members of the set. For this the database programmer has to specify a *compensating action*, i.e., a function that compensates for the insertion of a new *Cuboid* into a set of type *Workpieces*:

declare increase_total: Workpieces || Cuboid, float → float;

define increase_total (new_cuboid, old_total) **is**
    return old_total + new_cuboid.volume;

---

[4]The function *totalVolume* is associated with the type *Workpieces*.

Compensating actions may be specified only for update operations associated with argument types of materialized functions. It is not allowed to specify a compensating action for an update operation associated with a nonargument type, as this may lead to inconsistent GMR extensions. If, for example, a compensating action is specified for the materialized function $totalVolume$ and the update operation $Cuboid.scale$, the GMR $\langle\langle totalVolume \rangle\rangle$ could become inconsistent by an invocation of $Cuboid.scale$.

Based on Figure 21.2, assume $Cuboid\ id_1$ to be a member of the set id_{59} of type $Workpieces$. The invocation $id_{59}.remove(id_1)$ removes id_1 from the set id_{59}—due to our RRR maintenance algorithm, id_1 remains marked. Thus, if $Cuboid\ id_1$ is scaled subsequently, the invocation of the specified compensating action is triggered— leading to a wrong result, as id_1 is no longer a member of id_{59}.

Further, compensating actions can be specified only for *modified* update operations, i.e., update operations that are extended by statements to inform the GMR manager about updates. If an argument type t is strictly encapsulated only public update operations of t are modified. Otherwise, if t is not strictly encapsulated, the elementary update operations $t.A\leftarrow$ (if t is tuple structured) or $t.insert$ and $t.remove$ (if t is set structured) are modified.

21.8 Exercises

21.1 Complete the type definitions shown in Figure 21.1. In particular, outline the type definition of *Robot* that is needed to code the operation *distance* in the type definition frame of *Cuboid*. Further, outline the type definitions of *Valuables* and *WorkPieces* with the operations *totalValue* and *totalVolume*.

Then outline a sample object base extension that also contains some *Robot* instances.

21.2 Assume that the type *Cuboid* has a further operation *distanceToOtherCuboid* with the following signature:

declare distanceToOtherCuboid: Cuboid $\parallel$ Cuboid $\rightarrow$ float;

Outline the implementation of this function—assuming that the operation returns the distance between the centers of the two argument *Cuboids*.

Show the extension of the GMR $\langle\langle distanceToOtherCuboid \rangle\rangle$ for the example object base of Figure 21.2.

Why does this GMR contain redundancy—induced by the symmetry of the function *distanceToOtherCuboid*?

21.3 Specify the operations *forgetObject* and *newObject* that are provided by the GMR manager for ensuring the consistency of GMRs upon the creation of a new object and the deletion of an existing object, respectively.

21.4 Consider the operation *weightUnderDifferentGravitation*, whose signature and implementation are given as follows:

> **declare** weightUnderDifferentGravitation: Cuboid $\|$ float $\to$ float;

> **define** weightUnderDifferentGravitation(gravitation) **is**
>   **return self.**volume $*$ **self.**mat.specWeight $*$ gravitation/9.81;

What problems occur when we create $\langle\!\langle weightUnderDifferentGravitation \rangle\!\rangle$, which is caused by the (logically) infinite argument range of the *float* parameter?

Outline a solution by restricting the materialized results for particular argument values—i.e., gravitations—only. What problems occur in query processing when we use such a restricted GMR? Does an enumeration sort *Gravitations* help?

21.5 In the text, we outlined two approaches to reduce the invalidation and rematerialization overhead induced by database updates:

1. Recognition of irrelevant parts of an object

2. Marking those objects that are used in the materialization of particular function results

Devise the control structures of the GMR manager and the schema (operation) modifications necessary to exploit these optimization concepts.

21.6 Discuss the reason why *strict encapsulation* is required for controlling the invalidation of precomputed functions as specified in Section 21.7.3.

Illustrate your discussion on the *volume* and *weight* operations associated with *Cuboid*. Show, in particular, the problems possibly occurring when the *Vertex* instances referred to by the attributes $v1, \ldots, v8$ are accessible to outside clients. How can this be avoided?

21.7 An object of type *Company* consists of the name of the company, all departments of the company, and the set of projects that are carried out within the company. A *Department* is described by its name and the set of its employees. *Projects* are modeled by the project name, the status of the project, the size of the project and the set of programmers that are involved in the project. We consider only software projects. Thus, the size of a project is modeled by the lines of code it comprises. The status of a project is given by a decimal value that ranges between -1000 and 1000. A negative status denotes a delay and a loss that is caused by the project, a positive status denotes that the project is profitable.

The type *Employee* is a subtype of the type *Person*. Each *Employee* has a unique employee number, a salary, and a history of jobs, which is modeled as a set of jobs. A *Job* describes the part of a *Project* that has been delegated to a particular *Employee*. Therefore, each *Job* contains a reference to the *Project*, the number of lines of code that have been written by the *Employee* and two Boolean values that denote the status of the *Employee*. The attributes of the type *Job* are used to compute an assessment value—the *ranking* of an *Employee* is then given by the average of the assessment values of all jobs in the employee's job history. The function *ranking* is associated with type *Employee*.

The second function whose materialization should be considered is the function *matrix* associated with the type *Company*. This function computes the department project matrix for a company. A department project matrix is defined as a set of tuples of the type *MatrixLine*, defined as

$$[Dep : Department, Proj : Project, Emps : SetofEmployee]$$

One tuple τ of type *MatrixLine* states that the employees contained in the set $\tau.Emps$ are employed in department $\tau.Dep$ and work in project $\tau.Proj$. The matrix contains only *MatrixLine* instances τ with $\tau.Emps \neq \{\ \}$.

The reference graph depicted in Figure 21.8 elucidates the involved types. An arrow leading from type t_1 to type t_2 means that objects of type t_1 contain an attribute of type t_2. Double-pointed arrows denote that objects of the first type contain set-valued attributes with elements of the second type. Arrows leading from t_1 to t_2 are labeled with the name of the appropriate attribute.

1. Outline the type definitions of the above-described schema.

2. Sketch a sample object base extension for this schema.

3. Show the extension of the GMR materializing the *ranking* operation.

4. Outline the GMR that materializes the function *matrix*.

5. The function *matrix* is computationally very complex. Therefore, consider the concept of *compensating* actions to recompute an invalidated *matrix* result.

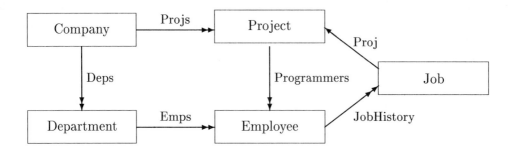

Figure 21.8: Reference Graph of Types in the Company Example

21.9   Annotated Bibliography

The function materialization discussed here is—in its basic ideas—similar to materialization of views in the relational context. The most important work is reported in [BCL89] and [BLT86]. Lindsay proposed so-called relational *snapshots* [AL80]—which, however, are not guaranteed consistent with the actual state of the database. The snapshots are only periodically recomputed and, thereby, brought back into a consistent state. Thus, a snapshot can be used only for certain applications, e.g., browsing, that do not require a completely consistent information contents.

Further work in precomputing queries and database procedures was done in the extended relational database project POSTGRES [SR86]. Here, the so-called QUEL as a Datatype attributes—see Section 6.3—are precomputed and cached in separate data structures. The control concepts are discussed in [Han87, Han88, Jhi88, Sel88, SAH87a, SJGP90].

The function materialization concepts discussed in this chapter were developed by the authors and C. Kilger. The basic ideas are described in [KKM91]; a more detailed discussion is given in [KKM90]. Some of the detailed algorithms for controlling the consistency of materialized results were developed by M. Steinbrunn in his Master's thesis [Ste91].

22

Pointer Swizzling

In this chapter, we classify and describe different approaches to optimizing the access to main memory resident persistent objects—techniques that are commonly referred to as *"pointer swizzling."* In order to speed up the access along interobject references, the persistent pointers in the form of unique object identifiers (OIDs) are transformed (swizzled) into main memory pointers (addresses). Thus, pointer swizzling avoids the indirection of a table lookup to localize a persistent object that is already resident in main memory.

22.1   Mechanisms for Localizing Objects

We investigate a class of techniques for managing references between main memory resident persistent objects, which is commonly referred to as *"pointer swizzling."* Pointer swizzling is a measure to optimize the access to persistent objects in main memory via converting references to main memory pointers. In object-oriented systems, objects are referenced by their unique object identifier (OID). Each time an object is referenced on the basis of its OID the system has to localize the object in main memory by a table lookup—if it is already memory resident; otherwise, it has to be brought into the buffer.

The mechanisms for localizing objects are depicted in Figure 22.1. The top part of the picture constitutes the main memory, the bottom part the secondary storage. Thus, some of the persistent objects—e.g., the *Cuboid* id_1 and the *Vertex* instances id_{11}, id_{12}, and id_{13}—are in main memory, called *main memory resident*. Other objects, like the *Material* instance id_{77}, are not (yet) in main memory and only resident on secondary storage. Let us now investigate how an object reference is actually processed by the database system. Consider the evaluation of the following expression (assuming that the variable *myCuboid* refers to the *Cuboid* id_1):

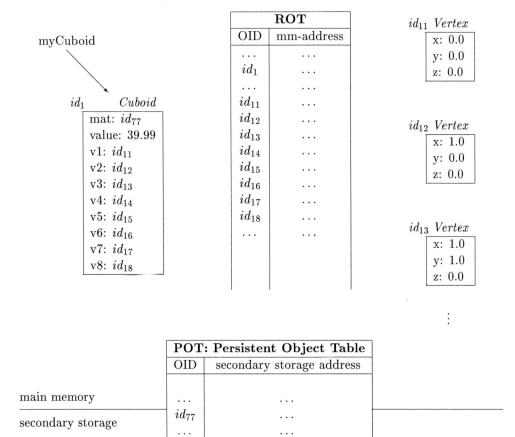

Figure 22.1: Mechanisms to Localize Persistent Objects

myCuboid.v1.x;

The object base system first determines that *myCuboid.v1* refers to the *Vertex* object identified by id_{11}. Then it has to determine whether this object id_{11} is already resident in main memory. For this purpose, the so-called *resident object table (ROT)* is inspected. In this particular state, the *Vertex* id_{11} is actually main memory resident as indicated by the entry for id_{11} in the ROT table. It can be localized by following the pointer obtained from the *mm-address* field of the ROT. The ROT table contains an entry for every object resident in main memory. It can thus grow into quite a large table. Therefore, in most systems it is maintained as a hash table in order to guarantee an efficient lookup when localizing a particular object.

The localization of an object that is not yet main memory resident is much more time-consuming, as the following discussion indicates. Consider the dereferencing induced by

myCuboid.mat.specWeight;

Again, the system determines id_{77} as the OID of the object that needs to be localized. Once again, the object base system tries to localize the object id_{77} via the ROT table. However, this time no entry id_{77} is found in the ROT, which indicates that the object is not yet resident in main memory and only resides on secondary storage. Therefore, a lookup in the *persistent object table (POT)* is required to find the secondary storage address of the object identified by id_{77}. Then the object is transferred into main memory whereupon its main memory address is inserted into the ROT table. Actually, the POT table is only required in object base systems supporting *logical* OIDs as opposed to *physical* OIDs, which refer to the secondary storage address without the additional indirection incurred by the POT table. We refer the reader to Section 7.5 for the discussion of logical versus physical object identity.

Typically the resident object table ROT is completely resident in main memory (as shown in Figure 22.1), whereas the persistent object table POT is only partially main memory resident—as indicated in Figure 22.1 by drawing the "borderline" between main memory and secondary storage through the POT table. Note that the POT table contains an entry for every object in the object base. Thus, it can assume a tremendous size. It is typically maintained as an extensible hash table on secondary storage. Thus, the table lookup in the POT may cause additional page faults.

22.2   The Goal of Pointer Swizzling

The goal of pointer swizzling is to speed up the localization of objects that are already main memory resident. In terms of our preceding example, pointer-swizzling techniques are designed to expedite the traversal from the *Cuboid* object id_1 to the *Vertex* object id_{11} during the execution of the expression

    myCuboid.v1.x;

The basic idea of pointer swizzling is to materialize the address of main memory resident persistent objects in order to avoid the lookup in the table ROT to localize the object each time it is accessed. Pointer swizzling converts database objects from an external (persistent) form containing OIDs into an internal (main memory) form replacing the OIDs by the main memory addresses of the referenced objects. This is, of course, particularly important in *computation-intensive* applications—as experienced in, e.g., the CAD/CAM application domain.

In terms of our example, this means that the object base system replaces the value id_{11} of the *Cuboid*'s attribute $v1$ by a main memory address. This main memory address constitutes a pointer leading to the main memory location of the *Vertex* object identified by id_{11}. Thus, the traversal from the Cuboid id_1 to the *Vertex* id_{11}—i.e., dereferencing of $id_{11}.v1$—does not incur the cost of looking up an entry in the ROT table.

In order to appreciate the virtues of pointer swizzling, consider a computer geometry application that successively rotates the *Cuboid* id_1—perhaps 1000 times for visualization on a computer screen. Each rotation involves dereferencing of the attributes $v1$, $v2$, ..., $v8$—possibly several times, depending on how "smart" the *rotate* operation was implemented. This example scenario indicates that the avoidance of the indirection via the ROT table promises a tremendous increase in performance of such computation-intensive applications.

22.3   Classification of Pointer Swizzling

We classify the pointer-swizzling techniques along three dimensions:

1. *In place/copy*
   Here, we distinguish whether the objects in which pointers are swizzled remain on their pages (*in place*) on which they are resident on secondary storage or whether they are copied (*copy*) into a separate object buffer.

2. *Eager/lazy*
   Along this dimension we differentiate between techniques that will swizzle *all* pointers that are detected versus those swizzling techniques that will only swizzle on demand, i.e., when the particular reference is dereferenced.

3. *Direct/indirect*
   Under direct pointer swizzling, the swizzled attribute (reference) contains a

Classification of Pointer-Swizzling Techniques		
In Place/Copy	Eager/Lazy	Direct/Indirect
in place	eager	direct
		indirect
	lazy	direct
		indirect
copy	eager	direct
		indirect
	lazy	direct
		indirect

Figure 22.2: The Three Dimensions of Pointer-Swizzling Techniques

direct pointer to the referenced in-memory object. Under indirect swizzling there exists one indirection; that is, the attribute contains a pointer to a so-called *descriptor*, which then contains the pointer to the referenced object.

The three dimensions are summarized in tabular form in Figure 22.2. In the subsequent sections, we will discuss those three dimensions in more detail.

22.4   In Place and Copy Swizzling

The secondary storage is organized in pages whose typical size is 4 K bytes. All objects are mapped onto these pages. Usually, objects are not allowed to go beyond page boundaries; that is, an object has to fit on one page. Actually, this requirement is relaxed for systems that support large objects, e.g., multimedia or image objects. In this case, a special customized organizational structure for segmenting these objects onto several pages is needed. In this discussion, however, we will assume that all objects reside on a single page.

The pages are the transport units between secondary memory and the main memory—as discussed before in Chapter 19. When an object is requested by an application—e.g., by dereferencing of an object-valued attribute—the system determines the page on which the requested object resides. This is determined by the lookup in the POT table—as indicated in Figure 22.1.[1] Then the entire page is brought into main memory, i.e., into an empty frame of the *page buffer*.

Two different possible architectures exist:

---

[1]In this picture, however, we omitted to show the page boundaries.

1. All objects remain on their original pages in the page buffer of the main memory.

2. The main memory is separated into the page buffer and an *object buffer*. The objects that are used in the application are copied from their original page into the object buffer. Thereafter, the page is subject to being swapped out of the page buffer—in order to free main memory space.

The first architecture avoids copy costs for transferring objects from the page buffer into the object buffer. The second alternative does incur these copy costs for the sake of achieving a better reference locality of the applications. If only a few objects resident on a particular page are needed in an application, then leaving the entire page in the page buffer wastes main memory space. In this case, copying the objects from the page into the object buffer increases the locality of the reference behavior of the object base application because the page with the unneeded objects will ultimately be swapped out of the page buffer.

Let us now investigate the effects of these two different main memory organizations on pointer swizzling. Under the first organizational alternative, pointers are swizzled *in place*. That is, objects whose pointers are swizzled remain on their original pages. This requires that *all* swizzled pointers contained in an entire page be restored— called *unswizzled*—when the page is written back to secondary storage. Thus, even swizzled pointers in otherwise unmodified objects have to be unswizzled at some time—provided that at least one object residing on the particular page is modified.

The object buffer architecture avoids this overhead of unswizzling pointers in otherwise unmodified objects. At the end of an application (transaction), only those objects that have been modified in the course of processing are written back to secondary storage. All other objects can be discarded or remain in the object buffer (for further access). Thus, only the swizzled pointers contained in modified objects have to be unswizzled.

In Figure 22.3, the architecture consisting of page buffer and object buffer is depicted. In this picture, we show two pages currently residing in the page buffer:

- One page inhabited by the *Cuboid* id_1 and the *Vertex* id_{11}

- The other containing—possibly among others—the *Vertex* instance id_{12}

Under *copy swizzling* the objects are moved (that is, copied) from their pages into the object buffer, which is shown in the right upper part of the picture. Pointers are swizzled only in these copies—which, in our picture, is indicated by the values $\rightarrow (id_{11})$ and $\rightarrow (id_{12})$ for the *Cuboid*'s attributes *v1* and *v2*, respectively. Here, we assume that $\rightarrow (id_{11})$ and $\rightarrow (id_{12})$ represent swizzled pointers, i.e., the main memory addresses of the *Vertex* objects id_{11} and id_{12}.

Let us reconsider the advantage of copy swizzling. Assume the only modification of the objects is the invocation of geometrical transformations, i.e., *scale*, *translate*, and

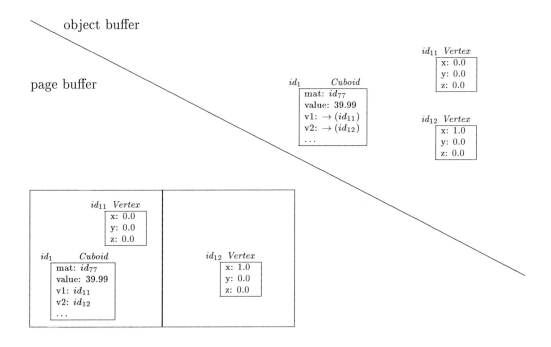

Figure 22.3: Architecture of Copy Swizzling

rotate. Then the *Cuboid* id_1, which contains the swizzled pointers, remains invariant. Thus, at the end of the application only the modified *Vertex* instances have to be written back to secondary storage. For this purpose, they have to be copied back onto their pages—which may still be in the page buffer or have to be brought back into the page buffer from the secondary storage. Only swizzled pointers in modified objects have to be restored. In our example, the modified objects, i.e., the *Vertex* instances identified by id_{11} and id_{12}, do not even contain any swizzled pointers since they comprise only atomic attributes.

Let us contrast this with *in-place swizzling*. In this case, the objects remain on their pages and pointers are swizzled in place. Then the references id_{11} and id_{12} are swizzled into main memory pointers in the original copy of the *Cuboid* object id_1. At the end of the application, all pages that contain modified objects have to be written back to secondary storage. This makes it necessary—in our example—for the swizzled pointers contained in the *Cuboid* id_1 to be unswizzled, since the page inhabiting id_1 contains also the modified *Vertex* id_{11}. Only after unswizzling the pointers can the page be transferred to secondary storage.

22.5   Eager and Lazy Swizzling

Eager swizzling guarantees that all the pointers in main memory are swizzled. An object is an aggregation of pointers and other bytes representing atomic values. When an object is loaded from disk, the object is "scanned through" and all pointers the object contains are immediately swizzled.

In contrast to eager swizzling, *lazy* swizzling swizzles pointers only on demand; i.e., a pointer is not swizzled until the object it refers to is accessed via this particular pointer. The advantage of lazy swizzling is that no pointers are swizzled in vain.On the negative side, lazy swizzling has to handle two different kinds of pointers at run time—swizzled and nonswizzled pointers. Checks need to be included that determine the state of a pointer (swizzled or not swizzled) each time an object is accessed. Hence, eager swizzling performs more efficient object accesses, whereas lazy swizzling is preferable in case the application has poor locality; i.e., a low percentage of pointers that are loaded into main memory are actually used to access the objects they reference.

22.6   Direct and Indirect Swizzling

Depending on whether it is possible to swizzle a pointer that refers to an object that is not (yet) main memory resident, we distinguish *direct* from *indirect* swizzling. Direct swizzling requires that the referenced object is resident. A directly swizzled pointer contains the main memory address of the object it references. The problem with direct swizzling is that in case an object is displaced from the page or object buffer—i.e., is no longer resident—all the directly swizzled pointers that reference the displaced object need to be unswizzled. In order to unswizzle these pointers, they are registered in a list called *reverse reference list (RRL)*.[2] Figure 22.4 illustrates the scenario of direct swizzling.

Note that in case of eager direct swizzling, we are not allowed to simply unswizzle the pointers, as eager swizzling guarantees that all pointers in the buffer are swizzled—instead, we have to displace those pointers (i.e., their "home objects"), too. This may result in a snowball effect—however, in this presentation we will not investigate this effect in detail.

Maintaining the RRL can be very costly; especially in case the degree of sharing of an object is very high. In our context, the degree of sharing can be specialized to the *fan-in* of an object that is defined as the number of swizzled pointers that refer to the object. Assume, for example, an attribute of an object is assigned a new value. First, the RRL of the object the old value of the attribute referenced needs

---

[2]In the RRL the OID of the object and the identifier of the attribute, in which the pointer appears, is stored—we say that the *context* of the pointer is stored.

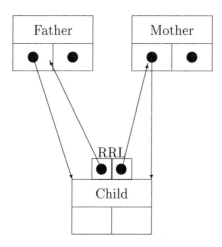

Figure 22.4: Direct Swizzling and the RRL

to be updated. Then the attribute needs to be registered in the RRL of the object it now references. Maintaining the RRL in the sequence of an update operation is demonstrated in Figure 22.5, in which an attribute, say, *spouse*, of the object *Mary* is updated due to a divorce from *John* and subsequent remarriage to *Jim*. First, the reverse reference to the object *Mary* is deleted from the RRL of the object *John*; then a reverse reference is inserted into the RRL of the object *Jim*.

Indirect swizzling avoids this overhead of maintaining an RRL for every resident object by permitting to swizzle pointers that reference nonresident objects. In order to realize indirect swizzling, a swizzled pointer materializes the address of a *descriptor*— i.e., a placeholder of the actual object. In case the referenced object is main memory resident, the descriptor stores the main memory address of the object; otherwise, the descriptor is marked invalid. In case an object is displaced, the swizzled pointers that reference the object need not be unswizzled—only the descriptor is marked invalid. Figure 22.6 illustrates this (the dashed box marks the descriptor invalid).[3]

In order to collect garbage, every descriptor keeps a counter, that counts the number of indirectly swizzled pointers pointing to the descriptor. Maintaining this counter is much cheaper than maintaining the RRL when executing updates or swizzling and unswizzling of pointers. On the other hand, indirect swizzling induces an additional overhead over direct swizzling when it comes to simple object lookups. Indirect swizzling leads to an additional indirection due to the descriptor and a residency check—a check whether the descriptor is valid whereas for direct swizzling the information that an object is resident is coded in the swizzled pointer.

---

[3]In case of eager indirect swizzling, we need to provide a special pseudo-descriptor for *NULL* and dangling references. This pseudo-descriptor is always marked invalid.

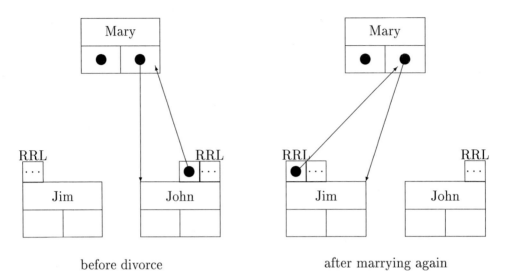

before divorce   after marrying again

Figure 22.5: Updating an Object under Direct Swizzling

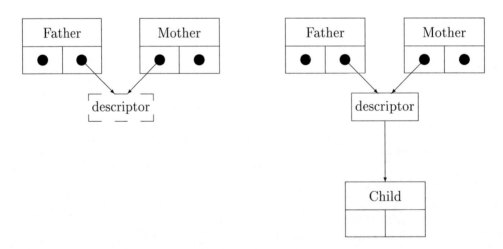

Figure 22.6: The Scenario of Indirect Swizzling

22.7   Exercises

22.1   Discuss the eight different pointer-swizzling techniques on our *Cuboid* example
of Figure 22.1. Illustrate your discussion by graphically sketching the respective organization of the main memory and the swizzled pointers.

22.2   Try to characterize application profiles that are particularly amenable to the
different pointer-swizzling techniques. That is, try to come up with a metric
that determines which of the eight pointer-swizzling techniques is superior for
given application profiles.

22.3   In Exercise 22.2 you experienced that no *one* pointer-swizzling technique is
superior for *all* application profiles. Therefore, it appears advantageous to design an *adaptable* system that can "switch" between different pointer-swizzling
strategies. Sketch the architecture of such an adaptable parameterized system.

22.8   Annotated Bibliography

There exist strong dependencies between pointer swizzling and other system components. Moss [Mos92] is the first who abstracted from system and/or object model characteristics and undertook a systematic—though incomplete—classification of pointer-
swizzling techniques. The classification of pointer-swizzling techniques used in this
chapter was devised by Kossmann in his Master's thesis [Kos91]. An adaptable system that allows to switch between different pointer swizzling strategies was proposed
by Kemper and Kossmann [KK93].

Various forms of pointer swizzling have—by now—been implemented in several
persistent object managers. Among the first that employ certain forms of pointer
swizzling are LOOM [KK83] and PS-Algol [CAC+84]. Also, many recently developed
object-oriented database systems incorporate pointer swizzling. Indirect swizzling is
employed in LOOM [KK83] and in the Orion object manager [KBC+88]. A similar
technique called node marking can be found in [HM90]. In Orion [KBC+88], an architecture is proposed that divides the system buffer into two parts: a page buffer
and an object buffer. Main memory resident persistent objects are copied from the
page buffer into the object buffer and, in the process, the interobject references are
converted (swizzled) into main memory pointers with one indirection. That is, the
pointer refers to a so-called resident object descriptor (ROD), which, in turn, points
to the referenced in-memory object. This method is referred to as indirect swizzling
in our terminology. EXODUS supports two forms of pointer swizzling: In version
EPVM 1.0 [SCD90] swizzling is restricted to local variables only. In the most recent
version, EPVM 2.0 [WD92], a more elaborate swizzling technique is incorporated

and is similar to Orion with respect to copying swizzled objects into a separate object buffer. However, the indirection via the ROD is omitted in EXODUS—thereby, however, eliminating the possibility of replacing a swizzled object from the buffer before the end of the application. In ObjectStore [LLOW91], all references are persistently stored similar to their swizzled form. As soon as a page is brought into main memory, all the pointers on the pages are swizzled. Eager direct swizzling is used, but the swizzled references may not be valid main memory addresses. Instead, the memory management unit of the underlying workstation is used to detect invalid pointers. Upon dereferencing an invalid pointer, an interrupt occurs for which ObjectStore provides a special interrupt handler that then brings the necessary page into main memory. After swizzling the pointers on that page, the original dereferencing can continue. Obviously, in this approach, the addresses used have to correlate to virtual memory addresses. That is why this approach is called *memory mapping*.

Part VI

Sample Systems

23

GemStone: An Object System Based on Smalltalk

GemStone is a system consisting of an "object server" and multiple client processes—which may be running on remote workstations—accessing the objects managed by the server. The GemStone system supplies tools that enable the following:

- Definition and creation of objects in the server

- Communication of client stations with the central server

- Concurrency control of parallel client processes accessing shared data

Among others, GemStone supports interfaces for client applications written in the programming languages C and Smalltalk-80.

GemStone's integrated data definition, access, and manipulation language, called OPAL, is the subject of this chapter.

23.1   Class Creation

The GemStone object model is very closely related to the Smalltalk-80 model. The *class* concept plays the predominant role in data modeling. Every GemStone object is a (direct) instance of exactly one *class*. The *class* concept combines the two concepts *object type* and *extension*—treated separately in the generic object model (GOM). A class specifies

- The internal representation of its instances

- The methods (operations) that are provided to query and manipulate the internal representation

The general format for defining a new class is outlined below:

(1)   ⟨SuperClass⟩ **subclass:** '⟨SubClass⟩'
(2)   **instVarNames:** #(⟨instance variable names⟩)
(3)   **classVars:** #(⟨class variable names⟩)
(4)   **poolDictionaries:** #[ ]
(5)   **inDictionary:** ⟨dictionary-name⟩
(6)   **constraints:** #(⟨list of constraints⟩)
(7)   **isInvariant:** true | false.

The first line (1) specifies the superclass of which the created class is to be a subclass.
Every user-defined class is the subclass of exactly one superclass; if no other class is
suitable as a superclass, the most general class *Object* has to be chosen as superclass.
Line (2) lists the *named* instance variables—attributes in our terms—that every instance of the class possesses. Aside from the named instance variables subclasses of
the built-in class *Collection* (cf. Section 23.8) provide two other kinds of storage slots,
called *unnamed* instance variables:

- *Indexable instance variables*
   These instance variables can be accessed by their position very much like elements of a list or array.

- *Anonymous instance variables*
   They can only be accessed *associatively*, that is, by predicates based on their
   value.

Unnamed instance variables are used to aggregate an unknown number of objects
within an instance. The number of unnamed instance variables can dynamically
grow and—due to deletion—shrink. In line (3) class variables are listed if the class
provides any. Class variables—contrary to instance variables—are shared by all instances of the respective class. Line (4) contains an array of optionally specified
pool dictionaries. These dictionaries hold so-called pool variables that can be shared
across different classes. They serve as global variables to maintain information that
is relevant to many otherwise unrelated classes and objects. Line (5) determines the
dictionary by which the class definition is accessible. Typically, there exists a system dictionary for those class definitions that are shared by all users—mostly called
UserGlobal—and, in addition, for each user a private dictionary containing more specialized class definitions. The **constraints** clause in line (6) contains any constraints
that limit the type of a particular instance variable. By default, GemStone does not
statically restrict the types of any data components. In certain situations, however, it
is necessary for the sake of performance to require instance variables to be instances
of a certain class (cf. Section 23.11). The last line (7) states whether instances can

be modified once they have been created (**isInvariant:** *false*) or not (**isInvariant:** *true*).

Let us now demonstrate the class creation on a well-known example: a two-dimensional *Vertex*.

> Object **subclass:** 'Vertex'
> **instVarNames:** #('xValue' 'yValue')
> **classVars:** #('dimension')
> **poolDictionaries:** #[ ]
> **inDictionary:** UserGlobal
> **constraints:** #( )
> **isInvariant:** false.

It is a GemStone—and Smalltalk—convention to denote classes with identifiers starting with a capitalized letter; instance and class variables as well as method names and parameters start with a small letter. *Vertex* is being defined as a subclass of *Object*, the most basic built-in class. Instances of type *Vertex* contain two instance variables: *xValue* and *yValue*. In addition, *Vertex* contains one class variable: *dimension*. While each instance of *Vertex* will have storage slots for *xValue* and *yValue*, the variable *dimension* exists just once; it is associated with the class and not with every instance. The class variable *dimension* is intended to be set to 2—once and for all.

The instance variables of an object, whether named or unnamed, are not accessible from outside the object. GemStone requires absolute information hiding in this respect.[1] Therefore, an instance of *Vertex*, which could be created as

> someVertex := Vertex new.

is of no use. The instantiation "*Vertex new*" merely returns an empty "container" with all instance variables set to **nil**, the GemStone symbol for an undefined reference. With the class definitions provided so far it is not possible to modify any of the internal instance variables of *someVertex*, or even to read them.

Before we can define operations to modify and query objects, we have to outline the general idea of message passing—operation invocation in our terms—in the next section.

23.2 Message Passing

In GemStone all actions are invoked by *message passing*. Message passing requires that some object, say, o_1, sends a message to another object o_2. The object o_1 is

---

[1]There is one exception to this rule, though. When using indexes to efficiently retrieve members of a collection one directly accesses the instance variables, as will be demonstrated in Section 23.11.

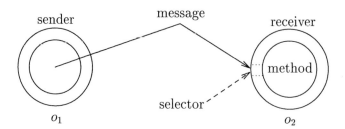

Figure 23.1: Message Passing in GemStone

naturally called the *sender*, o_2 is the *receiver*. Figure 23.1 highlights this description graphically. The message that o_1 sends has to be understood by the receiver: o_2 has to contain a *selector* that is compatible with the received message. Depending on the class of which o_2 is an instance a number of methods (operations) are applicable on o_2. Each method has a header, called the *selector*, by which it is uniquely determined (within o_2). We say that object o_2 can *respond* to a message if it contains a method with a selector that matches the message format. This is roughly analogous to the operations in the generic object model (GOM); an operation on some instance could only be invoked if the interface of the corresponding type supplied this operation. One could compare the sending of a message with the invocation of an operation and the method with the implementation of an operation.

All methods have a return value that is sent back to the sender. If none is specified, the receiver object is implicitly determined as the return value. In this respect a method is comparable to a function that may, however, produce some side effects. The side effects are limited to the receiver, though. Parameters that are passed within the message cannot be modified directly—only by explicitly sending a message to the according parameter allows for its modification.

23.3   Methods

23.3.1   Method Definition

The methods are, as pointed out before, the concrete implementations that can be invoked by a message sent to an instance. A method definition has the following syntactical framework:

> **method**: ⟨classname⟩
>   ⟨selector⟩
>   "|" ⟨local variables⟩ "|"
>   ⟨method implementation⟩
> "%"

Here, ⟨classname⟩ determines the class to which the subsequently specified method is added. The ⟨selector⟩ specifies the message format by which this method can be invoked. ⟨local variables⟩ is a list of variables whose scope is restricted to the implementation of the method. The special symbol "%" marks the end of the method implementation.

We will demonstrate the implementation of a method on two easy examples: The methods x and y of class *Vertex* return the *xValue* and the *yValue*, respectively. Their implementation could be specified as follows:

```
method: Vertex                    method: Vertex
    x                                 y
   | |                               | |
   ↑xValue                           ↑yValue
%                                 %
```

These two methods have no parameter. Their only argument is the receiver object. The return value of a method follows the special symbol "↑". For example, the clause "↑*xValue*" states that the instance variable *xValue* is the return value of the method x. We see that instance variables can be accessed inside the method implementation—but only those that belong to the class within which the method is defined.

These methods are invoked by merely sending a message of the appropriate name to an instance of *Vertex*, e.g.,

    someNumber := someVertex x.

The receiver, in this case *someVertex*, always stands to the left of the message, here x. Since *someVertex* was—so far—not initialized, the return value would be **nil** to which *someNumber* will subsequently refer.

Let us now consider methods that require arguments in addition to the receiver. In order to set the instance variables *xValue* and *yValue* to (numeric) objects that are passed as arguments, one could include the following two methods in the class *Vertex*:

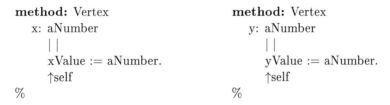

```
method: Vertex                    method: Vertex
   x: aNumber                        y: aNumber
     | |                               | |
     xValue := aNumber.                yValue := aNumber.
     ↑self                             ↑self
%                                 %
```

Like in GOM, the variable *self* is implicitly defined in every method and refers to the receiver of the message that selects the respective method. A GemStone method with arguments can be viewed as having no name. Rather, the formal parameter names serve as the selector of the method. Each formal parameter has a name followed by a colon (:) followed by an identifier that refers to the actual argument value that will be passed in the message. In our example, *x:* is the selector of the method and *aNumber* is the identifier referring to the actual parameter. It is the convention in Smalltalk-like languages to characterize the "expected" type of an argument by an appropriately chosen name, e. g, *aNumber*, *aDog*, *aVertex*, etc.

The assignment of 3 to the *x*-coordinate of *someVertex* can be invoked by the following message:

    someVertex x: 3.

The entire set of methods defined for a particular class is often called the class's *protocol*.

23.3.2   Message Formats

The two types of methods, i.e., no-parameter and *n*-parameter methods, lead to two different message formats. The no-parameter methods are invoked by their name, i. e.:

    ⟨receiver object⟩ ⟨method name⟩

A method with parameters does not really have a name. It is identified by its parameter names. A method with *n* formal parameters is invoked by sending a message of the following format:

    ⟨receiver object⟩ ⟨param$_1$⟩: ⟨object$_1$⟩ ··· ⟨param$_n$⟩: ⟨object$_n$⟩.

⟨param$_1$⟩, ..., ⟨param$_n$⟩ are the parameter names which are followed by the actual arguments ⟨object$_1$⟩, ..., ⟨object$_n$⟩. Binary operations, like $+$, $-$, etc., are treated in a special way. For the sake of simplicity of usage—and to conform to the conventional programming language and mathematical notation—the binary operations are defined as "normal" infix operators, e.g.,

    myNumber + yourNumber.

adds the two numerical objects. However, one should keep in mind that the GemStone message-passing paradigm leads to an execution of an expression from left to right; all operators have lower precedence than the left to right rule. For example,

    myNumber + yourNumber * hisNumber.

has the meaning

    ((myNumber + yourNumber) * hisNumber).

23.4 Class Methods

Aside from messages that are sent to instances of a class we can also have methods associated with the class. These are invoked by sending the corresponding message to the class name. For example, the initialization of the class variable *dimension* should be done at some time. This is achieved by invoking the method *initializeDimension* that may have been implemented as follows:

```
classmethod: Vertex
    initializeDimension
        ||
        dimension := 2.
        ↑self
%
```

The following message passing protocol will carry out the initialization

```
Vertex initializeDimension.
```

23.5 Instantiation

Having gained an understanding of the difference between methods that operate on instances and class methods, we can now review the instantiation of an object. The message *new* is always sent to a class—it couldn't possibly be sent to the instance that one has yet to create. Every class inherits a built-in method *new*, which returns just an appropriate empty container as an uninitialized instance. Therefore, in Section 23.1 the message

```
someVertex := Vertex new.
```

could be responded to without having explicitly defined the method *new*.

23.5.1 Initialization during Instantiation

Often it is disadvantageous to instantiate an object without—at the same time—initializing at least some of its instance variables. Consider, for example, the creation of the vertex (2.5, 3.5). Using our class protocol defined so far we could proceed like

```
myVertex := Vertex new.
myVertex x: 2.5.
myVertex y: 3.5.
```

This procedure has two disadvantages: It is lengthy, and inadvertently the user may leave one coordinate uninitialized.

A more elegant way that always guarantees to initialize the instance variables would be the definition of a class method that integrates the instantiation and initialization of the new object. The following class method "*newX: y:*" provides this functionality (comments are bounded by quotation marks):

```
classmethod: Vertex
    newX: aNumber1 y: aNumber2            "the selector"
        | newVertex |
        newVertex := self new.
        newVertex x: aNumber1.        "instance method x:"
        newVertex y: aNumber2.        "instance method y:"
        ↑newVertex
        "Possible Abbreviation: ↑(((self new) x: aNumber1) y: aNumber2)."
%
```

Now we could create the object *myVertex* as follows:

```
myVertex := Vertex newX: 2.5 y: 3.5.
```

23.6   Object Identity versus Object Equality

GemStone—like GOM—enforces a strict object identity concept. Once an object has been created through instantiation, it is assigned an invisible identifier that remains invariant throughout its lifetime.

Based on the object identity concept two categories of comparison predicates are distinguished in GemStone:

- Equality testing using the predicates = (equal) and ~= (not equal)

- Identity testing using the predicates == (identical) and ~~ (not identical)

Let us explain the difference on an example application:

```
myVertex := Vertex newX: 2 y: 3.
yourVertex := Vertex newX: 2 y: 3.
```

Here, *myVertex* and *yourVertex* are really two different objects that happen to have the same value. In this situation the comparison

```
(myVertex = yourVertex)
```

returns *true*, while the identity test

(myVertex == yourVertex)

returns *false* because the two objects are distinct, which is tested by:

(myVertex ~~ yourVertex)

which—consequently—returns *true*. An assignment, such as

myVertex := yourVertex

always has the effect that the component on the left is associated with the object on the right of the ":=" sign. Therefore, the subsequent test for identity

(myVertex == yourVertex)

now returns *true* because *myVertex* and *yourVertex* refer to the same object. Therefore, a modification of this object through *myVertex*, e.g.,

myVertex x: 5.

becomes also visible through *yourVertex*, i. e.,

((yourVertex x) = (myVertex x))

returns *true* because both messages return 5.

23.7   Type Hierarchy

GemStone—just like Smalltalk—provides single inheritance for structuring the class hierarchy. Every user-defined class is a direct subclass of one (and only one) superclass. This is enforced by the definition procedure: A new class is defined by sending the message **subclass:** to its superclass. The message **subclass:** is understood by (almost) all classes. If a user-defined class has no obvious superclass, it inherits at least all features of the built-in class *Object*. *Object* is the root of all super/subclass hierarchies.

Thus, the GemStone type structure consists of one expandable tree with the root *Object*. Every user-defined class has to be fitted into this tree as a child of some existing node (class). This is highlighted in Figure 23.2. A class, say, *BasicStuff*, is created as the subclass of *Object* as follows:

Object **subclass**: 'BasicStuff'
    ...   "remaining class definition"

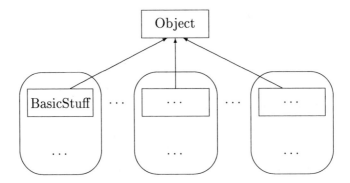

Figure 23.2: *Object* as the Root of All Class Hierarchies

This leads to the creation of a new subtree below the node *Object*.

A subclass inherits not only all instance, class, and pool variables of its superclasses but also the methods defined in its ancestors. This holds for class methods as well as for instance methods. Just like in GOM an inherited method may be redefined. Whereas type-safe models with static type consistency verification, like GOM, impose restrictions on the legal refinements, the GemStone system allows any refinement. An incompatibility is thus detected only at run time when, e.g., a return value of a message does not conform to the expected object—potentially leading to all kinds of catastrophic run-time errors.

23.8   Collection Classes

GemStone provides built-in support for managing collections of objects by the predefined *Collection* class and its subclasses. A *Collection* instance can store groups of objects—not necessarily of the same type—in indexable or anonymous storage slots. The diagram of Figure 23.3 outlines (part of) the class hierarchy of collection classes. *Collection* itself is an abstract (virtual) class in the sense that it factors out the common protocol of all collections. However, *Collection* itself does not give rise to useful direct instances, because some essential methods are provided only by its subclasses. *Bag*, *Set*, *Array*, and *String* are classes that yield meaningful instances with all the functionality needed to access the individual objects. A *Bag* instance—contrary to a *Set* instance—may contain the same object several times.

Instances of, e.g., *Array* and *Set* are created as follows:

```
myVertexArray := Array new.
myVertexSet := Set new.
```

Let us briefly survey the most important methods that the abstract class *Collection*

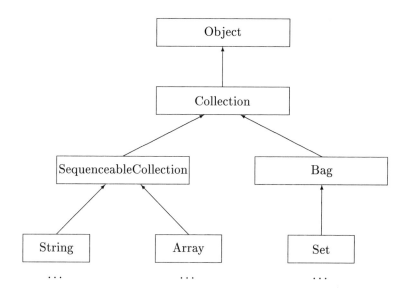

Figure 23.3: Class Hierarchy of Collection Classes

implements, i.e., the set of basic methods, that is common to all of its descendants.

23.8.1   Testing of Membership

The method *includes*: *anObject* tests whether the receiver collection contains the object *anObject*. For example,

myVertexSet includes: myVertex.

returns *true* if *myVertex* is contained in *myVertexSet*, otherwise *false*.
   An empty *Collection* instance may be tested for with the method *isEmpty*, e.g.,

myVertexSet isEmpty.

   The method *occurrencesOf*: *anObject* returns the number of times the argument object is contained in the collection, e.g.,

timesContained := myVertexArray occurrencesOf: myVertex.

Applied to a *Set* instance, such as *myVertexSet*, *occurrencesOf* always returns either 0 or 1.

23.8.2   Enumerating the Elements

There are several methods that allow to iterate over all elements of a collection and execute a block—i.e., a sequence of statements—on each member object.

For example, the method *do: aBlock* performs the program specified in *aBlock* on every member. A block is a GemStone program enclosed in square brackets. In general, a block has the syntactical form

"[" ⟨parameter list⟩ "|" ⟨code⟩ "]"

The parameter list may contain parameters—each one preceded by a colon—that are passed to the block when invoked. All blocks that are applied in the enumeration messages of *Collection* instances have exactly one parameter. This parameter serves as "cursor" over the *Collection* instance; i.e., it references the elements of the receiver collection one by one.

For example,

myVertexSet do: [:aVertex | aVertex x: 0].

sets the *xValue* of each member of the set *myVertexSet* to 0. Here, *aVertex* is the parameter of the block by which the elements of *myVertexSet* are referenced.

Another subcategory of enumeration methods loops though the elements of a collection and returns those objects that satisfy a specified predicate. These enumeration methods are used for querying the object base.

The method *detect: aBlock* iterates through the receiver (a *Collection*) and returns the first member found for which *aBlock* evaluates to true. For example,

myVertexSet detect: [:aVertex | (aVertex x) = 0]

returns the (first) *Vertex* element whose *xValue*—that is retrieved by *aVertex x*— equals 0. If none of the elements of the *Collection* satisfies the condition the statement generates an error. This can be avoided by using the method

detect: aBlock ifNone: exceptionBlock

where the *exceptionBlock* is executed only if *aBlock* evaluated to *false* for all members of the receiver.

A typical database application is to retrieve all elements of a *Collection* that satisfy a search condition. The *select: aBlock* method is used for this purpose. For example, the selection of all vertices of *myVertexSet* that have a distance of less than 10 from the origin is achieved as follows (*square* and *sqrt* are built-in operations with obvious semantics):

nearToOrigin :=
    myVertexSet select: [:v | ((v x square) + (v y square)) sqrt < 10]

The *select:* method always returns a *Collection* instance of the same kind as the receiver. Thus, *nearToOrigin* is a *Set* of *Vertex* instances that are close to the origin. If applied to an *Array* instance, e.g.,

nearToOrigin :=
    myVertexArray select: [...].

the *select:* method would return an *Array* of *Vertex* instances.

23.8.3   Arrays

The subclass *SequenceableCollection* and, in particular, its subclass *Array* furnish additional methods for indexed access to its elements. *Array* is, like *String*, a subclass of *Collection*'s subclass *SequenceableCollection*, as can be observed in Figure 23.3.

We will briefly describe the most important methods to insert elements into an *Array* and to retrieve particular elements by their position from an *Array*. Let us create a new *Array* associated with the variable *myVertexArray*:

> myVertexArray := Array *new.*

A GemStone *Array* is dynamic in the sense that it can dynamically grow and shrink depending on the number of elements inserted (there is no fixed upper bound).

An object is added to the end of an *Array* instance using the *addLast: anObject* method, e.g.:

> myVertexArray addLast: myVertex.

This causes the size of the array to be increased by one.

An entry at position *n* of an *Array* may be overwritten as follows:

> myVertexArray at: n put: myVertex.

This inserts (a reference to) the object *myVertex* at the nth position of the *Array*.

Elements of an *Array* can be accessed by their position by sending the message *at: aPositiveInteger*. For example, the statement

> fifthVertex := myVertexArray at: 5.

returns the fifth element and associates it with the variable *fifthVertex*.

The method *deleteObjectAt: aPositiveInteger* is used to remove an element from an *Array*. For example,

> myVertexArray deleteObjectAt: 5

causes the removal of the fifth element and a left shift of all indexes starting at 6 by one position. Also, the *size* of the receiver, which is a method to query the number of elements in a *Collection*, is reduced by 1.

23.8.4   Bags and Sets

Contrary to *SequenceableCollection* instances nonsequenceable *Collections* (*Bag* and *Set* instances) do not maintain any order on their elements. The difference between a *Bag* and a *Set* is that instances of *Bag* may contain the same object in several occurrences, whereas a *Set* contains an element just once—even though it might have been inserted several times.

Elements are inserted using the *add:* method, e.g.,

> myVertexSet add: myVertex

inserts (a reference to) *myVertex* into the respective *Set*.

23.9   Typing

Every object in a GemStone database belongs to exactly one *direct* class. Each object contains information that allows to deduce its direct type. This may (internally) be realized by maintaining a reference from the object to its direct class. An object's type can be queried by the method *class*. For example,

> myVertex class.

returns *Vertex*.

However, GemStone is not strongly typed; rather, the type of a database component is dynamic and unconstrained and has to be determined at run time. Later on, in Section 23.11, we will see that it is, however, possible and sometimes necessary to state type constraints within class definitions that are verified dynamically.

Dynamic typing allows to associate any object to any database component. For example, the named instance variable *xValue* of the *Vertex myVertex* could be assigned differently typed objects. For example,

> myVertex x: 2.

causes *xValue* of *myVertex* to be assigned the *SmallInteger* (a built-in class) instance 2. A subsequent assignment, e.g.,

> myVertex x: 2.978.

may associate a *Float* instance with the *xValue* of *myVertex*. These two assignments of differently typed instances are quite meaningful. However, GemStone also allows the assignment

> myVertex x: aDog.

where *aDog* may be an instance of a user-defined class *Dog*. This assignment can then lead to a run-time type error; e.g., when trying to evaluate ($myVertex\ y\ +$ $myVertex\ x$).

In spite of such contrived examples highlighting the danger of dynamically typed models, the flexibility of dynamic typing is sometimes very beneficial. It allows the association of database components with any kind of class instance. To illustrate the usefulness of this flexibility consider the following example. We want to store, in some *Set allAssets*, the valuables that some organization possesses. Then this set *allAssets* may contain instances of entirely unrelated classes such as *Car*, *House*, *Boat*, *RealEstate*, *Plane*, *Machinery*, etc. But no predictions can be made as to whether all members of the set can actually respond to a particular message. For example, the following operation is intended to total the estimated value of all assets:

```
|total|
total := 0.
allAssets do: [:anAsset | total + (anAsset estimatedValue)].
```

It cannot be guaranteed statically—that is, without looking at the current database state—that all elements of the set *allAssets* have a selector—and thus an implementation—for *estimatedValue*. For example, one could insert *myVertex* into *allAssets*—even though it strains the authors' professional ethics:

```
allAssets add: myVertex.
```

Now the program to total the *estimatedValue* of all assets leads to a run-time error because the element *myVertex* of the set cannot respond to the message.

Dynamic typing places a heavy burden on the shoulders of the database users to always make sure that they associate database components with instances of proper classes. The typing system of GemStone provides only limited help in this respect (cf. Section 23.11). Dynamic typing is particularly dangerous if persistent database components (e.g., sets) are shared among many different users. Then, the caution exercised by one user can completely be nihilated by some other, less experienced user.

Another consequence of the flexible dynamic typing scheme is that most methods have to be bound dynamically at run time. This is necessary because there is too little static type information. Therefore, dynamically typed models are inherently harder to optimize.

23.10   Example Application

So far, we have illustrated the concepts of GemStone on a rather trivial example. Let us now expand this application in order to get an appreciation of the data modeling constructs of GemStone on a more realistic "real–life" model.

Let us introduce some of the data types and their associated functionality of very large scale integration (VLSI) design applications. As was outlined in Section 3.3 each layer of a chip consists of a collection of paths, whose intersection on different layers leads to transistors. Let us first introduce the two most fundamental classes *Polygon* and its subclass *Box*, which are used to model the geometric layout of a chip.

```
Object subclass: 'Polygon'
    instVarNames: #( 'vertices' 'function' )
    classVars: #( )
    poolDictionaries: #[ ]
    inDictionary: UserGlobal
    constraints: #( )
    isInvariant: false.
```

The definition of *Polygon* specifies that an instance possesses two named instance variables:

- *vertices*, a *SequenceableCollection* instance to hold the bounding vertices of the *Polygon* in sequential order

- *function*, a *String* detailing the use of the particular geometric shape

Having introduced *Polygon*, one could define *Box*—i.e., a rectangle—as a subclass thereof.

Polygon **subclass:** 'Box'
   instVarNames: #('center' 'layer')
   classVars: #('numberVertices')
   poolDictionaries: #[ ]
   inDictionary: UserGlobal
   constraints: #( )
   isInvariant: false.

In addition to the inherited instance variables, *vertices* and *function*, a *Box* consists of the instance variables

- *center*, intended to refer to a *Vertex* that coincides with the center of the rectangle

- *layer*, a *String* denoting the layer in the chip (i.e., diffusion, metal, silicon) in which the particular *Box* instance forms a path

To place predefined shapes on a chip design one needs to provide methods for translating and rotating two-dimensional geometrics.

```
method: Vertex
    translateByX: aNumber1 andY: aNumber2
        | |
        xValue := xValue + aNumber1.
        yValue := yValue + aNumber2.
    ↑self
%

method: Vertex
    rotateByAngle: aNumber
        | xOld |
        xOld := xValue.
        xValue := (xValue * (aNumber cos)) − (yValue * (aNumber cos)).
        yValue := (xOld * (aNumber sin)) − (yValue * (aNumber cos)).
    ↑self
%
```

Depending on the angle, the *xValue* and the *yValue* are appropriately updated.

> **method:** Polygon
>   translateByX: aNumber1 andY: aNumber2
> | |
>   vertices **do**: [:aVertex | aVertex translateByX: aNumber1
>   andY: aNumber2].
>   ↑self
> %

> **method:** Box
>   translateByX: aNumber1 andY: aNumber2
> | |
>   "first translate the bounding vertices"
>   **super** translateByX: aNumber1 andY: aNumber2.
>   "now translate the center"
>   center translateByX: aNumber1 andY: aNumber2.
>   ↑self
> %

This method implementation in class *Box* requires some explanation. Obviously it is a refinement of the method *translateByX: andY:* inherited from *Polygon*. In this implementation it turns out that it is quite useful to invoke the overwritten method that was inherited from *Polygon* in order to do part of the work, i.e., translating the bounding vertices that are associated with the *Collection vertices*. This can be achieved in GemStone by using the keyword **super**, which causes the system to start the search for an appropriate method in the direct superclass of the receiver. That is, the statement

> **super** translateByX: aNumber1 andY: aNumber2.

invokes the method defined in *Polygon* rather than the one defined in *Box*—which, by the way, would have led to an infinite recursion. Having taken care of translating the boundary points of the *Box* the *center Vertex* remains to be translated, which is achieved by passing the message *translateByX: aNumber1 andY: aNumber2* to the *Vertex* representing the center. Note that *Vertex* contains a method with the same selector, which, however, can and should not be overwritten in *Box* because the two classes are unrelated.

Following the "liberal" dynamic typing philosophy, refinement of operations is not constrained in GemStone. Thereby, a refinement can easily be defined that conflicts with the originally inherited method by, e.g., returning a differently typed object. This can then lead to a run-time error.

23.11   Indexed Associative Access

The search for an object by value, which is typically called *associative access* as opposed to accessing an object by its object identifier, may cause a high overhead on the system. Consider the following example application in which the set *myVertexSet* may contain in the order of several thousands of objects of type *Vertex*:

```
inOriginVertices :=
    myVertexSet select [:aVertex | aVertex x absolute < 0.001 &
                        aVertex y absolute < 0.001]
```

This "query" requires that for every *Vertex* instance seven messages are sent: *x*, *absolute*, $<$, &, *y*, *absolute*, and $<$. Each of the messages invokes a method that has to be dynamically looked up at run time because of the interpretative execution mode of GemStone.

In some cases, the database designer can take special care to speed up associative access in GemStone. This requires a very careful database design because several provisions have to be taken:

- The *named* instance variables on the basis of which the associative access is performed have to be constrained to a particular object type.

- The *Collection* of instances over which the query is stated is constrained to contain only instances of a specified class (or its descendants).

- An index on the named instance variable that is accessed in the search predicate has to be created.

- The query (*select:* method) has to be explicitly marked as an index-supported associative access.

23.11.1   Constraining Named Instance Variables

It is infeasible to construct an index over an attribute (named instance variable) that holds different incomparable objects for different members of the collection.

Consider, for example, the perfectly legitimate GemStone scenario that two *Vertex* instances *v1* and *v2* are created as

```
v1 := Vertex newX: true y: false.
v2 := Vertex newX: 1 y: 2.
```

These vertices could be included in the set *myVertexSet*, i.e.,

```
(myVertexSet add: v1) add: v2.
```

It is now impossible to construct an index over the x–coordinate of all instances of *myVertexSet* because (v1 x), which returns *true*, and (v2 x), which returns 1, are incomparable values.

Even though the preceding example has a contrived flavor, there are plenty of situations in which the assignment of incomparable objects to the same instance variable is quite natural. Consider, for example, a class *Person* with the named instance variable *mostValuedPossession*. Without any further restrictions, *mostValuedPossession* could be associated—for different *Person* instances—with objects of type *Dog*, *Car*, *House*, or even *Person* (for *Persons* who value their spouses as the most valuable possession).

In order to assure that named instance variables hold only comparable instances, one can specify type constraints in the class definition. This expands the class *Vertex* to

```
Object subclass: 'Vertex'
    instVarNames: #( 'xValue' 'yValue' )
    classVars: #( 'dimension' )
    poolDictionaries: #[ ]
    inDictionary: UserGlobal
    constraints: #[ #[#xValue, Float],
                    #[#yValue, Float] ]
    isInvariant: false.
```

These constraints restrict the named instance variables *xValue* and *yValue* of *Vertex* to be instances of class *Float*—or any descendant thereof. Now it is ensured that all *Vertex* instances have comparable *xValue* and *yValue* values.

23.11.2   Constraining Collection Elements

It is, however, not yet guaranteed that the set *myVertexSet* contains only *Vertex* elements. For example, it is still legitimate to include an instance of, say, *Dog*, in the set, e.g.,

```
myVertexSet add: aDog.
```

This has to be avoided in order to facilitate index creation over all elements of a collection. GemStone allows the creation of subclasses of any nonsequenceable collection class for which the element type is constrained. A nonsequenceable collection class is either the class *Bag* or any subclass thereof, such as *Set*. In our case, we could define

```
Set subclass: 'VertexSet'
    instVarNames: #( )
    classVars: #( )
    poolDictionaries: #[ ]
```

> **inDictionary:** UserGlobal
> **constraints:** Vertex
> **isInvariant:** false.

This definition restricts an instance of *VertexSet* to contain only *Vertex* instances, or—if applicable—instances of a descendant of *Vertex*. Now, *myVertexSet* could be defined as an instance of *VertexSet*:

> myVertexSet := VertexSet *new*.

The attempt to insert an object whose type is not *Vertex* (or one of its subtypes) will be rejected—but only at run time and not at compile time as in GOM.

23.11.3   Index Creation

An index is constructed by sending either of two messages:

- *createIdentityIndexOn:*
  This message initiates the creation of an index that supports predicates based on object identity.

- *createEqualityIndexOn:*
  An index for predicates comparing the value (state) of the object is created.

For example, two equality indexes can be created for the *Collection myVertexSet* as follows:

> myVertexSet createEqualityIndexOn: 'xValue'.
> myVertexSet createEqualityIndexOn: 'yValue'.

From now on the access to elements of the set *myVertexSet* that is based on comparing values of *xValue* and *yValue* is supported under the condition that the queries are stated by obeying a special syntax (cf. Section 23.11.5).

23.11.4   Indexing over a Path

In the foregoing example the index was created over an atomic attribute of members of the collection. It is also possible to index over a path expression. A path expression links an object o_1 via a named instance variable v_1 with another object o_2. From o_2 another instance variable v_2 may lead to a third object o_3, and so on. Then the path expression has the form

$$o_1.v_1.v_2.\cdots.v_n$$

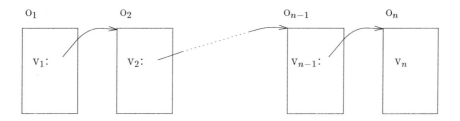

Figure 23.4: Graphical Representation of a Path Expression

Here, v_i for $(1 \leq i \leq n)$ has to be a named instance variable of the object referenced by $o_1.v_1. \cdots. v_{i-1}$. Such a path is pictured in Figure 23.4.

To clarify this concept we will introduce the new class *Square*:

    Object **subclass:** 'Square'
       instVarNames: #('center' 'corner')
       classVars: #('dimension')
       poolDictionaries: #[ ]
       inDictionary: UserGlobal
       constraints: #[#[#center, Vertex],
                  #[#corner, Vertex]]
       isInvariant: false.

A square is described by its *center* and one *corner*. Both instance variables are of type *Vertex*, which is listed in the **constraints** clause. In order to store instances of *Square* in an indexed collection, we could create a subclass of *Set* that is constrained to contain members of type *Square* only:

    Set **subclass:** 'SquareSet'
       instVarNames: #( )
       classVars: #( )
       poolDictionaries: #[ ]
       inDictionary: UserGlobal
       constraints: Square
       isInvariant: false.

An instance, say *manySquares*, can be created as

    manySquares := SquareSet new.

Access support on the *center* coordinates of *Square* requires an index over a path expression. In our case, it could be created as:

> manySquares createEqualityIndexOn: 'center.xValue'.
> manySquares createEqualityIndexOn: 'center.yValue'.

The creation of an index over a path expression requires that all instance variables occurring in the path have been constrained in their respective class. The *Square* example also allows to demonstrate the use of an *identity index*. Equality indexes support the retrieval of objects based on comparing the *value* of the last-named instance variable in the path expression, e.g., the *xValue* in "*center.xValue*".

Identity indexes, on the other hand, speed up the retrieval of objects based on an identity comparison in the search condition. For example, one could create the following identity index on *Square*:

> manySquares createIdentityIndexOn: 'center'.

This index would support a search of the form (the GemStone query is formulated in the next section):

> "Find the elements of *manySquares* whose center is identical to some given *Vertex*."

23.11.5 Formulating the Search

An index-supported query has to be explicitly marked in GemStone. Let us work out the example: Find the *Squares* whose *center* is in the origin.

> manySquares select: {:s | (s.center.xValue = 0.0) &
>   (s.center.yValue = 0.0)}.

Syntactically, the argument block of the *select:* message is enclosed in braces—as opposed to square brackets—in order to let the system know that an index should be used to execute the search condition. Also, contrary to a "normal" select method's condition block, the index-supported block directly accesses the named instance variables in the path. Somehow the information-hiding concept of GemStone that allows it to access objects only via messages is circumvented. That is, GemStone treats the block in an index-supported *select:* query as a simple, nonprocedural expression that performs no message passing. The normal *select:* expression for our query could be stated as

> manySquares select: [:s | (s getCenter x = 0.0) &
>   (s getCenter y = 0.0)].

Here, we assume that *getCenter* is an appropriately defined method in *Square* that returns the *center* variable. This statement cannot utilize the index, though.

The query that utilizes the identity index on *manySquares* is formulated analogously:

manySquares select: {:s | s.center == myVertex}.

where *myVertex* is some *Vertex* instance. This *select:* expression returns only those *Squares* whose center was set to *myVertex*; i.e., the *center* and *myVertex* have to refer to the same object. Having identical *xValue* and *yValue* coordinates is not sufficient to satisfy the search condition.

Note that requiring a different syntax for index-supported query evaluation violates data independence. Ideally, the database user should not have to worry about any existing indexes. Moreover, the GemStone strategy requires rewriting existing applications when an index is dropped from or added to the database. Consequently, application program are not independent of the underlying physical object base representation.

23.12 GemStone's Transaction Control

Since GemStone objects can be shared among different users, some concurrency control mechanism has to be provided that prevents an inconsistent state of the database due to concurrent actions by different users. GemStone's primary concurrency control mechanism is an optimistic one—which is briefly described here (for more details we refer to Chapter 15, in particular Section 15.3).

Sets of actions, read and write operations, on the object base can be encapsulated in a transaction. A transaction can always be **aborted** without creating any side effects. This is achieved by establishing a private *workspace* for each transaction. The workspace constitutes a (logical) private copy of the shared object base which is created at the begin of the transaction (BOT). Therefore, any changes of the object base caused by other, concurrent transactions will not become visible by your own transaction. Any modifications are performed on the transaction's workspace. Only the initiation of a **commit** of a transaction attempts to merge the private workspace with the global object base. A **commit** may fail if a conflict with some other, already committed transaction is detected (recall Section 15.3).

The following messages are issued to abort and commit a transaction, respectively:

System abortTransaction.
System commitTransaction.

These messages are sent to the built-in kernel class *System*, which implements the functionality to detect any conflicts with other transactions.

Transactions are automatically (and implicitly) begun after any of the following:

1. The user logs in

2. A previous transaction is aborted

3. A transaction is committed successfully

Therefore, no message has to be sent to *System* to begin a new transaction. For the appropriateness of optimistic concurrency control for engineering applications the reader is again referred to Section 15.6.

23.13 Exercises

23.1   Expand the example object base schema for VLSI design that was started in the text by further object class definitions.

23.2   Illustrate the differences among *loose typing* (as in LISP), *dynamic typing* (as in GemStone and Smalltalk-80), *strong typing* (as in GOM), and strict *static typing* (as in Pascal). Illustrate your discussion by way of examples. Discuss the pros and cons of the four different typing philosophies.

23.3   Why is it impossible to detect the violations of the GemStone type constraints statically—that is, already at compile time—in order to avoid run-time checking and, in the case of a violation, a run-time error.

Hint: Consider database components that do not have any type constraints.

Again, illustrate your discussion by examples.

23.4   Compare the GemStone indexing capabilities with the access support relations introduced in Chapter 20. Investigate, in particular, what extensions the GemStone indexing scheme supports. Why doesn't it makes sense to distinguish between an *identity* index and an *equality* index in GOM?

23.5   Devise a realistic example—within your area of expertise—of a conflict between two parallel transactions in which one of the transactions reads (but does not write) objects modified by the other transaction. Show that this could potentially lead to inconsistencies within the object base.

23.6   Consider a transaction that reads objects only—but does not write any objects. Show the difference between optimistic and pessimistic concurrency control for such a transaction.

23.14 Annotated Bibliography

GemStone adopted the Smalltalk-80 object model. The book [GR83] provides an in-depth description of the Smalltalk class concept. [Gol84] details the Smalltalk

programming environment. A review of the historical development and some retro-
spective comments about Smalltalk may be found in [Kra83]. The direct predecessor
of Smalltalk was Simula-67, the first object-oriented language that was well ahead of
its time [DMN70, ND81].

The GemStone discussion in this chapter is based on the OPAL Programming
Manual, Version 1. 4 [SL88]. More research oriented topics of the GemStone development are discussed in the following papers:

- [MS87, MOP85, CM84, MSOP86] discuss general issues related to the development of the Smalltalk object server.

- [MS86] outlines the index support facilities of GemStone.

- [PSM87b] deals with the architectural design of the coupling between the Gem-
  Stone *server* and *client* applications.

- [PS87, PSM87a] discuss some implementation issues of the storage manager of
  GemStone.

24

Ontos: An Integrated Object System

Ontos is an object-oriented database management system that is integrated in an extended C++ programming environment. Persistent objects, i.e., entities that are stored in the database, can be accessed and manipulated within a C++ application program. The Ontos object base contains objects that are instances of user-defined types. Each object encapsulates as an entity the describing data and the operations to access and manipulate the internal state of the instance. All the public features that are provided in the form of operations and attributes associated with a database entity can be invoked by the C++ application program.

24.1   Basic Architecture of Ontos Applications

Ontos is an object-oriented database that provides an interface to C++ applications. Database objects can be "activated," that means transferring an object's state from the secondary-storage database into the virtual memory of the C++ application, and conversely objects can be "deactivated," which involves their transfer back into the database. Between activation and deactivation, processing is performed via standard C++ programming.

The C++ interface to Ontos does not involve syntactic extensions to the language. There exists a tool, called schema loader, that translates C++ class definitions into database classes and loads the resulting schema into the database. Because of this utility, there is no need for a separate schema definition language in Ontos.

The Ontos object database is a distributed database and is based on the client/server architecture for data interaction. The server side manages the data store, and the client side provides the interface to user processes and manages the mapping of

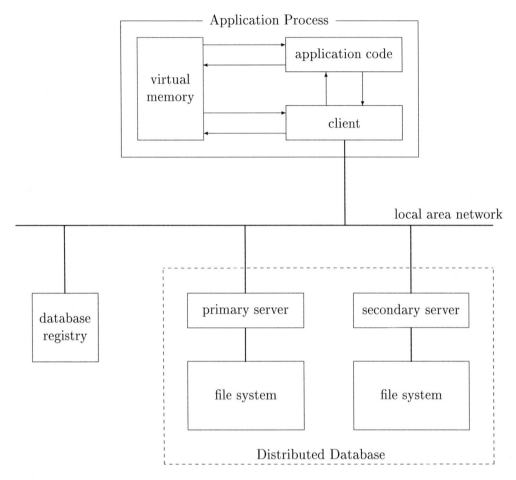

Figure 24.1: Ontos's Client-Server Architecture

data to the application processes' virtual memory space. The client and the server can communicate over a local area network. The overall architecture of the Ontos system is depicted in Figure 24.1.

For implementing large data-intensive applications in Ontos one can rely on three major components of the integrated object system:

The kernel database: In its internal architecture the Ontos database is object-oriented. Therefore, the Ontos system is delivered with a set of basic classes, the kernel classes. The classes can be divided into two categories:

- Classes that describe the Ontos system and are called the "meta-schema"

- Classes that facilitate the development of applications by providing some basic functionality

The class libraries and the header files: The kernel database contains the database objects that describe the kernel classes. The class libraries and the header files provide the support needed to write an application using instances of the kernel classes. Each class has an associated header file that must be included—using the C++ **#include** statement—in an application using that class.

The utility programs: They supply support for developing large applications. They consist of various software engineering tools, such as a browser, a debugger, etc. Since we are more concerned with the object-oriented language and database concepts, these utilities will not be discussed in this presentation.

24.2   Type Definition in Ontos / C++

As pointed out before, in Ontos C++ is used instead of a separate schema definition language to specify the representation of a class of objects and—equally important in an object-oriented system—the signatures of the associated operations.

In C++ the class definition statement (which is comparable to the type definition frame of the generic object model (GOM)) has the following basic form:

```
class ClassName : public SuperClassName {
public:
    MemberList      /* public part */
protected:
    MemberList      /* protected part */
private:
    MemberList      /* private part */
};
```

In the first line we specify the location of the new type in the overall inheritance hierarchy of the type system that has been created for a particular application. In C++ terminology, *SuperClassName* is called *base class*, and *ClassName* is called *derived class*. The **public** label before *SuperClassName* indicates that all properties and functions available to clients of the *SuperClassName* are also available for clients of *ClassName*. Subsequently three clauses follow, each of which can be repeated in any order within the class definition. After the **public** label all attributes—called *members* in C++—and operations—called *member functions* in C++—available to clients of *ClassName* are listed or declared. These are also accessible by possible subclasses of *ClassName*. The next clause—called **protected**—lists those members

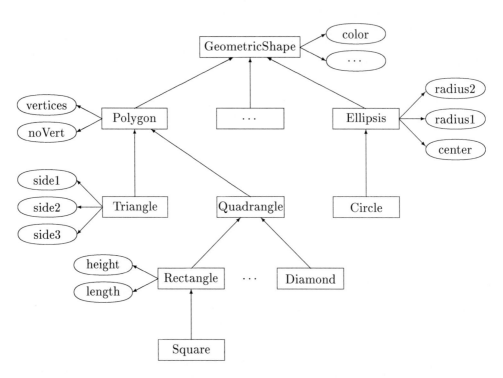

Figure 24.2: Type Hierarchy of Two-Dimensional Geometric Shapes

and member functions that are visible only to subclasses of *ClassName*. They are not visible to general clients of it. The last clause labeled **private** is the most restrictive. The members listed there can only be used within the implementation of member functions of *ClassName*. Their use outside of these, e.g., in general clients or subclasses of *ClassName* is impossible.

In C++ there is no constraint to separate the structure and the behavior of an object. For example, it is possible to declare an attribute of an object within the public part and, therefore, to allow the access to this attribute for everyone. In our understanding of object-oriented programming, it is necessary to distinguish between structure and behavior of an object and to protect the structure of an object from uncontrolled access by every user. Therefore, for the rest of this chapter, we will use the protected part of a class to define the structure of objects and the public part to define the behavior of the class that should be visible to clients.

We will illustrate the type definition capabilities of Ontos on the conceptual schema sketched in Figure 24.2. This conceptual representation of two-dimensional geometric objects involves *generalization*, which can be handled in Ontos by the subclass/superclass concept, as we will see shortly. A *GeometricShape* is either an *Ellipsis*, a *Polygon*, or .... *Ellipsis* itself can be specialized to a *Circle*. *Polygons* are distinguished in *Triangle*, *Quadrangle*, etc. A *Quadrangle* itself is either a *Rectangle*, a *Diamond*, or .... Finally, a *Rectangle* could be a *Square*. The generalization hierarchy discriminates two-dimensional geometric objects according to their different geometric shapes, which leads to variant representation and behavior. For example, *Polygons* can most naturally be modeled by their bounding *Vertices*, whereas a natural representation of an *Ellipsis* stores the *Center* and the two *Radius* values. At each of these generalization levels the object types provide some properties, which are inherited by its descendants.

We start by sketching the class *GeometricShape* in C++ as used in Ontos:

```
class GeometricShape : public Object {
protected:
    char *color;
    . . .
public:
    . . .
};
```

The first line specifies that *GeometricShape* is a subtype of *Object*, which is a built-in type in Ontos. *Object* is the parent of all persistent classes; that is, only objects of (direct or indirect) subclasses of *Object* can possibly reside in the Ontos database. Among others, *Object* defines a member function for storing objects in the database and a delete function for removing objects from the database. It also defines properties for an object name and for the clustering of objects in the database.

24.2.1 Inheritance

To map our conceptual geometric model into an Ontos type definition, one would code each generalization abstraction by introducing subclasses for the specialization types. These subclasses inherit all the features of the supertype, i.e., the generic type. Thus, we code:

```
class Polygon : public GeometricShape {
protected:
    int noVert;
    List *vertices;
public:
    . . .
};
```

This type definition makes *Polygon* a subtype of *GeometricShape*, which induces that *Polygon* inherits the properties of GeometricShape such as, e.g., *color*. In addition, each *Polygon* obtains the properties *noVert*, which stores the number of vertices the modeled polygon possesses, and *vertices*, which is a reference to a *List* aggregate containing the bounding vertices.

The attribute *vertices* is of particular interest: It is an *Aggregate* property in the sense that more than one object is associated with this property. In this case, it consists of a *List* of *Vertex* instances or, more precisely, a list of references to *Vertex* instances. For more details on aggregate properties refer to the section on *Aggregate* types below (Section 24.6).

24.2.2   Further Example Type Definitions

Let us briefly sketch the remaining class definitions of our sample class hierarchy *GeometricShape*—Exercise 24.1 asks the reader to complete these rudimentary class specifications:

```
class Ellipsis : public GeometricShape {
protected:
    float radius1, radius2;
    Vertex *center;
public:
    ...
};


class Triangle : public Polygon {
protected:
    float side1, side2, side3;
public:
    ...
};

class Quadrangle : public Polygon {
    ...
};

class Rectangle : public Quadrangle {
protected:
    float length, height;
public:
    ...
};
```

```
class Diamond : public Quadrangle {
   . . .
};


class Square : public Rectangle {
   . . .
};
```

24.2.3   Static Members

A class is a type, not a data object, and each object (instance) of the class has its own copy of the data members of the class. However, some types are most elegantly implemented if all objects of that type share some data. Preferably, such shared data are declared as part of the class. In C++ it is possible to declare such shared data in the form of *static members* of a class. For example, it will be useful for all *DisplayedObjects* to know the origin and the orientation of the coordinate system on the screen:

```
class DisplayedObject : public Object {
protected:
   . . .
   static Vertex *origin;
   static Vector *orientation;
public:
   . . .
};
```

Declaring the members *origin* and *orientation* as *static* ensures that there will be only one copy of them, and not one copy for each instance of *DisplayedObject*.

24.3   Operations in Ontos / C++

While attributes implement the more static aspects of the instances of a particular type, operations code the dynamic behavior of the types' instances. In Ontos there is no distinction between operations that modify an instance and side-effect free functions; there is only one category, called *function*. A C++ function may behave like a function inasmuch as it returns a result. But, at the same time, the function invocation may induce some side effects.

24.3.1   Member Functions

In C++, functions that are attached to a class are called *member functions* of the class. These functions are usually declared within the public part of the class and represent a possibility to access the properties of the protected part of the class.

```
class Polygon : public GeometricShape {
protected:
    int noVert;
    List *vertices;
public:
    float perimeter();
    ...
};
```

However, the reader should not be misled in believing that all functions are necessarily in the **public** section. It is also possible to include functions in the **protected** and/or **private** part of the class definition—even though this is not demonstrated in this presentation.

Since different classes can have member functions with the same name, one must specify the class name when defining a member function. Therefore, the definition of the member function *perimeter* looks like this:

```
float Polygon::perimeter() {
    ...      /* some application code */
};
```

The double colon "::" delimiter serves to separate the class to which the function belongs (preceding the "::") from the function name. Such a class-associated operation is invoked by using the following syntax:

```
somePolygon→perimeter();
```

Here, we assume that *somePolygon* is a C++ variable referring to (containing a pointer to) a *Polygon* instance—or a subclass thereof.

24.3.2   Friends

In C++ there also exists a possibility to allow a nonmember function access to private members of a class. Such functions are called *friends* of the class. A function is made a friend of a class by a friend declaration in that class. It is also possible to make all functions of one class friends of another class:

```
class X {
    friend class Y;
    . . .
};
```

This friend declaration makes all of X's members—also the private ones—accessible to instances of Y.

24.3.3 Virtual Functions

The operation *perimeter* that was defined for *Polygon* is inherited by *Rectangle* and, hence, provides the same code; i.e., it sums up the distances between adjacent vertices. But for *Rectangle* and also for *Diamond* there is an easier, more efficient (and possibly more accurate) way to implement *perimeter* by just returning the value:

$$2 * height + 2 * length$$

Therefore, we would want to redefine *perimeter* in *Rectangle* and allow for its dynamic binding. In C++, this is possible by declaring member functions in a base class as *virtual*. Hence, we change the declaration of the member function *perimeter* in the class *Polygon* to the following form:

```
class Polygon : public GeometricShape {
protected:
    . . .
public:
    virtual float perimeter();
    . . .
};
```

The keyword **virtual** indicates that the function *perimeter* can have different implementations for different subclasses and that these different implementations are bound dynamically.[1]

The signature of the member function is declared in the base class *Polygon* and cannot be redeclared in a subclass. A virtual function may be redefined (re-coded) for any class that is a subclass of the class in which the virtual function was first declared. Then the most specific version of the function is dynamically bound—depending on the direct type of the receiver object.

Note that this usage of **virtual** functions does not entirely match the terminology we used earlier in the GOM model—recall Section 11.1.

---

[1] In C++ it is also possible to refine nonvirtual functions. However, these are not dynamically bound.

24.4   Accessing Persistent Objects

24.4.1   Naming of Objects

Ontos's class *Object* provides the functionality to (optionally) name any of the persistent objects in the database. In a way, this is analogous to persistent variables in GOM. The name of an object can serve as an entry point to the database. From there, the application can access further objects either by name or by traversing an object reference emanating from the retrieved object. Rather than enforcing a single flat name space, Ontos provides the possibility to create a hierarchy of name spaces, called *Directory*. An object's name is then specified by a search path starting with the *RootDirectory*.

An example of accessing a particular *Ellipsis* object called "myFavoriteEllipsis" is as follows:

```
    . . .
    Ellipsis *myEllipsis = (Ellipsis*) OC_lookup("myFavoriteEllipsis")
    . . .
```

In this statement the variable *myEllipsis* is assigned (a reference to) the *Ellipsis* object obtained from the database by executing the *OC_lookup* routine. This routine takes as a parameter the name of the object to be retrieved—including, possibly, the search path within the hierarchical directory structure. The *OC_lookup* invocation is prefixed with a so-called *cast* operation—i.e., (*Ellipsis*) in this case—which ensures that a C++ virtual memory pointer to an *Ellipsis* object is returned by *OC_lookup*. This becomes necessary because *OC_lookup* is an operation abstractly defined for class *Object*. Thus, when a more specific object, such as an *Ellipsis*, is expected it has to be explicitly "casted."

The name of an object is typically set during the initialization; that is, it is part of the constructor operation. For example, the *Ellipsis* could be defined as follows:

```
    class Ellipsis: public GeometricShape{
    protected:
        float radius1, radius2;
        Vertex *center;
    public:
        Ellipsis(float r1, float r2, Vertex *c, char *theName);
        float getRadius1();
        float getRadius2();
        . . .
    };
```

The public member function *Ellipsis* is the initializer that initializes the fields *radius1* to *r1*, *radius2* to *r2*, and *center* to *c*—the (straightforward) implementation is left as an exercise to the reader. Furthermore, the *Name* of the *Ellipsis* is set to *theName*, where *Name* is an inherited attribute from the root class *Object*.

An *Ellipsis* can then be created as follows:

myEllipsis = **new** Ellipsis(5.0, 6.5, someVertex, "myFavoriteEllipsis");

Here, we assume that *someVertex* is a variable referencing a *Vertex* object representing the *center* of the *Ellipsis*.

24.4.2   Activation and Deactivation of Persistent Objects

Objects are transferred from the database to the application's (virtual) memory and vice versa by operations that achieve *activation* and *deactivation*. One such operation was already introduced in the preceding subsection: *OC_lookup*.

Activation of Objects

Activation initiates transferring the object's state from the database to main-memory and, thereby, transforming all object references contained in the activated object to direct virtual memory-based pointers if the referenced object has already been activated. Furthermore, all objects resident in virtual memory (i.e., all previously activated objects) containing a reference to the newly activated object are updated such that their reference is translated into a new virtual memory pointer.

Activating objects can critically affect the application's performance. If too many objects are activated a lot of disc I/O is performed in vain for those objects never accessed. On the other hand, checking whether a referenced object has yet to be brought into memory each time an object reference is traversed imposes a rather high overhead on computationally complex operations. Therefore, activation of objects is delegated to the programmer in Ontos—under the assumption that the programmer can best decide which objects are most likely accessed in an application program.

Objects can be activated in three different granularities:

1. A single object can be activated using the *OC_lookup* operation. Furthermore, a single object is activated by traversing a direct or abstract reference (cf. Section 24.5).

2. All objects contained in an *Aggregate*—e.g., a *List* or a *Set*—can be activated by a single invocation of the operation *OC_getCluster*, which requires the name of the Aggregate as a parameter.

3. All objects reachable from a particular object can be activated by the operation *OC_getClosure*. In this case, one can be certain that all object references
   can safely be traversed because all are guaranteed to be valid virtual memory
   pointers.

The three operations' signatures are given as follows:

```
Object *OC_lookup(
    char *objectName,
    LockType lock = readLock,
    ... );
Object *OC_getCluster(
    char *clusterName,
    LockType lock = readLock,
    ... );
Object *OC_getClosure(
    char *objectName,
    LockType lock = readLock,
    ... );
```

The first two techniques—activation of single objects and of logical clusters (*Aggregates*)—are more efficient in most applications. They allow to restrict the (expensive) activation to those objects that are likely to be used in the processing phase.
However, these two techniques do have the disadvantage of leaving at least some references unresolved (i.e., nontranslated to virtual memory pointers). This problem,
will be discussed in more detail in Section 24.5, below.

Conversely, activating the closure of all the entry points of an application is the
safest technique because it can ensure that all accessible objects are in virtual memory and, therefore, all references are resolved to virtual memory pointers. However,
reading the closure may be prohibitively expensive in terms of disc I/O and translating the objects' state. In the worst case, it involves reading the entire database.
Figure 24.3 depicts the scenario of activating the closure of the black object. All
shaded objects are implicitly activated by *OC_getClosure*, since they are reachable
by interobject reference chains (marked as arrows) originating at the black object.
The white objects cannot be reached by traversing references originating at the black
object and, therefore, they are not activated.

Deactivation of Objects

An object is *deactivated* with the operation

```
void Object::putObject();
```

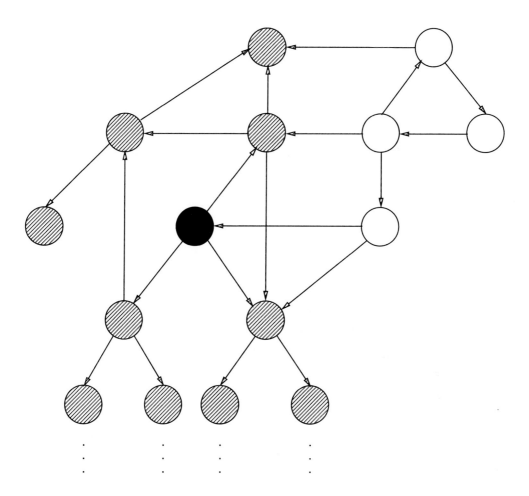

Figure 24.3: Activating the Closure of the "Black Object"

For example, the *Ellipsis* referenced by the C++ variable *myEllipsis* is deactivated as follows:

    myEllipsis→putObject();

This creates a (new) copy of the *Ellipsis* object in the database when and if the current transaction completes successfully. It is the programmer's responsibility to invoke *putObject* on all objects that need to be saved in the database at the end of the transaction. Forgetting to invoke *putObject* results in losing the effects of the application program on this particular object, if there were any.

Note that in Ontos invoking *putObject* on the *Ellipsis* does not (at all) induce the saving of the referenced *Vertex* object. This has to be done explicitly, e.g.:

    myEllipsis→getCenter→putObject();

It is, however, possible to deactivate an entire *Aggregate* by *putCluster*, which obeys the following signature:

    void Aggregate::putCluster(...);

Deactivation is also implicitly achieved by deleting an object or an entire cluster from the database. The operations' signatures are as follows:

    void Object::deleteObject(...);
    void Aggregate::deleteCluster(...);

In this case, however, the objects are not written back to the database; rather, their original database states are also deleted.

24.5   Reference Mechanisms in Ontos

The object model Ontos facilitates two vastly different types of inter-object references:

1. Direct C++ references

2. Abstract references

In the first case, the maintenance of the reference—e.g., taking precautions that the referenced object is brought into the main memory, etc.—is delegated to the programmer. The second kind of reference, called *abstract* references, is much more convenient to handle. For example, upon traversing an abstract reference the referenced object is automatically brought into main memory. Also, maintaining the reference when the abstractly referenced object is moved within main memory is taken care of by the system—whereas under direct referencing this is, again, the task of the user (programmer).

24.5.1 Direct References

The concept of *directly referencing* another persistent object was already used in our *Ellipsis* example. Recall the part of the class definition:

```
class Ellipsis: public GeometricShape
{
protected
    . . .
    Vertex *center;
public
    . . .
    void setCenter (Vertex *c) {center = c;}
    Vertex *getCenter() {return center;}
};
```

This way of directly referencing the *center Vertex* object has the advantage that the traversal along the reference can be performed at (about) the same speed as a normal C++ virtual memory pointer navigation. However, on the negative side, it is the programmer's responsibility to ensure that all referenced objects are activated prior to traversing the reference. We have seen before that loading the closure of all entry points of an application is one safe way to do so. If this cannot be afforded (because of an abundance of reachable objects), referenced objects have to be activated individually. For our example we could perform the following activation of a particular *Ellipsis* instance:

```
myEllipsis = (Ellipsis*) OC_lookup ("myFavoriteEllipsis");
```

Now the *Ellipsis* is activated. But at this point in time the reference to the *Vertex* via the variable *center* is *INACTIVE* and, therefore, cannot be traversed. For this purpose, we have to activate the associated *Vertex* first using the free function *OC_directActivateObject* with the signature

```
Entity *OC_directActivateObject(Entity ** fieldAddress,
                                LockType lock = readLock);
```

on the *center* field of the Ellipsis.

   In order to always be on the safe side, one could, of course, program such that each reference traversal is preceded by a test whether the reference is *ACTIVE* or *INACTIVE*—in the latter case, one would first activate the object. However, this procedure would nihilate the performance gain obtained by direct references altogether. The speed-up of direct references can be obtained only when the programmer can (somehow) avoid such costly tests.

24.5.2   Abstract References

In the last paragraph of the preceding section, we sketched a way of ensuring that each
reference traversal is preceded by a check whether the referenced object is already
read into virtual memory, i.e., activated. This is implicitly done when using *abstract
references*. Let us modify our *Ellipsis* example to employ an abstract reference to the
center Vertex, e.g.:

```
class Ellipsis : public Geometric Shape {
protected:
    . . .
    Reference center;
public:
    . . .
    void setCenter(Vertex *c) {center.Reset(c, this);}
    Vertex *getCenter() {return (Vertex*) center.Binding(this);}
};
```

In this (modified) class definition, we now utilize Ontos's built-in class *Reference*,
which is used to maintain interobject abstract references. We invoked two built-in
operations provided on the class *Reference*. In the implementation of *setCenter* we
invoked

```
center.Reset(c, this);
```

in order to associate the *Vertex c* with the *center* field of **this** object (the keyword
this denotes a pointer to the receiver object—analogous to **self** in GOM).

Further, in the implementation of *getCenter* we invoked

```
(Vertex*) center.Binding(this);
```

which obtains a virtual memory pointer to the object associated with the *center* field
of **this**. In this case, the test whether the associated *Vertex* is ACTIVE or INACTIVE
is implicit. In the latter case (INACTIVE), it is implicitly activated.

On the negative side of abstract references—as present in the Ontos model—is the
lack of static type checking. As can be seen in the example, the type definition states
only that the particular attribute is an abstract reference, denoted by *Reference*.
However, it cannot be ensured statically that only *Vertex* instances are associated
with the *center* attribute (it might as well be a reference to any other object, e.g., a
Banana instance).

24.5.3   Discussion

One of the tenets of the Ontos design philosophy seems to have been to provide a direct
mechanism in all cases in which an abstract mechanism forces a performance tradeoff. Therefore, Ontos provides two reference types. One, the direct references are very

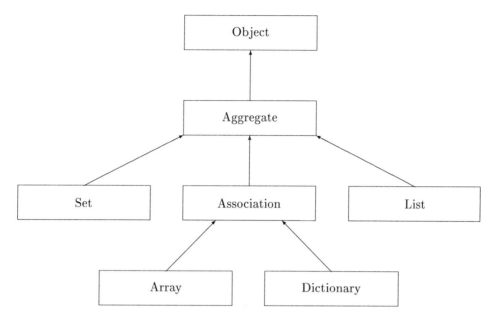

Figure 24.4: Hierarchy of Ontos's Aggregate Classes

efficient but place the burden of ensuring reference consistency on the shoulders of the programmer. Two, abstract references implicitly initiate the activation of referenced objects. It is a tradeoff between efficiency on the one side and safety on the other side. However, even Ontos's abstract reference mechanism bears considerable danger of encountering type violations at run time—as pointed out above.

By providing such direct mechanisms as, e.g., direct references with the necessity of explicitly activating and deactivating persistent objects, the Ontos designers (deliberately or not) gave up on a "seamless" integration, which demands that persistent objects are accessed in the same way as transient C++ objects.

24.6 Aggregates

Ontos provides some useful predefined classes for handling collections—called *Aggregates* in Ontos terminology. The class hierarchy of Aggregates is shown in Figure 24.4. The diagram shows the four *Aggregate* types supported:

- *Array*, an indexed collection where the index range is given by integer numbers

- *Dictionary*, which is similarly indexed as an *Array*—however, the keys are arbitrary *Entities*, e.g. strings, not necessarily integer numbers

- *List*, which is an ordered collection of objects

- *Set*, which provides for sets in the mathematical sense, i.e., disallowing multiple entries

The other two classes, shown in the diagram—i.e., *Aggregate* and *Association*—are abstract (virtual) classes not intended for instantiation. They merely comprise some abstract functionality inherited by their subclasses.

24.6.1 Instantiation of Aggregates

The class *Aggregate* has the built-in constructor

Aggregate::Aggregate (Type *memberSpec, char *name = (char *) NULL);

which is used to create a new *Aggregate* with a particularly constrained element type, i.e, elements of type *memberSpec*.

24.6.2 Element Insertion and Removal

The *Aggregate* subclass *Array* facilitates the insertion—and possibly overwriting of the old element—of an element at a particular index position. This is achieved by the operation *setElement* with the following signature:

Array::setElement(long index, Argument data);

Here, the parameter *index* specifies the position and the formal parameter *data* specifies the new element which has to be a subclass element of the built-in *Argument* class.

Elements are inserted into a *Dictionary* by the operation

Dictionary::Insert(Argument tag, Argument element);

where both, *tag* and *element*, are subtypes of *Argument*.

There are two operations to insert elements into a *List*:

(1) List::Insert(Argument element, long pos = −1);

(2) List::setElement(long index, Argument data);

The first operation inserts the *element* at a specified position *pos* (default at the beginning, i.e., position −1) and shifts all existing elements by one position to the right.

The second operation, *setElement*, overwrites a possibly existing element at position *index* and inserts the *data* at this position.

Insertion into *Set* objects is, of course, invoked without any position specification, i.e.:

Set::Insert(Argument member);

24.6.3   Cardinality and Membership

There is a built-in operation for obtaining the cardinality of an *Aggregate* object:

> long unsigned Aggregate::cardinality();

Further, the membership of an element in a collection is tested by the built-in operation *isMember*, which has the following signature:

> OC_Boolean Aggregate::isMember (Argument element);

This operation returns an *OC_Boolean* value *true* if the member is contained, *false* otherwise.

24.6.4   Activation of Aggregates

The free operation *OC_lookup* can be used to look-up (and activate) a named *Aggregate* object. However, this procedure only activates the single *Aggregate* object; not the members of the *Aggregate*. For this purpose the operations *getCluster* and *OC_getCluster*—already partly introduced in Section 24.4.2—are employed. Their signatures are as follows:

> **void** Aggregate::getCluster(LockType lock);
> Object* OC_getCluster(char *clusterName, LockType lock);

The *getCluster* operation is invoked on an already active *Aggregate* object and activates all its members. Since the *Aggregate* object, on which *getCluster* is invoked, is already active the operation returns no result; i.e., its result type specification is **void**.

The free operation *OC_getCluster* first obtains (i.e., reads and activates) the *Aggregate* named *clusterName* and subsequently activates all its members. The (reference to the) newly activated *Aggregate* is returned as the result of invoking this operation.

24.6.5   Iterators

Ontos provides special *Iterator* classes for iterating through the elements of an *Aggregate*. The best way to do so is to ask the *Aggregate* to be iterated through to generate its own *Iterator*, e.g.:

> *AggregateIterator* *anIterator = anAggregate→getIterator();

Here, *anAggregate* is a C++ variable referring to an *Aggregate* object, e.g., a *Set* object. The invocation of *getIterator()* on *anAggregate* returns a reference to an iterator that is assigned to the variable *anIterator*. This iterator can than be used to access all elements of the collection *anAggregate*, e.g.:

```
while (anIterator→moreData()) {
    Object *anObject = (Object *) (Entity *) (anIterator→operator()());
        ...    // do something with anObject
}
delete anIterator;
```

The variable *anObject* is looping through all objects contained in *anAggregate*. This is achieved by invoking *operator()()* on the iterator that retrieves the next element in the collection. This is done as long as the iterator has still more elements to access, which is verified by the Boolean function *moreData()* invoked on *anIterator*.

24.7   Transaction Control

The activation and deactivation of objects implicitly assumed a single user. In a database with many concurrent users, conflicts between users must be detected and resolved. Ontos handles such conflicts through a transaction mechanism discussed below. In common with most transaction mechanisms, it ensures atomicity of change; i.e., either all changes comprising a transaction are committed or none become manifest. An application writes objects to the database over time. These changes are not propagated to the database immediately. Rather, they are kept in a *transaction data pool*. When the transaction is successfully committed, they are propagated to the database in their entirety. However, within the transaction this behavior is not visible. The application can therefore work with its own private database state that is invisible outside the transaction until the transaction is committed.

24.7.1   Basic Transactions

The Ontos lock mechanism provides three graduations for locking objects. All object activation calls—e.g., *OC_lookup*, *OC_getCluster*, etc.—have a lock argument. This argument may specify a read lock, a write lock, or no lock at all. The *LockMode* argument is optional; if none is specified the default *LockMode*, i.e., the *readLock*, is implicitly assumed. The granting of locks follows the usual lock management protocol.
    The interface to the transaction mechanism consists of three functions:

- *TransactionStart*: The *TransactionStart* call defines the beginning of a transaction scope. It also defines several parameters. For example, it specifies:

- A transaction name (used for sharing transactions, see below).

- A function for dealing with any conflicts that may arise in reading or writing objects during the transaction. Because the action that should be taken in the event of a lock conflict varies in different situations (a conflict might lead the application to abort the current transaction, wait and try to obtain the lock later or simply continue processing without having obtained the object) this function can be one of the system-supplied functions or a function provided by the user.

- *TransactionCommit*: The *TransactionCommit* call ends the transaction, and an attempt is made to write all deactivated objects to the database.

- *TransactionAbort*: The *TransactionAbort* call also ends the transaction, but no changes are made to the database.

24.7.2   Nesting Transactions

Transactions may be nested (i.e., multiple *TransactionStart* calls without intervening commits or aborts are activated) to create units of atomicity within a longer transaction. The initial state of a child transaction is exactly the state of its parent transaction at the time it is initiated. At the end of a child's transaction the parent's state includes either all the changes made by the child, in case of a commit, or none of these changes, in case of an abort. Furthermore, a child's changes are not visible outside the parent until the parent commits successfully.

Nested transactions make long transactions more practical. A single long transaction can be divided into a series of many small nested transactions. Then, if a data conflict occurs upon activation, only the nested child transaction needs to be aborted and the work accomplished so far in the long enclosing transaction is unaffected.

24.7.3   Shared Transactions

Cooperating processes can share a single transaction. This facility allows dealing with tasks that are atomic from the standpoint of changes to the database but are implemented more easily by a number of cooperating processes. For example, think of a multiple window application in which a separate process operates on each window.

The transaction names are used to make a transaction that is already in progress "joinable" by another process. The first process to issue a *TransactionStart* call with a particular name argument initiates the transaction. Subsequent processes issuing a *TransactionStart* call with the same name simply join the existing transaction.

A shared transaction may be thought of as one in which the transaction data pool is visible to all processes participating in the shared transaction. The pool stores the latest state of each object, as established by the last deactivation of an object initiated by any of the cooperating processes. When a process activates an object that is already in the pool, it obtains this copy rather than a copy from the database. A process can also refresh its cache copy of an object from the pool at any time and thus get immediate access to the latest work done on objects by other cooperating processes.

The shared transaction "bundles" the work of all participating processes into a single atomic unit. A shared transaction is committed, and thereby its changes are made visible only if all processes sharing it are committed successfully. A *TransactionAbort* by any process aborts the entire shared transaction.

24.8   Versioning Support

The Ontos versions and alternatives mechanism allows a single object to exist in any number of versions. All objects of the same version are collected into a *configuration* object. Configurations are related to each other through *derivation links*. Each configuration (except the first) has a parent and may have any number of children. Thus, both alternatives and serial versions are supported. Objects in leaf configurations may be modified freely; objects in inner configurations are immutable. Inner configurations can be deleted, however. When an inner configuration is deleted, its parent inherits all the children of the deleted configuration.

Applications create new configurations by specifying a derivation link between the desired new configuration and some existing previous configuration. When created, a new configuration consists of the state of all objects as they exist in its parent configuration. Objects are written into a configuration implicitly. Upon deactivation, an object is propagated into the application's currently active configuration. Thus, versions of individual objects are not denotable; rather, configurations are denotable.

24.9   ObjectSQL

Ontos provides an embedding of an (extended) SQL query facility in the C++ programming environment. Since SQL is a set-oriented query language whereas C++ obeys the one-record-at-a-time processing paradigm, the interface of the two (incompatible) systems utilizes a particular *Iterator* class. The reader may recall our discussion of the so-called *impedance mismatch* problem encountered when coupling a relational database system with a procedural language, such as C (see Sections 4.5 and 5.3).

Let us base our discussion of Ontos's Object SQL interface on the example query, stated in pseudo SQL:

> **select** ∗
> **from** Ellipsis
> **where** getColor() = "red"

This query obtains all red *Ellipsis* instances.

Generally, such an SQL query should be braced by an *ExceptionHandler* to "catch" SQL errors—however, the details of exception handling in Ontos cannot be covered here. Thus, the format of specifying such a query in the Ontos/C++ environment is as follows:

```
OC_startQuerySession();
ExceptionHandler sql_handler("SQLproblem");    // handles exceptions
QueryIterator *myIter;
if (sql_handler.doesNotOccur())
    { ... }  // code to query the database
else
    { ... }  // code to handle SQL error
delete myIter;
OC_endQuerySession;
```

In this example we declared—aside from the exception handling features that we will ignore here—the variable *myIter* as a pointer to a *QueryIterator* instance. A *QueryIterator* is created (instantiated) by passing the SQL query as a string to the constructor, e.g.:

```
myIter = new QueryIterator("(select ∗
                            from Ellipsis
                            where getColor() = "red")";
```

The resulting collection of *Ellipsis* objects satisfying the SQL query can then be obtained by stepping through the iterator. The following code fragment sketches the principle:

```
while (myIter→moreData())
    {
        ...   //get each Ellipsis
        ...   // and process it somehow
    }
```

In principle, this embedding of the ObjectSQL language in the Ontos/C++ environment resembles very much the *cursor* concept used to couple (standard) relational SQL with programming languages—as discussed in Section 4.5 for the Oracle system.

24.10   Exercises

24.1   Complete the class definitions of the *GeometricShape* class hierarchy. In particular, outline sample member functions for each class definition. Verify the
correctness of your class definitions either on an Ontos system (if you have access to one) or as an (ordinary) C++ application. In the latter case, you have
to take out the Ontos-specific operations for storing objects and retrieving
them from the database.

24.2   Discuss the problems that may occur under direct referencing when a main-
memory resident object is inadvertently deactivated from the virtual C++
memory. What happens when such a reference is traversed? Is it possible to
"catch" such an error by preceding the reference traversal by some **if**-statement
to test the validity (consistency) of the pointer?

24.3   Directly referencing objects is very similar to the pointer-swizzling techniques
that were discussed in Chapter 22. Discuss the main difference that is caused
by the fact that pointer swizzling is done implicitly and automatically by the
run-time system, whereas maintaining the consistency of direct references is
the programmer's task. Why is pointer swizzling inherently less efficient?

24.11   Annotated Bibliography

In this presentation we assumed that the reader has some working knowledge of the
C++ programming language. If this is not the case, the C++ language description
by Stroustrup [Str90b] is a good starting point.

The Ontos database system was developed by Ontologic Inc. (see Chapter 27
for addresses). This company developed a predecessor, called Vbase, which was an
object-oriented database system with its own object model (unlike Ontos, which is
based on the C++ object model). The Vbase model is described in [AH87b]. The
Object SQL interface of Vbase and Ontos is presented in [HD91].

The guiding design principle behind the Ontos development is that the user can
choose between convenience and safety on the one hand and mere performance on
the other hand. It is the programmer who decides about the appropriate tradeoff
between safety and performance, since Ontos does not prevent the programmer from
"breaking the rules," as, for example, substituting safe abstract references by direct
references. In this respect Ontos can be viewed as a "lower-level" product compared
to its predecessor Vbase, which achieved a seamless and safe integration—for the sake
of performance, as the Ontos designers argue. The presentation in this book is based
on the Ontos user's manual [Ont92] and the concise overviews of Ontos prepared by
Andrews, Harris, and Sinkel [AHS90, AHS91].

25

O₂

In this chapter, we briefly describe the essentials of the object-oriented database system O₂. O₂ was originally developed at Altaïr and is now marketed (and further developed) by O₂ Technology.

The data structuring concepts of O₂ are quite similar to the ones incorporated in the generic object model (GOM). There is, however, one essential difference between O₂ and GOM. The O₂ model is designed to be embedded into existing programming languages—most important C and C++. Therefore, the model and its implementation language for methods do not quite form one seamless, integrated, and orthogonal framework—as in GOM. Rather, the O₂ data model stands somewhat outside the implementation language—that is, the host language C or C++. Thus, all O₂ constructs have to be explicitly marked in order to be detected by the C compiler and initiate appropriate actions, e.g., data transformations.

25.1 The Basics of Data Modeling in O₂

Out of all the data models of the commercial products described in this book, O₂ has the one closest to the GOM object model. Therefore, the subsequent description of the O₂ data modeling concepts can be very brief.

25.1.1 The Type Constructors

The O₂ model provides four type constructors:

1. *Tuple* constructor

2. *Set* constructor

3. *Unique set* constructor

4. *List* constructor

The *tuple* and *list* constructor are analogous to the corresponding type constructors in the GOM model. The O_2 *set* constructor, however, should actually have be termed a *multiset* or *bag* constructor, since it allows the same element to be contained several times. Thus, it is not a set in the mathematical sense. The *unique set* constructor provides the functionality of the mathematical set—it precludes multiple containment of an element.

25.1.2   Types

The O_2 terminology is different from the GOM notation: An O_2 *type* corresponds to a *sort* in GOM and a GOM *object type* is called a *class* in O_2. O_2 provides the usual built-in atomic types. Values have a type and may be atomic, i.e., elements of one of the built-in atomic types, or complex structured. Just like in GOM, values in O_2 do not possess an identity and, consequently, cannot be shared. Here are some example values:

> **o2** integer myAgeIn92;
> **o2 tuple**(day: integer, month: integer, year: integer) myBirthDay;
>
> myBirthDay.day = 16;
> myBirthDay.month = 4;
> myBirthDay.year = 1958;
>
> myAgeIn92 = 1992 − myBirthDay.year;

A few words are needed to explain the syntax of O_2 applications. The O_2 variable declarations are explicitly marked by the prefix **o2**. This is needed because O_2 constructs are embedded in a host language—in this presentation, we will always assume C as the host language. The variable *myAgeIn92* is marked as an O_2 variable of type integer. The variable *myBirthDay* can assume complex-structured values consisting of a tuple with the three attributes *day*, *month*, and *year*—all being constrained to an integer value. An attribute of a tuple value is accessed by the dot notation as exemplified above—later on we will see that attributes of a tuple object are accessed in a different form. As in C, an assignment has the syntactical form $a = b$ where a is assigned the value of b.

25.1.3   Classes and Objects

In O_2 terminology objects belong to a class. A class combines the structural representation that consists of a type and the behavioral specification consisting of a collection of operations. The extension of a class is not (yet) automatically maintained. If the user wishes to maintain the extension, he or she has to explicitly create a set-structured object into which all the newly generated instances are inserted during initialization. For this purpose, the initializer—called *init* in O_2—has to be appropriately defined.

25.1.4   Operations

Operations are associated with classes. The signature of operations is specified in the class definition. The implementation is coded in one of the interface languages; the most important one is O_2C, an embedding of O_2 in the programming language C. Other possible host languages are Basic and C++.

25.1.5   Inheritance

O_2 provides single and multiple inheritance. Operations and structural properties are inherited by the subtype. Operations may be refined—*overridden* in O_2 terminology. The refined operations are dynamically bound.

25.1.6   Declarative Query Language

For formulating declarative queries O_2 provides the language O_2SQL (formerly called O_2Query), which is a "dialect" of SQL—in parts already introduced in Section 14.3. O_2SQL can be used interactively—e.g., for browsing the object base—or embedded within the host programming language, e.g., the language O_2C. However, the embedding in the host programming language is (still) rather simplistic: The query is passed as a text string to the database component.

25.1.7   Type Safety

The O_2 model does require type constraints on database components. Nevertheless, the O_2 model provides some features that possibly violate type safety. Examples thereof are:

- Exceptional attributes

- Relaxed refinement conditions

- Subtyping on collection classes even if the element types are subtypes of each other

- Embedding O_2SQL queries, which cannot always be type checked, in the host language (see below).

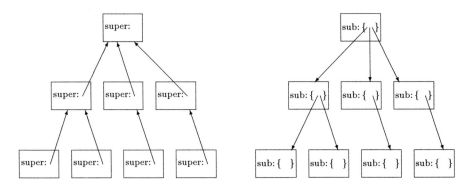

Figure 25.1: Alternative Representation of a Part Hierarchy

25.2   Example Application

Recall the conceptual schema for modeling part hierarchies—which was introduced in Chapter 9. For convenience, the entity-relationship diagram is iterated below:

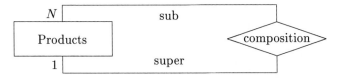

This schema allows two different representations of a part hierarchy, which are outlined in Figure 25.1—this figure was also already introduced in Chapter 9. The two representations of the relationship *composition* are visualized at the object level. Each box represents a *Product* instance; the arrows represent references. On the left-hand side we depict a sample object base state in which the type *Product* contains the attribute *super*. On the right-hand side the same object base state is shown for the case that the type *Product* contains the set-valued attribute *sub*.

25.3   Product Hierarchy with *super* References

Let us first implement the schema for a *Product* hierarchy under the assumption that the references are directed from sub- to superpart, i.e., the scenario shown on the left-hand side of Figure 25.1.

25.3.1   Class Specification

Let us define the class *Product* including the most important operations:

```
class Product
    public type tuple
    (
        name: string,
        costs: real,
        price: real,
        weight: real,
        super: Product
    )
    method
        init(name: string, costs: real, price: real,
             weight: real, super: Product): Product,
        public partList: unique set(Product),
        public isPartOf(theSuperPart: Product): boolean
end;
```

The above class definition specifies that a *Product* object has the structural representation of a tuple with four atomic attributes—*name*, *costs*, *price*, and *weight*—and one attribute *super* referring to another object. The keyword **public** preceding **tuple** states that all attributes can be read and written by clients; i.e., no protection is enforced. It is also possible to specify the access rights selectively for each attribute. The possible modes are:

- *public*: The attribute can be read and written by clients,

- *private*: The corresponding attribute is invisible by clients

- *read*: The attribute can only be read by clients. In this case, all write access has to be performed by operations associated with the class.

Our example class *Product* has only three associated operations whose signatures have to be specified as part of the class definition. The operation *init* is the initializer, which requires values for all attributes as parameters. The other two operations, *partList* and *isPartOf*, correspond to the equally named operations introduced in Chapter 9: *partList* computes the set of all constituent *Products* of which the object **self**—i.e., the object on which the operation is invoked—is composed. The Boolean operation *isPartOf* determines whether the receiver object (i.e., **self**) is a subpart of the composite *Product* passed as the parameter called *theSuperPart*. Just like for the structural components the O$_2$ model allows to specify selective access rights for the operations. The applicable modes are restricted to *public* and *private*—since the mode *read* does not make sense for an operation.

25.3.2 Class Extensions

We mentioned before that O_2 does not maintain any implicit class extensions. This is the user's task. For this purpose, we create a named database set object called *AllProducts* as follows:

> **name** AllProducts: **unique set**(Product);

During the initialization a new *Product* has to be inserted into this named database set.

The named database objects form the (only) entry points via which applications can access persistent objects or persistent values. It is also possible to create single named objects or values—as opposed to the named set *AllProducts*.

25.3.3 Coding of the Methods

Let us first code the initializer *init* in the host language O_2C:

```
method body init(name: string, costs: real, price: real,
                    weight: real, super: Product): Product
    in class Product
    {
        self→name = name;
        self→costs = costs;
        self→price = price;
        self→weight = weight;
        self→super = super;
        AllProducts += unique set(self);
        return self;
    };
```

As required, the initializer inserts the newly created *Product* instance into the set *AllProducts*. In the context of sets, the overloaded operator "+=" means set union. Analogously, the operator "−="denotes set difference—when applied to two sets. Note that the attributes of the tuple object are accessed using the right-arrow ($\rightarrow$) notations—as opposed to the dot notation used for accessing attributes of tuple values.

We can now code the two operations *partList* and *isPartOf*. Their implementation can easily be derived from the corresponding GOM realization of Chapter 9.

```
method body partList: unique set(Product)
    in class Product
    {
        o2 unique set(Product) resultSet;
        o2 Product part;
```

```
        resultSet += unique set(self);
        for (part in AllProducts) {
            if (part→super == self) {
                resultSet += part→partList;
            }
        }
        return resultSet;
    };
```

```
method body isPartOf(theSuperPart: Product): boolean
    in class Product
    {
        o2 Product part;
        part = self;
        while (part != nil) {
            if (part == theSuperPart) {
                return true;
            }
            else {
                part = part→super;
            }
        }
        return false;
    };
```

25.3.4 Declarative Query Embedded in O_2C

The language O_2Query provides an SQL-like declarative query capability. It can be used interactively or within the host language O_2C. Let us concentrate on the embedded query facility. An example is given as follows:

```
o2 set(Product) rightWeightProducts;
o2 integer lowerBound, upperBound;
...
o2query (rightWeightProducts,
            "select p
             from p in AllProducts
             where p.weight > $1 and p.weight < $2",
            lowerBound, upperBound);
```

We defined a set variable *rightWeightProducts*, which is subsequently used to hold the result of the query. The query is prefixed with the keyword **o2query** to indicate that the query has to be passed to the database component. In this case, we formulated a parameterized query that has two formal parameters—denoted \$1 and \$2 in the **select** ... **from** ... **where** ... query. The parameters are substituted at run time by the actual values of the variables *lowerBound* and *upperBound*, respectively.

It should be obvious that this kind of parameterized query language embedded in a host language can easily lead to type inconsistencies. The parameters are replaced at run time by the actual values. Now consider an embedded query in which the named set over which the query is evaluated is passed as a string as exemplified by

```
    ...
    o2 string namedSet:
    ...
    o2 query (rightWeightProducts,
       "select p
        from p in $1
        where p.weight > $2 and p.weight < $3",
       namedSet, lowerBound, upperBound);
```

Since there is only a rather loose coupling between database and host language—consisting of passing strings—it cannot be ensured statically that the *string* variable *namedSet* has a meaningful and/or type-consistent value. The compiler merely enforces that the variable *namedSet* can only assume a string value—which, however, may as well be "hello world".

25.4   Alternative Representation of Part Hierarchies

For completeness we will also include the O_2 realization of the alternative representation of part hierarchies. In this representation the class *Product* contains a set-valued attribute *sub* that contains (references to) those *Product* objects that are (direct) constituent parts.

25.4.1   The Class Definition

The class Product is defined as follows:

```
    class Product
       public type tuple
       (
            name: string,
            costs: real,
            price: real,
            weight: real,
```

```
    sub: unique set(Product)
)
method
    init(name: string, costs: real, price: real,
         weight: real, sub: unique set(Product)): Product,
    public partList: unique set(Product),
    public isPartOf(theSuperPart: Product): boolean
end;
```

name AllProducts: **unique set**(Product);

25.4.2   The Coding of the Most Important Methods

Let us restrict the coding of the operations to the two most important ones: *partList*
and *isPartOf*. The initializer *init* is implemented as shown before.

```
method body partList: unique set(Product)
    in class Product
    {
        o2 unique set (Product) resultSet;
        o2 Product part;
        resultSet += unique set(self);
        for (part in self→sub) {
            resultSet += part→partList;
        }
        return resultSet;
    };

method body isPartOf(theSuperPart: Product): boolean
    in class Product
    {
        o2 boolean isUsed = false;
        o2 Product part;

        for (part in theSuperPart→partList) {
            if (part == self) {
                isUsed = true;
            }
        }
        return isUsed;
    }
```

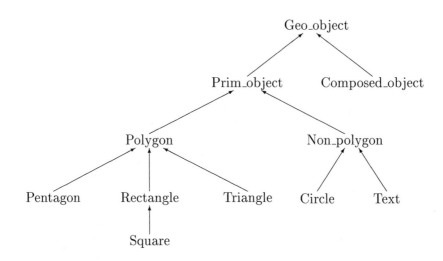

Figure 25.2: Type Hierarchy of the MINI-CAD System

25.5 Exercises

25.1 If you own an O_2 system implement—and extend—the two alternative representations of part hierarchies. Populate the object base with some realistic part hierarchies, e.g., a bicycle part hierarchy. Measure the efficiency of the operations *isPartOf* and *partList* for the two alternative representations and analyze your benchmark findings.

If you don't have access to an O_2 system, carry out the implementation in any other accessible object-oriented database system or in an object-oriented programming language coupled with a file system.

The remaining exercises are part of a student database project that was carried out during a database lab course for students at the Technical University Aachen and the University of Passau. The goal of the lab course was to develop a small, stand-alone drawing system—called the MINI-CAD system—under X windows and O_2.

25.2 The two-dimensional drawing objects of the MINI-CAD system are to be classified into the type hierarchy shown in Figure 25.2.

Define O_2 classes for the geometrical object types shown in the diagram. Incorporate methods for translation, rotation, and scaling in the class definition (see Figure 25.3).

```
class Geo_object
    type tuple
        (...)

    method
        init: Geo_object,          /* initialize object */
        public move(v: vector),    /* translate object */
        public rotate(angle: real, rotpoint: point),
                                   /* rotate the object about rotpoint */
        public scale(factor: real)  /* resize the object by factor */
end;
```

Figure 25.3: Skeleton of the Class Definition *Geo_object*

25.3   It is useful to provide an operation that allows to select a geometric figure on the screen by a mouse click. For this purpose, realize a function *contains* that—given the coordinates of the mouse click—determines whether the mouse click point lies inside the *Geo-objects* instance. An object of the class *Text* should always return false for this particular Boolean function.

Hint: There are different possibilities to realize this function. The standard method for polygons is the "beam algorithm": send a "beam" from the mouse-click point and count the number of times it cuts a polygon edge. If this number is odd, the mouse click point lies inside the polygon.

25.4   Write a program that allows mouse-controlled rotation, translation and scaling of two-dimensional objects on the screen.

25.5   For polygons with a large number of edges the function contains—as realized in Exercise 25.3—is rather inefficient. Enhance the program such that a *pretest*ing is done first by using the *bounding box* method. The bounding box is a rectangle with edges parallel to the coordinate axes and completely encompasses the polygon. Realize this pretesting using O_2C and O_2SQL. Then the actual *contains* test is done only if the polygons satisfied the *pretest*. Visualize the outcome of the *pretest* and the *contains* test graphically on the screen.

25.6   Implement an operations *intersects* that can be invoked on a set of *Polygons* and finds those pairs of polygons that intersect. A *Polygon* should, by convention, not intersect with itself and a once determined pair of intersecting polygons (P_1, P_2) should not be output again as (P_2, P_1).

Hint: The implementation should be based on the set of edges of a polygon. Use a function *crossing* that determines whether two edges cross.

25.7   Realize manipulation operations for composed objects. These include the composition of (atomic and composed) objects, the disconnection of composed objects into its constituent atomic objects, rotation and translation of an entire composed object. Furthermore, take care that the operations realized in foregoing exercises—such as *contains* and *pretest*—also work on composite objects.

25.8   To illustrate the functionality of your MINI-CAD system construct a drawing of your own liking.

25.6   Annotated Bibliography

The development of O_2 was very influential for the area of object-oriented database research because of the numerous papers published by the O_2 developers. Most of these papers are contained in a book edited by Bancilhon, Delobel, and Kanellakis [BDK92]. A survey article on O_2 appeared in a special issue on database research of the Communications of the ACM [Deu91]. Query languages for object models are reviewed in [Ban89]; the declarative query language of O_2 is described in [CDLR89, CD92].

26

Two Other Commercial Systems

26.1 Itasca

The object-oriented database system Itasca is based on the series of the Orion prototypes developed at the Microelectronics and Computer Technology Corporation (MCC) Austin, Texas. It employs a fully distributed architecture with private and public databases. It does not exhibit any central component like a central data or central name server. It supports the coding of applications in Common Lisp and—in the near future—in C. The dynamically typed data model supports most of the features presented in this book.

26.1.1 Class Definitions in Itasca

Just to give a flavor of the Itasca Common Lisp interface we introduce some examples and briefly overview the most important features a type (class) definition exhibits in Itasca.

A type—in Itasca called *class*—definition for *Vertex* looks as follows:

```
(make-class 'Vertex
            :attributes
            ((X :domain number)
             (Y :domain number)
             (Z :domain number)))
```

A *Vertex* consists of three attributes with domain *number* as indicated by the keyword **:domain**. A *Vertex* with coordinates in the origin is created by **make**, e.g.:

```
(setq V (make 'Vertex :X 0.0 :Y 0.0 :Z 0.0))
```

The variable V now refers to the newly created *Vertex* instance.

The attribute value of the X coordinate of V can be changed to 1.0 by (**setf** (X V) 1.0)

Subsequently reading the *X* attribute of *V*, e.g.,

(X V)

yields the expected number 1.0.

The abstract syntax for the most important entries in the class definition is

> (**make-class** *ClassName* [**:superclasses** *ListOfSuperclasses*]
>   [**:documentation** *Documentation*]
>   [**:abstract** *TorNIL*]
>   [**:versionable** *TorNIL*]
>   [**:attributes** *ListOfAttributes*])

The components of this syntactical frame have the following meaning:

- *ClassName* is the name of the newly defined class.

- *ListOfSuperclasses*, which is associated with the keyword **:superclasses**, is a list of the superclasses (supertypes) of the newly created class. All the features of all the superclasses are then inherited by the class *ClassName*. Thus, Itasca supports multiple inheritance.

- *Documentation* is a comment to the class definition in the form of a string.

- **:abstract** specifies whether or not the class is virtual—in the generic object model (GOM) terminology. In the former case *TorNIL* must be *T* (true); in the latter case it is set to *NIL*.

- **:versionable** specifies whether the instances of the class are versionable (*T*) or not (*NIL*).

- *ListOfAttributes* is a list of attribute specifications. An example has already been specified above. Besides the **:domain** key many more keys specifying default values and the like exist for individual attributes.

Class-associated methods can be defined by using **def-method**. What follows is the definition of the method *scale* for a Vertex (all dimensions are scaled by the same *scaleFactor*):

```
(def-method scale (self Vertex) (scaleFactor)
    (setf (X self) (* (X self) scaleFactor))
    (setf (Y self) (* (Y self) scaleFactor))
    (setf (Z self) (* (Z self) scaleFactor))
    self)
```

The name of the newly defined method is *scale*. The type (class) it is associated with is *Vertex* as specified in the list (**self** Vertex) following *scale*. The only argument to *scale* is *scaleFactor* specified in the second list following the method's name. The body of the method consists of three statements. Each statement sets one coordinate to the result of multiplying the value of the argument by the *scaleFactor*. Last, the method *scale* returns the scaled *Vertex* still referred to by **self**.

26.1.2   Summary

The predominant features of Itasca are as follows:

- It exhibits a fully distributed architecture with multiple servers and multiple clients. No central component is needed—neither a central data (object) server nor a central name server.

- New sites can be created, deleted, or migrated by the users at any time.

- It allows for public and private databases.

- The transaction management provides classical transaction control as well as support for long duration transactions by following the checkIn/checkOut paradigm.

- Recovery is provided for CPU failure and optionally for media failure.

- Its authorization concepts comprise most of the features introduced in Chapter 16.

- The schema management of Itasca is the most advanced of all commercial systems. It allows for schema evolution by invoking schema modification operations even from running applications. Schema evolution in Itasca provides the full range of evolution operations covering the whole taxonomy discussed in Chapter 18. It further provides methods to query the current schema from any application; e.g., it is possible to ask for the list of attributes of a certain class.

- It provides specific support for *composite objects*. A composite object consists of a hierarchy of objects referencing each other via their attributes. This hierarchy, if defined as a composite object, can be treated as a single unit. This allows, for example, the user to delete a composite object by only deleting its root object.

26.2   ObjectStore

ObjectStore is an object-oriented database system developed by ObjectDesign. Its data model is the C++ programming language. In this respect, it falls in line with Ontos and several other object-oriented database management systems. Some additional operations to create, open, close, and delete databases are introduced. Beyond this, class libraries are provided, e.g., to support collection types.

26.2.1   General Issues

The system itself is a multiple client/multiple server architecture. It runs on several hardware platforms covering all major workstations. Classical transactions as well as a checkIn/checkOut protocol for the client/server architecture are supported. The latter includes a notification mechanism. Versioning is used to enhance the degree of concurrency allowed. No additional support for design transactions is provided.

The support of schema evolution is very weak compared to the possibilities described in Chapter 17 of this book. The main operations are adding and modifying schema components. A special notion of schema is not supported.

There exist two features unique to ObjectStore. First, it provides persistence for C++ programs by a *memory mapping architecture*. Second, it provides a so-called *low-cost migration path* to convert a program whose persistence of data is based on file handling into a program using ObjectStore for persistent data store. The low-cost migration path relies heavily on the memory mapping architecture.

The basic idea of the memory mapping architecture is the same as for virtual memory management in operating systems. There, main memory pointers are used to point to data items. All data items reside on pages. These pages may either be in main memory or may be swapped onto disk. In the former case, all pointers to data items on the page present in main memory are valid and can be dereferenced without any problems. When a page is swapped out of main memory, the pointers to data items on it become invalid. Dereferencing an invalid pointer results in raising an exception. For this, hardware support exists in today's operating systems—the *memory management unit*. The result of the exception is that the missing page is transferred into main memory again. Independence of the location is a crucial issue, since the original place in main memory could be occupied by a page not to be swapped out. Providing location independence by a table converting virtual memory addresses into physical memory addresses is also the task of the memory management unit.

The memory mapping architecture of ObjectStore just exploits this idea that originated in virtual memory operating systems. References to objects are realized by virtual memory addresses. If the page an object to be dereferenced resides on is within main memory, no extra penalty for dereferencing is necessary. Thus, in this case dereferencing is as fast as for any C or C++ program. In case of a page fault, i.e., dereferencing an object present on a page not in main memory, this page must be loaded. Only thereafter can dereferencing take place.

Although, by applying a memory mapping architecture, ObjectStore surely provides a fast mechanism, there exist disadvantages induced by this approach. The main disadvantage is that the object references are physical. Moving of objects is not possible without the penalty of changing all references to it. But this is usually an impossible task, since all objects would have to be inspected or expensive backward references have to be maintained. By this, clustering becomes a severe problem, since the location of an object must be determined at the time of its creation and cannot be dynamically changed.

Considering object references being realized as virtual memory pointers, it becomes obvious how the fast migration path works. Any virtual memory pointer is interpreted as an object reference and hence can be handled by ObjectStore. The adaptations to the program handling its persistent data by reading and writing files are the following:

- The creation of a database where the data is to be stored

- The insertion of transaction boundaries

- The insertion of special commands to indicate the types whose instances are to be persistent

- The deletion of the read and write code within the application program

While this mechanism works for a single application, it fails to integrate different applications in which the class definitions differ. Applications treated this way are able to communicate only via instances of identical types. Thus, it is easy to change programs to use ObjectStore instead of file systems for persistence management; but it is impossible to integrate applications using different class definitions this way.

26.2.2   Example and Query Facilities

The main issue of this subsection is to give a flavor of the ObjectStore query capability. For this, we rely on the following example class definition for *Employee*s:

```
class Person {                    class Employee: public Person {
public                            public
    string name;                      float salary;
    int age;                      };
};
```

Note that we purposely made all attributes **public**—thereby allowing the clients of the classes *Person* and *Employee* direct access.

In order to have a set of objects against which queries can be run, we have to utilize ObjectStore's library facilities for set-structured classes. A persistent set containing *Employee* objects is declared as follows:

> **persistent**<db> **os_Set**<Employee *> myEmps;

where *db* identifies a database to store the persistent set to which *myEmps* is referring. Now assume that the set referred to by *myEmps* has been "populated" with *Employee*s.

A query in ObjectStore always returns a subset of the set against which the query is executed. The resulting set can then be assigned to an accordingly declared variable pointing to a set-structured object. Let us state an easy example in which we retrieve all *Employee*s earning more than 100000:

> myEmps [: salary > 100000 :];

Note that we directly used the attribute name *salary*. For query evaluation, it is implicitly evaluated for each member of the set *myEmps*. Within the query, the set members iterated through can be referred to by **this**. In fact, the above query is an abbreviation of

> myEmps [: **this**→salary > 100000 :];

As the query above, this query returns a set of all *employees* contained in *myEmps* who earn more than 100000. Of course, this query is useless in its stand-alone form, since it gets lost directly after returning the resulting set. Hence, we assign the result to a variable called *richEmps* as follows:

> **os_Set**<Employee * > richEmps = myEmps [: salary > 100000 :];

Note that the result set is transient, since we omitted the keyword **persistent** in the declaration of the variable *richEmps*.

More complex queries can be stated by using the standard C or C++ Boolean connectors. Here are two examples:

> myEmps [: salary > 50000 && salary < 100000 :];
> myEmps [: salary > 250000 || age > 60 :];

The first query asks for employees who earn between 50000 *and* 100000; the second query asks for those earning more than 250000 *or* those of age greater than 60.

Of course, in between "[:" and ":]" any legal Boolean expression of C++ is allowed. In particular, member function calls and occurrences of variables are legal.

26.3 Exercises

26.1 Redo the exercises of Chapter 8 for Itasca.

26.2 Restate the queries of the exercises of Chapter 14 in ObjectStore's query language. Which queries can be expressed, which can't? Why?

26.4 Annotated Bibliography

The research prototype Orion, which evolved into the Itasca product, is described in a series of articles [BCG+87, KCB88, KBC+88] and a book by Kim [Kim90].

A survey on ObjectStore can be found in [LLOW91]. Its query-processing capabilities are described in [OHMS92].

27

Selected Systems

This chapter contains a short survey of selected object-oriented database management systems currently available on the market. It is by no means complete. First of all, due to space limitations, we could not describe all available systems. Second, also for included systems the survey does not necessarily contain all the informations that might be of interest. The main idea of this chapter is to describe how the ideas and concepts presented in this book found their way into products. The goal was **not** to provide a top-ten list of object-oriented database management systems. Note that all features described are based on information the authors of the book received upon request from the vendors. Systems are described in alphabetical order.

27.1   GemStone

GemStone is also described in more detail in Chapter 23.

Address: Servio Logic Corporation
                  1420 Harbor Bay Parkway, Suite 100
                  Alameda, CA 94501
                  U.S.A.

The *GemStone* data management system was one of the first systems entering the market. It exhibits a *client/server* architecture. The GemStone system is distributed over two processes: the *Gem* and the *Stone* processes. The *Stone* process delivers the data management capabilities performing disk I/O, concurrency control, recovery, and authorization. It runs on the server. The *Gem* process runs either on the server or on a client. It comprises compilation facilities, browsing capabilities, and user authentication. GemStone is implemented using the C programming language.

27.1.1   Data Model

The data model of GemStone is called *OPAL*, which is a derivative of *Smalltalk-80*. It is dynamically typed and, hence, type safety cannot be assured at compile time. Instead, types are checked at run time. As in Smalltalk, a class is described by its structure and behavior in the form of *instance variables*—attributes in our terms— and *methods*—type-associated operations in our terms. Further, *single inheritance* is supported. Currently, there is no support of multiple inheritance, but this is planned for future releases. Type extensions are not maintained automatically. There exists no explicit support for composite objects.

Object deletion is implicit; that is, objects are deleted as soon as no incoming reference exists any more—i.e., automatic garbage collection is provided. Since explicit deletion of objects is not possible, there is no danger of dangling references.

A nice feature is that classes and methods can be created dynamically. Hence, they are available to the entire system immediately after their definition. Together with the interactive nature of the system, this speeds up prototyping because new classes and methods can instantaneously be utilized.

As the only support for *queries*, a special method called *select* is provided and allows to retrieve a subset of a collection. However, this query facility does not provide for restructuring the result objects.

27.1.2   Control Concepts

By default, GemStone provides *optimistic concurrency control*. Optional pessimistic concurrency control with locking at the object level is also available. The locks supported are *read*, *write*, and *exclusive* locks, where the *write* lock allows for concurrent reads. If locks are explicitly requested by a transaction, then pessimistic concurrency control takes place. Whenever a conflict is detected, GemStone sends a notification to the user, who then can save his or her work before aborting the transaction.

In order to protect data against losses due to disk failure, GemStone allows the user to replicate data onto up to six disk replicates—called *mirror disks*. These are automatically maintained. Committed data are fully recoverable.

In order to support authorization, GemStone provides read/write control and authorization at user, group, and world levels.

GemStone supports multiple name spaces that can be compared to different schemas as introduced in Chapter 18. No additional structuring of different schemas seems possible. On the schema evolution side, GemStone provides simple changes to class definitions as, e.g., adding and deleting an attribute. Automatic conversion of objects guarantees schema-object consistency after a class definition has been modified—as a response to changing system requirements, for example.

27.1.3   Performance Enhancements

GemStone supports *user-controlled clustering*. B-tree *indexes* on a single attribute defined for objects within a given collection are supported. This index can be freely defined for any collection. For more details see Section 23.11.

27.1.4   Miscellaneous

Many object-oriented systems do have problems with the input of large amounts of data within a single transaction. To overcome this deficiency, GemStone provides a *bulk loader*, which is capable of importing large amounts of data into the object base.

The set of available tools includes a source-level debugger.

GemStone provides foreign language interfaces for applications that are written in Smalltalk, C, Fortran, Pascal, Ada, C++, and Objective C. An interface library provides direct access to GemStone's DDL, DML, and query language. Extracting and utilizing data from relational database systems are also possible. In particular, there exists a relational gateway to SYBASE.

An extensive collection of system classes is provided in the form of a library.

27.1.5   Literature

A survey article on GemStone is contained in a special issue on database systems of the *Communications of the ACM* [BOS91]. The roots of GemStone can be found in [CM84]; later articles are [MSOP86], [MS87], and [BMO+89]. More background literature on GemStone is cited at the end of Chapter 23.

27.2   Itasca

Itasca is also described in some more detail in Section 26.1.

Address: Itasca Systems Inc.
                  2850 Metro Drive, Suite 300
                  Minneapolis, MN 55425
                  U.S.A.

The Lisp-based *Itasca* database management system was derived from the *Orion* system developed at MCC. It is implemented in Common-Lisp. The system provides for *multiple clients* and *multiple servers* where the nodes in a network can be dynamically reconfigured. No central server is required, which enhances availability. It runs on most UNIX workstations.

27.2.1   Data Model

The data model of Itasca is an extension of Common-Lisp. Within it, classes—types in our terms—are described by instance attributes—attributes in our terms—and methods—type-associated operations in our terms. Methods are stored within the database.

Multiple inheritance is supported. Itasca does not preclude conflicts due to multiple inheritance. Several conflict resolution strategies are available. Deviating from the object model presented in this book, classes are also objects, called *class objects*. They can receive messages in very much the same way as ordinary objects. These messages return schema information like attribute names, subclasses, and the like. Itasca does not support inverse attributes but provides composite objects to explicitly support the *part-of* relationship.

Itasca supports single target queries; i.e., a subset of a collection of objects can be queried.

27.2.2   Control Concepts

Traditional ACID transactions are called *short transaction* in Itasca. These transactions are organized within a *session*, where a session contains a sequence of transactions. Sessions themselves can be nested or even shared. Serializability for short transactions is guaranteed by a pessimistic locking protocol where locking takes place at the object granularity level. The locks supported by Itasca are IS, IX, S, SIX, and X. Lock requests have to indicate whether to abort or restart a transaction involved in the deadlock.

In order to support collaborative work, Itasca implements a database hierarchy consisting of *public*—in Itasca called *shared*—and *private databases*. Among these, a *checkOut/checkIn* mechanism also supports *long transactions*: Objects checked out into private databases remain there persistently, even after a crash. *Change notification* is supported in both forms, i.e., flag-based and message-based. Further, Itasca provides *hypothetical transactions* that always abort. These are useful to test transactions and, in this way, to answer questions of the "what-if" style.

Full recovery is provided. Itasca protects data against loss by CPU and disk failure. An UNDO log is used to restore a consistent database state after a crash occurred.

Itasca exhibits a very sophisticated authorization technique, very much like the one described in Chapter 16. Authorization is tied to the class hierarchy. At any level, positive and negative authorizations can be granted. Implicit authorization is supported. The levels where authorization can take place include the database level, classes, instances of classes, attributes, and methods.

Version control in Itasca supports the scheme of *transient, working, and released versions*. Both *static and dynamic binding* of versions to *design objects*—here called *generic versions*—are supported.

Itasca allows flexible dynamic schema evolution. That is, schema evolution operations can be invoked by the application at run time. For the taxonomy of possible schema evolution operations see Section 26.1.

27.2.3 Performance Enhancements

Composite objects can be clustered within the same segment. B-tree indexes on a single attribute defined at the class level are supported.

27.2.4 Miscellaneous

Graphical tools are available for editing the schema and the instances. Further, a particular DBA-Tool supports the database administrator in setting up authorizations, examining and modifying locks, and tuning the performance of the database.

Itasca allows applications to be written in several programming languages. Among the supported languages are C, C++, and Lisp. Calling an Itasca method from a foreign language is possible.

Another feature of Itasca is its support for multimedia data. More specifically, a long-data manager—supporting linear and spatial data—is incorporated into Itasca. As an example, the former is used for storing and retrieving audio data. Spatial data are supported in the form of bitmaps used for storing images.

27.2.5 Literature

The Itasca system evolved from the Orion prototype for which a large number of (very influential) publications exist. More specifically, there exist papers on topics like the object model [BCG+87, KBC+88], composite object support [KBC+87, KBG89], indexing techniques [BK89], clustering [BKKG88], schema evolution [BKKK87, KC88b], transaction management [GK88], and authorization [RBKW91]. [Kim90] covers most aspects of the Orion architecture in a single book.

27.3 MATISSE

MATISSE is a French development which is now also distributed in the United States—therefore two addresses are provided.

Address:

ODB (Object Databases)	Intellitic International SNC
238 Broadway	14, rue du Fort de Saint Cyr
Cambridge, MA 02139	BP 317 - Montigny le Bretonneaux
U.S.A.	78054 Saint Quentin en Yvelines Cedex
	France

27.3.1   Data Model

MATISSE is structured as a database dictionary with a semantic net as its underlying data model. This model is used to build higher-level data structures, e.g., the MATISSE object-oriented *MetaModel*. The *MetaModel* provides many of the concepts introduced in this book. Among the most important are *classes* (types in our terms), *attributes*, *relationships*, and *methods* (operations in our terms). Inverse relationships are automatically maintained—thereby, however, incurring the additional cost upon each reference update. Relationships are handled outside the objects and, hence, can easily be added and deleted allowing a very flexible relationship handling. This seems like a good feature when thinking of hypertext applications.

Besides its attractive dictionary architecture, the most prominent feature of MATISSE is its framework for defining and attaching consistency constraints to the database schema. These constraints include type constraints, cardinality constraints on relationships, pre- and postconditions. All the constraints are automatically checked upon access to the database. Also, MATISSE automatically maintains referential integrity. Both (the dictionary architecture and the consistency control) together allow the user to adjust the data model he or she is actually working with.

Another nice feature is that the DML and DDL of MATISSE are identical, which is due to consequently following the dictionary/MetaModel approach.

27.3.2   Control Concepts

On the safety side, MATISSE supports on-line incremental backup, transparent data replication, and dynamic disk recovery. There exists also some kind of concurrency control—the details were, however, not covered in the system overview accessible to the authors.

Dynamic schema evolution is supported. MATISSE supports versioning and historical data management.

27.3.3   Performance Enhancements

MATISSE provides for dynamic clustering and indexing.

27.3.4 Miscellaneous

MATISSE provides several tools for development and administration. Additionally, compression and encryption of data are available.

27.3.5 Literature

This overview of MATISSE is based on a product description obtained from the vendor.

27.4 O_2

O_2 is also described in more detail in Chapter 25.

Address: O_2 Technology
                7, rue du Parc de Clagny
                78000 Versaille
                France

O_2 exhibits a multiple client/multiple server architecture. Its data model is similar to the generic object model (GOM) presented in this book.

27.4.1 Data Model

O_2 has its own data model. It was designed to be hosted in several programming languages like C and C++. Therefore, for defining operations, a derivative of C called O_2C is used. The data model capabilities of O_2 are very similar to those presented in this book—abstracting from the syntax. For example, it provides for types and sorts. Multiple inheritance is supported. Late binding is provided.

Opposed to other systems, methods are stored in the database and, hence, are also under the control of the database. They are dynamically loaded and linked at run time.

Persistence itself is guaranteed by naming a component and making this name known to the schema. By inserting an object into a persistent, e.g., named, set, guarantees the persistence of this object. It is "deleted" by removing it from the set. The concept of persistence itself is bound to the notion of reachability from a named object: Objects no longer referenced by any persistent variable are implicitly deleted, i.e., automatic garbage collection is provided.

O_2 provides a powerful query language which exhibits a *select-from-where* style syntax (cf. Sections 25.1.6 and 14.3).

27.4.2   Control Concepts

Transaction control is optimistic in O_2. Read locks are acquired as soon as a page is accessed. If the contents of a page are modified, no further lock is acquired. Instead, granting write locks is deferred until the end of the transaction. O_2 guarantees serializability.

Schema evolution is possible, resulting in object conversion, if necessary.

27.4.3   Performance Enhancements

Run-time performance can be increased by clustering through the definition of placement trees and by creating indexes.

27.4.4   Miscellaneous

There exists support for including applications written in various languages.

O_2 comes with an extensive set of browsers: a schema browser, a class browser, an application browser, a function browser, and a persistent name browser. All browsers are based on the window system OSF/Motif. Additional tools provided by O_2 Technology are a program source editor, a debugger, and a graphical user interface generator.

27.4.5   Literature

The development of O_2 was very influential for the area of object-oriented database research because of the numerous papers published by the O_2 developers. Most of these papers are contained in a book edited by Bancilhon, Delobel, and Kanellakis [BDK92].

A survey article on O_2 appeared in a special issue on database research of the Communications of the ACM [Deu91].

Query languages for object models are reviewed in [Ban89]; the declarative query language of O_2 is described in [CDLR89, CD92].

27.5   Objectivity/DB

Address: Objectivity, Inc.
                    800 El Camino Real, 4th floor
                    Menlo Park, CA 940025
                    U.S.A.

The object-oriented database system Objectivity/DB provides full and transparent distribution. Multiple databases can be placed anywhere in a heterogeneous net. Objectivity/DB is one of the first object-oriented database management systems associated with a large distributor (Digital Equipment Corporation).

27.5.1   Data Model

The designers of Objectivity/DB chose C++ as its data model where *multiple inheritance* is supported. Furthermore support exists for *complex objects* and *binary relationships*—here called associations—of any cardinality type. Additionally, Objectivity/DB provides *VArrays*, which are arrays of variable size; i.e., the size of *VArrays* can be changed dynamically.

27.5.2   Control Concepts

Short transactions are supported by a two-phase locking protocol. To enhance the flexibility of the locking scheme, multiple granularity locking is provided. Deadlocks are automatically detected. The transaction requesting the latest lock involved in a deadlock is aborted. A checkIn/checkOut mechanism supports long transactions. Automatic recovery from hard and soft failures is provided. A password protection mechanism safeguards schemas. Objectivity/DB provides support for versioning and configuration management.

27.5.3   Miscellaneous

Through clustering, performance enhancements are possible.

A nice additional feature is that a graphical Hypertext view is provided in the database's contents. Other tools comprise a browser and a database debugger through which the database may be changed.

27.5.4   Literature

The description of Objectivity/DB is based on a product description obtained from the vendor.

27.6   ObjectStore

ObjectStore is also described in somewhat more detail in Section 26.2.

Address: Object-Design Inc.
                    1 New England Executive Park
                    Burlington, MA 01803
                    U.S.A.

ObjectStore exhibits a multiple server/multiple client system architecture. It has been implemented using C and C++. One of its most prominent features is the virtual memory mapping (cf. Section 26.2).

27.6.1   Data Model

ObjectStore's data model is an extension of the C++ object-oriented programming language. Object types are called *classes* in C++, the attributes *members*, and the type-associated operations *member functions*. In ObjectStore, member functions are stored in C++ binary files. To account for the database functionality, C++ is slightly extended providing a clean integration. Database functionality can also be accessed through a library interface. In this case, no extensions of C++ are needed.

Further, collections and queries on collections are supported. The query facility consists of a special member function, called *select*, taking a binary expression as its argument and returning a subset of the receiver collection.

ObjectStore supports binary relationships of any cardinality type—i.e., 1 : 1, 1 : N, and N : M relationships—in the form of inverse attributes. The integrity of these attributes is automatically maintained by the system.

Objects can be deleted explicitly by the database user. No automatic garbage collection is provided. If an illegal (dangling) reference is detected, ObjectStore provides two mechanisms to handle this. The first reaction is raising an exception; the second is just assigning NULL to this reference. The user can choose from these two possibilities.

Persistence is fully orthogonal to class definitions. At creation time, the application program has to indicate whether the object is to be persistent or transient. This is done by overloading the *new* operator, which takes a database as an argument, if the object to be created is persistent. This operator corresponds to the *create* operator in GOM.

27.6.2   Control Concepts

ObjectStore provides multiple granularity locking for objects in the database. The granularity levels are segments, pages, or groups of objects—called collections. A two-phase locking protocol guarantees serializability of all transactions. Following the automatic detection of a deadlock, the latest transaction involved is restarted by default. The restart is initiated until the transaction terminates successfully, or a specified limit of retrials is exceeded. Besides this default handling of deadlocks, many operations are available that let the user choose which transaction to abort.

Recovery is supported by a log-file containing after-images of modified pages. These redo records are held in a log-file at the server site in order to be able to restore the effects of a committed transaction after a system failure.

Long transactions are supported via a checkIn/checkOut mechanism. This mechanism is tied to the handling of versions and configurations. Configurations of objects can be checked out into private databases and—after possible modifications—checked back into the public database. The concurrency control protocol governing multi-user access can be either pessimistic or optimistic. For different versions, simultaneous access is granted.

ObjectStore also supports configurations that may even be nested. Through configurations, it is possible to group objects and use these groups for versioning, locking, and clustering purposes.

Authorization is supported similar to the access control provided by the operating system. The level of granularity is an entire database. Permissions assigned are read, write, and execute privileges. User identification and groups are protected by the operating system's password protection mechanism.

27.6.3 Performance Enhancements

ObjectStore essentially supports left-complete access support relations—as introduced in Chapter 20. Hence, indexing of paths is possible—these paths may even contain collection-structured objects. Indexes may be added and dropped at any time.

Run-time performance can be increased by user-defined clustering. At the objects' creation time, the user is able to specify an object close to which the newly created object is to be stored. This object has to be given as an optional parameter to the *new* operator.

Additional performance enhancement is achieved by the virtual memory mapping architecture, which immediately swizzles all references of a page as soon as the page enters the database buffer (cf. Section 26.2).

27.6.4 Miscellaneous

Large objects spanning many pages are supported. These can be used to store multimedia data. A graphical schema designer and interactive browser are available. A low-cost migration path to integrate C applications is provided (cf. Section 26.2).

27.6.5   Literature

A survey article on ObjectStore appeared in a special issue on database research of the Communications of the ACM [LLOW91]. The query-processing capabilities of ObjectStore are described in a recent paper [OHMS92].

27.7   Ontos

Ontos is also described in more detail in Chapter 24.

Address: Ontologic Inc.
                   3 Burlington Woods
                   Burlington, MA 01803
                   U.S.A.

Ontos is implemented in C++ and provides a multiple client/multiple server architecture. Each client process works with a logical database that may be distributed over several servers. One server—the primary server—is distinguished and channels the net traffic for the client. The distribution of data itself is transparent to the user. In order to cope with different application demands, Ontos supports the coexistence of several storage managers.

27.7.1   Data Model

As in many other systems, C++ is the data model of Ontos. Methods are stored in ordinary C++ binary files rather than in the database. They are linked dynamically with the object data when the objects are accessed. Methods can be invoked either classically by a compiled call or by a facility interpreting name strings at run time. Ontos operations are invoked by library calls.

Ontos supports inverse attributes whose integrity is maintained automatically. All possible cardinality types for binary relationships are supported. Ontos uses logical object identifiers. Objects are to be deleted explicitly, which may possibly lead to dangling references. Automatic garbage collection is not supported.

In Ontos, persistence is not orthogonal to class definitions. Instead, classes having only persistent instances are distinguished from classes having transient instances. Classes whose instances are persistent have to be direct or indirect subclasses of the persistent root class *Object*—see Chapter 24 for more details. The disadvantage of this approach is that if there is a class where transient as well as persistent instances are useful, two class definitions are necessary. To overcome this deficiency to some extent, the automatic generation of a database schema from C++ class definitions is possible.

As a query language ObjectSQL is provided. It is an almost full-fledged multitarget query language including quantifiers. Within the query, member functions can be invoked. Nesting of queries is excluded, so far. It is important to note that ObjectSQL bypasses the information hiding provided by C++. More specifically, all attributes become visible, even if they are specified as being private or protected in the class definition.

27.7.2 Control Concepts

In Ontos the two-phase locking protocol guarantees serializability of short transactions. The user may choose between optimistic and pessimistic concurrency control. Multiple granularity locking can be applied. The levels of locking are objects, aggregates—which are user-defined sets of objects—and types. Deadlocks are detected automatically. Nested transactions as well as notification are supported.

In order to enhance concurrency and support collaborative work, Ontos provides *shared transactions*. Here, several client processes may modify the same objects. The modification is committed only if all client processes agree. *Long transactions* are supported through a class library. Long transactions can be suspended and resumed later. All changes performed before a suspension are visible for the long transaction itself after resuming, but they become visible to the public only after the commit of the long transaction.

Ontos allows the definition of explicit safe points. Mirroring disks, i.e., disks having the same contents and, in this way, providing redundancy for data safety purposes, are maintained automatically by Ontos.

Versioning is supported through a class library. Instances are versioned by providing them with a name. Version spaces—or name spaces—are organized hierarchically. Version spaces can also be used to increase concurrency. Memory overhead is minimized by applying a delta technique where only the changes between different versions are stored.

Only simple schema evolution operations like adding or deleting an attribute are supported. Schema/object consistency is maintained by conversion. Masking is not possible.

Database security is realized through database file access rights.

27.7.3 Performance Enhancements

Performance enhancements are possible by clustering, indexing, and caching.

Persistent objects are in either of two states: *deactivated* or *activated*. Deactivated objects are stored on disk. After activation, the objects are transferred into main memory and can then be treated as ordinary C++ objects. The representations on disk and in main memory differ. Activation takes place under explicit programmer control or automatically upon dereferentiation—for more details see Chapter 24. At activation time, all direct references of objects pointing to the newly activated object are swizzled (the consistency maintenance of swizzled references is delegated to the user, however).

Ontos supports physical clustering of objects. Further, to enhance performance, logical clusters can be defined. Whenever one object within such a logical cluster—called *groups*—is activated, all the other objects in the group are activated, too. Regrouping of the objects during run time is possible. Grouping is independent of the physical distribution of the objects.

27.7.4   Miscellaneous

Interactive tools allow for schema editing and object browsing. An additional tool enables the creation of objects for which no class exists and for dynamically creating a new class. The interface builder allows for easy generation of user interfaces. The masks used by the interface builder and designed by the user are stored within Ontos. A fourth generation language (4GL) interface with report generator is available.

A *bulk loader* enables the loading of large amounts of data into the system. Further, there exists a tool for database administration. Via this tool, it is also possible to modify the physical clustering of objects. A *make* tool keeps application code synchronized with schema updates. The *DBATool* supports the tasks of the database administrator.

Language support exists for C, C++, Smalltalk, and Ada.

For multimedia data, Ontos provides support for large objects.

27.7.5   Literature

Ontos is described in an Ontologic technical report [AHS90], which also appeared in a slightly different form as [AHS91].

The book edited by Gupta and Horowitz [GH91] contains several chapters on the use of Ontos and its predecessor Vbase to CAx applications.

27.8   OpenODB

Address: OpenODB
                  Hewlett-Packard Laboratories
                  1501 Page Mill Road
                  Palo Alto, CA 94304
                  U.S.A.

27.8.1   Architecture

OpenODB came out of the Iris research prototype developed at the HP research laboratory in Palo Alto. It uses a client/server architecture. The server site consists of a storage manager—a relational DBMS—on top of which the object manager is placed.

27.8.2   Data Model

The data model of OpenODB is based on the functional data model. As a consequence, attributes as well as operations are modeled by functions. The data model and query language of OpenODB are called OSQL.

To give a flavor of the data model, we model the simple *Emp* and *Manager* type hierarchy and give the definition of the type-associated operation *increaseSalary*. The type definitions look as follows:

```
CREATE TYPE Emp
    FUNCTIONS (name Char,
                    salary Float);
CREATE TYPE Manager SUBTYPE OF Emp
    FUNCTIONS (backup Emp);
```

Attributes are called *functions*. Hence, the above type definition for *Emp* contains the two familiar attributes *name* and *salary*. Note that not only tuple-structured types are supported, but also bags, sets, and lists.

The specific type-associated operation *increaseSalary* is defined as follows:

```
CREATE FUNCTION increaseSalary (Emp e, Float inc) → FLOAT
    UPDATE salary(e) := (salary(e) + inc);
```

Single and multiple inheritance are supported. During their life cycle, objects can dynamically be associated to different types. Moreover, it is possible for an object to belong to several types at the same time—even to types that are not related via the sub/supertype relationship. This way, OpenODB supports a very flexible *role* concept that enables the use of the same object in many different roles.

27.8.3   Control Concepts

OpenODB provides for transaction and session control. A session is organized into a set of transactions. Transactions are guaranteed to exhibit the ACID properties. Within a transaction safe points can be defined and subsequently be restored.

OSQL exhibits a very orthogonal authorization concept. There exist two rights that can be granted for functions to a user or a group of users: CALL and UPDATE.

They grant the right to invoke and update a function, respectively. The owner, i.e., the creator, of a function is granted both rights by default. In addition, he or she can grant and revoke these rights to and from other users or groups of users.

The groups—constituting sets of users inheriting the rights granted to the group— are organized into a hierarchy. The root group is called PUBLIC. Every user is a member of this group. Groups inherit all the rights granted to parent groups. The set of rights a single user possesses is simply the union of all the rights of all groups the user is a member of.

27.8.4 Performance Enhancements

Clustering and indexing are supported.

27.8.5 Miscellaneous

The graphical browser of OpenODB allows the user to interactively browse the database schema and the database contents. It is also intended to ease finding code for reusability purposes.

27.8.6 Literature

OpenODB is based on the Iris database system prototype implementing the functional data model. It is described in [FBC+87, AD92, WLH90]. The access control concepts are described in [ADG+92].

27.9 Statice

Address: Symbolics, Inc.
8 New England Executive Park, East
Burlington, MA 01803
U.S.A.

The Statice database system with its multiple clients/single server architecture is based on Common-Lisp.

27.9.1 Data Model

The data model is similar in flavor to the Orion data model, which is also Lisp based. It includes multiple inheritance. For associative access *iterators* can be used.

Statice requires explicit deletion of objects. If an object is deleted, all the references pointing to this deleted objects are also deleted.

27.9.2   Control Concepts

Statice provides ACID transactions. Serializability is guaranteed through the two-phase locking protocol on a page granularity level. Deadlocks are detected automatically, where upon one of the involved transactions is aborted. Recovery is supported by log-files containing redo records.

27.9.3   Performance Enhancements

Performance enhancements through B-tree indexes on attribute values and clustering are possible.

27.9.4   Miscellaneous

A graphical browser is available as well as an extensive set of other development tools.

27.9.5   Literature

Statice is described in [WFGL91].

27.10   UniSQL

Address: UniSQL, Inc.
                 9390 Research Blvd.
                 Austin, TX 78759
                 U.S.A.

UniSQL is somewhat different from the other systems described, in that it tries to enhance a relational database by features pertinent to an object-oriented database. Hence, it can be classified as an extended relational system. On the one hand, it allows the user to stick fully to the relational model. On the other hand, it allows the user to incorporate as much "object-orientedness" as he or she wishes. This is supported by enhancing the strict relational model by allowing tables to be nested, attributes to contain procedures and pointers, and multiple inheritance. In this respect UniSQL is rather similar to the POSTGRES research prototype (and its commercial outcome called *Miro*).

27.10.1   Literature

The overview of UniSQL is based on a high-level product information obtained from the vendor.

27.11   Versant

Address: Versant Object Technology
                    4700 Bohannon Drive, Suite 125
                    Menlo Park, CA 94025
                    U.S.A.

The object-oriented database system Versant exhibits a multiple client/multiple server architecture.

27.11.1   Data Model

Versant relies on C++ with multiple inheritance as its data model.

Objects have to be deleted explicitly.

Persistence of objects is realized by a class *Persistent* from which other classes that ought to be persistent have to inherit.

Queries on instances of a class are possible via a *select* function, which can be invoked on a special meta class.

27.11.2   Control Concepts

Versant supports two types of transactions: *short* and *long transactions*. Short transactions obey the ACID rules. A two-phase locking protocol ensures serializability among short transactions.

A long transaction is organized into several short transactions, and each short transaction has to be attached to a long transaction. Locks can be transferred between the long transaction and its short transactions. Among the locks supported by Versant are S, X, and the browsing lock. Explicit safe-points are also supported.

A versioning concept is supported, too. It can be coupled via a checkIn/checkOut mechanism with public and private databases to increase concurrency.

Schema evolution is very limited in that addition and deletion of attributes are allowed only in leaves of the class hierarchy. Database security is supported through database file access rights.

27.11.3   Performance Enhancements

Indexing and clustering allow the user to enhance performance.

27.11.4   Miscellaneous

A graphical user interface providing a browser is included in the system. It allows retrieval at the schema level, i.e., of classes and their attributes, as well as retrieval and modification at the instance level.

27.11.5   Literature

The discussion of Versant is based on information obtained from the vendor.

27.12   Other Influential Systems and Prototypes

Besides the above mentioned commercial systems one should notice all the other influential systems and prototypes that in one or another way contributed to the development of object-oriented database management systems. Some of the more influential prototype developments are listed below. More discussions on object-oriented database systems can be found in [ASL90, AWSL92, Cat91, KMRC90].

A benchmark especially designed for object-oriented database systems together with some performance data for some systems can be found in [CDN93].

AIM-P [KDG87, DKA$^+$86, LKD$^+$88] is a database management system based upon the NF^2 data model. It has been developed at the IBM Scientific Research Center in Heidelberg. This project is tightly connected to the R^2D^2 [KLW87] prototype which builds upon AIM-P. The application considered in R^2D^2 was robot simulation.

AVANCE [BB88, BH89] (formerly called **OPAL** [ABB$^+$84]—not to be confused with the GemStone data model, which is also called OPAL), is an object-oriented database system for office information management.

Cactis [HK89, DKH90, HK88, HK87a] is a database system that implements a data model that combines features from the functional data model with semantic data modeling concepts.

ConceptBase [Jar92, MBJK90, HJEK90, JJR90, JJ91, RJG$^+$91] is a deductive object-oriented database system developed at the Universities of Passau and Aachen. Its main objective is to support collaborative work in design environments. The data model Telos integrates techniques from deductive, object-oriented, and temporal databases. Other problem areas attacked in this project are query optimization and efficient consistency checking.

CoReDB [BCFL90] is a complex relational database developed at the Universities of Dortmund and Bonn. Special emphasis is to be paid to the extensible object management system **OMS**, which has information on the type of the objects it stores. The main challenge of this system was to design it as open as possible; that is, it can be extended by hitherto not integrated types.

DAMOKLES [DGL86, DGL87b, DGL87a] is a structurally object-oriented database management system especially designed for software engineering applications. It was developed at the FZI (Computer Science Research Center) at Karlsruhe. Special emphasis was put on the development of versioning mechanisms. The successor of the DAMOKLES prototype is **OBST**, an object manager providing persistence to C++ objects.

DAMASCUS [DKM86, KDM88] is a structurally object-oriented database management system especially developed for very large scale integration (VLSI) design applications. Special emphasis was put on the development of versioning mechanisms and the support of activity. This project is tightly connected to the previous one.

DAPLEX [Shi81] is a functional data model that forms the basis of the **Probe** development. Furthermore, DAPLEX had a very strong influence on the *Iris* data model design.

DASDBS [PSS⁺87, SPSW90, SS91, SLT91] is an NF^2 database management system developed at the University of Darmstadt and the ETH Zürich. It is now used as a kernel of the fully object-oriented database management system **COCOON**. In this system special emphasis is put on view mechanisms and a more flexible association of objects to types.

DBPL [MS89] is the successor of PascalR [Sch77] and is based on Modula2. The aim in the development of DBPL was to provide a seamless integration of database functionality into a programming language.

EasyDB [SF93] is an object-oriented database system whose data model is based on the entity-relationship model. One-to-one and one-to-many relationships are supported bidirectionally. Entity types may be specified as a subtype of another type; thereby supporting single inheritance.

ENCORE [SZ86, ZH87, Zdo89, SZ89b, SZ90, SZ89a] is an object-oriented database management system built upon the *ObServer* object manager. Special emphasis is put onto optimization, evolution, and concurrency control.

The **EXODUS** [CDRS86, CDF⁺86, CDV88, RCS89, CDG⁺89, SC90, WD92] system was designed to support the development of database systems. It can be classified as a *database generator*. The core idea is to provide a toolkit that simplifies the development of database management systems. As one validation of this approach, a persistent object-oriented programming language called E was realized. E is an extension of C++. Special emphasis in developing the EXODUS system was put on performance and the support of large objects within the EXODUS storage manager. The **WiSS** storage manager can be seen as the predecessor of the EXODUS storage manager.

Galileo [ACO85] is a strongly typed, conceptual database language that had a strong influence on many object-oriented data model designs.

As EXODUS **GENESIS** [Bat87, Bat86, BLW88, Bat88, BBG⁺88] is also a database generator, but it does not rely on the toolkit approach. Instead, the core idea is to generate database management system modules automatically from a given specification.

GOM [KMWZ91] is a prototype of the data model presented in this book. Special emphasis was put on the development of optimization techniques [KM92a, KKM91, KMS92].

Gral [BG92] is an extensible relational database system that was designed to support geo-relational applications.

IDB [PDS] is a *fully distributed database* exhibiting a peer-to-peer architecture in which no central server is required. For the structural part of objects IDB provides three kinds of objects: *strings*, *nodes*, and *sequences*. Two types of sequences are supported: arrays where access is performed via an index, and double-linked lists. Single and multiple inheritance are supported. All nonleave classes are supposed to be virtual. Hence, no instance can be derived from a non-leave class.

As already mentioned, the **Iris** database management system [DKL85, FBC⁺87, DS89, WLH90, DFK⁺86, FAB⁺89] was the predecessor of the OpenODB product. It implements the functional data model upon a relational data manager.

Jasmine [ISK⁺93] is an object-oriented knowledge base management system developed at Fujitsu Laboratories. It enhances the core object-oriented model by multimedia capabilities.

KALA [SG91] is an object repository running on various hardware platforms; including PCs. Object placement is explicitly controlled by the user; i.e., logically related objects are placed in the same so-called *monad* that forms the unit of transportation between secondary and main memory. Unlike pages monads have variable size.

LOGRES/ALGRES [CCRZ⁺90, C⁺92] is a non-first normal form relational model that was implemented on top of a relational kernel database system.

Miro is the commercial outcome of the **POSTGRES** research effort—which is listed below.

Mneme [MS88, Mos92, HM90] is an object manager developed at the University of Massachusetts at Amherst. The main goal was to provide highest efficiency with a minimal orthogonal functionality. Special emphasis was put on the investigation of pointer-swizzling strategies.

ObServer is an object manager developed at the Brown University.

ODE is an object-oriented database management system developed at AT&T. Lately, the development puts special emphasis on object activity [AG89, GJS92b, GJS92a].

Orion is the prototype database system from which Itasca evolved. Literature on Orion is cited at the end of Chapter 26.

OSCAR [HFW90, Heu92, HS91] is an object-oriented database system developed at the University of Clausthal. The basis for its query and update language is an object algebra. A query language in the spirit of SQL is available. A remarkable difference to other object-oriented systems is that it provides support for rules.

OZ+ [WL89] is an object-oriented database system that was developed at the University of Toronto to support office information systems.

POSTGRES [SAHR84, SR86, SAH87b, SAH87a, SHP88] is an extended relational database system developed at the University of California at Berkeley. Its predominant features are the handling of the query language QUEL as a data type and the treatment of historical data. Recently, the POSTGRES development was turned into a commercial product, called **Miro**.

Prima [Mit88, HMWMS87, Mit89, GGH+92] is a database management system that builds upon the molecule atom data model (MAD) and was developed at the University of Kaiserslautern. Atoms are identifiers, references, records, arrays, sets, and lists. These can be dynamically tied together to form molecules. Special support for relationships is provided by links.

PROBE has been developed at the Computer Corporation of America. It implements an extension of the DAPLEX functional data model. Emphasis is put onto extensibility and optimization issues [DS85, GO87, OM88, DMB+87].

Rose [Har87, HS88] is an object-oriented database system primarily designed for computer graphics applications.

SIM [JGF+88, FGJ+90] is a commercially available DBMS developed by Unisys Corporation. It utilizes a semantic data model.

The extensible relational DBMS **Starburst** [S+86, LMP87, H+89, WSSH88] has been developed at IBM Almaden Research Center. It supports user-defined extensions for access methods and query optimization.

The **Trellis/Owl** prototype is described in [OBS86, OHK87, SCB+86].

The **VODAK** [DKT88, NS88, KNS90a, KNS90b, MRKN91, RGN90] system supports complex objects, abstract data types, inheritance, and message passing. It is designed to support an integrated information and publication processing system. Within the VODAK project emphasis was put on data mode issues, multimedia support, and new transaction models for supporting cooperation.

Volcano [GM93] is a database research project focusing on advanced query optimization and processing techniques.

Zeitgeist [FJL+88, WBT92] is an object-oriented database management system developed at Texas Instruments. Its successor is the OpenOODB project, which emphasizes an open object-oriented architecture. A beta release of **OpenOODB** is expected to be available for noncommercial use. Commercial possibilities are currently evaluated.

Note: Most of the system names are registered trademarks of their respective vendor companies.

Bibliography

[AB84]   S. Abiteboul and N. Bidoit. Non first normal form relations to represent
            hierarchically organized data. In *Proc. ACM SIGMOD/SIGACT Conf. on
            Princ. of Database Syst. (PODS)*, pages 191–200, 1984.

[AB87]   M. P. Atkinson and O. P. Buneman. Types and persistence in database
            programming languages. *ACM Computing Surveys*, 19(2):105–190, 1987.

[AB91]   S. Abiteboul and A. Bonner. Objects and views. In *Proc. of the ACM
            SIGMOD Conf. on Management of Data*, pages 238–248, Denver, CO, May
            1991.

[ABB+84]   M. Ahlsen, A. Björnerstedt, S. Britts, C. Hulten, and L. Soderlund. An
            architecture for object management in OIS. *ACM Trans. Office Inf. Syst.*,
            2(3), July 1984.

[ABC+76]   M. M. Astrahan, M. W. Blasgen, D. D. Chamberlin, K. P. Eswaran, J. N.
            Gray, P. P. Griffiths, W. F. King, R. A. Lorie, P. R. McJones, J. W. Mehl,
            G. R. Putzolu, I. L. Traiger, B. W. Wade, and V. Watson. System R: A
            relational approach to data. *ACM Trans. on Database Systems*, 1(2):97–137,
            June 1976.

[ABD+89]   M. Atkinson, F. Bancilhon, D. J. DeWitt, K. R. Dittrich, D. Maier, and
            S. Zdonik. The object-oriented database system manifesto. In *Proc. of the
            Conf. on Deductive and Object-Oriented Databases (DOOD)*, pages 40–57,
            Kyoto, Japan, Dec 1989.

[ACO85]   A. Albano, L. Cardelli, and R. Orsini. Galileo: A strongly-typed, interactive
            conceptual language. *ACM Trans. on Database Systems*, 10(2):230–260, June
            1985.

[AD92]   R. Ahad and D. Dedo. OpenODB from Hewlett-Packard: A commercial
            object-oriented database management system. *Journal of Object-Oriented
            Programming*, 4(9):31–35, 1992.

[ADG+92]   R. Ahad, J. W. Davis, S. Gower, P. Lyngbaek, A. Marynowski, and
            E. Onuegbe. Supporting access control in an object-oriented database lan-
            guage. In *Proc. Intl. Conf. Extending Database Technology (EDBT)*, pages
            184–200, Springer-Verlag, Heidelberg, 1992.

[AFS87]   S. Abiteboul, P. C. Fischer, and H.-J. Schek, editors. *Nested Relations and
            Complex Objects in Databases*, volume 361 of *Lecture Notes in Computer
            Science (LNCS)*. Springer-Verlag, Heidelberg, 1987.

[AG88]   S. Abiteboul and S. Grumbach. COL: A logic-based language for complex
            objects. In *Proc. Intl. Conf. Extending Database Technology (EDBT)*, pages
            271–293, Springer-Verlag, Heidelberg, 1988.

[AG89]   R. Agrawal and N. H. Gehani. ODE (object database and environment):
            The language and the data model. In *Proc. of the ACM SIGMOD Conf. on
            Management of Data*, pages 36–45, Portland, OR, 1989.

[AH87a]   S. Abiteboul and R. Hull. IFO: A formal semantic database model. *ACM
            Trans. on Database Systems*, 12(4):525–565, Dec 1987.

[AH87b]   T. Andrews and C. Harris. Combining language and database advances in
            an object-oriented development environment. In *Proc. of the ACM Conf.
            on Object-Oriented Programming Systems and Languages (OOPSLA)*, pages
            430–440, Orlando, FL, Oct 1987.

[AHS90]   T. Andrews, C. Harris, and K. Sinkel. The ONTOS object database. Ontos
            Technical Report, Ontos Inc., 3 Burlington Woods, Burlington, MA 01803,
            USA, 1990.

[AHS91]   T. Andrews, C. Harris, and K. Sinkel. ONTOS: A persistent database for
            C++. In R. Gupta and E. Horowitz, editors, *Object-Oriented Databases with
            Applications to CASE, Networks, and VLSI Design*, pages 387–406, Prentice
            Hall, Englewood Cliffs, NJ, 1991.

[AKMP85]   H. Afsarmanesh, D. Knapp, D. McLeod, and A. Parker. An extensible, object-
            oriented approach to databases for VLSI/CAD. In *Proc. of the Conf. on Very
            Large Data Bases (VLDB)*, pages 13–24, Stockholm, Sweden, Aug 1985.

[AKW90]   S. Abiteboul, P. Kanellakis, and E. Waller. Method schemas. In *Proc. ACM
            SIGMOD/SIGACT Conf. on Princ. of Database Syst. (PODS)*, pages 16–27,
            Atlantic City, NJ, 1990.

[AL80]   M. E. Adiba and B. G. Lindsay. Database snapshots. In *Proc. of the Conf.
            on Very Large Data Bases (VLDB)*, pages 86–91, Montreal, Canada, Aug
            1980.

[AM83]   H. Arisawa and K. Moriya. Operations and properties on non first normal
            form relational databases. In *Proc. of the Conf. on Very Large Data Bases
            (VLDB)*, pages 197–204, Florence, Italy, Oct 1983.

[ANSI81]   American National Standards Institute (ANSI). Digital representation for
            communication of product definition data (initial graphics exchange specifi-
            cation IGES). American Society of Mechanical Engineers (ANSI Y14.26M),
            1981.

[ANSI83]   American National Standards Institute. The programming language Ada
            reference manual. Available as: Lecture Notes in Computer Science No. 155,
            Springer-Verlag, Heidelberg 1983.

[ANSI86]   American National Standards Institute. Database language SQL. Document
            ANSI X3.135, 1986. Also available as: International Standards Organization
            Document ISO/TC 97/SC 21/WG 3 N 117.

[ASL90]   S. Ahmed, D. Sriram, and R. Logcher. A comparison of object-oriented
            database management applications for engineering applications. Research
            Report No R90-03, MIT, Cambridge, MA, Oct 1990.

[AWSL92]   S. Ahmed, A. Wong, D. Sriram, and R. Logcher. Object-oriented database
            management systems for engineering: A comparison. *Journal of Object-
            Oriented Programming*, 5(3):27–45, June 1992.

[Ban88]   F. Bancilhon. Object-oriented database systems. In *Proc. ACM SIG-
            MOD/SIGACT Conf. on Princ. of Database Syst. (PODS)*, pages 152–162,
            Austin, TX, 1988.

[Ban89]   F. Bancilhon. Query languages for object-oriented database systems: Anal-
            ysis and a proposal. In T. Härder, editor, *Datenbanksysteme in Büro, Tech-
            nik und Wissenschaft (GI/SI Fachtagung)*, Informatik Fachberichte Nr. 204,
            pages 1–18. Springer-Verlag, Heidelberg, 1989.

[Bat86]   D. S. Batory. Extensible cost models and query optimization in GENESIS.
            IEEE Database Engineering, 9(4), Dec 1986.

[Bat87]   D. S. Batory. A molecular database systems technology. Tech. Report TR-
            87-23, University of Austin, 1987.

[Bat88]   D. S. Batory. Concepts for a database system compiler. In *Proc. of the ACM
            SIGMOD Conf. on Management of Data*, pages 184–192, Chicago, IL, 1988.

[BB84]   D. S. Batory and A. P. Buchmann. Molecular objects, abstract data types,
            and data models: A framework. In *Proc. of the Conf. on Very Large Data
            Bases (VLDB)*, pages 172–184, Singapore, 1984.

[BB88]   A. Björnerstedt and S. Britts. AVANCE: An object management system. In *Proc. of the ACM Conf. on Object-Oriented Programming Systems and Languages (OOPSLA)*, pages 206–222, San Diego, CA, 1988.

[BBG+88]   D. S. Batory, J. R. Barnett, J. F. Garza, K. P. Smith, K. Tsukuda, B. C. Twichell, and T. E. Wise. GENESIS: An extensible database management system. *IEEE Trans. Software Eng.*, 14:1711–1730, 1988.

[BBG89]   C. Beeri, P. A. Bernstein, and N. Goodman. Concurrency control in distributed database systems. *Journal of the ACM*, 37(2):230–269, April 1989.

[BBKV87]   F. Bancilhon, T. Briggs, S. Khoshafian, and P. Valduriez. FAD, a powerful and simple database language. In *Proc. of the Conf. on Very Large Data Bases (VLDB)*, pages 97–105, Brighton, UK, Sep 1987.

[BCD89]   F. Bancilhon, S. Cluet, and C. Delobel. A query language for the O₂ object-oriented database system. In *Proc. Second Intl. Conf. on Database Programming Languages*, Glenedon Beach, OR, Morgan-Kaufmann, San Mateo, CA, 1989.

[BCFL90]   Th. Bode, A. B. Cremers, J. Freitag, and Th. Lemke. Coupling the complex-relational database CoReDB with the object management system OMS. In *Proc. Int. Conf. Database and Expert Systems Applications (DEXA)*, Vienna, Austria, 1990.

[BCG+87]   J. Banerjee, H. Chou, J. Garza, W. Kim, D. Woelk, N. Ballou, and H. Kim. Data model issues for object-oriented applications. *ACM Trans. Office Inf. Syst.*, 5(1):3–26, Jan 1987.

[BCL89]   J. A. Blakeley, N. Coburn, and P. A. Larson. Updating derived relations: Detecting irrelevant and autonomously computable updates. *ACM Trans. on Database Systems*, 14(3):369–400, Sep 89.

[BM72]   R. Bayer and E. M. McCreight. Organization and Maintenance of Large Ordered Indices. *Acta Informatica*, 1(3):173–189, 1972.

[BD90]   V. Benzaken and C. Delobel. Enhancing performance in a persistent object store: Clustering strategies in O₂. In A. Dearle, G. Shaw, and S. Zdonik, editors, *Implementing Persistent Object Bases*, pages 403–412, Martha's Vineyard, Morgan Kaufmann, San Mateo, CA, Sep 1990.

[BDK92]   F. Bancilhon, C. Delobel, and P. Kanellakis, editors. *Building an Object-Oriented Database System: The Story of O2.* Morgan Kaufmann, San Mateo, CA, 1992.

[BG81]   P. A. Bernstein and N. Goodman. Concurrency control in distributed database systems. *ACM Computing Surveys*, 13(2):185–221, June 1981.

[BG92]   L. Becker and R. H. Güting. Rule-based optimization and query processing in
             an extensible geometric database system. *ACM Trans. on Database Systems*,
             17(2):247–303, June 1992.

[BH89]   A. Björnerstedt and C. Hulten. Version control in an object-oriented archi-
             tecture. In W. Kim and F. H. Lochowsky, editors, *Object-Oriented Concepts
             and Databases*, chapter 18, pages 451–485. Addison-Wesley, Reading, MA,
             1989.

[BHG87]   P. A. Bernstein, V. Hadzilacos, and N. Goodman. *Concurrency control and
             recovery in database systems*. Addison-Wesley, Reading, MA, 1987.

[BK85]   D. S. Batory and W. Kim. Modeling concepts for VLSI CAD objects. *ACM
             Trans. Database Syst.*, 10(3):322–346, 1985.

[BK89]   E. Bertino and W. Kim. Indexing techniques for queries on nested objects.
             IEEE Trans. Knowledge and Data Engineering, 1(2):196–214, June 1989.

[BK91]   N. S. Barghouti and G. E. Kaiser. Concurrency control in advanced database
             applications. *ACM Computing Surveys*, 23(3):269–317, 1991.

[BKK85]   F. Bancilhon, W. Kim, and H. F. Korth. A model of CAD transactions. In
             Proc. of the Conf. on Very Large Data Bases (VLDB), pages 25–33, Stock-
             holm, Sweden, Sep 1985.

[BKK+90]   D. G. Bobrow, K. Kahn, G. Kiczales, L. Masinter, M. J. Stefik, and F. Zdybel.
             CommonLoops: Merging Lisp and object-oriented programming. In A. Car-
             denas and D. McLeod, editors, *Research Foundations in Object-Oriented and
             Semantic Database Systems*, pages 70–90, Prentice Hall, Englewood Cliffs,
             NJ, 1990.

[BKKG88]   J. Banerjee, W. Kim, S. J. Kim, and J. F. Garza. Clustering a DAG for CAD
             databases. *IEEE Trans. Software Eng.*, 14(11):1684–1699, Nov 1988.

[BKKK87]   J. Banerjee, W. Kim, H.-J. Kim, and H. Korth. Semantics and implemen-
             tation of schema evolution in object-oriented database systems. In *Proc.
             of the ACM SIGMOD Conf. on Management of Data*, pages 311–322, San
             Francisco, 1987.

[BLT86]   J. A. Blakeley, P. A. Larson, and F. W. Tompa. Efficiently updating materi-
             alized views. In *Proc. of the ACM SIGMOD Conf. on Management of Data*,
             pages 61–71, Washington, DC, 1986.

[BLW88]   D. S. Batory, T. Y. Leung, and T. E. Wise. Implementation concepts for an
             extensible data model and data language. *ACM Trans. Database Systems*,
             13(3):231–262, Sep 1988.

[BM88]   D. Beech and B. Mahbod. Generalized version control in an object-oriented
            database. In *Proc. IEEE Conference on Data Engineering*, Los Angeles, CA,
            pages 14–22, 1988.

[BMO+89]   R. Bretl, D. Maier, A. Otis, J. Penney, B. Schuchardt, J. Stein, and
            H. Williams. The GemStone data management system. In W. Kim and F. H.
            Lochowsky, editors, *Object-Oriented Concepts, Databases and Applications*,
            pages 283–308. Addison-Wesley, Reading, MA, 1989.

[BMS84]   M. L. Brodie, J. Mylopoulos, and J. Schmidt, editors. *On Conceptual Mod-
            elling*. Springer-Verlag, Heidelberg, 1984.

[Boc87]   S. Bocionek. Dynamic Flavors. Technical report, Institut für Informatik,
            Technical University Munich, Munich, Germany, 1987.

[Boo83]   G. Booch. *Software Engineering with Ada*. Benjamin/Cummings, Menlo
            Park, CA, 1983.

[Boo91]   G. Booch. *Object-Oriented Design with Applications*. Benjamin/ Cummings,
            Redwood City, CA, 1991.

[BOS91]   P. Butterworth, A. Otis, and J. Stein. The GemStone object database system.
            Communications of the ACM, 34(10):64–77, 1991.

[Bra83]   R. J. Brachman. What IS-A is and isn't: An analysis of taxonomic links in
            semantic networks. *IEEE Computer*, 16(10):30–36, Oct 1983.

[BTBO89]   V. Breazu-Tannen, P. Buneman, and A. Ohori. Can object-oriented databases
            be statically typed? In *Database Programming Languges Workshop*, pages
            226–237, Portland, OR, May 1989.

[Bud91]   T. Budd. *An Introduction to Object-Oriented Programming*. Addison-Wesley,
            Reading, MA, 1991.

[BW77]   D. G. Bobrow and T. Winograd. An overview of KRL, a knowledge repre-
            sentation language. *Cognitive Science*, 1(1):3–46, 1977.

[BW79]   D. G. Bobrow and T. Winograd. KRL, another perspective. *Cognitive Sci-
            ence*, 3(1), 1979.

[BW90]   K. B. Bruce and P. Wegner. An algebraic model of subtype and inheritance. In
            F. Bancilhon and P. Buneman, editors, *Advances in Database Programming
            Languages*, pages 75–96, Addison-Wesley, Reading, MA, 1990.

[C+92]   F. Cacace, S. Ceri, L. Tanca, and S. Crespi-Reghizzi. Designing and Proto-
            typing Data-Intensive Applications in the Logres and Algres Programming
            Environment. *IEEE Trans. Software Eng.*, 18(6):534–546, June 1992.

[CAC⁺84]   W. P. Cockshot, M. P. Atkinson, K. J. Chisholm, P. J. Bailey, and R. Morrison. Persistent object management system. *Software—Practice and Experience*, 14:49–71, 1984.

[Car84]   L. Cardelli. A semantics of multiple inheritance. In *Semantics of Data Types*, volume 173 of *Lecture Notes in Computer Science (LNCS)*, pages 51–69, Springer-Verlag, Heidelberg, 1984.

[Car88]   L. Cardelli. Types for data-oriented languages. In *Proc. Intl. Conf. Extending Database Technology (EDBT)*, volume 303 of *Lecture Notes in Computer Science (LNCS)*, pages 1–15. Springer-Verlag, Heidelberg, 1988.

[Cat91]   R. G. G. Cattell. *Object Data Management: Object-Oriented and Extended Relational Database Systems*. Addison-Wesley, Reading, MA, 1991.

[CCRZ⁺90]   S. Ceri, S. Crespi-Reghizzi, R. Zicari, G. Lamperti, and L. A. Lavazza. ALGRES: An advanced database system for complex applications. *IEEE Software*, 7:68–78, July 1990.

[CD92]   S. Cluet and C. Delobel. A general framework for the optimization of object-oriented queries. In *Proc. of the ACM SIGMOD Conf. on Management of Data*, pages 383–392, San Diego, CA, June 1992.

[CDF⁺86]   M. J. Carey, D. J. DeWitt, D. Frank, G. Graefe, J. E. Richardson, E. J. Shekita, and M. Muralikrishna. The architecture of the EXODUS extensible DBMS. In K. R. Dittrich and U. Dayal, editors, *Proc. IEEE Intl. Workshop on Object-Oriented Database Systems*, Asilomar, Pacific Grove, CA, pages 52–65, 1986.

[CDG⁺89]   M. J. Carey, D. J. DeWitt, G. Graefe, D. M. Haight, J. E. Richardson, D. T. Schuh, E. J. Shekita, and S. L. Vandenberg. The EXODUS extensible DBMS project: An overview. In S. Zdonik and D. Maier, editors, *Readings in Object-Oriented Databases*, pages 474–499. Morgan-Kaufmann, San Mateo, CA, 1989.

[CDLR89]   S. Cluet, C. Delobel, C. Lecluse, and P. Richard. Reloop, an algebra based query language for an object-oriented database system. In *Proc. of the Conf. on Deductive and Object-Oriented Databases (DOOD)*, pages 294–313, Kyoto, Japan, Dec 1989.

[CDN93]   M. J. Carey, D. J. DeWitt, and J. F. Naughton. The 007 benchmark. In *Proc. of the ACM SIGMOD Conf. on Management of Data*, pages 12–21, Washington, DC, 1993.

[CDRS86]   M. J. Carey, D. J. DeWitt, J. E. Richardson, and E. J. Shekita. Object and file management in the EXODUS extensible database system. In *Proc. of the*

Conf. on Very Large Data Bases (VLDB), pages 91–100, Kyoto, Japan, Aug 1986.

[CDV88]   M. J. Carey, D. J. DeWitt, and S. L. Vandenberg. A data model and query language for EXODUS. In *Proc. of the ACM SIGMOD Conf. on Management of Data*, pages 413–423, Chicago, IL, Jun 1988.

[CH91]   J. R. Cheng and A. R. Hurson. Effective clustering of complex objects in object-oriented databases. In *Proc. of the ACM SIGMOD Conf. on Management of Data*, pages 22–32, Denver, CO, May 1991.

[Che76]   P. P. S. Chen. The entity relationship model: Toward a unified view of data. *ACM Trans. Database Syst.*, 1(1):9–36, March 1976.

[CJ90]   W. Cellary and G. Jomier. Consistencies of versions in object-oriented databases. In *Proc. Int. Conf. on Very Large Data Bases (VLDB)*, pages 432–441, Brisbane, Australia, Aug 1990.

[CJK91]   W. Cellary, G. Jomier, and T. Koszlajda. Formal model of an object-oriented database with versioned objects and schema. In *Proc. Intl. Conf. on Database and Expert Systems Applications (DEXA)*, pages 239–244. Springer-Verlag, Heidelberg, 1991.

[CK86]   H.-T. Chou and W. Kim. A unifying framework for version control in a CAD environment. In *Proc. Int. Conf. on Very Large Data Bases (VLDB)*, pages 336–344, Kyoto, Japan, 1986.

[CM84]   G. Copeland and D. Maier. Making Smalltalk a database system. In *Proc. of the ACM SIGMOD Conf. on Management of Data*, pages 316–325, Boston, MA, 1984.

[CM90]   A. Cardenas and D. McLeod, editors. *Object-Oriented Database Systems*. Prentice Hall, Englewood Cliffs, NJ, 1990.

[Cod70]   E. F. Codd. A relational model for large shared data banks. *Comm. ACM*, 13(6):377–387, 1970.

[Cod79]   E. F. Codd. Extending the relational database model to capture more meaning. *ACM Trans. Database Syst.*, 4(4):397–434, Dec 1979.

[Cod90]   E. F. Codd. *The Relational Model for Database Management Version 2*. Addison-Wesley, Reading, MA, 1990.

[Com79]   D. Comer. The ubiquitous B-tree. *ACM Computing Surveys*, 11(2): 121–137, 1979.

[Cox86]   B. J. Cox. *Object Oriented Programming: An Evolutionary Approach*. Addison-Wesley, Reading, MA, 1986.

[CR90]   P. K. Chrysanthis and K. Ramamritham. ACTA: A framework for specifying and reasoning about transaction structure and behavior. In *Proc. of the ACM SIGMOD Conf. on Management of Data*, pages 194–203, Atlantic City, NJ, 1990.

[CR91]   P. K. Chrysanthis and K. Ramamritham. A formalism for extended transaction models. In *Proc. Int. Conf. on Very Large Data Bases (VLDB)*, Barcelona, Spain, pages 103–112, 1991.

[CW85]   L. Cardelli and P. Wegner. On understanding types, data abstraction, and polymorphism. *ACM Computing Surveys*, 17(4):471–522, Dec 1985.

[CY91a]   P. Coad and E. Yourdan. *Object-Oriented Analysis*, 2nd edition. Prentice Hall, Englewood Cliffs, NJ, 1991.

[CY91b]   P. Coad and E. Yourdan. *Object-Oriented Design*, 2nd edition. Prentice Hall, Englewood Cliffs, NJ, 1991.

[Dat86]   C. J. Date. *A Guide to DB2*. Addison-Wesley, Reading, MA, 1986.

[Dat87]   C. J. Date. *A Guide to the SQL Standard*. Addison-Wesley, Reading, MA, 1987.

[Dat90]   C. J. Date. *An Introduction to Database Systems*, volume I, 5th edition. Addison-Wesley, Reading, MA, 1990.

[DB82]   U. Dayal and P. Bernstein. On the correct translation of update operations on relational views. *ACM Trans. on Database Systems*, 7(3)381–416, 1982.

[DDKL87]   P. Dadam, R. Dillmann, A. Kemper, and P. C. Lockemann. Objektorientierte Datenhaltung für die Roboterprogrammierung. *Informatik: Forschung und Entwicklung*, 2(4):151–170, 1987.

[Deu91]   O. Deux. The O_2 system. *Communications of the ACM*, 34(10):34–48, 1991.

[DFK$^+$86] N. P. Derret, D. H. Fishman, W. Kent, P. Lyngbaek, and T. A. Ryan. An object-oriented approach to data management. In *Proc. of the COMPCON 31st IEEE Computer Society Intl. Conf.*, pages 330–335, San Francisco, CA, 1986.

[DGL86]   K. R. Dittrich, W. Gotthard, and P. C. Lockemann. DAMOKLES: A database system for software engineering environments. In *Proc. of the IFIP 2.4 Workshop on Advanced Programming Environments*, Trondheim, Norway, June 1986.

[DGL87a]   K. R. Dittrich, W. Gotthard, and P. C. Lockemann. Complex entities for engineering applications. In *Proc. of the Intl. Conf. on Entity-Relationship Approach*, pages 421–440. North-Holland, 1987.

[DGL87b]   K. R. Dittrich, W. Gotthard, and P. C. Lockemann. DAMOKLES: A data-
            base system for software engineering applications. In *Lecture Notes in Com-
            puter Science No. 244*, pages 353–371. Springer-Verlag, Heidelberg, 1987.

[DH55]   J. Denavit and R. S. Hartenberg. A kinematic notion for lower-pair mecha-
            nisms based on matrices. *ASME J. Appl. Mech.*, 22:215–221, 1955.

[DH86]   R. Dillmann and M. Huck. A software system for the simulation of robot
            based manufacturing processes. *Robotics* (North-Holland), 2(1), March 1986.

[DHP89]   K. R. Dittrich, M. Härtig, and H. Pfefferle. Discretionary Access Control in
            Structurally Object-Oriented Systems. In C. E. Landwehr, editor, *Database
            Security II: Status and Prospectus*, pages 105–121. Elsevier Science Publ.,
            1989.

[DK88]   M. Dürr and A. Kemper. Transaction control mechanism for the object cache
            interface of R^2D^2. In *Proc. of the Third Intl. Conf. on Data and Knowledge
            Bases: Improving Usability and Responsiveness*, pages 81–89, Jerusalem, Is-
            rael, Morgan-Kaufmann, June 1988.

[DKA$^+$86] P. Dadam, K. Küspert, F. Andersen, H. Blanken, R. Erbe, J. Günauer,
            V. Lum, P. Pistor, and G. Walch. A DBMS prototype to support extended
            NF^2 relations: An integrated view on flat tables and hierarchies. In *Proc. of
            the ACM SIGMOD Conf. on Management of Data*, pages 376–387, Washing-
            ton, DC, 1986.

[DKH90]   P. Drew, R. King, and S. Hudson. The performance and utility of the cactis
            implementation algorithms. In *Proc. of the Conf. on Very Large Data Bases
            (VLDB)*, pages 135–147, Brisbane, Australia, Aug 1990.

[DKL85]   N. Derret, W. Kent, and P. Lyngbaek. Some aspects of operations in an
            object-oriented database. *IEEE Database Engineering*, 8(4):66–74, 1985.

[DKL89]   K. R. Dittrich, A. Kemper, and P. C. Lockemann. Databases for Planning and
            Manufacturing. In U. Rembold, editor, *Robot Technology and Applications*,
            pages 485–552, Marcel Dekker, New York, NY, 1990.

[DKM86]   K. R. Dittrich, A. M. Kotz, and J. A. Mülle. An event/trigger mechanism to
            enforce complex consistency constraints in design databases. *ACM SIGMOD
            Record*, 15(3):22–36, 1986.

[DKT88]   H. Duchene, M. Kaul, and V. Turau. VODAK kernel data model. In K. R.
            Dittrich, editor, *Proc. 2nd. Int. Workshop on Object-Oriented Database Sys-
            tems*, pages 242–261, Lecture Notes in Computer Science No. 334, Springer-
            Verlag, Heidelberg, 1988.

[DL88]   K. R. Dittrich and R. A. Lorie. Version support for engineering database
             systems. *IEEE Trans. Software Eng.*, 14(4):429–437, April 1988.

[DMB⁺87]   U. Dayal, F. Manola, A. Buchmann, U. Chakravarthy, S. Heiler, J. Orenstein,
             and A. Rosenthal. Simplifying complex objects: The PROBE approach to
             modeling and querying them. In H.-J. Schek and G. Schlageter, editors,
             Informatik Fachberichte No. 136, pages 17–38, Springer-Verlag, Heidelberg,
             1987.

[DMN70]   O. J. Dahl, B. Myrhaug, and K. Nygaard. Simula 67: Common base language.
             Publication NS 22, Norsk Regnesentral (Norwegian Computing Center), Oslo,
             Norway, Oct 1970.

[DN66]   O. Dahl and K. Nygaard. Simula, an Algol-based simulation language. *Com-
             munications of the ACM*, 9:671–678, 1966.

[DS85]   U. Dayal and J. M. Smith. PROBE: A knowledge oriented database man-
             agement system. In *Proc. Islamorada Workshop on Large Scale Knowledge
             Base and Reasoning Systems*, Islamorada, 1985.

[DS89]   N. Derret and M.-C. Shan. Rule-based query optimization in Iris. In *Proc.
             of the Annual ACM Computer Science Conference*, pages 78–86, Louisville,
             KY, Feb 1989.

[DT88]   S. Danforth and C. Tomlinson. Type theories and object oriented program-
             ming. *ACM Computing Surveys*, 20(1):29–72, March 1988.

[DZ89]   C. Delcourt and R. Zicari. Preserving Structural Consistency in an Object-
             Oriented Database. Technical report, GIP Altair, July 1989.

[DZ91]   C. Delcourt and R. Zicari. The design of an integrity consistency checker
             (ICC) for an object-oriented database system. In *Proc. Europ. Conf. on
             Object-Oriented Programming (ECOOP)*, pages 97–117, 1991.

[EDIF84]   EDIF Technical Committee. Electronic design interchange format (EDIF).
             Preliminary Specification, Version 0.8, May 1984.

[EGLT76]   K. P. Eswaran, J. N. Gray, R. A. Lorie, and I. L. Traiger. On the notion of
             consistency and predicate locks in a relational database system. *Communi-
             cations of the ACM*, 19(11):624–633, 1976.

[Elm92]   A. K. Elmagarmid, editor. *Database Transaction Models For Advanced Ap-
             plications*. The Morgan Kaufmann Series in Data Management Systems.
             Morgan Kaufmann, San Mateo, CA, 1992.

[EN89]   E. Elmasri and S. Navathe. *Fundamentals of Database Systems*. Ben-
             jamin/Cummings, Redwood City, CA, 1989.

[FAB⁺89]   D. H. Fishman, J. Annevelink, D. Beech, E. Chow, T. Connors, J. W. Davis,
            W. Hasan, C. G. Hoch, W. Kent, S. Leichner, P. Lyngbaek, B. Mahbod, M. A.
            Neimat, T. Risch, M. C. Shan, and W. K. Wilkinson. Overview of the Iris
            DBMS. In W. Kim and F. H. Lochowsky, editors, *Object-Oriented Concepts
            and Databases*, chapter 10, pages 219–250. Addison-Wesley, Reading, MA
            1989.

[FBC⁺87]   D. H. Fishman, D. Beech, H. P. Cate, E. C. Chow, T. Connors, J. W. Davis,
            N. Derret, C. G. Hoch, W. Kent, P. Lyngbaek, B. Mahbod, M. A. Neimat,
            T. A. Ryan, and M. C. Shan. Iris: An object-oriented database management
            system. *ACM Trans. Office Inf. Syst.*, 5(1):48–69, Jan 1987.

[FGJ⁺90]   B. L. Fritchman, R. L. Guck, D. Jagannathan, J. P. Thompson, and D. M.
            Tolbert. SIM: Design and implementation of a semantic database system.
            In A. F. Cardenas and D. McLeod, editors, *Research Foundations in Object-
            Oriented and Semantic Database Systems*, pages 241–266. Prentice Hall, En-
            glewood Cliffs, NJ, 1990.

[Fis79]   W. E. Fischer. PHIDAS: A database management system for CAD/CAM
            application software. *Computer-Aided Design*, 11(3):146–150, 1979.

[FJL⁺88]   S. Ford, J. Joseph, D. E. Langworthy, D. F. Lively, G. Pathak, E. R. Perez,
            R. W. Peterson, D. M. Sparacin, S. M. Thatte, D. L. Wells, and S. Agar-
            wala. ZEITGEIST: Database support for object-oriented programming. In
            K. R. Dittrich, editor, *Advances in Object-Oriented Database Systems: Proc.
            Second Int. Workshop on Object Oriented Database Systems*, pages 23–42.
            Springer-Verlag, Lecture Notes in Computer Science (LNCS) No. 334, 1988.

[FSC75]   E. Fernandez, R. Summers, and C. Coleman. An authorization model for a
            shared database. In *Proc. of the ACM SIGMOD Conf. on Management of
            Data*, pages 23–31, 1975.

[FSL75]   E. Fernandez, R. Summers, and T. Lang. Definition and evaluation of access
            rules in data management systems. In *Proc. Int. Conf. on Very Large Data
            Bases (VLDB)*, Framingham, MA, 1975.

[FT83]   P. C. Fischer and S. J. Thomas. Non first normal form relations. In *Proc
            IEEE Computer Software and Applications Conference*, pages 464–475, 1983.

[Fv83]   J. D. Foley and A. van Dam. *Fundamentals of Interactive Computer Graphics*.
            Addison-Wesley, Reading, MA, 1983.

[GCG⁺89]   Rajiv Gupta, W. H. Cheng, Rajesh Gupta, I. Hardonag, and M. A. Breuer.
            An object-oriented VLSI CAD framework. *IEEE Computer*, 22(5):28–37,
            May 1989.

[GD72]   G. Graham and P. Denning. Protection: Principles and practice. In *AFIPS Conference Proceedings 40*, pages 417–429, AFIPS Press, Montvale, NJ, 1972.

[GGF93]   N. Gal-Oz, E. Gudes, and E. B. Fernandez. A Model of Methods Access Authorization in Object-Oriented Databases. In *Proc. of the Conf. on Very Large Data Bases (VLDB)*, pages 52–61, Dublin, Ireland, Aug 1993.

[GGH+92]   M. Gesmann, A. Grasnickel, T. Härder, C. Hübel, W. Käfer, M. Mitschang, and H. Schöning. PRIMA – A database system supporting dynamically defined composite objects. In *Proc. of the ACM SIGMOD Conf. on Management of Data*, page 5. June 1992.

[GH91]   R. Gupta and E. Horowitz, editors. *Object-Oriented Databases with Applications to CASE, Networks, and VLSI Design*. Prentice Hall, Englewood Cliffs, NJ, 1991.

[GJS92a]   N. H. Gehani, H. V. Jagadish, and O. Shmueli. Composite event specification in active databases: Model & implementation. In *Proc. Int. Conf. on Very Large Data Bases (VLDB)*, pages 327–338, Vancouver, Canada, 1992.

[GJS92b]   N. H. Gehani, H. V. Jagadish, and O. Shmueli. Event specification in an active object-oriented database. In *Proc. of the ACM SIGMOD Conf. on Management of Data*, pages 81–90, San Diego, CA, 1992.

[GK88]   J. Garza and W. Kim. Transaction management in an object-oriented database system. In *Proc. of the ACM SIGMOD Conf. on Management of Data*, pages 37–45, Chicago, IL, 1988.

[GKKM92b]   C. Gerlhof, A. Kemper, C. Kilger, and G. Moerkotte. Partition-based clustering in object bases: From theory to practice. In *Proc. of the Intl. Conf. on Foundations of Data Organization and Algorithms (FODO)*, Chicago, IL, Oct 1993.

[GKP92]   P. M. D. Gray, K. G. Kulkarni, and N. W. Paton. *Object-Oriented Databases: A Semantic Approach*. Prentice Hall, Englewood Cliffs, NJ, 1992.

[GM93]   G. Graefe and W. J. McKenna. The Volcano optimizer generator: Extensibility and efficient search, In *Proc. IEEE Conf. on Data Engineering*, pages 209–218, Vienna, Austria, April 1993.

[GMB+81]   J. Gray, P. R. McJones, M. W. Blasgen, B. Lindsay, R. A. Lorie, T. G. Price, G. R. Putzolu, and I. L. Traiger. The recovery manager of the System R database manager. *ACM Computing Surveys*, 13(2):223–242, June 1981.

[GO87]   D. Goldhirsch and J. Orenstein. Extensibility in the PROBE database system. *IEEE Database Engineering*, 10(2):24–31, June 1987.

[Gol84]   A. Goldberg. *Smalltalk-80: The Interactive Programming Environment.* Addison-Wesley, Reading, MA, 1984.

[GR83]   A. Goldberg and D. Robson. *Smalltalk-80: The Language and its Implementation.* Addison-Wesley, Reading, MA, 1983.

[GR93]   J. Gray and A. Reuter. *Transaction Processing: Concepts and Technology.* Morgan Kaufmann, San Mateo, CA, 1993.

[Gra78]   J. Gray. Notes on a database operating system. In R. Bayer, R. Graham, and G. Seegmüller, editors, *Operating Systems*, pages 393–481. Springer-Verlag, Heidelberg, Lecture Notes in Computer Science (LNCS) No. 60, 1978.

[Gra81]   J. N. Gray. The transaction concept: Virtues and limitations. In *Proc. of the Conf. on Very Large Data Bases (VLDB)*, pages 144–154, Cannes, France, 1981.

[GW76]   P. Griffith and B. Wade. An authorization mechanism for a relational database system. *ACM Trans. on Database Systems*, 3(3):242–255, Sept 1976.

[GZ84]   M. Groover and E. Zimmers. *CAD/CAM: Computer Aided Design and Manufacturing.* Prentice Hall, Englewood Cliffs, NJ, 1984.

[H$^+$89]   L. M. Haas et al. An extensible processor for an extensible relational query language. In *Proc. of the ACM SIGMOD Conf. on Management of Data*, Portland, OR, 1989.

[Han87]   E. Hanson. A performance analysis of view materialization strategies. In *Proc. of the ACM SIGMOD Conf. on Management of Data*, pages 440–453, San Francisco, CA, May 1987.

[Han88]   E. Hanson. Processing queries against database procedures. In *Proc. of the ACM SIGMOD Conf. on Management of Data*, pages 295–303, Chicago, May 1988.

[Här78]   T. Härder. Implementing a generalized access path structure for a relational database system. *ACM Trans. on Database Systems*, 3(3):285–298, Sep 1978.

[Har87]   M. Hardwick. Why Rose is fast: Five optimizations in the design of an experimental database system for CAD/CAM applications. In *Proc. of the ACM SIGMOD Conf. on Management of Data*, pages 292–298, San Francisco, CA, May 1987.

[HD91]   C. Harris and J. Duhl. Object SQL. In R. Gupta and E. Horowitz, editors, *Object-Oriented Databases with Applications to CASE, Networks, and VLSI Design*, pages 199–215. Prentice Hall, Englewood Cliffs, NJ, 1991.

[Hei87]   H. Heiß. Theorie und Anwendung der Koordinatentransformation bei Roboterkinematiken. *Informatik – Forschung und Entwicklung*, 2:19–33, 1987.

[Her90]   M. Herlihy. Apologizing versus asking permission: Optimistic concurrency control for abstract data types. *ACM Trans. on Database Systems*, 15(1):96–124, March 1990.

[Heu92]   A. Heuer. *Objektorientierte Datenbanken*. Addison-Wesley, Bonn, Germany, 1992.

[HFW90]   A. Heuer, J. Fuchs, and U. Wiebking. OSCAR: An object-oriented database system with a nested relational kernel. In H. Kangassalo, editor, *Proc. of the Intl. Conf. on Entity-Relationship Approach*, pages 95–110, Lausanne, Switzerland, North-Holland, Oct 1990.

[HJEK90]   U. Hahn, M. Jarke, S. Eherer, and K. Kreplin. CoAUTHOR: A hypermedia group authoring environment. In J. Benford and S. Bowers, editors, *Studies in Computer-Supported Cooperative Work*, pages 79–100. North-Holland, 1990.

[HK87a]   S. E. Hudson and R. King. Object-oriented database support for software environments. In *Proc. of the ACM SIGMOD Conf. on Management of Data*, pages 491–503, San Francisco, CA, May 1987.

[HK87b]   R. Hull and R. King. Semantic database modeling: Survey, applications, and research issues. *ACM Computing Surveys*, 19(3):201–260, 1987.

[HK88]   S. E. Hudson and R. King. The Cactis project: Database support for software environments. *IEEE Trans. Software Eng.*, 14(3):291–321, Sep 1988.

[HK89]   S. E. Hudson and R. King. Cactis: A self-adaptive, concurrent implementation of an object-oriented database management system. *ACM Trans. Database Systems*, 14(3):291–321, Sep 1989.

[HL82]   R. L. Haskin and R. A. Lorie. On extending the functions of a relational database system. In *Proc. of the ACM SIGMOD Conf. on Management of Data*, pages 207–212, Orlando, FL, June 1982.

[HM81]   M. Hammer and D. McLeod. Database description with SDM: A semantic database model. *ACM Trans. on Database Systems*, 6(3):351–386, 1981.

[HM90]   A. L. Hosking and J. E. B. Moss. Towards compile-time optimizations for persistence. In *Proc. of the Fourth International Workshop on Persistent Object Systems*, pages 17–27. Morgan Kaufmann, San Mateo, CA, Sep 1990.

[HMWMS87] T. Härder, K. Meyer-Wegener, B. Mitschang, and A. Sikeler. PRIMA – a DBMS prototype supporting engineering applications. In *Proc. of the Conf. on Very Large Data Bases (VLDB)*, pages 433–442, Brighton, UK, Sep 1987.

[Hof89]   C. M. Hoffmann. *Geometric and Solid Modeling*. Morgan Kaufmann, San Mateo, CA, 1989.

[HR87]   T. Härder and K. Rothermel. Concepts for transaction recovery in nested transactions. In *Proc. of the ACM SIGMOD Conf. on Management of Data*, pages 239–248, San Francisco, CA, 1987.

[HS76]   E. Horowitz and S. Sahni. *Fundamentals of Data Structures*. Computer Science Press, Rockville, MD, 1976.

[HS87]   M. Hardwick and D. L. Spooner. Comparison of some data models for engineering objects. *IEEE Computer Graphics and Applications*, 7(3):56–66, March 1987.

[HS88]   M. Hardwick and D. L. Spooner. ROSE: An object-oriented database system for interactive computer graphics applications. In K. R. Dittrich, editor, *Proc. 2nd. Int. Workshop on Object-Oriented Database Systems*, pages 242–261. Springer-Verlag, Heidelberg, Lecture Notes in Computer Science No. 334, 1988.

[HS91]   A. Heuer and P. Sander. Preserving and generating objects in the LIVING IN THE LATTICE rule language. In *Proc. IEEE Conference on Data Engineering*, pages 562–569, Kobe, Japan, 1991.

[HZ87]   M. Hornick and S. Zdonik. A shared, segmented memory system for an object-oriented database. *ACM Trans. Office Inf. Syst.*, 5(1):70–95, Jan 1987.

[ISK+93]   H. Ishikawa, F. Suzuki, F. Kozakura, A. Makinouchi, M. Miyagishima, Y. Izumida, M. Aoshima, and Y. Yamane. The model, language, and implementation of an object-oriented multimedia knowledge base management system. *ACM Trans. on Database Systems*, 18(1):1–50, March 1993.

[Jar92]   M. Jarke. ConceptBase V3.1 User Manual. Aachener Informatik Berichte 92-17, RWTH Aachen, Fachgruppe Informatik, D-5100 Aachen, Germany, 1992.

[JGF+88]   D. Jagannathan, R. L. Guck, B. L. Fritchman, J. P. Thompson, and D. M. Tolbert. SIM: A database system based on the semantic data model. In *Proc. of the ACM SIGMOD Conf. on Management of Data*, pages 46–55, Chicago, IL, 1988.

[Jhi88]   A. Jhingran. A performance study of query optimization algorithms on a database system supporting procedures. In *Proc. of the Conf. on Very Large Data Bases (VLDB)*, pages 88–99, Los Angeles, CA, Sep 1988.

[JJ91]   M. Jarke and M. Jeusfeld. From relational to object-oriented integrity simplification. In *Proc. Int. Conf. on Deductive and Object-Oriented Databases (DOOD)*, pages 460–477, Munich, Germany, 1991.

[JJR90]   M. Jarke, M. Jeusfeld, and T. Rose. A software process data model for knowledge engineering in information systems. *Information Systems*, 15(1):85–116, 1990.

[Kat85]   R. H. Katz. *Information Management for Engineering Design*. Surveys in Computer Science. Springer-Verlag, Heidelberg, 1985.

[Kat90]   R. H. Katz. Toward a unified framework for version modeling in engineering databases. *ACM Computing Surveys*, 22(4):375–408, Dec 1990.

[KBC⁺87]   W. Kim, J. Banerjee, H. T. Chou, J. F. Garza, and D. Woelk. Composite object support in an object-oriented database system. In *Proc. of the ACM Conf. on Object-Oriented Programming Systems and Languages (OOPSLA)*, pages 118–125, Orlando, FL, Oct 1987.

[KBC⁺88]   W. Kim, N. Ballou, H. T. Chou, J. F. Garza, D. Woelk, and J. Banerjee. Integrating an object-oriented programming system with a database system. In *Proc. of the ACM Conf. on Object-Oriented Programming Systems and Languages (OOPSLA)*, pages 142–152, San Diego, CA, Sep 1988.

[KBG89]   W. Kim, E. Bertino, and J. F. Garza. Composite objects revisited. In *Proc. of the ACM SIGMOD Conf. on Management of Data*, pages 337–347, Portland, OR, May 1989.

[KC86]   S. N. Khoshafian and G. P. Copeland. Object identity. In *Proc. of the ACM Conf. on Object-Oriented Programming Systems and Languages (OOPSLA)*, pages 408–416, Portland, OR, Nov 1986.

[KC88a]   R. H. Katz and E. Chang. Managing change in a computer-aided design database. In *Proc. Int. Conf. on Very Large Data Bases (VLDB)*, pages 400–407, Los Angeles, 1988.

[KC88b]   W. Kim and H.-T. Chou. Versions of schema for object-oriented databases. In *Proc. Int. Conf. on Very Large Data Bases (VLDB)*, pages 148–159, Los Angeles, 1988.

[KCB86]   R. H. Katz, E. Chang, and R. Bhateja. Version modeling concepts for computer-aided design databases. In *Proc. of the ACM SIGMOD Conf. on Management of Data*, pages 379–386, Washington, DC, 1986.

[KCB88]   W. Kim, H. T. Chou, and J. Banerjee. Operations and implementation of complex objects. *IEEE Trans. Software Eng.*, 14(7):985–996, July 1988.

[KD91]   U. Keßler and P. Dadam. Auswertung komplexer Anfragen an hierarchisch strukturierte Objekte mittels Pfadindexen. In *Proc. der GI-Fachtagung Datenbanksysteme für Büro, Technik und Wissenschaft (BTW)*, pages 218–237. Informatik-Fachberichte No. 270, Springer-Verlag, Heidelberg, 1991.

[KDG87]   K. Küspert, P. Dadam, and J. Günauer. Cooperative object buffer manage-
           ment in the Advanced Information Management Prototype. In *Proc. of the
           Conf. on Very Large Data Bases (VLDB)*, pages 483–492, Brighton, UK, Sep
           1987.

[KDM88]   A. M. Kotz, K. R. Dittrich, and J. A. Mülle. Supporting semantic rules by a
           generalized event-trigger mechanism. In *Proc. Intl. Conf. Extending Database
           Technology (EDBT)*, pages 76–92, Venice, Italy, March 1988.

[Kee89]   S. Keene. *Object-Oriented Programming in Common Lisp: A Programmer's
           Guide to CLOS*. Addison-Wesley, Reading, MA, 1989.

[Kem86]   A. Kemper. CAM databases: Requirements and survey. In *Proc. Nineteenth
           Ann. Hawaii Intl. Conference on System Sciences*, pages 363–378, Honolulu,
           Jan 1986.

[Kem92]   A. Kemper. *Zuverlässigkeit und Leistungsfähigkeit objektorientierter Daten-
           banken*, volume 298 of *Informatik Fachberichte*. Springer-Verlag, Heidelberg,
           1992.

[Ken79]   W. Kent. Limitations of record-based information models. *ACM Trans.
           Database Syst.*, 4(1):107–131, 1979.

[Kim90]   W. Kim. *Introduction to Object-Oriented Databases*. MIT Press, Cambridge,
           MA, 1990.

[KK83]   T. Kaehler and G. Krasner. LOOM—large object-oriented memory for
           Smalltalk-80 systems. In G. Krasner, editor, *Smalltalk-80: Bits of History,
           Words of Advice*. Addison-Wesley, Reading, MA, 1983.

[KK93]   A. Kemper and D. Kossmann. Adaptable pointer swizzling strategies in
           object bases. In *Proc. IEEE Conf. on Data Engineering*, pages 155–162,
           Vienna, Austria, April 1993.

[KKB88]   H. F. Korth, W. Kim, and F. Bancilhon. On long-duration CAD transactions.
           Information Science, 46:73–107, 1988.

[KKM90]   A. Kemper, C. Kilger, and G. Moerkotte. Materialization of Functions in
           Object Bases: Design, Realization, and Evaluation. To appear in: *IEEE
           Trans. Knowledge and Data Engineering*, 1994.

[KKM91]   A. Kemper, C. Kilger, and G. Moerkotte. Materialization of functions in
           object bases. In *Proc. of the ACM SIGMOD Conf. on Management of Data*,
           pages 258–268, Denver, CO, May 1991.

[KKS92]   M. Kifer, W. Kim, and Y. Sagiv. Querying object-oriented databases. In
           Proc. of the ACM SIGMOD Conf. on Management of Data, pages 393–402,
           San Diego, CA, June 1992.

[KL70]   B. Kernighan and S. Lin. An efficient heuristic procedure for partitioning
             graphs. *Bell System Technical Journal*, 49(2):291–307, Feb 1970.

[KL84]   R. H. Katz and T. Lehman. Database support for versions and alternatives
             of large design files. *IEEE Trans. Software Eng.*, 10(2):191–200, March 1984.

[KL89a]   M. Kifer and G. Lausen. F-logic: A higher-order language for reasoning about
             objects, inheritance, and scheme. In *Proc. of the ACM SIGMOD Conf. on
             Management of Data*, pages 134–146, Portland, OR, 1989.

[KL89b]   W. Kim and F. H. Lochovsky, editors. *Object-Oriented Concepts, Databases,
             and Applications*. ACM Press, Frontier Series. Addison-Wesley, Reading,
             MA, 1989.

[KLMP84]   W. Kim, R. Lorie, D. McNabb, and W. Plouffe. A transaction mechanism
             for engineering design databases. In *Proc. of the Conf. on Very Large Data
             Bases (VLDB)*, pages 355–362, Singapore, Aug 1984.

[KLW87]   A. Kemper, P. C. Lockemann, and M. Wallrath. An object-oriented database
             system for engineering applications. In *Proc. of the ACM SIGMOD Conf. on
             Management of Data*, pages 299–311, San Francisco, CA, May 1987.

[KM89]   A. Kemper and G. Moerkotte. Typing in object bases. In *Proc. Advanced
             Database Symposium*, pages 19–31, Kyoto, Japan, Dec 1989.

[KM90a]   A. Kemper and G. Moerkotte. Access support in object bases. In *Proc. of
             the ACM SIGMOD Conf. on Management of Data*, pages 364–374, Atlantic
             City, NJ, May 1990.

[KM90b]   A. Kemper and G. Moerkotte. Correcting anomalies of standard inheri-
             tance—a constraint based approach. In *Proc. of the Intl. Conf. on Database
             and Expert Systems Applications (DEXA)*, pages 49–55, Vienna, Austria,
             Springer-Verlag, Aug 1990.

[KM92a]   A. Kemper and G. Moerkotte. Access Support Relations: an indexing method
             for object bases. *Information Systems*, 17(2):117–146, 1992.

[KM92b]   A. Kemper and G. Moerkotte. A framework and a type inference system for
             strong typing in (persistent) object models. In *Proc. of the Intl. Conf. on
             Database and Expert Systems Applications (DEXA)*, Springer-Verlag, Heidel-
             berg, pages 257–263, Aug 1992.

[KM93]   A. Kemper and G. Moerkotte. Query optimization in object bases: Exploit-
             ing the relational techniques. In J.-C. Freytag, D. Maier, and G. Vossen,
             editors, *Query Processing for Advanced Applications*, pages 63–94. Morgan
             Kaufmann, San Mateo, CA, 1993.

[KMP93]   A. Kemper, G. Moerkotte, and K. Peithner. A blackboard architecture for query optimization in object bases. In *Proc. of the Conf. on Very Large Data Bases (VLDB)*, pages 543–554, Dublin, Ireland, Aug 1993.

[KMRC90]   M. A. Ketabchi, S. Mathur, T. Risch, and J. Chen. Comparative analysis of RDBMS and ODBMS: A case study. In *Proc. COMPCON IEEE International Conference*, San Francisco, CA, Feb 1990.

[KMS92]   A. Kemper, G. Moerkotte, and M. Steinbrunn. Optimizing Boolean expressions in object bases. In *Proc. of the Conf. on Very Large Data Bases (VLDB)*, pages 79–90, Vancouver, BC, Canada, Aug 1992.

[KMWZ91]   A. Kemper, G. Moerkotte, H.-D. Walter, and A. Zachmann. GOM: A strongly typed, persistent object model with polymorphism. In *Proc. der GI-Fachtagung Datenbanken in Büro, Technik und Wissenschaft (BTW)*, pages 198–217, Kaiserslautern, Springer-Verlag, Informatik-Fachberichte Nr. 270, March 1991.

[KNS90a]   W. Klas, E. Neuhold, and M. Schrefl. Metaclasses in VODAK and their application in database integration. Technical Report 462, GMD, Darmstadt, Germany, July 1990.

[KNS90b]   W. Klas, E. Neuhold, and M. Schrefl. Using an object-oriented approach to model multimedia data. *Computer Communications*, 13(4):204–216, Special Issue on Multimedia Systems, May 1990.

[Kos91]   D. Kossmann. Entwurf und Implementierung von Laufzeitoptimierungs-maßnahmen im GOM-Prototyp. Master's thesis, Universität Karlsruhe, Fakultät für Informatik, D-7500 Karlsruhe, Germany, Oct 1991.

[KR81]   H. T. Kung and J. T. Robinson. On optimistic methods for concurrency control. *ACM Trans. on Database Systems*, 6(2):213–226, June 1981.

[Kra83]   G. Krasner. *Smalltalk-80: Bits of History, Words of Advice*. Addison-Wesley, Reading, MA, 1983.

[Kru56]   J. B. Kruskal. On the shortest spanning subgraph of a graph and the travelling salesman problem. *Proc. of the Amer. Math. Soc.*, 7:48–50, 1956.

[KS91]   H. F. Korth and A. Silberschatz. *Database System Concepts*, 2nd edition. McGraw-Hill, New York, 1991.

[KSUW85]   P. Klahold, G. Schlageter, R. Unland, and W. Wilkes. A transaction model supporting complex applications in integrated information systems. In *Proc. of the ACM SIGMOD Conf. on Management of Data*, pages 388–401, Austin, TX, 1985.

[KSW86]   P. Klahold, G. Schlageter, and W. Wilkes. A general model for version man-
            agement in databases. In *Proc. of the ACM SIGMOD Conf. on Management
            of Data*, pages 319–327, Washington, DC, 1986.

[KW84]   R. H. Katz and S. Weiss. Design transaction management. In *Proc. of the
            ACM IEEE Design Automation Conf.*, pages 692–693, Las Vegas, NV, 1984.

[KW87]   A. Kemper and M. Wallrath. An analysis of geometric modeling in database
            systems. *ACM Computing Surveys*, 19(1):47–91, March 1987.

[KWL86a]   A. Kemper, M. Wallrath, and P. C. Lockemann. Database support for
            robotics applications. In U. Rembold and K. Hörmann, editors, *NATO
            Intl. Advanced Research Workshop on Languages for Sensor-Based Control
            in Robotics*, Castelvecchio Pascoli, Italy. NATO ASI Series, Springer-Verlag,
            Heidelberg, Sep 1986.

[KWL86b]   A. Kemper, M. Wallrath, and P. C. Lockemann. Ein Datenbanksystem für
            Robotikanwendungen. *Robotersysteme*, 1(2):177–187, 1986.

[LAB+81]   B. Liskov, R. Atkinson, T. Bloom, E. Moss, J. Craig Schaffert, R. Scheifler,
            and A. Snyder. *CLU Reference Manual*. Lecture Notes in Computer Science
            No. 114. Springer-Verlag, Berlin, Heidelberg, New York, 1981.

[LAB+85]   P. C. Lockemann, M. Adams, M. Bever, K. R. Dittrich, B. Ferkinghoff,
            N. Gotthard, A. M. Kotz, R.-P. Liedtke, B. Lüke, and J. A. Mülle. An-
            forderungen technischer Anwendungen an Datenbanksysteme. In *Proc. der
            GI-Fachtagung Datenbanken für Büro, Technik und Wissenschaft (BTW)*,
            pages 1–26, Springer-Verlag, Heidelberg, Informatik Fachberichte No. 94,
            1985.

[Lam71]   B. Lampson. Protection. In *Proc. of the 5th Annual Princeton Conference
            on Information Sciences and Systems*, 1971.

[Lan88]   D. E. Langworthy. Evaluating correctness criteria for transactions. *SIGPLAN
            Notices*, pages 139–141, 1988.

[LF83]   Y. C. Lee and K. S. Fu. A CSG based DBMS for CAD/CAM and its sup-
            porting query language. In *Proceedings Engineering Design Applications,
            Database Week*, San Jose, 1983. (In conjunction with the ACM SIGMOD
            Conf. on Management of Data).

[LG86]   B. Liskov and J. Guttag. *Abstraction and Specification in Program Develop-
            ment*. The MIT Electrical Engineering and Computer Science Series. MIT
            Press, Cambridge, MA, 1986.

[LKD+88]   V. Linnemann, K. Küspert, P. Dadam, P. Pistor, R. Erbe, A. Kemper,
            N. Südkamp, G. Walch, and M. Wallrath. Design and implementation of an

extensible data base management system supporting user defined data types and functions. In *Proc. of the Conf. on Very Large Data Bases (VLDB)*, pages 294–305, Long Beach, CA., Sep 1988.

[LKM+85]   R. A. Lorie, W. Kim, D. McNabb, W. Plouffe, and A. Meier. Supporting complex objects in a relational system for engineering databases. In W. Kim, D. Reiner, and D. S. Batory, editors, *Query Processing in Database Systems*, pages 145–155, Springer-Verlag, Heidelberg, 1985.

[LLOW91]   C. Lamb, G. Landis, J. Orenstein, and D. Weinreb. The ObjectStore database system. *Communications of the ACM*, 34(10):50–63, 1991.

[LMP87]   B. Lindsay, J. McPherson, and H. Pirahesh. A data management extension architecture. In *Proc. of the ACM SIGMOD Conf. on Management of Data*, pages 220–227, San Francisco, May 1987.

[LP83]   R. A. Lorie and W. Plouffe. Complex objects and their use in design transactions. In *Proceedings Engineering Design Applications, Database Week*, pages 115–121, San Jose, May 1983. (In conjunction with the ACM SIGMOD Conf. on Management of Data).

[LS79]   B. Liskov and A. Snyder. Exception handling in CLU. *IEEE Trans. Software Eng.*, 5(6):546–558, Nov 1979.

[Män88]   M. Mäntylä. *An Introduction to Solid Modeling*. Computer Science Press, Rockville, MD, 1988.

[Mai89]   D. Maier. Making database systems fast enough for CAD applications. In W. Kim and F. H. Lochovsky, editors, *Object-Oriented Concepts, Databases, and Applications*, pages 573–582, Addison-Wesley, Reading, MA, 1989.

[MBJK90]   J. Mylopoulos, A. Borgida, M. Jarke, and M. Koubarakis. Telos: A language for representing knowledge about information systems. *ACM Trans. Office Inf. Syst.*, 8(4):327–362, 1990.

[MC80]   C. Mead and L. Conway. *Introduction to VLSI Systems*. Addison-Wesley, Reading, MA, 1980.

[Mei86]   A. Meier. Applying relational database techniques to solid modeling. *Computer-Aided Design (CAD)*, 18(6):319–326, 1986.

[Mey86]   B. Meyer. Genericity versus inheritance. In *Proc. of the ACM Conf. on Object-Oriented Programming Systems and Languages (OOPSLA)*, pages 391–405, Portland, OR, ACM SIGPLAN Notices, Vol. 21, No. 11, 1986.

[Mey87]   B. Meyer. Reusability: The case for object-oriented design. *IEEE Software*, 4(2):43–53, Sep 1987.

[Mey88]   B. Meyer. *Object-Oriented Software Construction*. International Series in Computer Science. Prentice Hall, Englewood Cliffs, NJ, 1988.

[Mic88]   J. Micallef. Encapsulation, reusability, and extensibility in object-oriented programming languages. *Journal of Object-Oriented Programming Languages*, 1(1):12–36, May 1988.

[Mil78]   R. Milner. A theory of type polymorphism. *Journal of Computer and System Sciences*, 17:348–375, 1978.

[Mit88]   B. Mitschang. *Ein Molekül-Atom-Datenmodell für Non-Standard-Anwendungen*, volume 185 of *Informatik-Fachberichte*. Springer-Verlag, Heidelberg, 1988.

[Mit89]   B. Mitschang. Extending the relational algebra to capture complex objects. In *Proc. of the Conf. on Very Large Data Bases (VLDB)*, pages 297–306, Amsterdam, The Netherlands, Aug 1989.

[MM92]   D. Mandrioli and B. Meyer, editors. *Advances in Object-Oriented Software Engineering*. Object-Oriented Series. Prentice Hall, Englewood Cliffs, NJ, 1992.

[Moo86]   D. A. Moon. Object-oriented programming with Flavors. In *Proc. of the ACM Conf. on Object-Oriented Programming Systems and Languages (OOPSLA)*, pages 1–8, Portland, OR, 1986.

[MOP85]   D. Maier, A. Otis, and A. Purdy. Object-oriented database development at Servio Logic. *IEEE Database Engineering*, 8(4):58–65, 1985.

[Mos81]   J. E. Moss. *Nested Transactions: An Approach to Reliable Distributed Computing*. PhD thesis, MIT, Dept. of Electrical Engineering and Computer Science, Boston, MA, April 1981.

[Mos92]   J. E. B. Moss. Working with persistent objects: To swizzle or not to swizzle. *IEEE Trans. Software Eng.*, 18(8):657–673, Aug 1992.

[MRKN91]   P. Muth, T. Rakow, W. Klas, and E. Neuhold. A transaction model for an open publication environment. *IEEE Database Engineering*, Special Issue on Transaction Models for Advanced Applications, March 1991.

[MS86]   D. Maier and J. Stein. Indexing in an object-oriented DBMS. In K. R. Dittrich and U. Dayal, editors, *Proc. IEEE Intl. Workshop on Object-Oriented Database Systems*, Asilomar, Pacific Grove, CA, pages 171–182. IEEE Computer Society Press, Sep 1986.

[MS87]   D. Maier and J. Stein. Development and implementation of an object-oriented DBMS. In B. Shriver and P. Wegner, editors, *Research Directions in Object-Oriented Programming*, pages 355–392, MIT Press, Cambridge, MA, 1987.

[MS88]   J. E. B. Moss and S. Sinofsky. Managing persistent data with Mneme: De-
            signing a reliable shared object interface. In K. R. Dittrich, editor, *Advances
            in Object-Oriented Database Systems: 2nd Int. Workshop on Object-Oriented
            Database Systems*, pages 298–316, Bad Münster, Lecture Notes in Computer
            Science 334, Springer-Verlag, Heidelberg, Sep 1988.

[MS89]   F. Matthes and J. W. Schmidt. The type system of DBPL. In *Proc. Sec-
            ond Intl. Conf. on Database Programming Languages*, Glenedon Beach, OR,
            Morgan-Kaufmann, San Mateo, CA, 1989.

[MSOP86]   D. Maier, J. Stein, A. Otis, and A. Purdy. Development of an object-oriented
            DBMS. In *Proc. of the ACM Conf. on Object-Oriented Programming Systems
            and Languages (OOPSLA)*, pages 472–482, Portland, OR, 1986.

[MZ92a]   G. Moerkotte and A. Zachmann. Multiple substitutability without affecting
            the taxonomy. In *Proc. Intl. Conf. Extending Database Technology (EDBT)*,
            pages 120–135, Vienna, Austria, March 1992.

[MZ93]   G. Moerkotte and A. Zachmann. Towards more flexible schema management
            in object bases. In *Proc. IEEE Conf. on Data Engineering*, pages 174–181,
            Vienna, Austria, April 1993.

[ND81]   K. Nygaard and O. J. Dahl. Simula 67. In R. W. Wexelblatt, editor, *History
            of Programming Languages*, Academic Press, New York, 1981.

[NHS84]   J. Nievergelt, H. Hinterberger, and K. C. Sevcik. The grid file: An adapt-
            able, symmetric multikey file structure. *ACM Trans. on Database Systems*,
            9(1):38–71, 1984.

[NR88]   K. Narayanaswamy and K. Bapa Rao. An incremental mechanism for schema
            evolution in engineering domains. In *Proc. IEEE Conf. on Data Engineering*,
            pages 294–301, Los Angeles, CA, 1988.

[NR89]   G. Nguyen and D. Rieu. Schema evolution in object-oriented database sys-
            tems. *Data & Knowledge Engineering*, pages 43–67, 1989.

[NS88]   E. Neuhold and M. Schrefl. Dynamic derivation of personalized views. In
            Proc. Int. Conf. on Very Large Data Bases (VLDB), pages 183–194, Los
            Angeles, 1988.

[OBBT89]   A. Ohori, P. Buneman, and V. Breazu-Tannen. Database programming in
            Machiavelli: A polymorphic language with static type inference. In *Proc. of
            the ACM SIGMOD Conf. on Management of Data*, Portland, OR, 1989.

[OBS86]   P. O'Brien, B. Bullis, and C. Schaffert. Persistent and shared objects in
            Trellis/Owl. In K. R. Dittrich and U. Dayal, editors, *Proc. IEEE Intl. Work-
            shop on Object-Oriented Database Systems*, pages 113–123, Asilomar, Pacific
            Grove, CA, IEEE Computer Society Press, 1986.

[OHK87]   P. O'Brien, D. Halbert, and M. Kilian. The Trellis programming environment. In *Proc. of the ACM Conf. on Object-Oriented Programming Systems and Languages (OOPSLA)*, pages 91–102, Orlando, FL, Oct 1987.

[OHMS92]   J. Orenstein, S. Haradhvala, B. Margulies, and D. Sakahara. Query processing in the ObjectStore database system. In *Proc. of the ACM SIGMOD Conf. on Management of Data*, pages 403–412, June 1992.

[Oho88]   A. Ohori. Semantics of types for database objects. In *Proc. of the Intl. Conf. on Database Theory (ICDT)*, pages 239–251, Bruges, Belgium, Aug 1988.

[OM88]   J. A. Orenstein and F. A. Manola. PROBE spatial data modeling and query processing in an image database application. *IEEE Trans. Software Eng.*, 14(5):611–629, May 1988.

[Ont87]   Ontologic. *Vbase Integrated Object System: System Documentation*. Ontologic Corp., The Structure of Possibility, Billerica, MA 01821, USA, Sep 1987.

[Ont92]   Ontos Inc. Ontos DB 2.2: First Time User's Guide. Ontos Inc., 3 Burlington Woods, Burlington, MA 01803, 1992.

[Ora85]   Oracle. *Pro∗C User's Guide*, Version 1.0. Oracle Corporation, Belmont, CA, 1985.

[Ora86]   Oracle. *SQL∗Plus User's Guide*, Version 1.0. Oracle Corporation, Belmont, CA, 1986.

[OY87]   Z. M. Ozsoyoglu and L. Y. Yuan. A new normal form for nested relations. *ACM Trans. on Database Systems*, 12(1):111–136, March 1987.

[PA86]   P. Pistor and F. Andersen. Designing a generalized NF^2 data model with an SQL-type language interface. In *Proc. of the Conf. on Very Large Data Bases (VLDB)*, pages 278–285, Kyoto, Japan, 1986.

[Pap86]   C. H. Papadimitriou. *The Theory of Database Concurrency Control*. Computer Science Press, Rockville, MD, 1986.

[PDS]   Persistent Data Systems, Inc. IDB: Product overview. P.O. Box 38415, Pittsburgh, PA 15238, U.S.A.

[PS85]   F. P. Preparata and M. I. Shamos. *Computational Geometry: An Introduction*. Springer-Verlag, Heidelberg, New York, 1985.

[PS87]   D. J. Penney and J. Stein. Class modification in the GemStone object-oriented DBMS. In *Proc. of the ACM Conf. on Object-Oriented Programming Systems and Languages (OOPSLA)*, pages 111–117, Orlando, FL, Oct 1987.

[PSM87a]   D. J. Penney, J. Stein, and D. Maier. Is the disk half full or half empty? Combining optimistic and pessimistic concurrency mechanisms in a shared, persistent object base. In *Proc. of the Workshop on Persistent Object Systems*, Appin, Scotland, Aug 1987.

[PSM87b]   A. Purdy, B. Schuchardt, and D. Maier. Integrating an object server with other worlds. *ACM Trans. Office Information Syst.*, 5(1):27–47, Jan 1987.

[PSS⁺87]   H. B. Paul, H.-J. Schek, M. H. Scholl, G. Weikum, and U. Deppisch. Architecture and implementation of the Darmstadt database kernel system. In *Proc. of the ACM SIGMOD Conf. on Management of Data*, pages 196–207, San Francisco, CA, June 1987.

[PT86]   P. Pistor and R. Traunmüller. A database language for sets, lists, and tables. *Information Systems*, 11(4):323–336, 1986.

[RBD87]   U. Rembold, C. Blume, and R. Dillmann. *Computer-Integrated Manufacturing Technology and Systems*. Marcel Dekker, New York, NY, 1987.

[RBKW91]   F. Rabitti, E. Bertino, W. Kim, and D. Woelk. A model of authorization for next-generation database systems. *ACM Trans. on Database Systems*, 16(1):88–131, March 1991.

[RBP⁺91]   J. Rumbaugh, M. Blaha, W. Premerlani, F. Eddy, and W. Lorensen. *Object-Oriented Modeling and Design*. Prentice Hall, Englewood Cliffs, NJ, 1991.

[RCS89]   J. E. Richardson, M. J. Carey, and D. T. Schuh. The design of the E programming language. Computer Sciences Technical Report #824, University of Wisconsin-Madison, Feb 1989.

[Req80]   A. A. G. Requicha. Representations for rigid solids: Theory, methods, and systems. *ACM Computing Surveys*, 12(4):437–464, Dec 1980.

[Reu84]   A. Reuter. Performance analysis of recovery. *ACM Trans. on Database Systems*, 9(4):526–559, Dec 1984.

[RGN90]   T. C. Rakow, J. Gu, and E. J. Neuhold. Serializability in object-oriented database systems. In *Proc. IEEE Conf. on Data Engineering*, pages 112–121, Los Angeles, CA, Feb 1990.

[RJG⁺91]   T. Rose, M. Jarke, M. Gocek, C. Maltzahn, and H. Nissen. A decision-based configuration process environment. *Software Engineering Journal*, Special Issue on Software Environments and Factories, July 1991.

[RKB87]   M. A. Roth, H. F. Korth, and D. S. Batory. SQL/NF: A query language for non-1NF relational databases. *Information Systems*, 12(1):99–114, 1987.

[Rub87]   S. Rubin. *Computer Aids for VLSI Design*. Addison-Wesley, Reading, MA, 1987.

[Rum87]   J. Rumbaugh. Relations as semantic constructs in an object-oriented language. In *Proc. of the ACM Conf. on Object-Oriented Programming Systems and Languages (OOPSLA)*, pages 466–481, Orlando, FL, Oct 1987.

[Rum88]   J. Rumbaugh. Controlling propagation of operations using attributes on relations. In *Proc. of the ACM Conf on Object-Oriented Programming Systems and Languages (OOPSLA)*, pages 285–296, San Diego, CA, 1988.

[RWK88]   F. Rabitti, D. Woelk, and W. Kim. A model of authorization for object-oriented and semantic databases. In *Proc. Intl. Conf. on Extending Database Technology (EDBT)*, Springer-Verlag, Heidelberg, 1988.

[S$^+$86]   P. Schwarz et al. Extensibility in the Starburst database system. In K. R. Dittrich and U. Dayal, editors, *Proc. IEEE Intl. Workshop on Object-Oriented Database Systems*, Asilomar, Pacific Grove, CA, 1986.

[SAH87a]   M. Stonebraker, J. Anton, and E. Hanson. Extending a database system with procedures. *ACM Trans. Database Systems*, 12(3):350–376, Sep 1987.

[SAH87b]   M. Stonebraker, J. Anton, and M. Hirohama. Extendability in POSTGRES. *IEEE Database Engineering*, 10(2):16–23, June 1987.

[SAHR84]   M. Stonebraker, E. Anderson, E. Hanson, and B. Rubinstein. QUEL as a data type. In *Proc. of the ACM SIGMOD Conf. on Management of Data*, pages 208–214, Boston, MA, June 1984.

[Sam90]   H. Samet. *The Design and Analysis of Spatial Data Structures*. Addison-Wesley, Reading, MA, 1990.

[SB86]   M. Stefik and D. G. Bobrow. Object-oriented programming: Themes and variations. *The AI Magazine*, 6(4):40–62, 1986.

[SC89]   E. J. Shekita and M. J. Carey. Performance enhancement through replication in an object-oriented DBMS. In *Proc. of the ACM SIGMOD Conf. on Management of Data*, pages 325–336, Portland, OR, May 1989.

[SC90]   E. J. Shekita and M. J. Carey. A performance evaluation of pointer-based joins. In *Proc. ACM SIGMOD Int. Conf. on Management of Data*, pages 300–311, Atlantic City, NJ, 1990.

[SCB$^+$86]   C. Schaffert, T. Cooper, B. Bullis, M. Kilian, and C. Wilpolt. An introduction to Trellis/Owl. In *Proc. of the ACM Conf. on Object-Oriented Programming Systems and Languages (OOPSLA)*, pages 9–16, Portland, OR, 1986.

[SCD90]   D. Schuh, M. Carey, and D. DeWitt. Persistence in E revisited—
            implementation experiences. In *Proc. of the Fourth International Workshop
            on Persistent Object Systems*, pages 345–359. Morgan Kaufmann, San Mateo,
            CA, Sep 1990.

[Sch77]   J. W. Schmidt. Some high level language constructs for data of type relation.
            ACM Trans. Database Systems, 2(3):248–261, Sep 1977.

[Sch87]   H.-J. Schek. DASDBS: A kernel DBMS and application-specific layers. *IEEE
            Database Engineering*, 10(2):62–64, June 1987.

[Sel88]   T. K. Sellis. Intelligent caching and indexing techniques for relational data-
            base systems. *Information Systems*, 13(2):175–186, 1988.

[SF93]   Objective Systems SF. Product overview of EasyDB–release 3.2. P.O.Box
            1128, S-16422 Kista, Sweden, June 1993.

[SG91]   S. S. Simmel and I. Godard. The KALA basket. In *Proc. of the ACM Conf.
            on Object-Oriented Programming Systems and Languages (OOPSLA)*, pages
            230–246, 1991.

[Shi81]   D. Shipman. The functional data model and the data language DAPLEX.
            ACM Trans. on Database Systems, 6(1):140–173, March 1981.

[SHP88]   M. Stonebraker, E. N. Hanson, and S. Potamianos. The POSTGRES rule
            manager. *IEEE Trans. Software Eng.*, 14(7):897–907, July 1988.

[SJGP90]   M. Stonebraker, A. Jhingran, J. Goh, and S. Potamianos. On rules, proce-
            dures, caching and views in data base systems. In *Proc. of the ACM SIGMOD
            Conf. on Management of Data*, pages 281–290, Atlantic City, NJ, May 90.

[SL88]   Servio Logic. GemStone: Product overview. Servio Logic Corporation, 1420
            Harbor Bay Parkway, Suite 100, Alameda, CA 94501, USA, March 1988.

[SLT91]   M. H. Scholl, C. Laasch, and M. Tresch. Updatable views in object-oriented
            databases. In *Proc. of the Conf. on Deductive and Object-Oriented Databases
            (DOOD)*, LNCS 566, Springer-Verlag, Heidelberg, pages 189–207, 1991.

[SLU89]   L. A. Stein, H. Lieberman, and D. Ungar. A shared view of nothing:
            Thetreaty of orlando. In W. Kim and F. H. Lochovsky, editors, *Object-
            Oriented Concepts, Databases, and Applications*, pages 31–48, Addison-
            Wesley, Reading, MA, 1989.

[Sny86]   A. Snyder. Encapsulation and inheritance in object-oriented programming
            languages. In *Proc. of the ACM Conf. on Object-Oriented Programming Sys-
            tems and Languages (OOPSLA)*, pages 38–45, Portland, OR, 1986.

[SO90]   D. D. Straube and T. Öszu. Queries and query processing in object-oriented database systems. *ACM Trans. Office Inf. Syst.*, 8(4):384–430, Oct 1990.

[SP82]   H.-J. Schek and P. Pistor. Data structures for an integrated database management and information retrieval system. In *Proc. of the Conf. on Very Large Data Bases (VLDB)*, pages 197–207, Mexico City, 1982.

[SPSW90]   H.-J. Schek, H.-B. Paul, M. H. Scholl, and G. Weikum. The DASDBS project: Objectives, experiences, and future prospects. *IEEE Transactions on Knowledge and Data Engineering*, 2(1):25–43, March 1990.

[SR86]   M. Stonebraker and L. Rowe. The design of POSTGRES. In *Proc. of the ACM SIGMOD Conf. on Management of Data*, pages 340–355, Washington, DC, 1986.

[SRG83]   M. Stonebraker, B. Rubinstein, and A. Guttman. Application of abstract data types and abstract indices to CAD databases. In *Proceedings of Database Week*, San Jose, May 1983. (In conjunction with the 1983 ACM SIGMOD Conf.)

[SS77a]   J. M. Smith and D. C. P. Smith. Database abstractions: Aggregation. *Communications of the ACM*, 20(6):405–413, June 1977.

[SS77b]   J. M. Smith and D. C. P. Smith. Database abstractions: Aggregation and generalization. *ACM Trans. on Database Systems*, 2(2):105–133, June 1977.

[SS86]   H.-J. Schek and M. H. Scholl. The relational model with relation-valued attributes. *Information Systems*, 11(2):137–147, 1986.

[SS90]   M. H. Scholl and H.-J. Schek. A relational object model. In *Proc. of the Intl. Conf. on Database Theory (ICDT)*, pages 89–108, Paris, France, Lecture Notes in Computer Science (LNCS), Springer-Verlag, Dec 1990.

[SS91]   H.-J. Schek and M. Scholl. From relations and nested relations to object models. In *Proc. 9th British National Conf. on Databases*, pages 202–225, Wolverhampton, UK, 1991.

[Sta84]   J. Stamos. Static grouping of small objects to enhance performance of a paged virtual memory. *ACM Trans. Comp. Syst.*, 2(2):155–180, May 1984.

[Ste91]   M. Steinbrunn. Entwurf und Realisierung der Funktionenmaterialisierung in GOM. Master's thesis, Universität Karlsruhe, Fakultät für Informatik, D-7500 Karlsruhe, Feb 1991.

[Sto85]   M. Stonebraker, editor. *The INGRES Papers: Anatomy of a Relational Database System*. Addison-Wesley, Reading, MA, 1985.

[Str90b]   B. Stroustrup. *The C++ Programming Language*, 2nd edition. Addison--Wesley, Reading, MA, 1990.

[SW87]   B. Shriver and P. Wegner, editors. *Research Directions in Object-Oriented Programming*. MIT Press, Cambridge, MA, 1987.

[SWKH76]   M. Stonebraker, E. Wong, P. Kreps, and G. Held. The design and implementation of INGRES. *ACM Trans. Database Syst.*, 1(3):189–222, Sep 1976.

[SZ86]   A. H. Skarra and S. B. Zdonik. The management of changing types in an object-oriented database. In *Proc. of the ACM Conf. on Object-Oriented Programming Systems and Languages (OOPSLA)*, pages 483–495, Portland, OR, Sep 1986.

[SZ87]   A. H. Skarra and S. B. Zdonik. Type evolution in an object-oriented database. In B. Shriver and P. Wegner, editors, *Research Directions in Object-Oriented Programming*, pages 393–415. MIT Press, Cambridge, MA, 1987.

[SZ89a]   G. M. Shaw and S. B. Zdonik. Object-oriented queries: Equivalence and optimization. In *Proc. of the Conf. on Deductive and Object-Oriented Databases (DOOD)*, pages 264–278, Kyoto, Japan, Dec 1989.

[SZ89b]   A. H. Skarra and S. B. Zdonik. Concurrency control and object-oriented databases. In W. Kim and F. H. Lochowsky, editors, *Object-Oriented Concepts, Databases, and Applications*, pages 395–421, Addison-Wesley, Reading, MA 1989.

[SZ90]   G. M. Shaw and S. B. Zdonik. A query algebra for object-oriented databases. In *Proc. IEEE Conf. on Data Engineering*, pages 154–162, Los Angeles, CA, Feb 1990.

[TF82]   T. J. Teorey and J. P. Fry. *Design of Database Structures*. Prentice Hall, Englewood Cliffs, NJ, 1982.

[Tic88]   W. Tichy. Tools for software configuration mangement. In *Proc. Int. Workshop on Software Version and Configuration Control*, pages 1–27, 1988.

[TK89]   L. Tan and T. Katayama. Meta operations for type management in object-oriented databases. In *Proc. Int. Conf. on Deductive and Object-Oriented Databases (DOOD)*, pages 58–75, Kyoto, Japan, Dec 1989.

[TN91]   M. M. Tsangaris and J. F. Naughton. A stochastic approach for clustering in object bases. In *Proc. of the ACM SIGMOD Conf. on Management of Data*, pages 12–21, Denver, CO, May 1991.

[TYF86]   T. J. Teorey, D. Yang, and J. P. Fry. A logical design methodology for relational databases using the extended entity-relationship model. *ACM Computing Surveys*, 18(2):197–222, June 1986.

[Ull88]   J. D. Ullman. *Principles of Data and Knowledge Bases*, volumes I and II. Computer Science Press, Woodland Hills, CA, 1988.

[Val87]   P. Valduriez. Join indices. *ACM Trans. on Database Systems*, 12(2):218–246, June 1987.

[VD91]   S. L. Vandenberg and D. J. DeWitt. Algebraic support for complex objects with arrays, identity, and inheritance. In *Proc. of the ACM SIGMOD Conf. on Management of Data*, pages 158–167, May 1991.

[VR77]   H. B. Voelcker and A. A. G. Requicha. Geometric modelling of mechanical parts and processes. *Computer*, 10(12), Dec 1977.

[Wal84]   B. Walter. Nested transactions with multile commit points: An approach to the structuring of advanced database applications. In *Proc. Int. Conf. on Very Large Data Bases (VLDB)*, pages 161–171, Singapore, 1984.

[Wal91]   E. Waller. Schema updates and consistency. In *Proc. Int. Conf. on Deductive and Object-Oriented Databases (DOOD)*, pages 167–188, Munich, Germany, 1991.

[WBT92]   D. L. Wells, J. A. Blakeley, and C. W. Thompson. Architecture of an open object-oriented database management system. *IEEE Computer*, 25(10):74–82, Oct 1992.

[WBWW90]   R. Wirfs-Brock, B. Wilkerson, and L. Wiener. *Designing Object-Oriented Software*. Prentice Hall, Englewood Cliffs, NJ, 1990.

[WD92]   S. J. White and D. J. DeWitt. A performance study of alternative object faulting and pointer swizzling strategies. In *Proc. of the Conf. on Very Large Data Bases (VLDB)*, pages 419–431, Vancouver, Canada, 1992.

[WFGL91]   D. Weinreb, N. Feinberg, D. Gerson, and C. Lamb. An object-oriented database system to support an integrated programming environment. In R. Gupta and E. Horowitz, editors, *Object-Oriented Databases with Applications to CASE, Networks, and VLSI Design*, pages 117–129, Prentice Hall, Englewood Cliffs, NJ, 1991.

[Wed88]   H. Wedekind. Die Problematik des Computer Integrated Manufacturing (CIM)—zu den Grundlagen eines "strapazierten" Begriffs. Informatik-Spektrum, 11:22–39, 1988.

[Wei89]   W. E. Weihl. Local atomicity properties: Modular concurrency control for abstract data types. *ACM Trans. Programming Languages and Systems*, 11(2):249–282, April 1989.

[Wei88]   G. Weikum. Transaktionen in Datenbanksystemen. Addison-Wesley, Bonn, Germany, 1988.

[WL89]   S. P. Weiser and F. H. Lochovsky. OZ+: An object-oriented database system.
            In W. Kim and F. H. Lochowsky, editors, *Object-Oriented Concepts and
            Databases*, chapter 13, pages 309–337. Addison-Wesley, Reading, MA, 1989.

[WLH90]   K. Wilkinson, P. Lyngbaek, and W. Hasan. The Iris architecture and im-
            plementation. *IEEE Trans. Knowledge and Data Engineering*, 2(1):63–75,
            March 1990.

[WSSH88]   P. F. Wilms, P. M. Schwarz, H.-J. Schek, and L. M. Haas. Incorporating data
            types in an extensible database architecture. In *Proc. Third Intl. Conf. on
            Data and Knowledge Bases: Improving Usability and Responsiveness*, pages
            180–192, Jerusalem, Israel, Morgan Kaufmann, San Mateo, CA, June 1988.

[YEEK87]   S. Yeh, C. Ellis, A. Ege, and H. Korth. Performance analysis of two concur-
            rency control schemas for design environments. Tech. rep. stp-036-87, MCC,
            Austin, Texas, 1987.

[Zan83]   C. Zaniolo. The database language GEM. In *Proc. of the ACM SIGMOD
            Conf. on Management of Data*, pages 207–218, San Jose, CA, May 1983.

[Zdo89]   S. B. Zdonik. Query optimization in object-oriented database systems. In
            Proc. of the Hawaii Int. Conf. on System Science, Hawaii, Jan 1989.

[ZH87]   S. B. Zdonik and M. Hornick. A shared, segmented memory system for an
            object-oriented database. *ACM Trans. Office Inf. Syst.*, 5(1):70–95, Jan 1987.

[Zic91]   R. Zicari. A Framework for O_2 Schema Updates. In *Proc. IEEE Conf. on
            Data Engineering*, Kobe, Japan, April 1991.

[Zic89b]   R. Zicari. Schema Updates in the O_2 Object-Oriented Database System.
            Technical Report 89-057, Dipartimento di Elettronica - Politecnico di Milano,
            1989.

[Zic91]   R. Zicari. Building a Toolset for Schema Modifications for an OODBMS
            (Position Paper). Technical report, Dipartimento di Elettronica - Politecnico
            di Milano, 1991.

[ZM89]   S. Zdonik and D. Maier. Fundamentals of object-oriented databases. In
            S. Zdonik and D. Maier, editors, *Readings in Object-Oriented Databases*,
            pages 1–32. Morgan Kaufmann, San Mateo, CA, 1989.

Index